Derelict Mines

Mine areas left behind by companies that no longer exist are defined as derelict mines – those that were operated and closed at a time when most countries did not have adequate regulations requiring rehabilitation of the impacted mine areas. This book provides unique information on the extent and severity of derelict mines' impact on environmental degradation and human and environmental health. It examines the nature of derelict mines, short-term and long-term risks to sensitive receptors, tools for monitoring and prioritizing risks, and technological advances for rehabilitation. This book considers a risk-based approach to managing derelict mines, which is reliable and cost-effective.

FEATURES

- Provides fundamental information on derelict mines and their inventory in different countries
- Explains risk-based management of derelict mines and the importance of community perspectives as a reliable and cost-effective method
- Identifies ownership and liability issues through many case studies in Australia and other countries that must deal with the remediation of derelict mines
- Presents remediation, assessment, and predictive tools for managing pit lakes
- Helps readers set standards, regulatory measures, and policies related to mine closures

This book is for engineers and professionals who work in mining, geology, chemistry, mineralogy, geotechnics, and hydrogeology and deal with industrial site management, waste management, mine closures, mine site reclamation, derelict mine remediation, and mine revegetation. It is also an insightful resource for graduate students, academics, and researchers focused on these courses.

Derelict Mines

Environmental Risk Assessment and Management

Edited by Ravi Naidu

CRC Press
Taylor & Francis Group
Boca Raton London New York

CRC Press is an imprint of the
Taylor & Francis Group, an **informa** business

Designed cover image: © Ravi Naidu

First edition published 2025
by CRC Press
2385 NW Executive Center Drive, Suite 320, Boca Raton FL 33431

and by CRC Press
4 Park Square, Milton Park, Abingdon, Oxon, OX14 4RN

CRC Press is an imprint of Taylor & Francis Group, LLC

ISBN: 978-1-138-30662-2 (hbk)
ISBN: 978-1-032-77756-6 (pbk)
ISBN: 978-1-315-14185-5 (ebk)

DOI: 10.1201/9781315141855

Typeset in Times
by Deanta Global Publishing Services, Chennai, India

While planning and working on this book, my thoughts were with my dad who was a minerals prospector and managed an open cast manganese mine in Fiji until early 1970s. An outstanding prospector as he used biogeochemical tracing of minerals. I do recall seeing his notebook which included plant leaves against symbols of certain minerals, including Cu, Mn and Zn. When in high school, I realized the meaning of these symbols and even before biogeochemical prospecting was in vogue, my dad was using plants to successfully identify mining areas. I also vividly remember his words when the manganese mine he was managing was closed – "this land is our life and we must repair this damage by rehabilitating and by planting trees", which he did. This book is dedicated to my wonderful Dad and Mum, Paidaiya and Parvathi Naidu. As the Chair of the Food and Agricultural Organization of the United Nations, I also dedicate this book to the International Network on Soil Pollution (INSOP).

FIGURE 0.0 Ravi Naidu's dad and mum, Paidaiya and Parvathi Naidu.

Contents

Preface

I hope that this book will stimulate more work on derelict mines especially from assessment perspectives and the viability of the existing and new approaches to managing and or remediating such sites. Mineral exploration and mining commenced centuries ago when the focus was on precious metals, including mining for metals and later for minerals that we commonly used for many purposes. Growth in the minerals industry boomed as the demand for all forms of minerals grew, especially in developed countries, and at a time when we did not have appropriate policies relating to mining and mine closure. Even today, there are numerous instances of mining in developed countries with such ventures by developed and certain developing countries that leave behind degraded landscape with limited investment to rehabilitate. This has resulted in many abandoned mines typically defined as derelict mines. The return on investment from mining for most companies and countries has been largely linked to the creation of jobs and contribution towards the national economy, and there has been limited attention given to planning for rehabilitation of the mine site once it was closed.

Globally we have hundreds of thousands of derelict mines which pose significant risks to the environment and human health. Our recent work in Australia has demonstrated the significant impact of toxic metal(loid)s to both terrestrial and aquatic ecosystems. The removal of topsoil with open cast mining leaves behind large areas of degraded land which just about eliminates carbon sequestration and, where there is sloping land, allows erosion of land, further impacting natural ecosystems. Rehabilitation of such sites could cost tens of millions of dollars though the key question is who underpins the cost of remediation/rehabilitation. Along with such impacts of mining to the environment, in some countries local Indigenous communities have also allowed access to their land and have failed to benefit economically or otherwise from past mining activities. However, in Australia now most if not all mining companies do seek input from local communities, noting that most Indigenous communities value the land as central to their culture, identity and daily lives.

This new book provides unique information on the extent and severity of derelict mines and their impact on environmental degradation and potential risks to human and environmental health. It examines the nature of derelict mines, short- and long-term risks to sensitive receptors, tools for monitoring and prioritising risks and technological advances for rehabilitation. It also touches on existing policies on mining and closure. The book considers a risk-based approach to managing derelict mines, which is reliable and cost-effective. In doing so, the book attempts to bring together the environmental impacts of the minerals industry in many countries and describes the policies in place and management strategies and challenges that constrain rehabilitation of such mines that may still include valuable resources. It also acknowledges that many decades ago we did not have appropriate technologies to mine, extract and utilise the products as well as we can today. The book helps readers set standards, regulatory measures and policies related to mine closure.

Editing this book proved quite challenging given the impact of COVID-19, with some contributors losing some of their family members as well. I am forever grateful to them for their continuing support.

I hope that this book will stimulate more work on derelict mines, especially from assessment perspectives, and the viability of the existing and new approaches to managing and or remediating such sites. I acknowledge the support that my loving wife, Shamila, provided me, allowing me much time to focus on this book, including reviewing and editing the book chapters.

Editor

Ravi Naidu is the Chief Executive and Managing Director of the Cooperative Research Centre for Contamination Assessment and Remediation of the Environment and Laureate Professor and Founding Director of the Global Centre for Environmental Remediation at the University of Newcastle. He has been a research scientist in environmental contaminants, toxicology, bioavailability and remediation for over 25 years. He has co-authored over 850 technical publications and co-edited 16 books in the field of soil and environmental sciences, including field remediation of contaminated sites. Since 1994 Ravi has worked with scientists from the United States, Europe and Asia on environmental contamination, including recent arsenic poisoning in Bangladesh, India and China, with this collaboration leading to extensive research as well as the training of 5,000 environmental managers. Ravi was awarded a Gold Medal in environmental science in 1998 by Tamil Nadu Agricultural University, elected as a Fellow of the Soil Science Society of America in 2000, a Fellow of the Soil Science Society of New Zealand in 2004, a Fellow of the American Society of Agronomy in 2006, a Fellow of the prestigious American Association for the Advancement of Science in 2013, a Fellow of the Royal Australian Institute of Chemistry in 2017, a Fellow of the Australian Academy of Technological Sciences and Engineering, a Foreign Fellow of the Indian Academy of Agricultural Science in 2021 and a Foreign Fellow of the New Zealand Academy of Science in 2021. In 2012 Ravi was chosen as a winner of the Soil Science Society of America's International Soil Science Award. In 2023, Ravi was presented with the European Geosciences Union Alina Kabata-Pendias Medal and also the Food and Agriculture of the United Nations World Soil Prize-the Glinka Award. He has also been Chair of the Standards Australia Technical Committee on Sampling and Analyses of Contaminated Soils (1999–2000), Chair of the International Union of Soil Sciences Commission for Soil Degradation Control, Remediation and Reclamation (2002–2010) and President of the International Society on Trace Element Biogeochemistry (2005–2007).

Contributors

Pedro Alberto Sifuentes Amez
Latin American Network for the Prevention and Management of Contaminated Sites – ReLASC
Magdalena del Mar, Lima, Perú

Harold Annegarn
Unit for Environmental Sciences and Management
North-West University
Potchefstroom, South Africa

Yudi Firmanul Arifin
Forestry Science Study Program
Faculty of Forestry
and
Center of Excellence for Innovation, Technology, Commercialization, Management: Forest and Wetland
and
Sustainable Development Goals (SDGs) Center
Lambung Mangkurat University
Banjarbaru, South Kalimantan, Indonesia

Mezbaul Bahar
CRC for Contamination Assessment and Remediation of the Environment (CRC CARE)
and
Global Centre for Environmental Remediation (GCER)
University of Newcastle
Callaghan, New South Wales, Australia

Siddharth Bari
CSIR – Central Institute of Mining and Fuel Research
Dhanbad, Jharkhand, India

Jo Cavanagh
Manaaki Whenua – Landcare Research
Lincoln, New Zealand

Girish Choppala
Global Centre for Environmental Remediation (GCER)
University of Newcastle
and
CRC for Contamination Assessment and Remediation of the Environment (CRC CARE)
Callaghan, New South Wales, Australia

Tamara S. Chibrik
Ural Federal University (named after first president of Russia, B.N. Yeltsin)
Ekaterinburg, Russia
Zhaoming Dong
Beihang University
Beijing, China

Jianhua Du
CRC for Contamination Assessment and Remediation of the Environment (CRC CARE)
and
Global Centre for Environmental Remediation (GCER)
University of Newcastle
Callaghan, New South Wales, Australia

Muthia Elma
Chemical Engineering Department
and
Material and Membrane Research Group (M2ReG)
Faculty of Engineering
Lambung Mangkurat University
Banjarbaru, South Kalimantan, Indonesia

Victor Andres Arias España
CES – Consulting Earth Scientists
Newcastle West, Australia

Elena I. Filimonova
Ural Federal University (named after first president of Russia, B.N. Yeltsin)
Ekaterinburg, Russia

John Frankish
Manager Operations Central
Department of Resources
City East, Queensland, Australia

Yonchao Gao
Ecology Institute
Qilu University of Technology (Shandong Academy of Sciences)
Shandong Provincial Key Laboratory of Applied Microbiology
Jinan, China

Cris Reven L. Gibaga
Department of Science and Technology
Philippine Nuclear Research Institute
Diliman, Quezon City, Philippines

Margarita A. Glazyrina
Ural Federal University (named after first president of Russia, B.N. Yeltsin)
Ekaterinburg, Russia

C.M. Hogan
NOAMI Secretariat
Natural Resources Canada
Ottawa, Canada

Oskar Kadletz
Department of Resources
City East, Queensland, Australia

Alvin Lal
CRC for Contamination Assessment and Remediation of the Environment (CRC CARE)
and
Global Centre for Environmental Remediation (GCER)
University of Newcastle
Callaghan, New South Wales, Australia

Dane Lamb
CRC for Contamination Assessment and Remediation of the Environment (CRC CARE)
and
Global Centre for Environmental Remediation (GCER)
University of Newcastle
Callaghan, New South Wales, Australia

Charles Lee
Newcastle Australia Institute of Higher Education
Singapore

S. Lucas
School of Environmental and Life Sciences
University of Newcastle
Callaghan, New South Wales, Australia

Natalia V. Lukina
Ural Federal University (named after first president of Russia, B.N. Yeltsin)
Ekaterinburg, Russia

M.C. Manna
ICAR – Indian Institute of Soil Science
Bhopal, India

Irdika Mansur
Department of Silviculture
Faculty of Forestry and Environment
IPB University
Bogor, West Java, Indonesia

R.E. Masto
CSIR – Central Institute of Mining and Fuel Research
Dhanbad, Jharkhand, India

S. Mukhopadhyay
Council of Scientific and Industrial Research (CSIR) – Central Institute of Mining and Fuel Research
Dhanbad, Jharkhand, India

Pumulo Mukube
Environment and Development
Copperbelt University
Kitwe, Zambia

Ravi Naidu
CRC for Contamination Assessment and Remediation of the Environment (CRC CARE)
and
Global Centre for Environmental Remediation (GCER)
University of Newcastle
Callaghan, New South Wales, Australia

Ramkrishna Nirola
CRC for Contamination Assessment and Remediation of the Environment (CRC CARE)
University of Newcastle
Callaghan, New South Wales, Australia

Ihsan Noor
PT Jorong Barutama Greston (Closed) Coal Mine Company
Karang Rejo, Jorong, Tanah Laut, South Kalimantan, Indonesia
and
Mining Engineering Study Program
and
Sustainable Development Goals (SDGs) Center
Lambung Mangkurat University
Banjarbaru, South Kalimantan, Indonesia

Sam Pegg
Department of Resources
City East, Queensland, Australia

Alfonso R. Rodriguez Pinilla
Pure Earth/Blacksmith Institute
New York, New York, USA
and
R3 Environmental Technology SAS
Bogotá, Cundinamarca, Colombia

H.A.B. Potter
Environment Agency
Bristol, UK

E. Pottie
School of Environmental and Life Sciences
University of Newcastle
Callaghan, New South Wales, Australia

Majeti N.V. Prasad
University of Hyderabad
Hyderabad, India

Bambang Joko Priatmadi
Soil Science Study Program
Faculty of Agriculture
Lambung Mangkurat University
Banjarbaru, South Kalimantan, Indonesia

Rico Neil M. Quierrez
Department of Science and Technology
Philippine Nuclear Research Institute
Diliman, Quezon City, Philippines

Eugene A. Rakov
Ural Federal University (named after first president of Russia, B.N. Yeltsin)
Ekaterinburg, Russia

Rony Riduan
Department of Environmental Engineering
Lambung Mangkurat University
Banjarbaru, South Kalimantan, Indonesia

Tim Roberts
School of Environmental and Life Sciences
University of Newcastle
Callaghan, New South Wales, Australia

Akhmad Rizalli Saidy
Soil Science Study Program
Faculty of Agriculture
Banjarbaru, South Kalimantan, Indonesia

Kenneth S. Sajwan
Department of Marine and Environmental Sciences
College of Sciences and Technology
Savannah State University
Savannah, Georgia, USA

Jessie O. Samaniego
Department of Science and Technology
Philippine Nuclear Research Institute
Diliman, Quezon City, Philippines

Peter Sanderson
Global Centre for Environmental Remediation (GCER)
University of Newcastle
Callaghan, New South Wales, Australia
and
Tetra Tech Coffey
Lambton, New South Wales, Australia

Lana Saria
Coal Business and Development
Energy and Mineral Resources Ministry Office
Jakarta, Indonesia

V.A. Selvi
CSIR – Central Institute of Mining and Fuel Research
Dhanbad, Jharkhand, India

Balaji Seshadri
CRC for Contamination Assessment and Remediation of the Environment (CRC CARE)
and
Global for Environmental Remediation (GCER)
University of Newcastle
Callaghan, New South Wales, Australia

Robyn Simcock
Manaaki Whenua – Landcare Research
Lincoln, New Zealand

A.K. Singh
CSIR – Central Institute of Mining and Fuel Research
Dhanbad, Jharkhand, India

P.K. Singh
CSIR – Central Institute of Mining and Fuel Research
Dhanbad, Jharkhand, India

Siddharth Singh
CSIR – Central Institute of Mining and Fuel Research
Dhanbad, Jharkhand, India

N.K. Srivastava
CSIR – Central Institute of Mining and Fuel Research
Dhanbad, Jharkhand, India

P. Stevens
School of Environmental and Life Sciences
University of Newcastle
Callaghan, New South Wales, Australia

Eko Suhartono
Medical Chemistry and Biochemistry Department
Lambung Mangkurat University
Banjarbaru, South Kalimantan, Indonesia

Suhernomo
PT Adaro Indonesia
Tanjung, South Kalimantan, Indonesia

Stephen Syampungani
Environment and Development
Copperbelt University
Kitwe, Zambia
and
Forest Science Postgraduate Programme
Department of Plant and Soil Sciences
University of Pretoria
Hatfield, Pretoria, South Africa

Alexandria M. Tanciongco
Department of Science and Technology
Philippine Nuclear Research Institute
Diliman, Quezon City, Philippines

Didik Triwibowo
Lambung Mangkurat Universityy
Banjarbaru, South Kalimantan, Indonesia

G.A. Tremblay
International Network for Acid Prevention (INAP)

Liang Wang
CRC for Contamination Assessment and Remediation of the Environment (CRC CARE)
and
Global Centre for Environmental Remediation (GCER)
University of Newcastle
Callaghan, New South Wales, Australia

Haibin Xu
CRC for Contamination Assessment and Remediation of the Environment (CRC CARE)
and
Global Centre for Environmental Remediation (GCER)
University of Newcastle
Callaghan, New South Wales, Australia

Xiaodong Yang
Department of Geography and Spatial Information Technology
Ningbo University
Ningbo, China

1 Minerals Industry
Risks, Impacts and Managing Derelict Mines – International Perspective

Ravi Naidu and Peter Sanderson

1.1 INTRODUCTION

Mining has occurred at scale globally since the Bronze Age, with exploitation of copper, tin, gold and lead evidenced by archaeology (Timberlake, 2017). For example, the Ross Island mine, which began operation in 2400 BC in southwestern Ireland, provided Britain with much of its early copper supply. Britain's first mining boom is said to have occurred between 1600 and 400 BC (Williams and Le Carlier de Veslud, 2019). This history has led to a legacy of environmental impact, with many thousands of mines now discharging metal-containing mine waters into adjacent waterways (Johnston et al., 2008).

With technological developments of the industrial revolution, mining increased substantially through the 19th and 20th centuries, with occasional "rushes" for several minerals (silver, gold, radium, etc.) which occurred in connection with booms in demand (Carvalho, 2017). These rushes drove exploration and large settlement and these communities went through booms and busts (BLM, 2013).

The California gold rush, which occurred between 1848 and 1855, saw an influx of some 300,000 people to the state (Starr and Orsi, 2000). The US Geological Survey estimates that as much as 3,300 tons of gold were mined during the gold rush and through the 1960s (USGS). In the later stage of the rush, hydraulic mining techniques were used, which used high-pressure jets of water to dislodge minerals in unconsolidated material; the resulting slurry was directed through a sluice to separate gold (Alpers et al., 2005). Mercury was used to enhance gold recovery from hard rock mining with an estimated 10–30 percent annual loss rate to the environment (Alpers et al., 2005). Consequently, due to environmental degradation and pollution from the practice, hydraulic mining was effectively banned in 1884 by the California courts (BLM, 2013). Hard rock mining was also undertaken for gold from quartz using density separation and mercury amalgamation (The Sierra Fund, 2008).

The legacy of such mining left behind unrecovered or subordinate minerals and other processing byproducts, such as arsenic, mercury, cyanide, lead and acids. Such wastes made their way into the soil and water (BLM, 2013). Historically, in the absence of environmental legislation for rehabilitation and closure when ore bodies were exhausted, miners would move on, leaving a legacy of abandoned mines, associated safety hazards and contaminated land and water (BLM, 2007). The environmental, social and economic problems associated with abandoned mine sites are serious and global (UNEP, 2001).

Another mining legacy comes in the form of the landscape impact of mine voids and pit lakes. Increasing demands for resources led to increasingly large open cut mines as resource extraction technologies became more sophisticated (Blanchette and Lund, 2016). Inadequate mine closure

DOI: 10.1201/9781315141855-1

planning did not provide for the restoration of the landscapes when operations ceased. The remaining void filled with water forming a pit lake presenting water quality issues such as acidity and toxicity.

The consequences of abandonment of mining activities are now well recognised. Issues associated with mining legacy include altered landscapes, acid mine drainage, metal leaching, and contamination of receiving environments, health and safety risks around mine shafts and openings, abandoned infrastructure and subsidence (IIED and WBCSD, 2002). These problems have social and economic impacts on countries and impacted communities due to loss of productive land; degradation of groundwater and pollution of surface water by sediments or salts; air pollution from dust or toxic gases; safety issues from open pits and shafts and landslides (UNEP, 2001).

Additionally, there are financial costs associated with reclamation and ongoing liabilities. Such legacies typically become the responsibility of governments to manage, which need mechanisms of funding and prioritisation due to the high costs associated with cleaning up the scale of mining legacies. Making progress towards dealing with abandoned mines has been hindered by a lack of clarity around the scale of the problem, key definitions, what are appropriate criteria and standards for rehabilitation and financial mechanisms that can be applied to fund needed measures. While Chapter 2 focusses on balancing sustainable development and assessing global environmental risks in the mineral industry operations, this book brings together an overview of abandoned/derelict mines in a number of countries together with policies and strategies in place for managing such mines – though the challenge to remediate and rehabilitate remains in most countries globally.

1.2 MINING LEGACIES

Since the early 2000s the problem of mining legacies has attracted increasing recognition internationally. Various conferences, workshops and working groups, such as the World Bank, the International Finance Corporation, the United Nations Environment Programme, the World Conservation Union and the International Council on Mines and Metals, have sought to identify key issues and coordinate a way forward (Pepper et al., 2014).

The issues related to abandoned mines are well documented; however, few systematic surveys and impact assessments have been carried out to fully appreciate the scope of the problem. As such, efforts have been directed towards conducting an inventory of mining legacies within jurisdictions, development of frameworks such as risk-based approaches and criteria. For example, a national approach has been developed in Canada – the National Orphaned/Abandoned Mine Initiative (NOAMI). In Australia, states have undertaken inventory to varying degrees, with the acknowledgement of the need for cooperation and information sharing between states and territories in Australia to tackle mining legacies most effectively. The following sections provide a brief inventory of abandoned mines in a number of countries with details presented in individual Chapters in the book.

1.3 INTERNATIONAL INVENTORIES OF ABANDONED MINES

1.3.1 South Korea

Most of the metal mines in Korea were established prior to 1940; however, by the 1980s these mines faced closure due to a decline in their economic feasibility (KMOE, 2017). By 2005, inactive or abandoned mines were reported to account for 64% of total mines (1,270 of 2,006 total), including cases where the responsible companies were no longer operating.

The Korean Ministry of the Environment (KMOE) began conducting soil contamination surveys on abandoned metal mines throughout Korea from 1992. Up until 2016, assessment of 1,536 abandoned metal mines had been completed and, of these, 672 were found to exceed guidance levels of soil contamination (KMOE, 2017). Heavy metals were also detected in river water, sediment and agricultural products (KMOE, 2017).

A mine reclamation corporation was established in 2006 to actively promote reclamation of abandoned mines. The Ministry of Environment also established a cooperative network for the assessment and management of abandoned mining areas, including determination of the status of soil contamination, the impact on the local residents and resolving the lack of sufficient post management after the completion of the project.

Reclamation work has been completed for 192 abandoned metal mines found to exceed environmental standards, and the work is continuing on the remaining mines (KMOE, 2017).

Community health impact assessments were conducted on residents living in areas near 142 abandoned metal mines between 2008 and 2017 reporting potential harmful level of As in humans (Chang et al., 2019)

Recognition of the negative impacts on the environment and human health due to abandoned mining operations, the Mine Pollution Prevention and Reclamation Act enacted in 2005 included provision for deposit by mine operators for each year of operation towards a fund for the rehabilitation of mining impacts after mine closure.

1.3.2 Canada

The National Orphaned/Abandoned Mines Initiative (NOAMI) in Canada is a notable initiative for managing abandoned mines, created in 2002 based on recommendations put forward at a multistakeholder workshop on abandoned mines. NOAMI involves cooperation amongst multiple stakeholders, including the mining industry; federal, provincial and territorial governments; NGOs; and indigenous peoples. Principles of the programme include collaboration, continuity, avoidance of duplication and ensuring government policy is based on sound science.

As detailed in Chapter 4, Task groups were established by the advisory committee towards the goal of:

- Information Gathering/inventory.
- Community involvement.
- Legislative and institutional barriers to collaboration.
- Funding approaches.
- Jurisdictional legislative review.
- Mine closure/return of mining lands.

NOAMI has developed a national inventory of orphaned and abandoned mines based on inventories from each province and territory. A web-based inventory tool provides access to the data, including interactive map display and a searchable database linking to the original data sources. Such a tool aids transparency, decision-making and planning.

The new national inventory ranks sites in classes (A, B, C and D) based on qualitative assessed level of risks to the environment, public health and public safety. It includes six categories for the status of each site ('inactive', 'neglected', 'abandoned', 'legacy', 'terminated' and 'unknown').

The total liability for contaminated mining sites across Canada is estimated to be well above 9.1 billion dollars CAD (Mining Watch Canada, 2017).

1.3.3 South Africa

Mining has been a major contributor to the South African economy, with large-scale mining starting in the mid- to late 19th century with the discovery of diamonds and gold (CSIR, 2019). However, it operated for a long time under limited regulation that left the responsibilities for mining impacts to the mine owners, and, in the process, many mines were abandoned (Mhlongo et al., 2016). An agreement in 1976 saw mines abandoned prior to 1976 become the responsibility of the state.

The Mines and Works Act 27 (1956) was concerned with the operation and works of mines, but did not contain requirements regarding environmental rehabilitation when mining operations are ceased, until amendment of the MWA and regulations on rehabilitation came into effect in 1980 (Hartzer, 2009). The Minerals Act 50 (1991) repealed the Mines and Works Act 27 (1956) and contained provisions for rehabilitation. Section 38 of the MA enforced rehabilitation on all holders of mining authorisations until a closure certificate was issued (Hartzer, 2009).

The Mineral and Petroleum Resources Development Act 2002 (MPRDA) came into effect in 2004 and became the major legislation regulating mineral resources and environmental impact. Section 46 of the act provides for the setting aside of funds to be retrieved from the rights holder for the rehabilitation of a prematurely abandoned site where it is urgently needed to prevent environmental degradation (Klopper and Wessels, 2017). The section classifies those mines for which a closure certificate has not been issued and no party can be traced to assume responsibility for their liabilities as abandoned and the government may provide funds for their rehabilitation (Cornelissen et al., 2019).

An integrated approach towards 'one environmental system' between government departments was adopted in 2014 for mine environmental management. Regulations were introduced under the National Environmental Management Act, 1998 (NEMA) for the financial provisioning for costs associated with the rehabilitation of mining impacts throughout the lifetime of operations and residual environmental impacts that may become known in the future. However, mine closure in South Africa remains elusive due to a lack of mine closure skill and knowledge, a lack of closure planning during feasibility stages, deficiencies in data and reporting on environmental impacts and financial liabilities (Van Druten and Bekker, 2017).

The Department of Mineral Resources (DMR) began inventory of abandoned mines in the early 2000s to understand the scope of the problem. By the end of 2008, almost 6,000 abandoned mines were known, but measures were not in place to ensure that abandoned mines were rehabilitated effectively and in a timely manner, and there was not an integrated information system to record and report on the status of mines to support this effort (DMR, 2009). Appropriate definitions of abandoned mine sites and features are needed for improving the inventory (Mhlongo et al., 2016).

The national strategy for abandoned mines finalised in 2009 included:

- Compilation and maintenance of an inventory of derelict and ownerless mines.
- A framework for the identification of priority sites in terms of public health and safety, environmental and social risks.
- Development of an action plan for the management of the risks related to derelict and ownerless mines.
- Ensure that all D&O sites are rehabilitated to acceptable levels where they can be habitable by humans.

Recent estimates for the cost for remediation of current officially listed abandoned mines in South Africa as of March 2019 were $48.7 billion (DMR, 2019). Estimated rehabilitation costs are affected by the lack of agreed criteria as to what needs to be remediated and what the goals for rehabilitation should be. To expedite progress on dealing with the legacy of abandoned mines in South Africa, the identification of rehabilitation criteria will assist in the prioritisation of rehabilitation efforts (Mhlongo et al., 2016). Additionally, going beyond qualitative risk ranking will further help prioritise funds (Mhlongo et al., 2016). Currently, the risk-ranking systems that have been developed are mostly internal to the organisations directly involved in their application, and there is little public or academic knowledge of their methodologies and application (Cornelissen et al., 2019).

Issues with the capacity of government agencies are evident in upholding regulation across Africa, and this is also the case in South Africa, which arguably has the most developed and sophisticated provisions for mine closure planning amongst African nations (Morrison-Saunders et al., 2016). Coordination between different agencies can lead to gaps and oversights in performance (Morrison-Saunders et al., 2016).

1.3.4 United States

The General Mining Act of 1872 gave federal authorisation to the prospecting and mining for minerals such as gold, platinum, silver, lead, zinc, uranium and tungsten. However, miners often abandoned mines, leaving behind structures, safety hazards and contaminated land and water (US GAO, 2008).

It has been estimated that as many as 500,000 abandoned mine sites exist, with an estimated clean-up cost of $54 billion dollars (House Committee on Natural Resources, 2016). These sites pose health and safety risks, particularly for recreational users of public lands due to mine shafts and abandoned toxic waste. There is also a substantial impact from AMD to surface and groundwater degrading water quality and ecosystems.

The BLM's abandoned mine lands (AML) programme (established in 1997) is a partnership that spans federal, state and local efforts to reduce the environmental and health risks of abandoned land mines. Several agencies are involved in reclaiming abandoned mines, including the federal agencies the Bureau of Land Management (BLM), the Office of Surface Mining Reclamation and Enforcement (OSMRE), the Forest Service and the Environmental Protection Agency (EPA).

The aim of BLM's Abandoned Mine Lands (AML) programme is to enhance public safety and improve water, soil and air quality by remediating the effects of legacy hard rock mines in the western United States.

These agencies are responsible for managing large areas of land, including abandoned mines on these lands (GAO, 2020):

- BLM maintains a national inventory of known abandoned mines (mostly hardrock mines). As of 2019 the focus is primarily tracking and reporting abandoned mine features rather than sites.
- The Forest Service does not have a national inventory (as of 2019) but tracks abandoned hard rock mines by features for physical safety hazards and by site for environmental hazards at regional and district levels.
- The Parks Service maintains a database of abandoned mines and features on its lands and manages them through a safety and environmental compliance and cleanup programme.
- The EPA maintains a database of mine sites on non-federal lands as part of its national database of contaminated sites. However, sites that pose only a physical safety hazard are not identified as they are outside the scope of the superfund programme.

As of May 2019, the Forest Service, BLM, the Park Service and EPA identified 140,652 abandoned hard rock mine features, of which 60% may pose environmental or physical safety hazards. Between fiscal years 1998 and 2007, BLM, the Forest Service, EPA and OSME spent at least $2.6 billion (in 2008 constant dollars) to reclaim abandoned hard rock mines. This increased to 2.9 billion from 2008 through 2017 (about 287 million dollars annually) (GAO, 2020). The BLM prioritises and takes appropriate action on these historic abandoned mine sites using a risk-based approach.

Federal law requires hard rock mining operations to reclaim disturbed land when operations cease and providing reclamation plans and bonds or other financial assurance to guarantee reclamation. Agency officials periodically review bonds held to ensure they will be adequate to cover expected reclamation costs. As of October 2017, BLM held over $3 billion in financial assurances for 1,425 hard rock mining operations, 4% of which had inadequate financial assurances (GAO, 2019). The Forest Service also holds $250 million in financial assurances, though lack a centralised mechanism to record the current estimated cost of reclamation to determine if financial assurances held are adequate (GAO, 2019).

For managing coal mines, the Surface Mining Control and Reclamation Act (SMCRA) of 1977 requires the coal mining areas to be restored to a condition capable of supporting the uses which it was capable of supporting prior to any mining, or higher or better uses. A reclamation plan is required for mining approval, and bonds must be posted to ensure funding is available for reclamation. The SMCRA also places a levy per ton on current coal mine operators to fund abandoned mines.

1.3.5 Europe

Although the mining industry in Europe has been declining over the last few decades, Europe has a long history of mining dating back more than 5,000 years ago (Žibret et al., 2020). Due to economic and environmental pressures, many mines in Europe closed through the 19th and 20th centuries without proper rehabilitation (Žibret et al., 2020). An estimated 1,000 to 1,500 km of water courses have been impacted by metal mine discharges in the EU member states. (Younger et al., 2002).

In Europe, the implementation of the Mining Waste Directive (2006/21/EC) includes requirements of provisions for proper closure of operating and future mines within the European Union. It also contains a mandate (Article 20) for production of an inventory of abandoned mining sites located on their territory which cause or have the potential for threat to human health and the environment in the medium or short term. This is covered by the 'Guidance document for a risk-based pre-selection protocol for the inventory of closed waste facilities as required by article 20 of the directive 2006/21/EC.'

Further, Article 21 addresses the need for the rehabilitation of those facilities identified under Article 20, including the establishment of the most appropriate risk assessment procedures and remedial actions. A support document (DHI, 2012) outlines the process to be undertaken by member states (Figure 1.1).

As of July 2017, over 3,400 closed and abandoned waste facilities were registered in 18 national inventories (Amec Foster Wheeler Environment et al., 2017). A study by Žibret et al. (2020) reported inventories of 9 member states totalling almost 6,000 sites, only a subset of which had undergone detailed assessment.

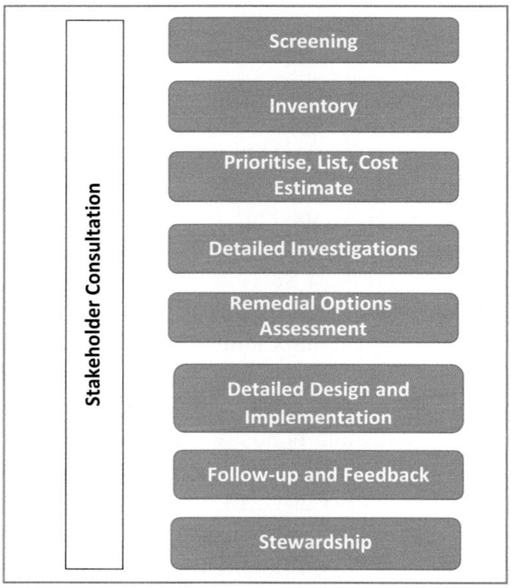

FIGURE 1.1 Outline of the process for addressing abandoned mines by EU member states.

1.4 CURRENT MINING PRACTICE

Globally mining has continued to expand in the modern era and remains economically important. According to world mining data, mining production has increased from 9.6 billion tons in 1985 to 17.2 billion tons in 2017. The biggest producers include China, the United States, Russia and Australia, and production is increasing in developing countries in Latin America and Africa (Reichl and Schatz, 2019). The global extent of land area impacted by mining and quarrying is estimated to range between 300,000 and 800,000 km (Cherlet et al., 2018).

Improperly regulated mining can disrupt the benefits that come from the sector through environmental harm and disruption of social and economic structures (IGF, 2020). There is a need to balance economic gains from mining with other economic, social, and environmental values. This requires strong and transparent institutional frameworks to regulate the mining sector and to ensure good governance practice (Cherlet et al., 2018).

Regulations implemented around the world with the aim to ensure safe and sustainable closure of mine sites (APEC, 2018). However, the development of these regulations has been uneven with some jurisdictions still having little or no regulation and others having a robust governance framework (APEC, 2018). Failure of the regulatory process undermines the social licence to operate (Unger et al., 2014) and mining legacies reflect poorly on the minerals industry (Pepper et al., 2014).

Chile, a country which holds the world's largest reserves of copper and lithium (USGS, 2020) is an example of recent development of a more robust governance framework. In 2012 laws regulating mines and mine closure came into force. Increased requirements for closure included an approved closure plan prior to the commencement of operations, including financial guarantees. Abandoned and inactive mine sites are recognised as a major environmental problem by the government with some 650 sites needing rehabilitation (Weeks et al., 2015).

1.4.1 Mine Closure

The potential for creating new mining legacies is reduced in areas where mining-related legislation is in place for mine closure, and financial assurance is provided for the proposed rehabilitation activities before a new mining project is approved (Bennet, 2016). Mine closure planning (including progressive rehabilitation) introduced over time has sought to reduce the likelihood of mine voids and pit lakes being left after the cessation of mining activities.

Clark and Clark (2005) note aspects of well-developed mining legislation such as is found in Japan, Australia, Canada, the United States, Germany and the United Kingdom. The legislative provisions include:

- EIA is required before leasing.
- Specific provisions for reclamation.
- Bonding procedures.
- Abandonment.
- Non-compliance.

Those countries with an extensive mining history, substantial numbers of abandoned mines and associated problems tend to have developed more comprehensive legislation. In the EU, Directive 2006/21/EC recognises the importance of guaranteeing that the taxpayer is not left with the financial burden of environmental cleanup and rehabilitation of mining liabilities, as has often been the case in the past. It requires a financial guarantee to be lodged by the mine operator prior to the commencement of deposition operations in the waste facility.

Developing countries in Africa, Asia and Latin/South America vary in the provisions for mine closure in mining law. Some of the features include:

- Only very general policy and legislation for mine closure.
- A high degree of state responsibility (abandoned and some operational mines).
- Mine closure negotiated and incorporated into individual mining agreements.
- Limited, if any bonding procedures to ensure comprehensive mine closure.

These provisions are often inadequate for comprehensive mine closure and provision for post-mining sustainable development is lacking, which presents the greatest problems for government (Clark and Clark, 2005).

The Intergovernmental Forum on Mining, Minerals, Metals and Sustainable Development has developed guidance on the legal framework for environmental and social impact assessment (ESIA) in mine planning and approval (IGF, 2020). ESIA is a key component in government decision-making regarding mine approval and provides a reference document for licence holders, government agencies, and other stakeholders during the life of the project.

Key aspects of the process include:

- Commitment to sustainable development.
- consistency and coordination of responsible authorities.
- Coverage of all stages of the life of mines.
- Community engagement and transparency.
- ESIA is part of project planning.
- Environmental and social management plans.
- Mine closure plans.
- Financial assurances.
- Oversight and enforcement of non-compliance.

With the aim of providing standards and guidance to mitigate long-term mining impacts and create environmental and socioeconomic value through mine closure and reclamation management, the International Organisation for Standardisation is developing standards for mine closure and reclamation management (ISO TC 82/SC 7), with the participation of 12 countries (Murphy et al., 2019).

The scope of ISO TC 82/SC 7 is to address all lifecycle stages of resource development, including design, development, mining, temporary suspension of operations, cessation of mining, closure and post-closure monitoring and management.

Key mine closure and reclamation themes from ISO TC 82/SC7:

- Mine closure knowledge.
- Management, risk and opportunity.
- Mine design for closure.
- Financial and economic aspects.
- Closure performance.
- Social aspects.
- Stakeholder engagement.
- Alternative post-closure land use.
- Legacy mines.

The principles of mining rehabilitation have shifted in recent years from a minimum standard of safety and stability and revegetation to restoration of the landscape to pre-mining conditions or providing future sustainability of land use for purposes such as agriculture, nature reserves, forests and grasslands (Sloss, 2013). In the EU, for example, 50% of reclaimed mined lands are in the form of forests and grasslands, whereas in China 70% are reclaimed for agriculture (Zhengfu et al., 2010).

1.4.2 Funding Mechanisms

It is not practical to rehabilitate all abandoned mine sites (Pepper et al., 2014); indeed, since resources are constrained it is desirable to prioritise addressing those sites which pose the most risk from a safety and human health perspective and impact on the environment. This is being increasingly recognised in the management of abandoned mines by different jurisdictions.

The typical funding mechanisms available for rehabilitation include state-funded remediation, 'polluter pays' or a mixed funding mechanism, based on state and private contributions. For historically abandoned mines 'polluter pays' is generally not applicable as no responsible party can be identified. Therefore, abandoned mines require state-funded or alternative funding mechanisms.

There is a view that the present mining industry should contribute to a fund that goes towards rehabilitating abandoned mines. In Western Australia, a Mining Rehabilitation Fund (MRF) was established by the Department of Mines, Industry Regulation, and Safety's (DMIRS). All tenement holders are required to report annually with respect to disturbance and contribute to the fund. Interest earned on money in the fund is available to rehabilitate abandoned mines across the State in circumstances where the original operator cannot be held responsible.

Similarly, since 2013 mine operators in the Northern Territory have also had to pay an annual 1% levy, based on the value of the financial security held for each mine site, into a Mining Remediation Fund, the proceeds of which are used by the government to address issues caused by abandoned mines. In the US, the Surface Mining Control and Reclamation Act (SMCRA) of 1977 imposes a levy on coal production, which is deposited in a fund for abandoned mine reclamation.

1.4.3 Social and Economic Considerations

Countries with economic dependence on mining and developing nations are vulnerable to price fluctuations on the minerals market. Sustainable rehabilitation of mining sites will only occur if the conditions for economic growth in the surrounding communities are restored and other associated risks are eliminated (Renner et al., 2009).

During mining, very few alternative and local economies develop, with dependence on mining and its value chain prevailing until mining ceases (Mthenjane, 2019). Diversifying economic activity which can continue to sustain the livelihoods and wellbeing of communities after mining is needed (Mthenjane, 2019).

Solutions are needed for sustainable post mining land use through socioeconomic regeneration, such as use of land for agriculture, energy or repurposing for tourism. Beneficial reuse of pit lakes may include aquaculture, irrigation, recreation, storage of harmful substances (e.g., tailings) and water supply (McCullough and Lund, 2006).

The Eden project in Cornwall is a botanical garden, located in a reclaimed clay mine. In Chile the old mining towns of Sewell and Humberston are UNESCO World Heritage Sites and have been extensively restored and redeveloped to support tourism. In Australia, the old Kidston gold mine, northwest of Townsville, Queensland, is being developed into 250 MW Pumped Hydro. A similar project is planned for an abandoned mine in Bendigo, Victoria.

1.5 CONCLUSIONS

There is a strong desire for environmental sustainability of the mining industry, to continue to realise the economic benefits and provide ongoing productive land use for the community. The strengthening of governance of mining through legislation, planning and compliance is needed to realise this aim. Considerable progress has been made in the implementation of robust governance of the mining industry, and strides are being made in developing nations as well. The provision of bonding requirements for rehabilitation provides important insurance to ensure the rehabilitation of mined lands.

Undertaking to deal with mining legacies is also progressing, though understanding the scope of the problem by abandoned mine inventories and the development of frameworks to assess, prioritise and manage risks at these sites. Some governments are implementing funding mechanisms, such as mining levies, to provide the funding needed to address mining legacies.

REFERENCES

Alpers, C. N., Hunerlach, H. J., May, J. T., & Hothem, R. L. (2005). Mercury contamination from historical gold mining in California Fact Sheet 3014. US Geological Survey.

Amec Foster Wheeler Environment & Infrastructure UK Ltd, BiPRO and Milieu. (2017). "Assessment of Member States' performance regarding the implementation of the Extractive Waste Directive" for Directorate-General for Environment. © European Union.

Asia-Pacific Economic Cooperation (APEC). (2018). Mine closure checklist for governments. APEC Mining Task Force, Singapore. https://www.apec.org/Publications/2018/03/Mine-Closure—Checklist-for-Governments.

Bennett, K. (2016). Abandoned mines — environmental, social and economic challenges. In A. B. Fourie & M. Tibbett (eds.), *Proceedings of the 11th International Conference on Mine Closure*, Australian Centre for Geomechanics, Perth, 241–252. https://doi.org/10.36487/ACG_rep/1608_16_Bennett

Blanchette, M. L., & Lund, M. A. (2016). Pit lakes are a global legacy of mining: An integrated approach to achieving sustainable ecosystems and value for communities. *Current Opinion in Environmental Sustainability*, 23, 28–34.

Bureau of Land Management (BLM), (2007), Abandoned Mine Lands: A Decade of Progress Reclaiming Hardrock Mines, Forest Service Publication (FS-891).

Bureau of Land Management (BLM). (2013). Abandoned mined lands – A new Legacy. BLM Publication Number: BLM/WO/GI-13/007-3720.

Carvalho, F. P. (2017). Mining industry and sustainable development: time for change. *Food and Energy Security*, 6(2), 61–77.

Chang, J. Y., Ahn, S. C., Lee, J. S., et al. (2019). Exposure assessment for the abandoned metal mine area contaminated by arsenic. *Environmental Geochemistry and Health*, 41(6), 2443–2458. https://doi.org/10.1007/s10653-019-00296-5

Cherlet, M., Hutchinson, C., Reynolds, J., Hill, J., Sommer, S., & von Maltitz, G. (eds.). (2018). World Atlas of desertification. Publication Office of the European Union, Luxembourg.

Clark, A. L., & Clark, J. C. (2005). An international overview of legal frameworks for mine closure. Environmental Law Alliance Worldwide: Energy and Biodiversity Initiative.

Cornelissen, H., Watson, I., Adam, E., & Malefetse, T. (2019). Challenges and strategies of abandoned mine rehabilitation in South Africa: The case of asbestos mine rehabilitation. *Journal of Geochemical Exploration*, 205,106354.

CSIR. (2019). A CSIR perspective on South Africa's post-mining landscape. CSIR Report No: stelgen16928. ISBN: 978-0-7988-5641-6. CSIR, Pretoria.

Department of Mineral Resources (DMR). (2009). The national strategy for the management of derelict and ownerless mines in South Africa.

Department of Mineral Resources (DMR). (2019). DMRE presentation on derelict and ownerless asbestos mine. September 28, 2019.

DHI. (2012). Annex 3 - Supporting document on closure methodologies for closed and abandoned mining waste facilities (April 2012). In 'Establishment of guidelines for the inspection of mining waste facilities, inventory and rehabilitation of abandoned facilities and review of the BREF document No. 070307/2010/576108/ETU/C2'.

Government Accountability Office (GAO). (2008). Hardrock mining: Information on abandoned mines and value and coverage of financial assurances on BLM land. GAO-08-574T.

Government Accountability Office (GAO). (2019). GAO-19-436R Hardrock mining financial assurances 2019 Hardrock mining: BLM and forest service hold billions in financial assurances, but more readily available information could assist with monitoring, Washington, DC.

Government Accountability Office (GAO). (2020). Abandoned hardrock mines: Information on number of mines, expenditures, and factors that limit efforts to address hazards. GAO-20-238 2020.

Hartzer, S. (2009). The liability of historical mine authorization holders for rehabilitation (Doctoral dissertation, North-West University).

IIED and WBCSD. (2002). Mining, Minerals and Sustainable Development, mining for the future appendix C: Abandoned mines working paper, 1–25.
Johnston, D., Potter, H., Jones, C., Rolley, S., Watson, I., & Pritchard, J. (2008). Abandoned mines and the water environment: Science project. SC030136-41, Environment Agency, the Scottish Environment Protection Agency (SEPA), the Coal Authority (The Environment Agency et al.). Rio House, Waterside Drive, Aztec West, Almondsbury, Bristol.
Klopper, D., & Wessels, J. A. (2017). Investigation of Western Australia's rehabilitation fund as a fiscal policy solution for South African abandoned mines. *Journal of the Southern African Institute of Mining and Metallurgy*, 117(11), 1081–1087.
KMOE (Ministry of the Environment in Korea). (2017). 2017 white paper of environment (pp. 342–345). KMOE, Republic of Korea Press.
McCullough, C. D., & Lund, M. A. (2006). Opportunities for sustainable mining pit lakes in Australia. *Mine Water and the Environment*, 25(4), 220–226.
Mhlongo, S. E., & Amponsah-Dacosta, F. (2016). A review of problems and solutions of abandoned mines in South Africa. *International Journal of Mining, Reclamation and Environment*, 30(4), 279–294.
Mining Watch Canada. (2017). Environmental liability for contaminated mine sites in Canada. https://miningwatch.ca/news/2017/5/10/finally-new-national-inventory-orphaned-and-abandoned-mine-sites
Morrison-Saunders,A., McHenry, M. P., Rita Sequeira, A., Gorey, P., Mtegha, H. & Doepel, D. (2016). Integrating mine closure planning with environmental impact assessment:challenges and opportunities drawn from African and Australian practice, Impact Assessment and Project Appraisal, 34:2, 117–128.
Mthenjane, M. I. (2019). Post-mining use of rehabilitated land – An opportunity for the South African mining industry for sustainable development. *Journal of the Southern African Institute of Mining and Metallurgy*, 119(9), 693–696.
Murphy, D. P., Nahir, M., & Didier, C. (2019). An update on the development of mine closure and reclamation standards by the International Organization for Standardization. In *Proceedings of the 13th International Conference on Mine Closure* (pp. 761–770). Australian Centre for Geomechanics.
Pepper, M., Roche, C. P., & Mudd, G. M. (2014). Mining legacies—Understanding life-of-mine across time and space. In *Proceedings of the Life-of-Mine Conference* (pp. 449–465). Australasian Institute of Mining and Metallurgy, Brisbane.
Reichl, C., & Schatzm, M. (2019). *World Mining Data 2019*. Volume 34. Federal Ministry for Sustainability and Tourism, Vienna.
Renner, S., Rankin, M., Ponce, R., Griffin, B., Moraes, R., Dalheimer, M., & Parot, M. E. (2009). *8th International Conference on Acid Rock Drainage (ICARD) and Securing the Future: Mining, Metals & the Environment in a Sustainable Society 2009*, 1542–1554. Swedish Association of Mines, Minerals and Metal Producers.
Sloss, L. (2013). Coal mine site reclamation. IEA Clean Coal Centre, CCC/216 ISBN, 978-92.
Starr, K., & Orsi, R. J. (Eds.). (2000). *Rooted in Barbarous Soil: People, Culture, and Community in Gold Rush California* (Vol. 3). University of California Press.
The House Committee on Natural Resources. (2016). Abandoned mines. https://naturalresources.house.gov/imo/media/doc/Abandoned%20Mines.pdf
The Intergovernmental Forum on Mining, Minerals, Metals and Sustainable Development (IGF). (2020). *Guidance for Governments: Improving Legal Frameworks for Environmental and Social Impact Assessment and Management*. IISD, Winnipeg.
The Sierra Fund. (2008). Mining's toxic legacy an initiative to address mining toxins in the Sierra Nevada.
Timberlake, S. (2017). New ideas on the exploitation of copper, tin, gold, and lead ores in Bronze Age Britain: The mining, smelting, and movement of metal. *Materials and Manufacturing Processes*, 32(7–8), 709–727.
UNEP. (2001). *Abandoned Mines. Problems, Issues and Policy. Challenges for Decision Makers*. Santiago, Chile. 18 June 2001. Summary report. United Nations.
Unger, C. J., Lechner, A. M., Walton, A., Glenn, V., Edraki, M., & Mulligan, D. R. (2014). Maturity of jurisdictional abandoned mine programs in Australia based on web-accessible information. In *Proceedings Life-of-Mine 2014*. The Australian Institute of Mining and Metallurgy, Brisbane.
U.S. Geological Survey (USGS). (1999). https://www.usgs.gov/news/gold-rush-over-environmental-effects-linger. Accessed 1/09/2020.
U.S. Geological Survey. (2020). Mineral commodity summaries 2020: U.S. Geological Survey, 200 p. https://doi.org/10.3133/mcs2020
Van Druten, E. S., & Bekker, M. C. (2017). Towards an inclusive model to address unsuccessful mine closures in South Africa. *Journal of the Southern African Institute of Mining and Metallurgy*, 117(5), 485–490.

Weeks, B. (2015). Mine closure in Chile – challenges and changes. *Geotechnical News March*, 42–45.
Williams, R., & Le Carlier de Veslud, C. (2019). Boom and bust in Bronze Age Britain: Major copper production from the Great Orme mine and European trade, c. 1600–1400 BC. *Antiquity*, 93(371), 1178–1196. https://doi.org/10.15184/aqy.2019.130
Younger, P. L., Banwart, S. A., & Hedin, R. S. (2002). *Mine Water. Hydrology, Pollution, Remediation*. Kluwer Academic Publishers, Dodrecht.
Zhengfu, B. I. A. N., Inyang, H. I., Daniels, J. L., Frank, O. T. T. O., & Struthers, S. (2010). Environmental issues from coal mining and their solutions. *Mining Science and Technology (China)*, 20(2), 215–223.
Žibret, G., Lemiere, B., Mendez, A. M., Cormio, C., Sinnett, D., Cleall, P., Szabó, K., & Carvalho, M. T. (2020). National mineral waste databases as an information source for assessing material recovery potential from mine waste, tailings and metallurgical waste. *Minerals*, 10(5), 446.

2 Balancing Sustainable Development

Assessing Global Environmental Risks in the Mineral Industry Operations

Ramkrishna Nirola, Ravi Naidu and Alvin Lal

2.1 INTRODUCTION

All mining processes are a form of temporary land usage, and their closure provides opportunities for rehabilitation (Glenn et al., 2014). Most anthropogenic sources of heavy metal(loid)s occur through mining, followed by industrial and agricultural activities. Soils are the major sinks for heavy metal(loid)s that do not undergo microbial or chemical degradation but, instead, oxidation to cations in the presence of oxygen and moisture (Kirpichtchkova et al., 2006). The natural source of heavy metal(loid)s pollution is the weathering of rocks and volcanic activities. Metallic elements such as copper, iron, mercury, zinc, silver, lead, chromium, etc., have high densities ranging from 3.5 to 7 gcm^{-3} (Lenntech, W., 2004, Inobeme et al., 2014). These high-density heavy metals are persistent toxic chemicals, and are bioaccumulative in nature (Nirola et al., 2016). Historically, in the absence of environmental legislation for rehabilitation and closure, when ore bodies were exhausted, miners would proceed to new sites, leaving a legacy of abandoned mines, associated safety hazards and contaminated land and water (BLM, 2007). Consequently, the environmental, social and economic problems associated with abandoned mine sites are serious and global (UNEP, 2001).

With technological advances made possibly by the Industrial Revolution, mining increased substantially through the 19th and 20th centuries, with occasional "rushes" for several minerals (silver, gold, radium, etc.) which coincided with booms in demand (Carvalho, 2017). According to the Bureau of Land Management (BLM), these "rushes" drove exploration and large settlements, giving rise to communities that, in turn, experienced "booms and busts" (BLM, 2013). For example, the California gold rush, which occurred between 1848 and 1855 saw an influx of some 300,000 people to the state (Starr and Orsi, 2000). The US Geological Survey estimates that as much as 3,300 tons of gold were mined during the gold rush and through the 1960s (Wanty et al., 1998). In the later stages of the rush, hydraulic mining techniques, which used high pressure jets of water, served to dislodge minerals in unconsolidated material. The resulting slurry was directed through a sluice to separate gold from other materials (Alpers et al., 2005). Mercury was used to recover gold from hard rock mining with an estimated 10–30% annual loss rate to the environment (Alpers et al., 2005). Consequently, due to environmental degradation and pollution from this practice, hydraulic mining was effectively banned in 1884 by the California courts (BLM, 2013). Hard rock mining was also undertaken for getting gold from quartz using density separation and mercury amalgamation, which simply led to more mercury pollution (The Sierra Fund, 2008). The legacy of such mining left behind unrecovered or subordinate minerals and other processing

DOI: 10.1201/9781315141855-2

by-products, such as arsenic, mercury, cyanide, lead and acids. Such wastes made their way into the soil and water (BLM, 2013).

The landscape legacy of mines is also evident in the form of mine voids and pit lakes. Increasing demands for resources led to increasingly large open cut mines as resource extraction technologies become more sophisticated (Blanchette and Lund, 2016). Inadequate mine closure planning did not provide enough scientific input for restoration of the landscapes when mining operations ceased. The remaining void filled with water, forming a pit lake presenting serious water quality problems, such as acidity and toxicity (Venkateswarlu et al., 2016). The consequences of abandonment of mining activities are well recognized because of a myriad of problems. Issues associated with mining legacy include altered landscapes, acid mine drainage, metal leaching and contamination of receiving environments, health and safety risks around mine shafts and openings, abandoned infrastructure and subsidence (IIED and WBCSD, 2002).

The metal pollution problems generate social and economic challenges, such as loss of productive land, degradation of groundwater and pollution of surface water by sediments or salts, air pollution from dust or toxic gases and safety issues from open pits and shafts and landslides (UNEP, 2001). Additionally, there are financial costs associated with reclamation and ongoing liabilities. Such legacies typically become the responsibility of governments or their agencies to manage, which need mechanisms of funding and prioritization due to the inflated costs associated with cleaning up. Making progress in dealing with abandoned mines has been hindered by a lack of clarity around the scale of the problem, key definitions, appropriate criteria and standards for rehabilitation and financial mechanisms that can be applied to fund needed measures (Mchaina, 2001). This chapter builds on Chapter 1 which touches on risks, impacts and managing derelict mines – International Perspective.

2.2 MINING LEGACIES AND ENVIRONMENTAL HEALTH RISKS

Environmental quality standards are used as objectives for the cleanup of pollutants globally. The goal of environmental compliance is to minimize as much as possible the risk of harming terrestrial and aquatic life. However, it is technically not feasible to return soil and water environments to a pre-mining state. In some rivers in the UK, for instance, life forms in the ecological community have adapted to centuries of exposure to elevated metals even after remediation (Environment Agency, 2012). For this reason, the standard European metrics are less effective for identifying metal pollution impacts since they were actually designed to assess the health of aquatic biology due to urban pollution (Environment Agency, 2008; Gozzard et al., 2011).

Elements such as iron, copper, zinc, nickel and other trace elements are required for biological systems' proper functioning, and concentrations exceeding safe threshold values in a body lead to toxicity and disorders (Lokeshwari and Chandrappa, 2006). The heavy metal(loid)s most associated with poisoning of humans are lead, mercury, arsenic and cadmium. Poisoning may occur because of industrial exposure, air or water pollution, foods, medicines, improperly coated food containers or the ingestion of lead-based paints. Most entries of toxic heavy metal(loid)s into the human food chain entry are through water (Sonone et al., 2020). According to the National Organization for Rare Disorders (NARD; see (https://rarediseases.org/rare-diseases/heavy-metal-poisoning/) there is a serious human health concern as chemicals at very toxic levels cause various diseases. The pollutants causing damage to people's health are anthropogenic and the main sources are metal borne as presented in the Table 2.1.

2.3 AWAKENING TO ENVIRONMENTAL PROBLEMS

Abandoned mined land (AML) as per the United States Environmental Protection Agency (USEPA) are defined as lands, waters and surrounding watersheds where the extraction, beneficiation and processing of ores and minerals have occurred (Hu et al., 2015). Since the early 2000s, environmental

TABLE 2.1
Contaminant Source and Associated Toxicity and Human Health Impacts from Exposure to Certain Pollutants

Metal	Use/Source	Health Impact
Arsenic	Pesticide, Groundwater	Encephalopathy, Peripheral neuropathy, Skin problems, Gastroenteritis, Muscle aches, chills and fevers, Hyperpigmentation, Cardiomyopathy, Jaundice.
Cadmium	Electroplating, Storage batteries, Vapour lamps, Soldering	Emphysema, Pulmonary oedema, Dyspnoea, Tachycardia, Anaemia, Cyanosis, Anosmia, Renal tubular dysfunction, Osteomalacia.
Chromium	Car manufacture, Glass, Pottery, Linoleum manufacture.	Lung and respiratory tract cancer, Kidney diseases, Acidosis, Sepsis, Myocardium.
Cobalt	Jet engines	Nausea, Vomiting, Anorexia, Tinnitus, Nerve damage, Goitre.
Lead	Battery, Welding	Lethargic, Anorexia, Dysarthria, Hyperproteinemia, Pallor, Anaemia, Ataxia, Encephalopathy, Papilledema.
Manganese	Purifying agent, Metallurgy	Damage to the central nervous system, Hallucinations, Gait, Dystonia.
Mercury	Chemical workers, Dentists	Lethargy, Dyspnoea, Fibrosis, Ataxia, Choreoathetosis, Polyneuropathy, Dysarthria.
Phosphorous	Fertiliser	Abdominal cramps, Ptosis, Miosis, Sensorimotor polyneuropathy, Atrophy.
Thallium	Cement, Coal, Pyrites.	Somnolence, Hematemesis, Alopecia, Optic atrophy, Retrobulbar neuritis, Ophthalmoplegia.
Copper, Zinc	Mining, Metallurgy	Metal fume flu disease, Blood poisoning, Liver dysfunction.

Source: Adapted from NORD (www.rarediseases.org)

problems of mining legacies have attracted increasing recognition and research internationally. Various conferences, workshops and working groups, such as the World Bank, International Finance Corporation, United Nations Environment Program, World Conservation Union and International Council on Mines and Metals, have sought to identify key issues and coordinate a way to make progress in solving the issues (Pepper et al., 2014).

The environmental issues related to abandoned mines are well documented, yet few systematic surveys and impact assessments have been carried out to fully appreciate the scope of the problem (Pepper et al., 2014). Efforts have been made to conduct inventories of mining legacies within jurisdictions and development of frameworks such as risk-based approaches and criteria for assessment. For example, a national approach called the National Orphaned/Abandoned Mine Initiative (NOAMI) has been developed in Canada. In Australia, states have undertaken mining inventories to varying degrees, with the acknowledgment of the need for cooperation and information sharing between states and territories to tackle legacy mine pollution (Gruenhagen et al., 2020; Environment Agency (2008); Venkateswarlu et al., 2016). Below we review the birth of environmental awakenings for selected countries as a result of mining legacies.

2.3.1 China

Mining has been an industry in China ever since the Bronze Age, dating back to 2000 BC during the Xia Dynasty (Ma, 1986). The modernization of mining in China occurred in Chairman Mao's rule during the 1970s with technical assistance from the USSR (Joseph, 1986). China initiated its market-oriented open-door policy in 1978, which began as a capitalistic approach and led to minerals trading becoming internationalized (Dorian 1994). It took a while in China to set up laws, and, in 1986 itself, the Mineral Resources Law governing the mineral exploration and exploitation commenced. In this law, the Ministry of Land and Resources was given the responsibility to supervise, approve, register and license mine mineral resources exploration. As well, the mine safety regulations were now supervised by the Ministry of Emergency Management. Provincial governments, such as Sichuan Mineral Resources Management, enacted regulations that oversee local mining activities. Recently, the National Mineral Resources Plan (NMRP) of 2016–2020 was designed to augment the modernization and globalization of the industry. As per the Ministry of Land Resources of PRC, the NMRP includes oil and natural gas besides metal mines with a provision to reclaim more than 500,000 ha of legacy mined land (Que et al., 2019).

Mining in China in recent decades has seen a steady growth of reported serious accidents and environmental disasters. One such case is Yuxi River basin in Hunan Province where the concentration of antimony in river sediments exceeded the regulatory limits and polluted adjacent farmlands. As a result, human exposure to antimony was detected in children (Fei et al., 2017). This is evidence of certain gaps in China's regulatory system. Globally, China followed by Australia have retained their dominance of global mining for the past decade. China has its own rising domestic demand and, for this, access to resources is needed. Exhaustion of domestic resources is minimized by importing major metals like iron from Australia, placing both countries at the top followed by the United States (Figure 2.1). According to the Analysis and Policy Observation, Australia; countries

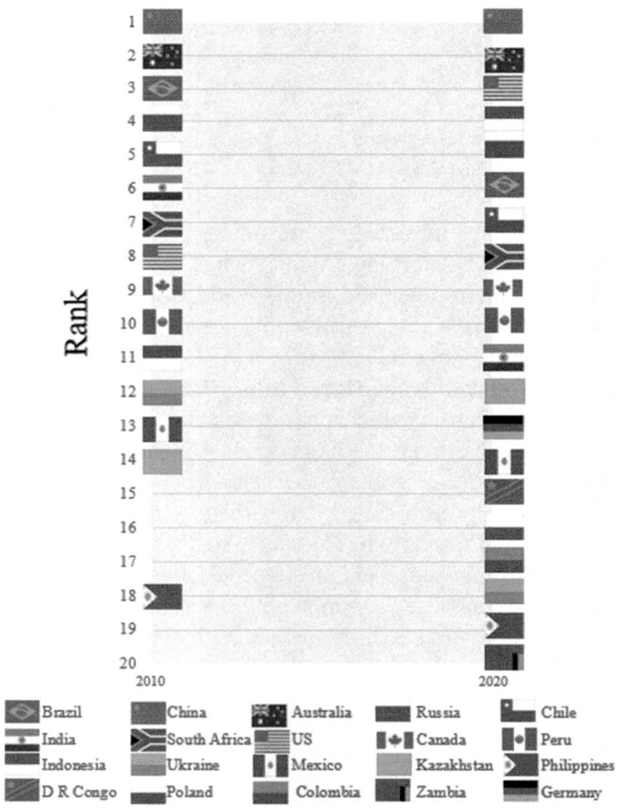

FIGURE 2.1 Mining contribution index ranking. Adopted from SNL Mine Economic database 2016 and Analysis and Policy Observatory (APO), Australia 2020.

such as India lost their position in the top ten mining contribution index ranking, dropping from 6th to 11th place, whereas Indonesia climbed from 11th to 4th position in 2020. Brazil, which was in third position in 2010, was pushed down to 6th position in 2020 with China and Australia retaining their first and second ranks, respectively (Cordes et al., 2016).

China passed its sole land reclamation law in 1986 to address the recovery and land utilization once mining operations had ceased. In 1989, the Land Reclamation Regulation (LRR) was formally promulgated. This was followed by the issuance of the Land Reclamation Ordinance in 2011 further defining the mining responsibility of companies through the establishment of reclamation plans. As per Liu (2011), the reclamation rate of China rose from 1% in 1987 to 25% in 2010. One of the major environmental problems caused by metal mining in China is acid mine drainage. To manage such problems, percolation leaching technology is recently employed mainly to produce gold, copper and uranium (Ilankoon et al., 2018). For instance, there were particularly three policies implemented in 31 provinces of the PRC which helped the reclamation of mines (Cheng and Skousen, 2017). The policies were:

- In 2005, the Comprehensive Rectification and Standardization notice of Mineral Resources Exploitation Order (NCRS) was released.
- The year 2006 saw the issuance of the Gradual Establishment of the Responsibility Mechanism of Mine Environment Management and Ecological Restoration (GERM) guidance.
- In 2009, the Protection of Mine Geological Environment (PPMGE) provisions were put forth.

China has more than 10,000 active mines and is the world's largest producer of coal and gold. China uses 50% of the world's coal, 56% of its iron ore and 64% of its copper (MLR, China 2017). Due to the exhaustion of resources as a result of decades of intensive extraction of coal, an increasing number of abandoned mines now exist (Hu et al., 2015). To reduce the environmental problems the PRC's 13th five-year plan recommends measures such as replacing coal-fired power stations with nuclear power (Xing et al., 2017). Moreover, the quality of mineral ores in China has been declining and increasingly depleted, raising the processing costs and scarcity of high-grade mineral ores. This has caused potential hazards of waste rocks due to over-exploitation in the quest to obtain better-quality high-volume ores (Norgate and Jahanshahi, 2010). Further information on mining and derelict mines in China is provided in Chapter 5.

2.3.2 South Korea and the Philippines Archipelago Belt

Exploring the mining legacy and its environmental problems further east, South Korea, and the Philippines and Archipelago belt is considered here for review due to their environmental vulnerability. South Korea's Ministry of the Environment (KMOE) began conducting soil contamination surveys on abandoned metal mines throughout the country in 1992. Recognition of the negative impacts on the environment and human health in South Korea due to abandoned mining operations was articulated through the Mine Pollution Prevention and Reclamation Act (MPPRA) – 2005. It included provisions for mine operators' paying money for each year's operation into a fund for rehabilitation after a mine had closed. By 2016 the assessment of 1,536 abandoned metal mines had been completed and, of these, 672 were found to exceed guidance levels of soil contamination (KMOE, 2017). Heavy metals were also detected in river water, sediment and agricultural products (KMOE, 2017). Since most of the metal mines in South Korea were established prior to 1940, by the 1980s these mines faced closure due to a decline in their economic feasibility (KMOE, 2017). By 2005, inactive or abandoned mines were reported to account for 64% of all mines (1270 of 2006 in total), including cases where the responsible companies were no longer operating.

A new entity, Mine Reclamation Corporation (MRC), was established in 2006 to actively promote reclamation of abandoned mines in South Korea. The Ministry of Environment also established a cooperative network for the assessment and management of abandoned mining areas, including determining the status of soil contamination, impact on local residents and resolving the lack of sufficient post-mining management after a project had finished. According to KMOE (2017) reclamation work has been completed for 192 abandoned metal mines found to exceed environmental standards, and work is continuing on the remaining mines. Community health impact assessments were conducted on residents living in areas near 142 abandoned metal mines between 2008 and 2017, reporting potential harmful levels of arsenic in humans (Chang et al., 2019).

The Philippines Archipelago environment is equally vulnerable to increasing heavy metal pollution as evident in the suspended nickel laterite mine in Manicani Island in the country's Eastern Samar province (Corales et al., 2019; Bueno, 2018; Holden, 2012). The outcome of a 2001 semi-detailed assessment of mining activity in the Philippines showed that all abandoned mine sites pose environmental risks from unstable tailings ponds and toxic contaminants in soil and water (Samaniego et al., 2019). The accumulation of heavy metals in marine sediments was also observed in Marcopper Mine on Marinduque Island where elevated levels of Cu, Mn, Pb and Zn were detected (David, 2002; Samaniego et al., 2019). The Philippines is an island country which has 27 abandoned and active mines of gold, copper, chromite, mercury, etc., that started operations in the 1960s (Aggangan et al., 2019). According to the records, the first Philippines Mining Act was passed in 1995, incorporating guidelines regarding mine decommissioning and rehabilitation (Corales et al., 2019).

2.3.3 Australia

The Aboriginal Australians utilized the earth's mineral resources for cultural, technological and trade reasons until the arrival of the settler economy and the Victorian gold rush which started in 1851 (McLean, 2012). Australia saw a rapid growth in the mining sector in the 19th century, mostly dominated by coal, gold, silver and base metals (Eklund, 2015). The mining industry in Australia contributes to about 8% of total GDP (Han et al., 2014). This was because of $400 billion investment between 2003 and 2014. Australian exports account for 60% of the total GDP where Australia is the largest exporter of iron ore and the second largest exporter of coal (Han et al., 2014). In Australia mining directly employs more than 200,000 people supporting the regional economies and exports aluminum, copper, gold, uranium and zinc besides the major ones – iron and coal (McMahon and Moreira, 2014).

The major income revenue source for the national, state and territory governments of Australia amounted to approximately $25 billion during the 2013–14 financial year. Exporting these commodities further generates employment to the importing countries, including infrastructure such as roads, schools and hospitals (McMahon and Moreira, 2014). The federal and state governments of Australia are proactively ensuring that the closure of mines is automatically linked to rehabilitation by developing all-encompassing operational plans (Pepper et al., 2014). As such, the Society for Ecological Restoration along with the Department of Trade, Investment, Regional Infrastructure and Services in 2013 recommended using the following attributes of mine site restoration (McManus and Connor, 2013):

- The site exhibits an ecosystem diversity and community structure akin to that of reference sites.
- Indigenous species are present, along with the necessary functional groups for ensuring long-term stability.
- The physical environment possesses the capacity to support reproducing populations. It operates normally and is seamlessly integrated within the landscape.

- Potential threats have been eliminated, and the site displays resilience to natural disturbances.
- Furthermore, it demonstrates self-sustainability.

The ability of a derelict or abandoned mine to achieve its rehabilitation targets rests on the scale and type of mining impacts. It is generally hard to restore its pre-existing ecosystems due to the soil nutrition quality and rainfall availability (Audet et al., 2013; Biswas et al., 2019). To solve soil nutrition problems, treatment criteria were implemented in Western Australia (Nichols, 2006) using the CGAP process. The CGAP process comprises the following: criteria and intent of remediation of soil; guidelines for acceptance; and accepted standard and potential corrective actions. Even after landscaping and safety actions have been taken, and vegetation on or near mines were established, the results have not been very satisfactory due to the lack of direct monitoring (Cristescu et al., 2013). Like most countries globally, rehabilitation of derelict mines poses significant challenges to both mining companies as well as state and territory governments and the Australian federal Government as discussed in Chapter 3.

2.3.4 Canada

Commercial mining started in Canada in 1738 with an iron mine operating at Forges du Saint-Maurice in Quebec followed by a copper mine in Ontario in 1848 (Sandlos and Keeling, 2015). Today, Canada hosts headquarters of approximately 75% of the world's mining companies, making Toronto the financial hub of the industry (Marques, 2016). As such, there are more than 10,000 orphaned and abandoned mines (OAM) in Canada, associated with scores of environmental liabilities, human health concerns and a huge financial burden to clean up the OAMs (Hogan and Tremblay, 2006). As per Mining Watch Canada, 2017 estimates, the total liability for contaminated mining sites throughout Canada is well above $9.1 billion. The National Orphaned/Abandoned Mines Initiative (NOAMI) in Canada is a notable initiative for managing abandoned mines, created in 2002 based on recommendations put forward at a multi-stakeholder workshop on abandoned mines.

NOAMI involves cooperation among key parties, including the mining industry; federal, provincial and territorial governments; NGOS and indigenous peoples. Principles of the program include Information Gathering/Inventory, Community Involvement, Legislative and Institutional Barriers to Collaboration, Funding Approaches, Jurisdictional Legislative Review and Mine Closure / Return of Mining Lands. NOAMI operates according to the following guiding principles (Cowan et al., 2013):

- Communication and Information Dissemination.
- Inclusiveness and Participation.
- Representation and Consistency of Involvement.
- Fostering Confidence in Decision-Making.
- Resources, Assistance and Facilitation.
- Integration and Respecting Local Cultures.

The six case studies undertaken in various mine sites across Canada (Cowan et al., 2013) gave rise to key lessons which are outlined below:

i. Clear and well-written legislation and policies are imperative in jurisdictions to facilitate the return of lands to their former state. Establishing a system for receiving and managing funds is equally crucial to ensure the viability needed for addressing long-term issues.
ii. In cases where planning and procedures for relinquishment are lacking, defaulting will eventually occur, and it will be at the taxpayers' expense.

iii. Only a few jurisdictions have a satisfactory funding regimen in place to handle surrendered, or about-to-be-surrendered, lands.
iv. In most jurisdictions that provide for rehabilitation releases and/or relinquishment of lands, there are no complete releases for environmental responsibility or liability.
v. It is evident that only certain jurisdictions have institutional control plans in place for relinquished lands. This means there is no administrative unit directly responsible for issues following surrender, such as records maintenance and land use planning inputs.
vi. Public consultation efforts are generally considered inadequate, especially in comparison to current expectations, particularly concerning Aboriginal communities. Effective consultation necessitates comprehensive information on the issues and understanding of what the affected communities need or want.
vii. The length of post-closure monitoring periods requires more extensive evaluation, especially in cases where documentation or identification of features is lacking.
viii. Well-defined risk assessment procedures for property returns are lacking in many situations. More information on quantitative risk assessment is needed to support determinations of chemical and physical stability.
ix. In situations involving cutting-edge technology, longer-term monitoring and scientific assessment may be anticipated, particularly in new schemes related to permafrost or climate change issues.
x. It is crucial to develop effective cost estimation procedures for long-term care and maintenance.
xi. The storage and maintenance of mine plans and records are essential for technical assessment. Several case studies were compromised due to the loss or unavailability of historic documentation.
xii. Providing a final closure report that details all completed decommissioning and reclamation work stands as an excellent best practice.

Recent research on comparison of mine closure plans in northern Canada highlighted the need to integrate indigenous knowledge with wider social, economic and cultural circumstances (Monosky and Keeling, 2021). This suggests that more needs to be done for effective and sustainable mine closure beyond NOAMI.

2.3.5 South Africa

Mining has been a major contributor to the South African economy, with large-scale mining starting in the mid-late 19th century, once diamonds and gold were discovered (CSIR, 2019). However, it operated for a long time under limited government regulation/oversight that left the responsibilities for mining impacts to the mine owners, and, in the process, many mines were abandoned (Mhlongo et al., 2016). An agreement in 1976 saw mines that had been abandoned prior to that year become the responsibility of the state to manage and rehabilitate. Such a timely intervention to manage mines has been economically positive for the ecosystem, in that there is now a better balance between the mining and agricultural industries. For instance, South Africa, Brazil and Lesotho share 2% of the total GDP from mining compared to 25% in Indonesia, DR Congo and Bolivia (Figure 2.2), indicating that strict regulations mean countries must find alternative sources of revenue other than mining (DMR, 2009, 2019; CSIR, 2019; Van and Bekker, 2017).

According to Hartzer, 2009, the Mines and Works Act 27 (1956) was concerned with the effects of metal mines operations on the environment. However, this statute did not have the requirements regarding rehabilitation when mining operations ceased until amendments to the MWA and regulations on rehabilitation came into effect in 1980 (Hartzer, 2009). The Minerals Act 50 (1991) repealed the Mines and Works Act 27 (1956) and contained a provision for rehabilitation. Section 38

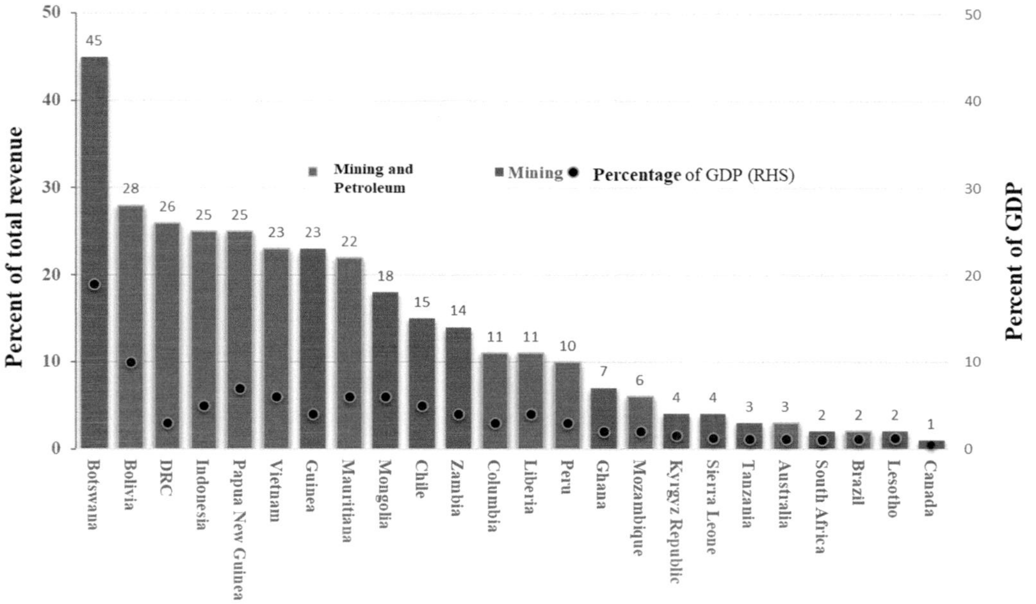

FIGURE 2.2 Government receipts from natural resources (averages 2000–2013). Adopted from IMF (2016) and Analysis and Policy Observatory Australia (2016).

of the MA enforced rehabilitation on all holders of mining authorizations until a closure certificate was issued (Hartzer, 2009).

The Mineral and Petroleum Resources Development Act-2002 (MPRDA) came into effect in 2004 and served as the major legislation regulating mineral resources and environmental impacts. Section 46 of the legislation sets aside funds to be retrieved from the rights holder for the rehabilitation of a prematurely abandoned site where it is urgently needed to prevent environmental degradation (Klopper and Wessels, 2017). The section 46 of the legislation from the MPRDA Act 2002 (Cornelissen et al., 2019) classifies those mines for which a closure certificate has not been issued and no party can be traced to assume responsibility for their liabilities as abandoned. Furthermore, the government may provide funds for their rehabilitation (Cornelissen et al., 2019).

An integrated approach toward "one environmental system" between government departments was adopted in 2014 for mine environmental management. Regulations were introduced under the National Environmental Management Act, 1998 (NEMA) for the financial provisioning for costs associated with rehabilitation of mine sites through the lifetime of operations and residual environmental outcomes that may become known in the future. However, mine closure in South Africa remains elusive due to the lack of mine closure expertise and knowledge, lack of closure planning during feasibility stages, deficiencies in data and reporting on environmental impacts and financial liabilities (Van and Bekker, 2017).

The Department of Mineral Resources (DMR) of South Africa began an inventory of abandoned mines in the early 2000s to understand how big this problem was. By the end of 2008, almost 6,000 abandoned mines were known, but measures were not in place to ensure that abandoned mines were rehabilitated effectively and in a timely manner and there was no integrated information system to record and report on the status of mines to support this effort (DMR, 2009). Appropriate definitions of abandoned mine sites and features are needed for improving the inventory (Mhlongo et al., 2016), and consequently the national strategy for abandoned mines in South Africa finalized in 2009 included:

- Compiling and maintaining an inventory of derelict and ownerless mines.
- Establishing a framework for identifying priority sites based on considerations of public health and safety as well as environmental and social risks.
- Creating an action plan to effectively manage the risks associated with derelict and ownerless mines.
- Guaranteeing the rehabilitation of all derelict and operational sites to meet acceptable standards for habitability and environmental soundness.

The cost for remediation of officially listed abandoned mines in South Africa as of March 2019 was R48.7 billion (DMR, 2019). Estimated rehabilitation costs are affected by, first, the lack of agreed criteria as to what needs to be remediated; and, second, what the goals for rehabilitation should be. To expedite progress on dealing with the legacy of abandoned mines in South Africa, the identification of rehabilitation criteria is expected to assist in the prioritization of rehabilitation efforts (Mhlongo et al., 2016). Additionally, going beyond qualitative risk ranking will help to further prioritize funds (Mhlongo et al., 2016). Currently, the risk-ranking systems that have been devised are mostly internal to the organizations directly involved in their application. Subsequently, there is little public or academic knowledge of their methodologies and application (Cornelissen et al., 2019).

Issues with capacity of government agencies are evident in upholding regulations throughout Africa and this is also the case in South Africa, which arguably has the most developed and sophisticated provisions for mine closure planning on the continent (Morrison-Saunders et al., 2016). An effective and risk-based approach to remediation and coordination between different agencies can lead to discrepancies and oversights in rehabilitation efforts (Morrison-Saunders et al., 2016).

2.3.6 United States of America

One of the incentives that European colonists and explorers were hoping to obtain in the Americas was mineral treasure along the Atlantic Ocean coast (Albert, 1879). Iron ore was sent to England from Jamestown in 1608, followed by Virginia in 1620 and Massachusetts in 1628. The first copper mining took place in Connecticut as early as 1709; however, the famous Lake Superior copper was first mined not until 1771. In this rush for fortune, the environment was least regarded primarily due to the state of ignorance about environmental issues (Albert, 1879). The General Mining Act of 1872 gave federal authorization to the prospecting and mining for minerals such as gold, platinum, silver, lead, zinc, uranium and tungsten. However, miners often abandoned mines, leaving behind structures, safety hazards and contaminated land and water (Seymour, 2004). It has been estimated that as many as 500,000 abandoned mine sites exist in the United States, with an estimated cleanup cost of $54 billion (House Committee on Natural Resources, 2016). These abandoned mine sites pose health and safety risks, particularly for recreational users of public lands due to the existence of undocumented mine shafts and abandoned toxic waste. Furthermore, acid mine drainage (AMD) wields a substantial impact on surface and groundwater, thus further degrading water quality and ecosystems (Venkateswarlu et al., 2016).

Established in 1997, the Bureau of Land Management (BLM) runs an abandoned mine lands (AML) program. It runs in partnership with federal, state and local governments to reduce the environmental and health risks of abandoned land mines. Several other agencies are also involved in reclaiming abandoned mines, such as the Office of Surface Mining Reclamation and Enforcement (OSME), Forest Service, and Environmental Protection Agency (EPA). The aim of the BLM program is to enhance public safety and improve water, soil and air quality in AML by remediating the effects of the legacy hard rock mines, as in the western United States. These agencies as per the Government Accountability Office (GAO, 2008) are responsible for managing large areas of land, including abandoned mines on mined lands (GAO, 2020). Furthermore, the BLM maintains a national inventory of known abandoned mines (mostly hard rock mines). As of 2019, the focus is primarily on tracking and reporting abandoned mine features rather than sites. The Forest Service does not have a national inventory as of 2019, but it tracks abandoned hard rock mine by features for physical safety hazards and by site for environmental hazards at regional and district levels. The Parks Service maintains a database of abandoned mines and features on its lands and manages them through a safety and environmental compliance and cleanup program. The EPA maintains a database of mine sites on non-federal lands as part of its national database of contaminated sites. However, sites which pose only a physical safety hazard are not identified because they are outside the scope of the superfund program (GAO, 2020). The Forest Service, BLM, the Park Service and EPA identified 140,652 abandoned hard rock mines as of 2019, of which 60% are expected to pose environmental or physical safety hazards. Between the fiscal years 1998 and 2007, BLM, the Forest Service, EPA, and OSME spent at least $2.6 billion (in 2008 constant dollars) to reclaim abandoned hard rock mines. This increased to $2.9 billion from 2008 through 2017 (about $287 million annually) (GAO, 2020). It is evident that BLM prioritizes and takes appropriate action on these historic abandoned mine sites using a risk-based approach (GAO, 2020).

Federal law requires hard rock mining operations to reclaim disturbed land when operations cease and provides reclamation plans and bonds or other financial assurance to guarantee reclamation. Agency officials periodically review bonds held to ensure they will sufficiently cover expected reclamation costs. As of October 2017, BLM held over $3 billion in financial assurances, for 1,425 hard rock mining operations, 4% of which had inadequate financial assurances (GAO, 2019). The Forest Service also holds $250 million in financial assurances, though it lacks a centralized mechanism to record the current estimated cost of reclamation to determine if financial assurances held are in fact adequate (GAO, 2019).

For managing coal mines, the Surface Mining Control and Reclamation Act (SMCRA) of 1977 requires coal mining areas to be restored to a condition capable of supporting the uses prior to any mining, or higher or better uses. A reclamation plan is required for mining approval, and bonds must be posted to ensure funding is available for reclamation. The SMCRA also places a levy per ton on current coal mine operators to fund the abandoned mine reclamation (Ziemkiewicz et al., 2021).

2.3.7 Europe

Many mines in Europe were shut down through the 19th and 20th centuries without any rehabilitation work due to economic and environmental pressures (Žibret et al., 2020). Although the mining industry in Europe has been declining over the last few decades, Europe has a long history of mining dating back more than 5,000 years ago (Žibret et al., 2020). Mining has been an occupation since the Bronze Age, with exploitation of copper, tin, gold and lead as evidenced by archaeological excavations in Europe (Timberlake, 2017). For example, the Ross Island mine which began operation in 2400 BC in southwestern Ireland (County Kerry) provided the inhabitants of the British Isles with much of their early copper supply. Britain's first mining boom is said to have occurred between 1600 and 1400 BC (Williams and Le Carlier de Veslud, 2019).

This history has led to a legacy of widespread environmental damage, with many thousands of mines now discharging metal-containing mine waters into adjacent waterways (Johnston et al., 2008). An estimated 1,000 to 1,500 km of water courses have been impacted by metal mine discharges in the EU member states (Younger et al., 2002). For instance, the United Kingdom is managing its metal pollution legacy by focusing on the most significant environmental impacts where more than 2,000 km of rivers and several major public water supply aquifers are polluted or threatened by abandoned mines (Mayes et al., 2015). As presented in Figure 2.3, technologies which function by adding no chemicals or energy, and reactions in the compost mixture bind up the metals flowing in from metal contaminated water as sulfides, are being implemented. This innovative treatment strategy has diminished the metal concentrations in the water that flows back into the Coledale Beck, a river running through Cumberland in the United Kingdom (Mayes et al., 2015).

As such, in Europe the implementation of the Mining Waste Directive (2006/21/EC) includes requirements for providing proper closure of operating and future mines within the European

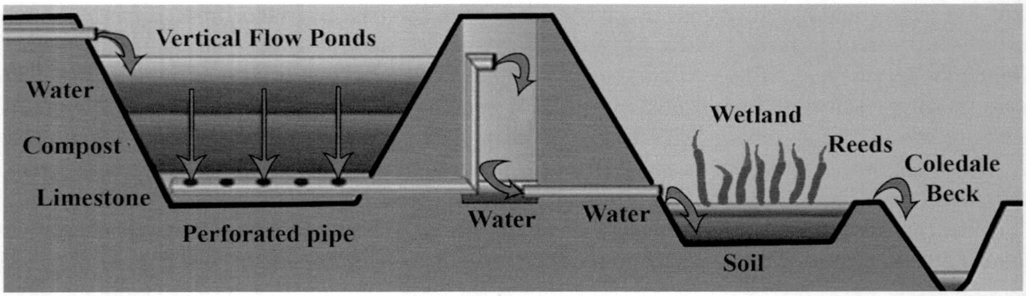

FIGURE 2.3 Water from mine workings on vertical flow ponds to treat metal from mines contaminated water in the Coledale Beck, UK. Adopted from Tamar Estuaries Consultative Forum, 2019.

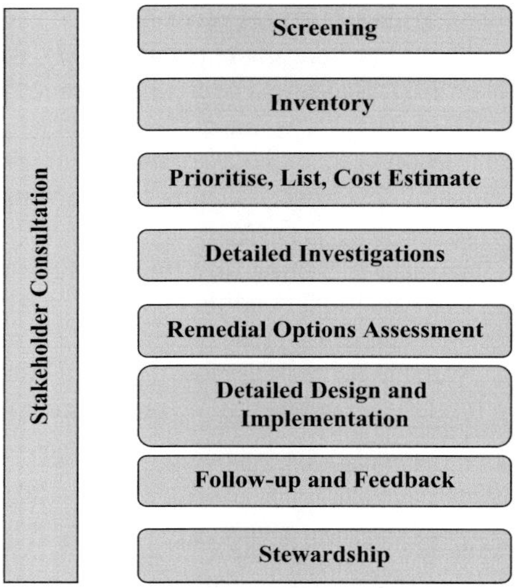

FIGURE 2.4 Outline of the process for addressing abandoned mines by EU member states (DHI, 2012).

Union. It also contains a mandate (Article 20) for production of an inventory of abandoned mining sites located on their territory which cause or have the potential to threaten human health and the environment in the short or medium term (Kulczycka et al., 2020). This is covered by the "Guidance document for a risk-based pre-selection protocol for the inventory of closed waste facilities as required by article 20 of the directive 2006/21/EC". Furthermore, Article 21 addresses the need for the rehabilitation of those facilities identified under Article 20, including establishing the best risk assessment procedures and remedial actions (Environment Agency, 2008, 2012). A support document (DHI, 2012) outlines the process to be undertaken by member states (Figure 2.4) involving the screening and inventory preparation leading to smart stewardship.

2.4 CURRENT MINING PRACTICE

Globally, mining has continued to expand in the modern era and remains economically important. According to world data, mining production rose from 9.6 billion tons in 1985 to 17.2 billion tons in 2017. The biggest producers were China, the United States, Russia and Australia and production is increasing in developing countries, including those in Latin America and Africa (Reichl and Schatz, 2019). The global extent of land area impacted by mining and quarrying is estimated to range between 300,000 and 800,000 km^2 (Cherlet et al., 2018). According to the Mining Contribution

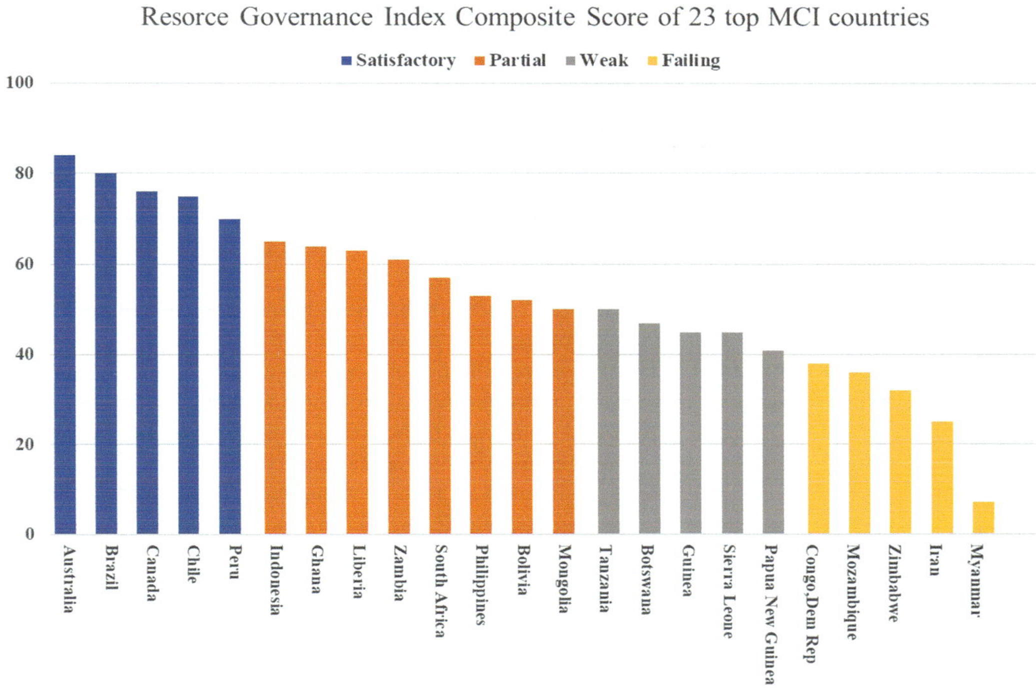

FIGURE 2.5 Resource Governance Index Composite Scores for 23 of the top Mining Contribution Index (MCI) countries. Adopted from NGRI and APO 2016 Australia.

Index (MCI) of the ICMM, Australia ranks number one in resource governance followed by Brazil and Canada (see Figure 2.5). This figure indicates the effectiveness of the emerging market economies compared to the United States, China and the United Kingdom.

As per the Asia-Pacific Economic Cooperation (APEC), mining regulations are implemented throughout the world with the aim to ensure safe and sustainable closure of mine sites (APEC, 2018). However, the evolution of these regulations has been uneven with some jurisdictions still having little or no meaningful regulation while others having a robust governance and enforcement framework in place (APEC, 2018). Failure of the regulatory process undermines the social license to operate (Unger et al., 2014) and mining legacies reflect poorly on the minerals industry (Pepper et al., 2014). According to the Intergovernmental Forum on Mining, Minerals, Metals and Sustainable Development, improperly regulated mining can disrupt the benefits that come from the sector because environmental harm and disruption of social and economic structures are not guarded against (IGF, 2020). There is a need to balance economic gains from mining with other economic, social and environmental values. This requires strong, honest and transparent institutional frameworks to regulate the mining sector and to ensure good governance practices (Cherlet et al., 2018). For instance, abandoned and inactive mine sites are recognized as a major environmental problem in Chile with some 650 sites needing rehabilitation (Weeks et al., 2015). Chile holds the world's largest reserves of copper and lithium (USGS, 2020). Chile has recently been successful in keeping its laws updated and viable by developing a more robust governance framework (ICMM, 2007; Weeks, 2015). In 2012, laws regulating mines and mine closure came into force. Increased requirements for closure included an approved closure plan prior to commencement of operations, including financial guarantees (Weeks. 2015).

2.4.1 Mine Closure Provisions

The potential for creating new mining legacies is less significant in areas where mining-related legislation is already in place for mine closure, with enough and guaranteed financial assurance for the proposed rehabilitation activities (Bennet, 2016). Mine closure planning, including progressive rehabilitation, reduces the likelihood of mine voids and pit lakes being left unattended after the cessation of mining activities. Clark and Clark (2005) note aspects of well-developed mining legislation such as those of Japan, Australia, Canada, the United States, Germany and the United Kingdom. The legislative provisions include EIA required before lease, specific provisions for reclamation, bonding procedures, abandonment and non-compliance.

In those countries with an extensive mining history, substantial numbers of abandoned mines and associated problems have generally compelled the development of more comprehensive legislation. The EU Directive 2006/21/EC recognizes the importance of guaranteeing taxpayers that they do not have to pay the financial burden of environmental cleanup and rehabilitation of mining liabilities. Ensuring this requires a financial guarantee to be lodged by the mine operator/owner prior to the commencement of deposition operations in the waste facility (Clark and Clark, 2005). However, developing countries in Africa, Asia and Latin/South America vary in their mining law provisions for mine closure. Some of the features include:

- There is a noticeable lack of specific policy and legislation regarding mine closure.
- The state bears a significant degree of responsibility, particularly for abandoned and certain operational mines.
- Negotiations for mine closure were integrated into individual mining agreements.
- There are minimal, if any, bonding procedures in place to guarantee thorough mine closure.

These provisions are often inadequate for comprehensive mine closure. A provision for post-mining sustainable development is lacking, which emerges as the greatest problem for governments to deal with and find a solution (Clark and Clark, 2005). It is evident from Figure 2.6 that revenue allocation

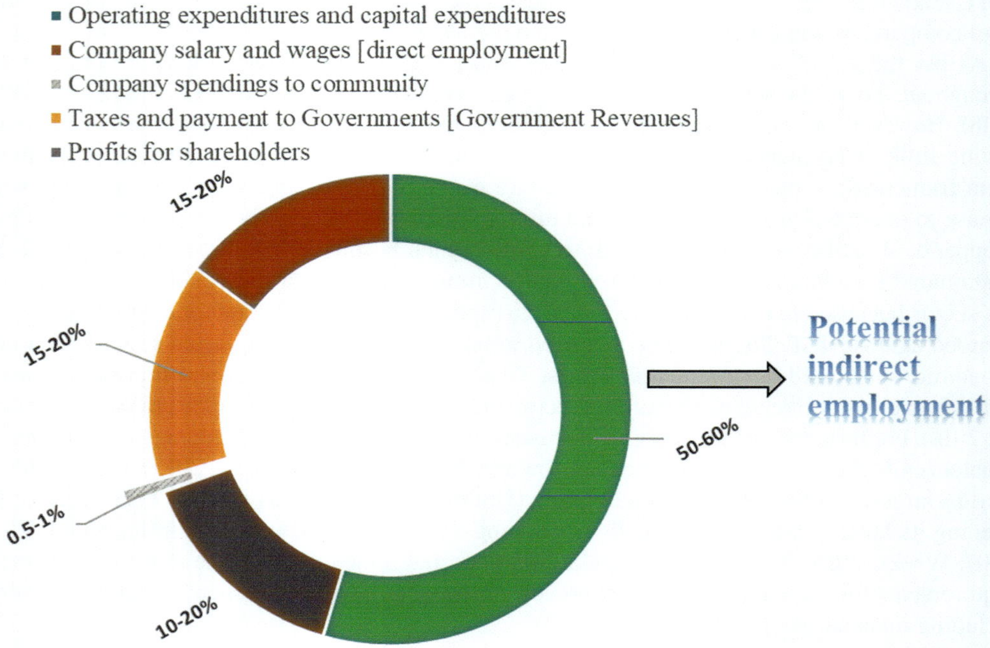

FIGURE 2.6 Revenue allocation to different stakeholders for a typical mine operation. Adopted from APO 2016, Australia.

for remediation or rehabilitation of mines is negligible, especially where a large slice of expenditure goes for operation and capital expenses (ICMM, 2016).

To manage capital expenses, the Intergovernmental Forum on Mining, Minerals, Metals and Sustainable Development has developed guidelines on the legal framework for Environmental and Social Impact Assessment (ESIA) in mine planning and approval (IGF, 2020). ESIA is a key component of government decision-making regarding mine approval and provides a reference document for license holders, government agencies and other stakeholders during the life of a mining project. The key aspects of the rehabilitation process of EISA include commitment to sustainable development; consistency and coordination of responsible authorities, including community engagement and transparency; environmental and social management plans; financial assurances; and oversight and enforcement of non-compliance.

With an aim of providing standards and guidelines to mitigate long-term mining damage and to create environmental and socioeconomic value through mine closure and reclamation management, the International Organization for Standardization is developing standards for mine closure and reclamation management (ISO TC 82/SC 7), with the participation of 12 countries (Murphy et al., 2019). The scope of ISO TC 82/SC 7 is to address all lifecycle stages of resource development, including design, development, mining, temporary suspension of operations, cessation of mining, closure and post-closure monitoring and management. With these in mind, the principles of mining rehabilitation have shifted in recent years from a minimum standard of safety, stability and revegetation to restoration of the landscapes to pre-mining conditions. The objective here is to ensure future sustainability of land use for purposes such as agriculture, nature reserves, forests and grasslands (Sloss, 2013). In the EU, for example, 50% of reclaimed mined lands are in the form of forests and grasslands, whereas in China 70% are reclaimed for agriculture (Bian et al., 2010) and the aim is to create more than 80% in the future (Murphy et al., 2019).

2.4.2 Funding Mechanisms for Managing Mines

It is not practical to rehabilitate all abandoned mine sites within a short period of time and to the acceptable level (Pepper et al., 2014). This is indeed true since resources are constrained, and countries do not have the revenues or budgets to make this possible. Therefore, it is desirable to prioritize addressing those sites which pose the most risk from the safety and human health perspective and impact on the environment (Unger et al., 2012). This is being increasingly experienced in the management of abandoned mines by different jurisdictions. A typical funding mechanism available for rehabilitation of mines includes state-funded remediation, "polluter pays" or a mixed funding mechanism, based on state and private sector contributions. For historically abandoned mines "polluter pays" policies are generally not applicable since no responsible party can be identified. Therefore, abandoned mines require state- funded or alternative funding mechanisms (Tremblay and Hogan, 2016). Since the aim of the mining industry is to make profits and develop communities, the value chain governance as presented in Figure 2.7 involves the stakeholders involved in the process. The funding mechanism for mining falls under revenue, whereas the environmental regulator operates as a producer of revenue. Therefore, the order of process for effective rehabilitation and regulation of mines must have a place for environment regulators at the policymaking level (Macfarlane, 2015).

There is a view that the mining industry as it currently operates should contribute to a fund that goes toward rehabilitating abandoned mines (Macfarlane, 2015). Since 2013, mine operators in the Northern Territory of Australia pay an annual 1% levy, based on the value of the financial security held for each mine site, into a Mining Remediation Fund, the proceeds of which are used by the government to address issues caused by abandoned mines. In the United States, the Surface Mining Control and Reclamation Act (SMCRA) of 1977 imposes a levy on coal production, which is deposited in a fund for abandoned mine reclamation (Macfarlane, 2015).

In Western Australia, a Mining Rehabilitation Fund (MRF) was established by the Department of Mines, Industry Regulation and Safety (DMIRS). All tenement holders are required by DMIRS

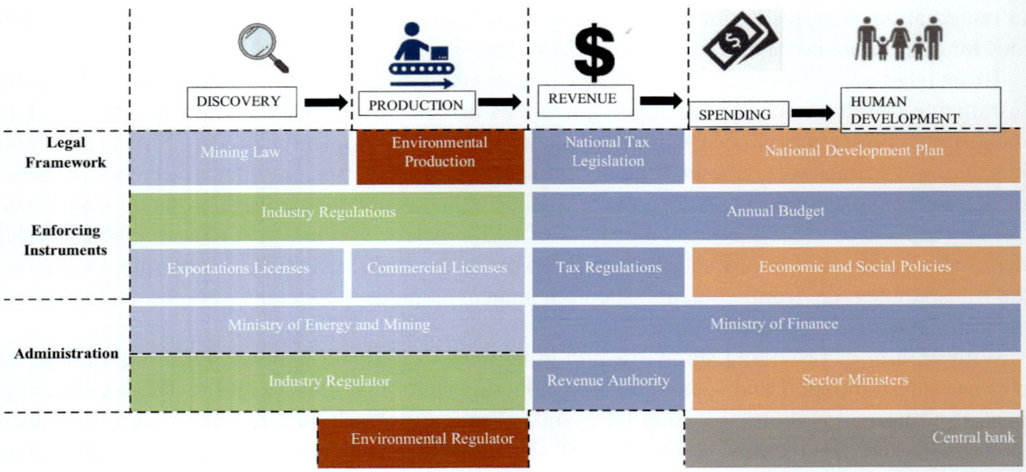

FIGURE 2.7 The value chain and governance. Adopted from Macfarlane, 2015 and APO Australia 2016.

to report annually with respect to disturbance and contribute to the fund. Interest earned on money in the fund is available to rehabilitate abandoned mines across the State in circumstances where the original operator cannot be held responsible (Mitchell et al., 2019). For instance, in the southwest of Western Australia (Figure 2.8), approximately 70% of the understory plant diversity has returned in Jarrah (*Eucalyptus marginata*) forest bauxite mines because of using the original topsoil (Koch, 2007). This is one of the ideal examples and a success story for the industry when it was implemented by Alcoa World Alumina Australia. This company operates two bauxite mines at Willowdale and Huntly in the Darling Range in Western Australia, 80–140 kilometers south of Perth (Grant and Gardner, 2005).

FIGURE 2.8 Jarrah Forest restoration by Alcoa World Alumina south of Perth, Western Australia. (Picture credit: Alcoa Website; Mine Site Remediation, WA).

2.5 SOCIAL AND ECONOMIC CONSIDERATIONS OF MINING

Farsighted planning strategies by government bodies are critical but also subject to the volatility of long-term economic development goals through mining and metal contributions. The impact on the national economy by revenue generated through mining may vary based on the circumstances of low- and middle-income countries. Generally, mining typically can contribute up to 15% of a country's total employment (Dorin et al., 2014). With an average annual salary of US$94,000, the American mining industry supports more than 1.1 million jobs. According to one ICMM report several years ago, an estimated 3.5 additional jobs are generated by each mining job in comparison to 2.6 additional jobs by the non-mining sector (ICMM, 2014).

The mining sector has an enormous potential to generate jobs, indirectly reducing poverty, boosting living standards, raising productivity and allowing individuals to invest in their health and education needs. Due to advances and changes in technology, the number of direct jobs in the mining sector has fallen and positions are limited. For example, a single computerized machine can do the jobs of several workers. However, the indirect positive revenue impact for the government by export income increases business for non-mining sectors of the economy, thereby creating jobs. Subsequently, indirectly the service industry flourishes (Dorin et al., 2014). Australia has been at the forefront of attracting foreign investment since the mid-1800s when the gold rush was in full swing. It shows that harnessing abundant resources is important for prosperity. Foreign investment in the mining sector in Australia during the fiscal year 2013–14 was 38%, which effectively meant $265 billion (Stevens et al., 2015).

The ICMM in its third edition of *Role of Mining in National Economies* (2016) reports that direct employment by the mining industry is only 1 to 2%, however, indirect employment is higher (ICMM, 2014). According to the World Bank in 2016, minerals production in the United States was 0.6% of the total world GDP PPP in 1996, which peaked at 1.9% of global GDP in 2011, making the mining sector an important contributor to GDP (Roe and Dodd, 2017). From 2004 onward, total minerals production surpassed the global GDP (Figure 2.9), indicating that all other goods and services were not economically viable and subject to swings in inflation.

Albeit, the African mining boom is an indication of the vision that metals and minerals are the backbone of a nation's economy (ICMM, 2007). The leading example of booms to their national economies induced by mining are found in West African countries (Ericsson and Lof, 2019).

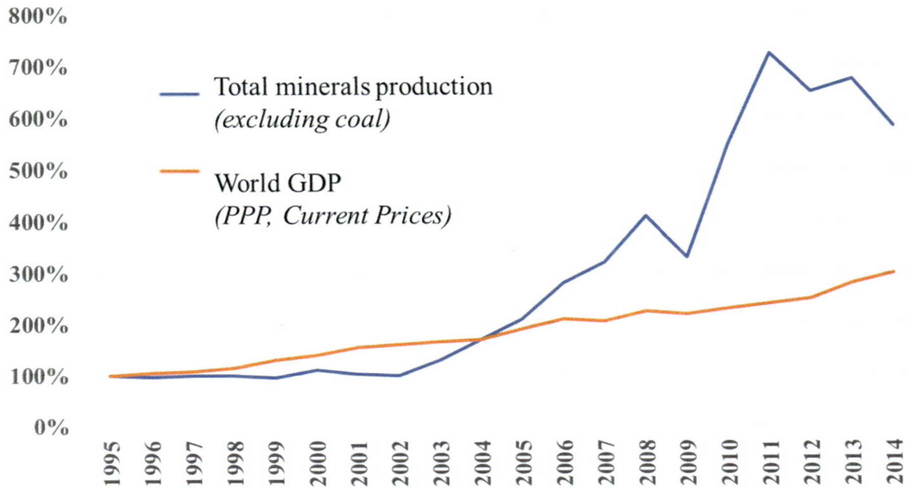

FIGURE 2.9 World GDP and total mineral production (1995 to 2014). Adopted from: SNL Database, World Bank and APO Australia.

The failure to rehabilitate and remediate former mine sites is possible and only becomes effective if the economy of a given locality and the stakeholders are stable and reliable. Often, the unstable and fluctuating nature of the mining business, given the changes in prices of minerals, ores, etc., hampers the rehabilitation process (Renner et al., 2009). Along with the dependence on mining activity, it is important to diversify economic activity so that there are more GDP sources apart from mining for achieving the goal of sustainable and environmentally friendly development (Mthenjane, 2019). An alternative solution to mine closure is to consider the promotion of tourism, as it is viable for generating income. Alternative uses such as aquaculture, irrigation, recreation and storage of waste chemicals, etc., can generate incomes and local economic livelihoods from abandoned and formerly mined lands. One such example of such an activity is the former Kidston Gold Mine in north Queensland, Australia. This site is located 270 km northwest of Townsville and being turned into a hydropower plant (McCullough and Lund, 2006). A comprehensive analysis underscores the critical role of innovation in empowering the mining sector to tackle its most pressing challenges, while emphasising the nuanced influence of diverse stakeholders on the diffusion and adoption process, paving the way for informed decision making and future research endavours (Gruenhagen and Parker, 2020) .

2.6 CONCLUSIONS

The fervent desire for environmental sustainability within the mining industry reflects a commitment to both reaping economic benefits and ensuring ongoing productive land use for communities. To realize this valuable objective, it is imperative to continually strengthen governance rules and procedures through legislation, planning and compliance efforts.

Significant strides have already been made in implementing robust governance within the mining industry, with progress also evident in developing nations. The introduction of bonding requirements for rehabilitation is viewed as crucial insurance for ensuring effective restoration of mined lands. Additionally, efforts to address mining legacies are showing satisfactory progress, characterized by a deeper understanding of issues associated with abandoned mines and the development of frameworks for assessing, prioritizing and managing risks at these sites.

Certain governments are taking proactive steps by implementing funding mechanisms, such as mining levies, to secure the necessary resources for tackling mining legacies. With the attainment of a zero-emission target, the future of mining is poised to be sustainable.

REFERENCES

Aggangan N.S., Anarna J.A., Cadiz N.M. (2019). Tree legume – Microbial symbiosis and other soil amendments as rehabilitation strategies in mine tailings in the Philippines. *Philippine Journal of Science,* 148(3): 481–491.

Alpers, C.N., Hunerlach, H.J., May, J.T., Hothem, R.L. (2005). Mercury contamination from historical gold mining in California Fact Sheet 3014. US Geological Survey.

Asia-Pacific Economic Cooperation (APEC). (2018). Mine closure checklist for governments. APEC Mining Task Force, Singapore. APEC Project: MTF 03 2016A. SOM Steering Committee on Economic and Technical Cooperation (SCE), Mining (Mining). https://www.apec.org/publications/2018/03/mine-closure---checklist-for-governments.

Audet, P., Arnold, S., Lechner, A.M., Baumgartl, T. (2013). Site-specific climate analysis elucidates revegetation challenges for post-mining landscapes in eastern Australia. *Biogeosciences*, 10(10): 6545–6557.

Bennett, K. (2016). Abandoned mines — Environmental, social and economic challenges. In A.B. Fourie & M. Tibbett (Eds.), *Proceedings of the 11th International Conference on Mine Closure*, Australian Centre for Geomechanics, Perth, pp. 241–252. https://doi.org/10.36487/ACG_rep/1608_16_Bennett.

Bian, Z., Inyang, H.I., Daniels, J.L., Otto, F., Struthers, S. (2010). Environmental issues from coal mining and their solutions. *Mining Science and Technology (China)*, 20(2): 215–223.

Biswas, B., Nirola, R., Biswas, J.K., Pereg, L., Willett, I.R., Naidu, R. (2019). Environmental microbial health under changing climates: State, implication, and initiatives for high-performance soils. In *Sustainable Agriculture Reviews*, vol. 29 (pp. 1–32). Cham: Springer.

Blanchette, M.L., Lund, M.A. (2016). Pit lakes are a global legacy of mining: An integrated approach to achieving sustainable ecosystems and value for communities. *Current Opinion in Environmental Sustainability*, 23: 28–34.

Bolles, A.S. (1879). Industrial history of the United States, volume IV. Norwich, CN: Henry Bill Publishing Company. https://www.legendsofamerica.com/we-mininghistory/#:~:text=The%20Schuyler%20Mine%2C%20near%20Belleville,in%20America%20in%201793%2D94.

Bueno, M. (2018). Evaluation of abandoned mine lands rehabilitation: The case of Bagacay mine, Western Samar, Philippines. Masters thesis, University of Tokyo.

Bureau of Land Management (BLM). (2007). Bureau of Land Management (BLM) standard Document Indentification Numbering System (DINS). US Department of the Interior. https://www.blm.gov/policy/im-2007-058.

Bureau of Land Management (BLM). (2013). Abandoned mined lands – A new Legacy. BLM Publication Number: BLM/WO/GI-13/007-3720. US Department of the Interior.

Carvalho, F.P. (2017). Mining industry and sustainable development: Time for change. *Food and Energy Security*, 6(2): 61–77.

Chang, J.Y., Ahn, S.C., Lee, J.S., et al. (2019). Exposure assessment for the abandoned metal mine area contaminated by arsenic. *Environmental Geochemistry and Health*, 41(6): 2443–2458. https://doi.org/10.1007/s10653-019-00296-5.

Cheng, L., Skousen, J.G. (2017). Comparison of international mine reclamation bonding systems with recommendations for China. *International Journal of Coal Science & Technology*, 4(2): 67–79.

Cherlet, M., Hutchinson, C., Reynolds, J., Hill, J., Sommer, S., von Maltitz, G. (Eds.). (2018). *World Atlas of Desertification*. Luxembourg: Publication Office of the European Union.

Clark, A. L., Clark, J. C. (2005). An international overview of legal frameworks for mine closure. Environmental Law Alliance Worldwide: Energy and Biodiversity Initiative.

Corales-Ultra, O.G., Peja, R.P., Casas, E.V. (2019). Baseline study on the levels of heavy metals in seawater and macroalgae near an abandoned mine in Manicani, Guiuan, Eastern Samar, Philippines. *Marine Pollution Bulletin,* 149: 110549. https://doi.org/10.1016/j.marpolbul.2019.110549.

Cordes, K., Östensson, O., Toledano, P. (2016). Employment from mining and agricultural investments: How much myth, how much reality? Columbia Center on Sustainable Investment, Columbia University.

Cornelissen, H., Watson, I., Adam, E., Malefetse, T. (2019). Challenges and strategies of abandoned mine rehabilitation in South Africa: The case of asbestos mine rehabilitation. *Journal of Geochemical Exploration*, 205: 106354.

Cowan, W.R., Mackasey, W.O., Robertson, J.G.A. (2013). Case studies and decision-making process for the relinquishment of closed mine sites. Prepared for NOAMI, viewed 16 March 2017. http://www.abandoned-mines.org/en/document/publication/.

Cristescu, R.H., Rhodes, J., Frére, C., Banks, P.B. (2013). Is restoring flora the same as restoring fauna? Lessons learned from koalas and mining rehabilitation. *Journal of Applied Ecology*, 50(2): 423–431.

CSIR. (2019). A CSIR perspective on South Africa's post-mining landscape. Authored by W. de Lange, N. Funke, S. Oelofse, K. Nortje. CSIR Report No.: stelgen16928. Pretoria: CSIR.

David, C.P. (2002). Heavy metal concentrations in marine sediments impacted by a mine-tailings spill, Marinduque Island, Philippines. *Environmental Geology*, 42(8): 955–965. https://doi.org/10.1007/s00254-002-0601-4.

Department of Mineral Resources (DMR). (2009). The national strategy for the management of derelict and ownerless mines in South Africa.

Department of Mineral Resources (DMR). (2019). DMR presentation on derelict and ownerless asbestos mine. September 28, 2019.

DHI. (2012). Annex 3 – Supporting document on closure methodologies for closed and abandoned mining waste facilities (April 2012). In Establishment of guidelines for the inspection of mining waste facilities, inventory and rehabilitation of abandoned facilities and review of the BREF. Document No. 070307/2010/576108/ETU/C2.

Dorian, J.P. (1994). *Minerals, Energy and Economic Development in China*. Oxford: Clarendon Press.

Dorin, I., Diaconescu, C., Topor, D.I. (2014). The role of mining in national economies. *International Journal of Academic Research in Accounting, Finance and Management Sciences*, 4(3): 155–160.

Eklund, E. (2015). Mining in Australia: An historical survey of industry–community relationships. *The Extractive Industries and Society*, 2(1): 177–188.

Environment Agency. (2008). Using science to create a better place: Assessment of Metal Mining-Contaminated River Sediments in England and Wales. Science Report: SC030136/SR4. https://www.gov.uk/government/uploads/system/uploads/attachment_data/file/291646/scho1108bozd-e-e.pdf.

Environment Agency. (2012). Evidence: Ecological indicators for abandoned mines. Report: SC090024/R2. https://www.gov.uk/government/uploads/system/uploads/attachment_data/file/291486/LIT_7657_ef99fa.pdf.

Ericsson, M., Löf, O. (2019). Mining's contribution to national economies between 1996 and 2016. *Mineral Economics*, 32(2): 223–250.

Fei, J.C., Min, X.B., Wang, Z.X. et al. (2017). Health and ecological risk assessment of heavy metals pollution in an antimony mining region: A case study from South China. *Environmental Science and Pollution Research*, 24: 27573–27586.

Glenn, V., Doley, D., Unger, C., McCaffrey, N., McKenna, P., Gillespie, M., Williams, E. (2014). Mined land rehabilitation-is there a gap between regulatory guidance and successful relinquishment? *AusIMM Bulletin*, (3): 48, 50–52, 54.

Government Accountability Office (GAO). (2008). Hardrock mining: Information on abandoned mines and value and coverage of financial assurances on BLM land. GAO-08-574T.

Government Accountability Office (GAO). (2019). GAO-19-436R Hardrock Mining financial assurances 2019 Hardrock Mining: BLM and forest service hold billions in financial assurances, but more readily available information could assist with monitoring. Washington, DC.

Government Accountability Office (GAO). (2020). Abandoned Hardrock mines: Information on number of mines, expenditures, and factors that limit efforts to address hazards. GAO-20-238.

Gozzard, E., Mayes, W.M., Potter, H.A.B., Jarvis, A.P. (2011). Seasonal and spatial variation of diffuse (non-point) source zinc pollution in a historically metal mined river catchment, UK. *Environmental Pollution*, 159: 3113–3122.

Grant, C., Gardner, J. (2005). Mainstreaming biodiversity in the mining industry: Experiences from Alcoa's Bauxite mining operations in Western Australia. http://citeseerx.ist.psu.edu/viewdoc/download?doi=10.1.1.502.761&rep=rep1&type=pdf.

Gruenhagen, J.H., Parker, R. (2020). Factors driving or impeding the diffusion and adoption of innovation in mining: A systematic review of the literature. *Resources Policy*, 65: 101540.

Han, X., Khan, H., Zhuang, J. (2014). Do governance indicators explain development performance? A cross-country analysis. Asian Development Bank Economics Working Paper Series No. 417.

Hartzer, S. (2009). The liability of historical mine authorization holders for rehabilitation. Doctoral dissertation, North-West University, South Africa.

Hogan, C.M., Tremblay, G.A. (2006). Abandoned mines in Canada. International Conference on Acid Rock Drainage, March 26–30, St. Louis, Missouri, pp. 774–784. American Society of Mining and Reclamation (ASMR).

Holden, W.N. (2012). Ecclesial opposition to large-scale mining on Samar: neoliberalism meets the church of the poor in a wounded land. *Religions*, 3(3): 833–861. https://doi.org/10.3390/rel3030833.

Hu, Z., Fu, Y., Xiao, W., Zhao, Y., Wei, T. (2015). Ecological restoration plan for abandoned underground coal mine site in Eastern China. *International Journal of Mining, Reclamation and Environment*, 29(4): 316–330.

ICMM. (2007). Chile. The challenge of mineral wealth: using resource endowments to foster sustainable development. Country case study. March 2007. London: ICMM.

ICMM. (2014). The role of mining in national economies (2nd edition). Mining's contribution to sustainable development. October 2014. London: ICMM.

ICMM. (2016). Role of mining in national economies, 3rd ed. https://apo.org.au/sites/default/files/resource-files/2016-10/apo-nid69735.pdf.

Institute for Environment and Development (IIED) and World Business Council for Sustainable Development (WBCSD). (2002). Mining for the future appendix C: Abandoned mines, No. 28 Working Paper. [Report commissioned by the MMSD project of IIED]. http://pubs.iied.org/pdfs/G00882.pdf.

Ilankoon, I.M.S.K., Tang, Y., Ghorbani, Y., Northey, S., Yellishetty, M., Deng, X., McBride, D. (2018). The current state and future directions of percolation leaching in the Chinese mining industry: Challenges and opportunities. *Minerals Engineering*, 125: 206–222.

IGF (The Intergovernmental Forum on Mining, Minerals, Metals and Sustainable Development). (2020). IGF guidance for governments: Improving legal frameworks for environmental and social impact assessment and management. Winnipeg: International Institute for Sustainable Development (IISD).

Inobeme, A., Ajai, A.I., Iyaka, Y.A., Ndamitso, M., Uwem, B. (2014). Determination of physicochemical and heavy metal content of soil around paint industries in Kaduna. *International Journal of Scientific and Technology Research*, 3(8): 221–225.

Johnston, D., Potter, H., Jones, C., Rolley, S., Watson, I., Pritchard, J. (2008). Abandoned mines and the Water Environment: Science project. SC030136-41, Environment Agency, the Scottish Environment

Protection Agency (SEPA), the Coal Authority (The Environment Agency et al.), Rio House, Waterside Drive, Aztec West, Almondsbury, Bristol.
Kirpichtchikova, T., Manceau A., Spadini, L., Panfili, F., Marcus, M.A., Jacquet, T. (2006). Speciation and solubility of heavy metals in contaminated soil using X-ray microfluorescence, EXAFS spectroscopy, chemical extraction, and thermodynamic modeling. *Geochimica et Cosmochimica Acta*, 70(9): 2163–2190.
Klopper, D., Wessels, J. A. (2017). Investigation of Western Australia's rehabilitation fund as a fiscal policy solution for South African abandoned mines. *Journal of the Southern African Institute of Mining and Metallurgy*, 117(11): 1081–1087.
KMOE (Korean Ministry of the Environment). (2017). *White Paper of Environment* (pp. 342–345). Republic of Korea Press.
Koch, J.M. (2007). Restoring a Jarrah Forest understorey vegetation after bauxite mining in Western Australia. *Restoration Ecology*, 15(S4): S26–S39.
Kulczycka, J., Dziobek, E., Szmiłyk, A. (2020). Challenges in the management of data on extractive waste—The Polish case. *Mineral Economics*, 33, 341–347.
Lenntech, W. (2004). *Water Treatment and Air Purification*. Rotterdam: Lenntech.
Liu, Z. (2011). To give land life and vigor - Interpretation of "Land reclamation ordinance" by the director of policy and regulations of the Ministry of Land and Resources. *Land Resources Communication*, 5: 12–13.
Lokeshwari, H., Chandrappa, G.T. (2006). Impact of heavy metal contamination of Bellandur Lake on soil and cultivated vegetation. *Current Science*, 91(5): 622–627.
Ma, C. Y. (1986). *Ancient Chinese Bronzes*, ed. by Shih, H. Y. Hong Kong: Oxford University Press.
Macfarlane, A.S. (2015). Reconciliation along the mining value chain. *Journal of the Southern African Institute of Mining and Metallurgy*, 115(8): 679–685.
Marques, J.C. (2016). Private regulatory fragmentation as public policy: Governing Canada's mining industry. *Journal of Business Ethics*, 135(4): 617–630. https://doi.org/10.1007/s10551-014-2377-3.
Mayes, W.M., Anton, A.D., Thomas, C., Potter, H.A.B., Rudall, S., Amezaga, J., Gandy, C.J., Jarvis, A.P. (2015). April. National assessment of sediment-related diffuse mining pollution in England and Wales. In *Proceedings of the 10th International Conference on Acid Rock Drainage (ICARD) and the International Mine Water Association (IMWA)*, Santiago, Chile.
McCullough, C. D., Lund, M. A. (2006). Opportunities for sustainable mining pit lakes in Australia. *Mine Water and the Environment*, 25(4): 220–226.
Mchaina, D.M. (2001). Environmental planning considerations for the decommissioning, closure and reclamation of a mine site. *International Journal of Surface Mining, Reclamation and Environment*, 15(3): 163–176.
McLean, I.W. (2012). *Why Australia Prospered: The Shifting Sources of Economic Growth The Princeton Economic History of the Western World, vol. 43*. Princeton University Press.
McMahon, G., Moreira, S. (2014). The contribution of the mining sector to socioeconomic and human development. Extractive Industries for Development Series No. 30. World Bank. April 2014.
McManus, P., Connor, L.H. (2013). What's mine is mine (d): Contests over marginalisation of rural life in the Upper Hunter, NSW. *Rural Society*, 22(2): 166–183.
Mhlongo, S. E., Amponsah-Dacosta, F. (2016). A review of problems and solutions of abandoned mines in South Africa. *International Journal of Mining, Reclamation and Environment*, 30(4): 279–294.
Mining Watch Canada. (2017). Finally! New national inventory for orphaned and abandoned mine sites, 10 May. https://miningwatch.ca/news/2017/5/10/finally-new-national-inventory-orphaned-and-abandoned-mine-sites
Ministry of Land and Resources (MLR). (2017). *Ministry of Land and Resources of the P.R. of China, China Mineral Resources*. Beijing: Geological Publishing House. http://www.mlr.gov.cn/sjpd/zybg/2017/201710/P020171017524608365416.pdf
Mitchell, I., Hryczyszyn, K., Read, T. (2019). A framework to prioritise high-risk abandoned mine features for rehabilitation in Western Australia. In A.B. Fourie & M. Tibbett (Eds.), *Mine Closure 2019: Proceedings of the 13th International Conference on Mine Closure*, Australian Centre for Geomechanics, Perth, pp. 123–132.
Monosky, M., Keeling, A. (2021). Planning for social and community-engaged closure: A comparison of mine closure plans from Canada's territorial and provincial North. *Journal of Environmental Management*, 277: 111324.

Morrison-Saunders, A., McHenry, M.P., Rita Sequeira, A., Gorey, P., Mtegha, H., Doepel, D. (2016). Integrating mine closure planning with environmental impact assessment: Challenges and opportunities drawn from African and Australian practice. *Impact Assessment and Project Appraisal*, 34(2): 117–128.

Mthenjane, M.I. (2019). Post-mining use of rehabilitated land - An opportunity for the South African mining industry for sustainable development. *Journal of the Southern African Institute of Mining and Metallurgy*, 119(9): 693–696.

Murphy, D.P., Nahir, M., Didier, C. (2019). An update on the development of mine closure and reclamation standards by the International Organization for Standardization. In A.B. Fourie & M. Tibbett (Eds.), *Mine Closure 2019: Proceedings of the 13th International Conference on Mine Closure, Australian Centre for Geomechanics*, Perth, pp. 761–770.

Nichols, O.G. (2006). Developing completion criteria for native ecosystem reconstruction——A challenge for the mining industry. In A.B. Fourie & M. Tibbett (Eds.), *Mine Closure 2006: Proceedings of the First International Seminar on Mine Closure*, Australian Centre for Geomechanics, Perth, pp. 61–74.

Nirola, R., Megharaj, M., Venkateswarlu, K., Aryal, R., Correll, R., Naidu R. (2016). Assessment of metal toxicity and bioavailability in metallophyte leaf litters and metalliferous soils using *Eisenia fetida* in a microcosm study. *Ecotoxicology & Environmental Safety*, 126: 264–272.

Norgate, T., Jahanshahi, S. (2010). Low grade ores–smelt, leach or concentrate? *Minerals Engineering*, 23(2): 65–73.

Pepper, M., Roche, C. P., Mudd, G. M. (2014). Mining legacies—Understanding life-of-mine across time and space. In *Proceedings of the Life-of-Mine Conference*. Australasian Institute of Mining and Metallurgy, Brisbane, pp. 449–465.

Que, S., Awuah-Offei, K., Demirel, A., Wang, L., Demirel, N., Chen, Y. (2019). Comparative study of factors affecting public acceptance of mining projects: Evidence from USA, China and Turkey. *Journal of Cleaner Production*, 237: 117634.

Reichl, C., Schatzm, M. (2019). World mining data 2019: Iron and ferro-alloy metals non-ferrous metals precious metals industrial minerals mineral fuels, vol. 34. Vienna: Federal Ministry for Sustainability and Tourism.

Renner, S., Rankin, M., Ponce, R., Griffin, B., Moraes, R., Dalheimer, M., Parot, M.E. (2009). 8th International Conference on Acid Rock Drainage (ICARD) and securing the future: Mining, metals & the environment in a sustainable society (pp. 1542–1554). Skelleftea Swedish Association of Mines, Minerals and Metal Producers, Skelleftea.

Roe, A.R., Dodd, S. (2017). Dependence on extractive industries in lower-income countries: The statistical tendencies. Working Paper 2017/98. Helsinki: UNU-WIDER.

Samaniego, J.O., Gibaga, C.R.L., Tanciongco, A.M., Rastrullo, R.M., Costa, M.A.V. (2019). Surface water characteristics in the vicinity of abandoned mercury mine site in Puerto Princesa City, Philippines. *Philippine Journal of Science*, 148(3): 493–498.

Sandlos, J., Keeling, A. (2015). Mining. In The Canadian Encyclopedia. Archived from the original on October 10, 2020. https://www.thecanadianencyclopedia.ca/en/article/mining.

Seymour, J.F. (2004). Hardrock mining and the environment: issues of federal enforcement and liability. *Ecology Law Quarterly*, 31: 795.

Sloss, L. (2013). Coal mine site reclamation. IEA Clean Coal Centre, CCC/216.

Sonone, S.S., Jadhav, S., Sankhla, M.S., Kumar, R. (2020). Water contamination by heavy metals and their toxic effect on aquaculture and human health through food chain. *Letters in Applied NanoBioScience*, 10(2): 2148–2166.

Starr, K., Orsi, R. J. (Eds.). (2000). Rooted in barbarous soil: People, culture, and community in gold rush California. California History Sesquicentennial series, vol. 3. Berkley, California: University of California Press.

Stevens, P., Lahn, G., Kooroshy, J. (2015). The resource curse revisited [research paper]. Chatham House: The Royal Institute of International Affairs.

The House Committee on Natural Resources. (2016). Abandoned mines. https://naturalresources.house.gov/imo/media/doc/Abandoned%20Mines.pdf

The Sierra Fund. (2008). Mining's toxic legacy: An initiative to address mining toxins in the Sierra Nevada. Nevada, CA: The Sierra Fund.

Timberlake, S. (2017). New ideas on the exploitation of copper, tin, gold, and lead ores in Bronze Age Britain: The mining, smelting, and movement of metal. *Materials and Manufacturing Processes*, 32(7–8): 709–727.

Tremblay, G.A., Hogan, C.M. (2016). Managing orphaned and abandoned mines–a Canadian perspective. Paper presented at the Dealing with Derelict Mines Summit, Singleton, NSW, Australia, December 2016.
Unger, C., Lechner, A.M., Glenn, V., Edraki, M., Mulligan, D.R. (2012). Mapping and prioritising rehabilitation of abandoned mines in Australia. *Life-of-Mine Conference 2012: Conference Proceedings, The Australasian Institute of Mining and Metallurgy*, pp. 259–266.
Unger, C.J., Lechner, A.M., Walton, A., Glenn, V., Edraki, M., Mulligan, D.R. (2014). Maturity of jurisdictional abandoned mine programs in Australia based on web-accessible information. In *Conference Proceedings: Life-of-Mine Conference 2014*, The Australian Institute of Mining and Metallurgy, Brisbane.
United Nations Environmental Programme (UNEP). (2001). Abandoned mines. Problems, issues and policy. Challenges for decision makers. Santiago, Chile. 18 June 2001. Summary Report. United Nations. UNEP, Division of Technology, Industry and Economics/Chilean Copper Commission.
US Geological Survey (USGS). (2020). Mineral commodity summaries 2020: US Geological Survey. https://doi.org/10.3133/mcs2020.
Van Druten, E.S., Bekker, M.C. (2017). Towards an inclusive model to address unsuccessful mine closures in South Africa. *Journal of the Southern African Institute of Mining and Metallurgy*, 117(5): 485–490.
Venkateswarlu, K., Nirola, R., Kuppusamy, S., Thavamani, P., Naidu, R., Megharaj, M. (2016). Abandoned metalliferous mines: Ecological impacts and potential approaches for reclamation. *Reviews in Environmental Science and Bio/Technology*, 15(2): 327–354.
Wanty, R.B., Wang, B., Vohden, J., Briggs, P.H., Meier, A.L. (1998). Regional baseline geochemistry and environmental effects of gold placer mining operations on the Fortymile River, eastern Alaska. Geologic Studies in Alaska by the US Geological Survey, 1998, p. 101.
Weeks, B. (2015). Mine closure in Chile – challenges and changes. In *Geotechnical News March*, pp. 42–45.
Williams, R., Le Carlier de Veslud, C. (2019). Boom and bust in Bronze Age Britain: Major copper production from the Great Orme mine and European trade, c. 1600–1400 BC. *Antiquity*, 93(371): 1178–1196. http://doi.org/10.15184/aqy.2019.130.
Xing, W., Wang, A., Yan, Q., Chen, S. (2017). A study of China's uranium resources security issues: Based on analysis of China's nuclear power development trend. *Annals of Nuclear Energy*, 110: 1156–1164.
Younger, P.L., Banwart, S.A., Hedin, R.S. (2002). *Mine Water. Hydrology, Pollution, Remediation.* Environmental Pollution, vol. 5. Dordrecht; Boston: Kluwer Academic Publishers.
Žibret, G., Lemiere, B., Mendez, A.M., Cormio, C., Sinnett, D., Cleall, P., Szabó, K., Carvalho, M.T. (2020). National mineral waste databases as an information source for assessing material recovery potential from mine waste, tailings and metallurgical waste. *Minerals*, 10(5): 446. https://doi.org/10.3390/min10050446
Ziemkiewicz, P.F., Skousen, J., White, K.D., Bruce Leavitt, B., Stiles, J. (2021). Guidelines for the design of abandoned mine land remediation and water treatment. https://wvwri.wvu.edu/files/d/cab72499-a80a-45a6-bdd1-21f801041b34/guidelines-for-aml-remediation-manual-final-15-march-2021.pdf.

3 Derelict Mine Sites – Australian Perspective

Girish Choppala and Ravi Naidu

3.1 INTRODUCTION

The mining industry plays a significant role in the development of our contemporary societies and, in many instances, serves as a significant economic driver at regional and national levels (Guerin, 2006; Weng et al., 2013). Mining for ores and metal smelting may have begun between 6,000 and 7,000 years ago (Goudie, 2018) and excavation of metals and minerals via formal and informal mining industries employs nearly 20 million people around the globe. More than 40,000 people are employed directly in the mining sector in New South Wales (NSW), Australia alone.

Global mining output was predicted to be worth US $1 trillion in 2016, with coal, iron ore, and gold contributing 50, 13 and 11%, respectively, of total mining output (Ericsson and Löf, 2019). Mining operations in Australia contribute 8% of the total gross domestic product (GDP) and the highest percentage of gross value added (10.2%) (Worrall et al., 2009). Australia is one of the world's top five producers of iron ore, lead (Pb), zinc (Zn), gold (Au), nickel (Ni), mine sands and coal (Britt et al., 2017).

Derelict mine sites are areas of land, water and surrounding water bodies where mineral extraction and ore processing have taken place, or ceased for a period of one year or more, with little or no remediation or rehabilitation work undertaken, causing significant environmental damage (Newton et al., 2000). These areas are frequently represented as stranded, inactive and unattended mines. Mine waste comprises metalliferous and nonmetalliferous materials generated from tailings, wastewater, waste sand, silt, clay, millings, drilled muds, crushed gravels and rooks, dust and powdery waste. This waste can overburden topsoil, which is then released into the atmosphere (Issaka and Ashraf, 2021).

Environmental contaminants are frequently released into the environment as a direct result of mining operations, such as grinding, concentrating areas and disposing of tailings and wastewater (Navarro et al., 2008). Abandoned mines pose real or potential threats to human safety, health and the environment, making them a primary global concern. These mine sites contain abundant metal sulphide minerals and, despite being surrendered and abandoned, these zones continue to discharge toxic metals from their build-up residue, posing a dangerous hazard (Mohapatra and Kirpalani, 2016).

In general, regular derelict mine sites comprise stripped areas (59%), open-pit mines (20%), tailing dams (13%), waste tipping areas (5%) and land affected by mining subsidence (3%) (Miao and Marrs, 2000). The geochemical properties and mineralisation rate of mine tailings determine the degree and extent of heavy metal contamination in and around abandoned mine sites (Johnson et al., 2000).

Elevated levels of heavy metals and metalloids in surface water and waterbodies in mining regions are directly associated with the oxidation of sulphide-bearing minerals that are subjected to weathering conditions, which results in acidity generation and high levels of dissolved metals, including copper (Cu), arsenic (As), Zn, cadmium (Cd) and carbonates and sulphates (Navarro et al., 2008).

The fate and transport of these metals is complex and depends on the degree of oxidation of sulphides and soil mineralogy. Heavy metal concentrations are high in and around abandoned mine sites as a result of mine waste discharge and dispersion into nearby agricultural soils and stream

 DOI: 10.1201/9781315141855-3

systems (Jung, 2001). The extent of heavy metal contamination is regulated by a series of precipitation, co-precipitation and sorption reactions (Berger et al., 2000).

3.2 DERELICT MINE SITES – GLOBAL CONTEXT

There are an estimated one million derelict mines worldwide, including shafts, adits and alluvial working mines (Coelho et al., 2011). There are up to 500,000 abandoned mine sites in the United States alone, many of which have been designated as a potential threat to biota and the environment (Struhsacker and Todd, 1998; Young et al., 2021). Australia's heavy reliance on mining is thought to have led to as many as 55,000 abandoned mines dispersed across the nation (Venkateswarlu et al., 2016).

In Europe, a study by the UK government's Department of Environment, Food and Rural Affairs reported that abandoned mines in England and Wales might have polluted up to 10% of water bodies across the United Kingdom (Jarvis et al., 2008). The mining industry in Africa employs an estimated half a million people. While mining for gold, diamonds, metals and coal accounts for more than 40% of the national economy, it has left a legacy of contamination affecting African society with abandoned mines, mainly abandoned gold mines, becoming hotspots for small-scale and illegal mining operations, promoting environmental and physical hazard issues (Mhlongo and Amponsah-Dacosta, 2016). Mining sites have already degraded around 40,000 km^2 of land in China, and abandoned mining land grows by 330 km^2 yearly (Zou et al., 2012).

3.3 MINE SITES IN AUSTRALIA

From the gold rushes that began in the early 1850s and continued for decades to present-day Australia, where mining is the most significant contributor to the country's GDP, Australia was built on mining. The number of derelict mine sites in Australia is estimated to be more than 50,000 (Figure 3.1); however, this is an underestimation due to a lack of data and different state-based

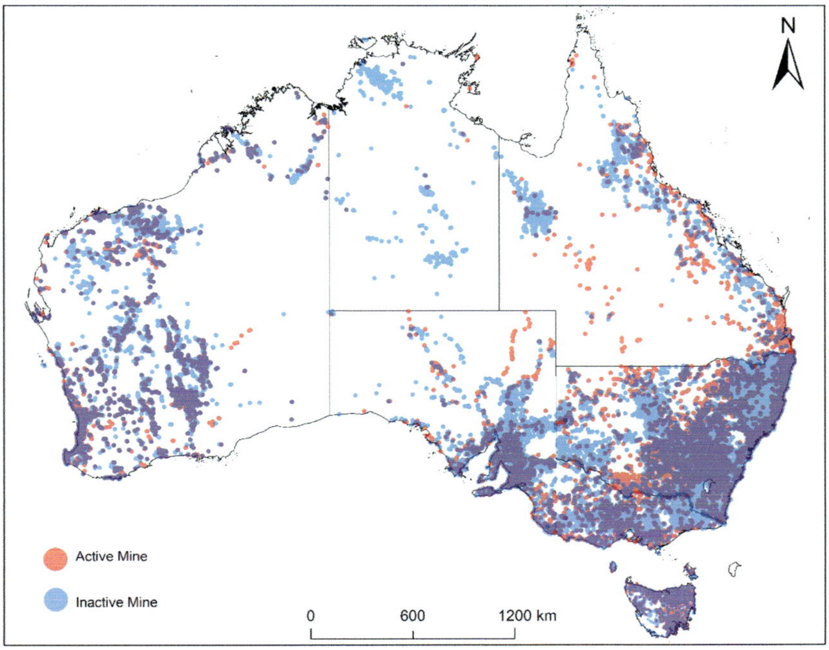

FIGURE 3.1 Australian map displaying the number of derelict mine sites, as reported by July 2011 (adopted from Unger et al. (2012).

approaches complicating efforts to quantify derelict mine sites as a national issue (Pepper et al., 2014).

These sites can range from a shallow excavation, costean, adit or shaft to a significant mining legacy site, such as Mount Morgan in Queensland, Redbank in the Northern Territory (NT) and Mount Lynell in Tasmania (Koehnken et al., 2003; Unger et al., 2003). According to the Queensland government, there are 17,000 abandoned mines in that one state alone, 3,500 reportedly on public land and 12,500 on private land.

Mining activities and operations in New South Wales (NSW) date back to the mid-1800s, and nearly 2,000 derelict mine sites exist in NSW (Grant et al., 2002).

3.4 ENVIRONMENTAL POLICIES TO REGULATE POLLUTION IN DERELICT MINE SITES

Australia has inherited a history of about 75% of mines undergoing premature or unplanned closure, resulting in mines left with no care and maintenance or simply abandoned with no attempt at formal closure of dangerous mine sites and unclaimed or unreclaimed lands (Laurence, 2006; Roche and Judd, 2016). In the past, mining corporations used reckless mining practices, ignoring the need for environmental protection and mine rehabilitation. In the absence of adequate legislative frameworks and controls, insufficient and inefficient mine closure policies, legislative controls and past mining practices have resulted in a legacy of abandoned and derelict mine sites, which significantly affects the ecosystem and mine rehabilitation costs. Governments may incur significant cleanup expenses and closure responsibilities as a consequence of abandoned mines (WorldBank, 2002).

Mining destroys natural ecosystems by removing soil and plants and burying waste disposal facilities underneath them. In particular, mining operations can substantially degrade the productivity of land and generate waste rock and tailings, subsequently altering the quality, level, availability and flow of surface and groundwater. Communities living near these derelict mine sites are becoming more conscious and critical of these socioeconomic and environmental repercussions, as well as mining legacies (Hilson, 2003). As derelict mine sites are geologically fixed and contamination type and extent varies among mine sites, pollution prevention and related management concepts, including waste management and minimisation of toxic waste generation, must be implemented for the mining industry prior to their incorporation into policies and laws.

Although mining operations in Australia occupy just 0.02% of the total geographical area (ABARES, 2019), their environmental and social consequences are increasingly significant to practitioners, regulators, academics and the general public. This is because negative legacies are so prevalent, such as the >50,000 abandoned or orphaned mining features. The deteriorated environmental and living conditions surrounding mining sites have disrupted livelihoods and subsequently affected the quality and commitment of the workforce, culminating in the mine's closure and a loss of revenue.

The Australian Constitution has the authority to create and govern mine site policies. Yet, because coal isn't included as a matter in Australia's Constitution, it is up to the individual states to enact laws governing coal mining. In addition, the Australian territories have a different legislative structure than the states, and, whilst still subject to Commonwealth regulation, they retain the right to adopt and enforce their own mining laws (Dwiki, 2018).

In Australia, the Indigenous inhabitants and traditional owners were vulnerable to mining operations. For example, Energy Resources Australia has been mining uranium at Ranger mine in Kakadu National Park, NT, since the 1970s, and they proposed to extend uranium mining to nearby Jabiluka. Since this proposed location is home to 32 Mirrar community members, the Jabiluka Action Group (JAG), in collaboration with Mirrar Aboriginal people living in Kakadu, has prevented the significant development of a uranium mine site in Jabiluka through shareholder protests, legal actions, direct action and direct appeals to international bodies, in addition to shareholder protests, legal actions, direct action and direct appeals to international bodies (Hintjens, 2000).

Mining policies in Australia are primarily governed by (I) environmental protection and (II) Native Title. Based on Commonwealth and Queensland law, the Environmental Protection Act 1994 (EPA) regulates the environmental impacts of mining. The Environmental Protection and Biodiversity Conservation Act of 1999 also governs this policy (EPBC Act). The Commonwealth regulates Native Title and Indigenous cultural heritage; consequently, when exploration and exploitation are conducted in certain areas, the Commonwealth can also impose legislation (Dwiki, 2018).

The Western Australia (WA) Mining Rehabilitation Fund Act (MRF Act) and regulations are used to define an abandoned mine site. Within WA, an abandoned mine site can be declared by the Minister, and the regulations state that the mining rehabilitation advisory panel may advise the minister of suitable land for this declaration. The WA policymakers defined abandoned mine sites as *"areas of land impacted by former mining activities for which no individual, company, or organisation can be held responsible for rehabilitation. An abandoned mine site may comprise multiple areas of land or site features (e.g., mine shafts, waste dumps, abandoned equipment) that may or may not be adjacent to each other"* (DMP 2015).

There is presently no national government body in Australia with formal duty for remediating abandoned mines, despite a non-binding national policy. With a few rare exceptions, abandoned mines on Crown property are the exclusive responsibility of the state or territory, while those on freehold land are the sole responsibility of the landowner. Nevertheless, there are a few abandoned mines that fall within the jurisdiction of the Australian government.

The Ministerial Council on Mineral and Petroleum Resources (MCMPR) was composed of state and territory ministers who were in charge of minerals and petroleum, as well as the ministers for resources, energy and tourism for the governments of Australia and New Zealand. The Standing Council on Energy and Resources (SCER) now performs the duties of MCMPR, which is no longer in existence. The Council of Australian Governments (COAG) formed SCER as a council to explore important national-level problems in the energy and resources sectors and advance major reform initiatives for the mining, oil and energy industries. At a national level, the Strategic Framework was prepared to encourage convergence of state and territory jurisdictions on the following aspects (Bennett and Lacy, 2012):

- Site inventories and data management.
- Improved understanding of liability and risk relating to abandoned mines.
- Improvement in performance reporting.
- Standardisation of processes and methodologies.
- Knowledge and skills sharing across jurisdictions.

Three jurisdictions (Queensland, NSW and Tasmania) had formal abandoned mine programmes prior to the completion of the 2010 Strategic Framework. Since 2010, Western Australia (WA) and the Northern Territory (NT) have introduced new policies to address mining legacies. The WA Draught Department of Mining and Petroleum (2015) policy (DMP) was developed to establish the principles that should be used in making decisions about the management and/or rehabilitation of these sites (Bennett and Lacy, 2012). Management decisions regarding abandoned mine sites need to balance risk, costs and benefits to both the environment and the Western Australian community, and also recognise the varying values of abandoned mines.

According to NSW's existing legacy mines programme regulations (NSW LMP), mine operators are in charge of rehabilitation under current law. This involves mining companies providing a security deposit at the start of operations for the entire cost of rehabilitation (https://www.resourcesregulator.nsw.gov.au/rehabilitation/legacy-mines-program). If the mine has financial trouble, the funds from the security deposit are utilised to pay for cleanup costs.

3.5 ENVIRONMENTAL IMPACT OF DERELICT MINE SITES

Mining processes such as exploration, construction, operation, maintenance, expansion, abandonment, decommissioning and repurposing of a mine may have a major positive and negative influence on social and environmental systems (Haddaway et al., 2019). Furthermore, both indirect and direct mining operations have an impact on the environment at the regional, district and global levels.

Mining may have major environmental consequences due to damage mechanisms such as mine dump erosion, sinkholes, contamination of surface water or groundwater and soil from chemicals emitted during mining operations and the extinction of species globally. Furthermore, the release of carbon from these processes has a negative impact on the climate, influencing biodiversity and the quality of human health. Derelict mine sites with an abundance of sulphide minerals in their waste materials represent severe environmental risks, causing contamination problems in streams, soils and ecosystems so these mine sites pose a significant environmental threat (Lottermoser, 2003).

3.5.1 Water Pollution – Acid Mine Drainage

The mining of minerals such as gold, copper and nickel is associated with excavation of sulfidic rich ores which can have a negative impact on waterways over the long term. Wastewaters generated by the mining industry contain substantial quantities of hazardous chemicals such as cyanides and heavy metals (Akcil and Koldas, 2006). The production of acid mine drainage waters, and the potential liberation of dissolved components into streams, are two of the most significant outcomes that can occur when mine sites containing sulphide minerals are not adequately remediated after they have been abandoned (Luo et al., 2020; Tomiyama et al., 2019).

Acid mine drainage (AMD) is the seepage of sulfuric acid produced when sulphide-bearing minerals are oxidised during exposure to atmospheric oxygen and water (Figure 3.2). The water released from AMD is typically highly acidic (pH between 2 to 4) with a high content of sulphate (1000 – 130,000 mg L^{-1}) and dissolved iron (200 – 1000 mg L^{-1}) (Figure 3.3). In addition to sulfuric acid generation, dissolved metals such as Cu, Ni, Mn, Pb, Zn, Cd, Hg, Ni and Co, and metalloids such as As and Sb are liberated from the breakdown of sulphide minerals (Figure 3.4) (Butler, 2009; Ferreira et al., 2021; Moncur et al., 2005; Naidu et al., 2019; Zhou et al., 2017).

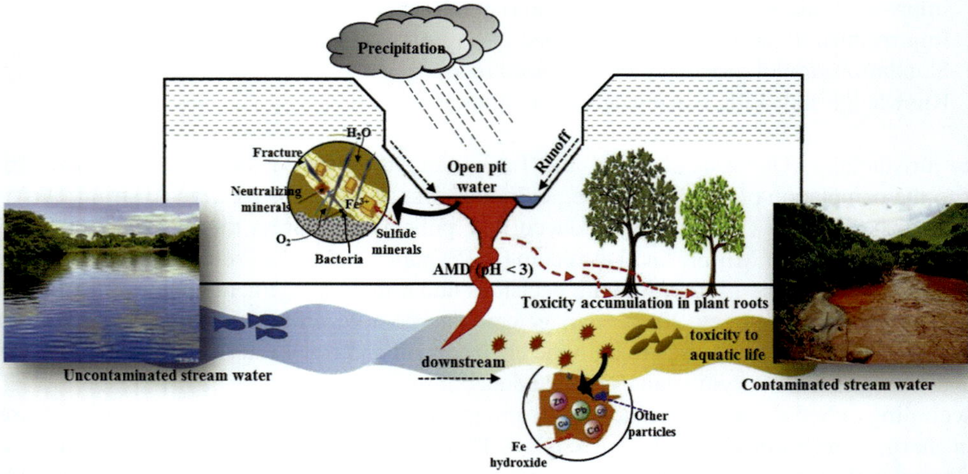

FIGURE 3.2 Schematic diagram representing the processes and generation of acid mine drainage and associate contaminants transport from open pit mine sites (adopted from Naidu et al. (2019)) (ABARES 2016).

FIGURE 3.3 Surface waters from AMD, enriched with Fe(III) flocs near a derelict coal mine site, Greta, NSW) (Photo courtesy: Girish Choppala).

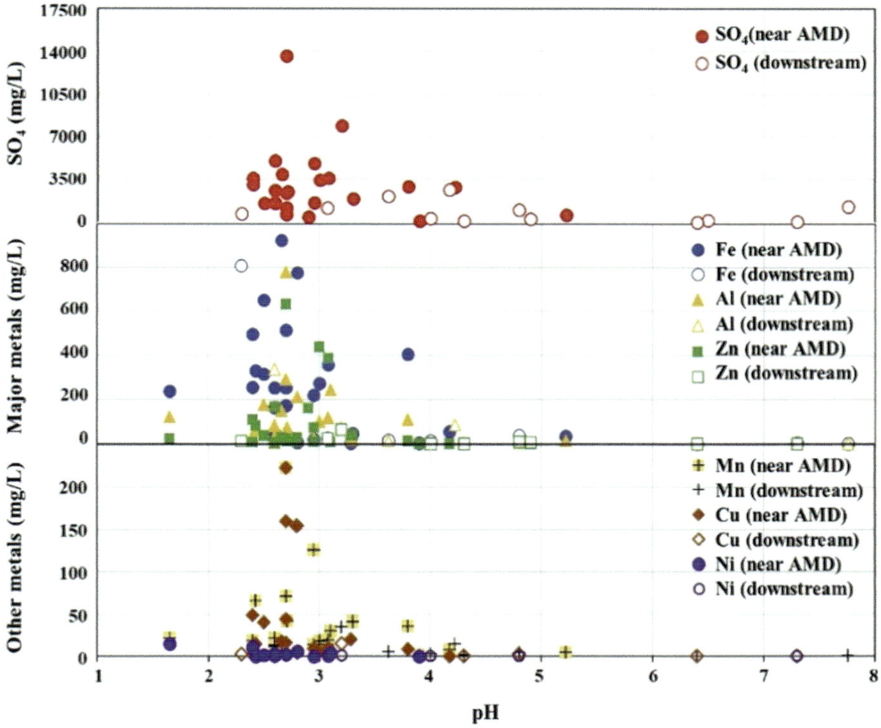

FIGURE 3.4 Distribution of metal contaminants and sulphate content in water streams collected near mine sites and down streams (adopted from Naidu et al. (2019).

Management of AMD produced by an operational mine is typically less expensive than after mine closure. However, in Australia, between the early 1980s and 2005, the closure of approximately 70% of mining sites was unexpected and unplanned due to factors such as steady decline in global commodity prices, decrease in grade or size of orebody, expensive operations, reduced revenues and difficult geotechnical settings (Laurence, 2006).

Investigation of surface waters from creeks up to 10 metres downstream of an abandoned mine site at Croydon (North QLD) showed elevated concentrations of As, Cd, Cu, Ni, Pb, S and Zn relative to local background concentrations, highlighting the importance of employing effective rehabilitation and remediation practices at these sites (Parbhakar-Fox et al., 2014). They reported that galena has undergone partial weathering and produced anglesite, suggesting a potential source of Pb in the surface water.

The extent of acid mine drainage contamination is correlated to the sulphur content of the coal (Johnson, 2003). A closed underground colliery can still act as a potential source of contaminants to the surrounding waters. Natural flooding of a closed mine in Berrima Colliery (NSW) discharges acid mine drainage into the Wingecarribee River and a study conducted by Wright et al. (2018) found hazardous levels of Ni (418 $\mu g\ L^{-1}$), Zn (1161 $\mu g\ L^{-1}$) and Mn (11.9 mg L^{-1}) in the acid mine drainage which is eventually polluting the water in the river. This study also casts doubt on the efficacy of the environmental regulatory regime imposed on this mine by the NSW EPA, which provided insufficient restrictions on the contaminants discharged by the mine through the acid mine drainage adit to the Wingecarribee River.

3.5.2 Impact on Soil

Derelict mine sites can be one of the most important sources of metals contamination if not properly remediated (Table 3.1). Therefore, it is critical to regularly monitor the areas surrounding them, in particular those through which the runoff and streams flow to feed lakes and reservoirs. Shamsoddini et al. (2014) investigated the heavy metal content of soils in the vicinity of a derelict mine site, which was a silver-lead-zinc mining area of Yerranderie, NSW. Mining activities were started when galena was found in the area and a number of small underground mines and four processing sites were established. Activities ceased in 1929 due to the construction of the Warragamba Dam, which served as a major water supply catchment for Sydney. The average elemental concentration (Ag, As, Cd, Pb, Zn) in sampled soils was found to be higher than the Australian and New Zealand Environment and Conservation Council (ANZECC, 2000) standards, with Pb and As showing significant heavy metal contamination (Table 3.2). Furthermore, 98% and 87% of the collected samples exceeded ANZECC guideline levels for Cu and Hg, respectively.

The size of the derelict mine in relation to the quantity of heavy metals discharged may be altered depending on the metals released and the subsequent pollution of soils. Ashley et al. (2004) studied the heavy metal contamination of soils surrounding small-scale mining and mineral processing at the Webbs Consols Pb-Zn-Ag deposit in northern NSW and reported very high levels, with median As, Pb, Ag and Zn concentrations exceeding those of background median values by 1–2 orders of magnitude, although both background and contaminated soils have the same mineralogy and contain jarosite type minerals and goethite. Their results further suggest that contaminated soils collected across the mine site have very similar As/Pb ratios, suggesting physical dispersion of fine-grained waste particles.

The trace elements in soils around the derelict mine sites pose a significant health risk that requires environmental assessment. Therefore, it is critical to assess the local contamination extent. Jafari et al. (2020) investigated the regional distribution of trace elements in 395 soil samples by using X-ray fluorescence (XRF) in the vicinity of the industrial complex that comprises former Pb and Cu smelters at Port Kembla, NSW.

The geographical distribution of Cu, Pb and Zn in soil samples shows high concentrations of these metals centred upon the Port Kembla copper smelter stack, with lower concentrations around the Kanahooka smelter (Figure 3.5 a, c, e). The copper content ranges between 21 to 6,240 mg kg^{-1}) with concentrations decreasing on moving farther from the smelter. The distribution of Pb shows high concentrations (1,569 mg kg^{-1}) within a distance of about 1,500 m around the copper smelter stack with an average value of 116 mg kg^{-1}. High nickel concentrations were reported in the copper slag dump and up to 53 mg kg^{-1} around the copper smelter. Overall, this study suggested a significant

TABLE 3.1
List of Abandoned Mine Sites across Australia with an Array of Contaminants

State	Location	Soil/Spoil/ Sediment	Contaminant	Concentration ($mg\ kg^{-1}$)	Reference
Victoria	Maldon	Soil	Arsenic (As) Lead (Pb) Copper (Cu) Cobalt (Co) Mercury (Hg)	185 76 59 25 9	Abraham et al. (2018)
Victoria	Blackwood	Sediment	Hg	130	Bycroft et al. (1982)
South Australia	Kapunda	Soil	Cu	65-9941 (Average: 4201)	Nirola et al. (2016)
South Australia	Port Pirie	Soils	Pb	6-7546	Fergusson (2017)
Queensland	Mount Carbine	Spoil	Tungsten (W) Cu	0.6 (0-5 cm) 0.9 (5-15 cm) 9005 (0-5 cm) 12604 (5-15 cm)	Pyatt and Pyatt (2004)
	Mungana	Spoil	W Cu	0.01 (0-5 cm) 0.01 (5-15 cm) 964 (0-5 cm) 1408 (5-15 cm)	
Queensland	Mount Morgan	Sediment	As Cu Pb Th (thorium) U (uranium) Zn	123-952 105-711 89-577 25-67 4.8-11.6 121-782	Lottermoser and Ashley (2011)
Queensland	Mount Isa	Soil	Cu Pb Zn	31-12,100 8-5800 26-11,100	Taylor et al. (2010)
NSW	Yerranderie	soil	Pb As Cd Zn	40-5000 311-20150 4-64.4 272-5049	Shamsoddini et al. (2014)
NSW	Mole River	Soil	As- Cu Pb Sb (antimony)	2,600-26,600 3040 609 514	Ashley and Lottermoser (1999)
NSW	Yerranderie	Soil	As Pb	24,000 128,000	Kavehei et al. (2021)
NSW	Gulf Creek		Cu Zn	900-2000 1100-1400	Ashelford and Gore (2020)
	Mole River	Tailings	As	87600	
NSW	Webbs Consols	Soils	As Pb Sb Cu	30,500 681 25 588	Bari et al. (2020)
	Mole River	Soils	As Pb Sb Cu	46600 209 1010 467	
	Halls Peak		As Pb Sb Cu	420 32,300 85 4900	
NSW	Sunny Corner	Sediment	Cu Zn Pb As	200 1700 1660 2800	Kavehei et al. (2021)

TABLE 3.2
Heavy Metal Content (mg kg^{-1}) in Soil Samples Collected from the Surroundings of a Derelict Ag-Pb-Zn Mine Site, Yerranderie, NSW

Heavy Metal Content	ANZECC Standard	ANZECC Limit Excess
Pb	50	× 358
Ag	1	× 14
As	20	× 232
Cd	1.5	× 12
Cu	65	× 10
Hg	0.15	× 3
Zn	200	× 8

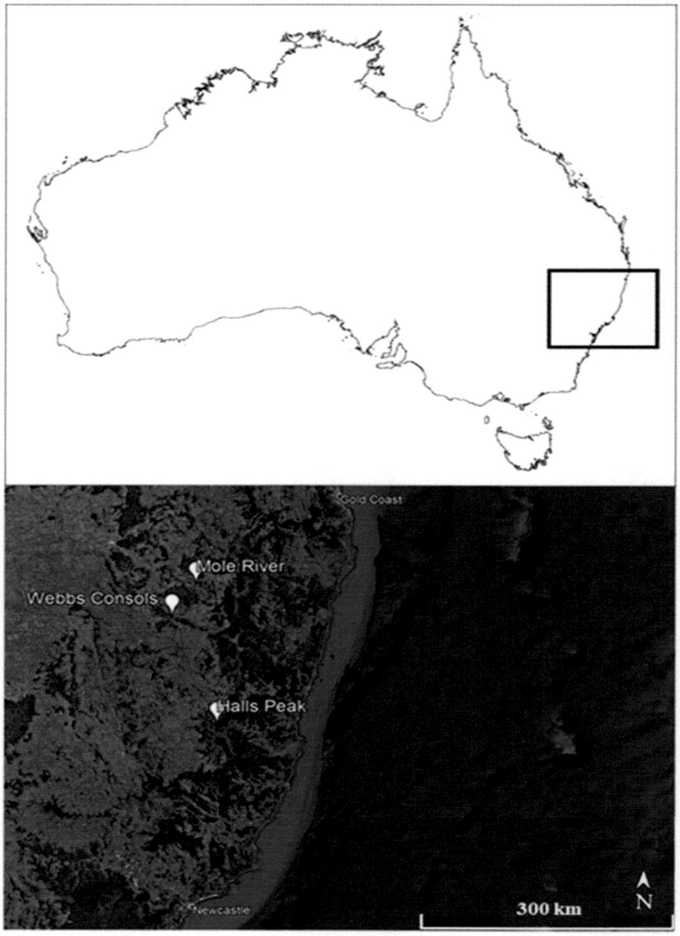

FIGURE 3.5 Locations of investigated derelict mine sites with abundant arsenic in their soils.

decrease in trace element content on moving away from smelter sites, which may be due to the deposition of trace elements in the soil by atmospheric transportation and fallout of larger-sized particles within a relatively short distance of the smelter. This study highlighted the potential sources of toxic metal contaminants in soils and negative impacts on the natural environment.

3.5.3 Loss of Biodiversity

Mining-affected areas often exhibit steep, unstable slopes, disruption of natural drainage systems and a rise in absence of vegetation brought on by nutrient deficiency, unfavourable soil composition, acidity and/or heavy metal induced toxicity (Batty, 2005).

The specific nature of the environmental issues brought on by mining varies, mostly dependent on the material extracted, the processing methods used, the local topography and geology and the time frames involved. It is also broadly acknowledged that mining has a detrimental impact on the native flora and fauna, ranging from soil microorganisms to large mammals, altering the native community structure and lowering biodiversity. This is especially true for water bodies contaminated with heavy metals, where rivers downstream of the discharge may experience a significant loss of flora and fauna (Kapoor and Singh, 2021).

Specific sources of erosion, including those from abandoned mining sites, can have detrimental impacts on aquatic biodiversity (Wager and Jackson, 1993). An investigation on the biodiversity downstream of the Dee River suggested that pollution-sensitive invertebrates, fishes and submerged plants were absent within 20 km of the Mount Morgan mine in central Queensland, which was attributed to acid rock drainage.

According to Woodside and O'Neill (1995), the impacts of mining-related pollution on biodiversity have been significant in the past. For instance, the Rum Jungle Mine in the NT released 130 tonnes of copper, 100 tonnes of manganese, 40 tonnes of zinc and 13,000 tonnes of sulphate into the Finniss River in one year (Saunders et al., 1996). A survey conducted in 1971 showed that the river was polluted with heavy metals, acidification and radiation to such an extent that it was virtually dead. Recent surveys of the same river (after rehabilitation of the site in the 1980s) indicate that macroinvertebrates in the waterways continue to be less common downstream of the mining activities.

3.6 RISK CHARACTERISATION

Derelict mining districts are often characterised by significant geological, ecological, environmental, industrial, cultural and archaeological values (Carbone et al., 2013). However, a number of environmental problems persist even when mining activities are stopped, including soil and water contamination, hydrogeological instability, subsidence, ecosystem harm and the erosion or destruction of landscapes. Mining activities may have a substantial negative influence on the environment even after ore extraction has finished, especially in places where sites have been abandoned and no cleanup of metals has been carried out (Passariello et al., 2002).

These metals and metalloids have significant toxic effects on ecosystems. In addition, this has a negative impact on the quality of the water and the sediments. Many different processes have the potential to liberate heavy metals and metalloids that are already present in contaminated stream sediments, thereby making them bioavailable (Siegel, 2002; Lottermoser, 2003). In addition, acid attacking rock-forming minerals mobilise elements, including Al and Mn, which can lead to the formation of new secondary minerals in the soil and sediments (Ashley et al., 2004).

Wastes from mining and processing industries may affect ecosystems by showing up in groundwater, surface water and soil. All living organisms, including flora, are susceptible to contamination and risks to human health may potentially derive from (i) inhalation of aerosols containing high levels of metals; (ii) absorption following skin contact; (iii) drinking of contaminated water; and (iv) consumption of food from contaminated areas.

Risk characterisation of derelict mine tailings can help us understand the extent of contamination in the surrounding environment and offers knowledge to evaluate the best remediation options to mitigate the contamination. Passariello et al. (2002) conducted a quantitative assessment of derelict mine wastes at various depths for metal concentrations including As, Cd, Hg, Pb Sb, Sn and Zn. Results found that Pb content remains almost constant in samples taken at various depths and As and Hg levels were also quite similar. However, they noticed that release of these elements via leaching was reasonably low.

3.6.1 Case Studies – Risk Characterisation in Australian Mines

3.6.1.1 Case Study I

Toxic heavy metals from derelict mine sites pose great threats around the world to human health and ecosystem integrity. Although several legislations have been imposed on reclamation and remediation of contamination in abandoned mine sites, there are multiple mine sites across Australia where no reclamation work has been undertaken (Ashley et al., 2004). Dump heaps, discarded low-grade ores in overburden materials and mine drainage water act as a potential source of trace metal contamination from derelict mine sites (Abraham et al., 2018; Park et al., 2014).

Particle size distribution can be used to identify relationships between toxic elements and their solubility, which is vital for the implementation of various remediation practices at abandoned mine sites. To fill this knowledge gap, Bari et al. (2020) analysed trace metal content and mineralogy in three contrasting mine site soils collected from Webbs Consols, Halls Peak and Mole River in NSW (Figure 3.6). Geochemical parameters including EC, CEC, total metal content and sequential extraction of As were measured. Particle size fractionation of these three different mine site soils was separated via dry sieving to yield six particle size fractions; (i) 4–2 mm, (ii) 2–1 mm, (iii) 1–0.25 mm, (iv) 0.25–0.125 mm, (v) 0.125–0.053 mm and (vi)<0.053 mm. Mineralogical and micromorphological characterisation, including XRD, scanning electron microscopy with energy-dispersive X-ray spectroscopy (SEM-EDX) and transmission electron microscopy energy-dispersive X-ray spectroscopy (TEM-EDX), were conducted.

This research indicates that the size of soil particles affected the redistribution of trace elements in soils. In four soil samples obtained from Webbs Consols and Halls Peak, the amounts of As and

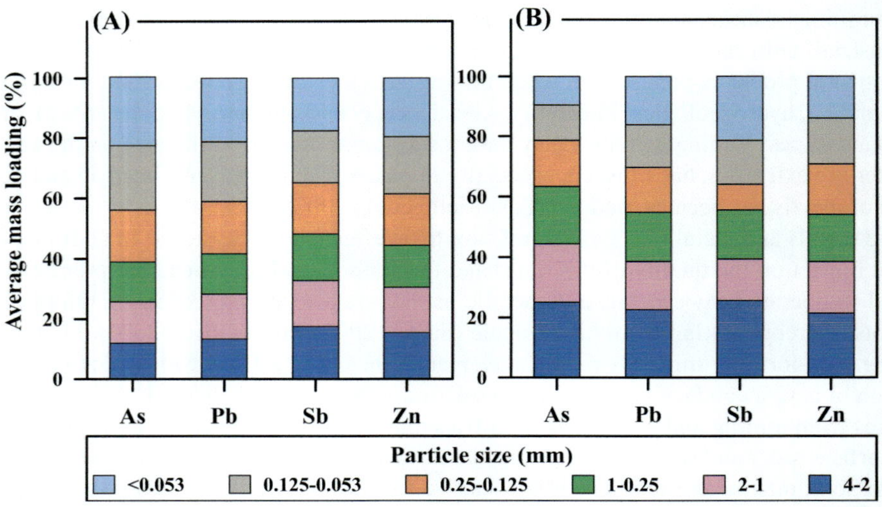

FIGURE 3.6 The mass loading (%) of trace elements-As, Pb, Sb and Zn in different particle size fractions in (A) Webbs Consols and Halls Peak, and (B) Mole River.

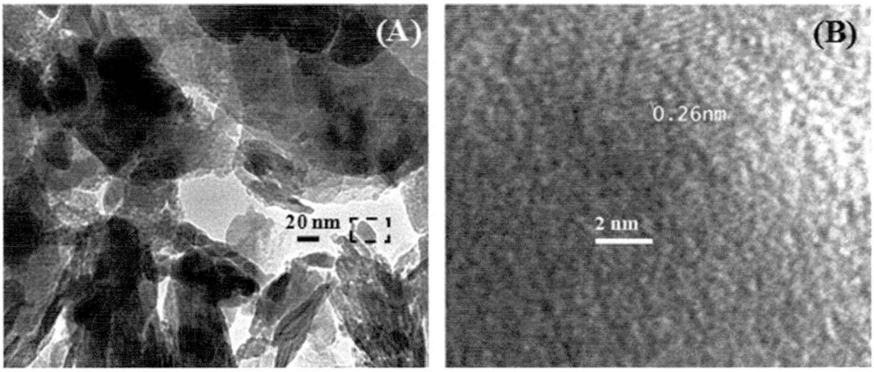

FIGURE 3.7 High resolution transmission electron microscopy (HR TEM) images of Mole River soils showing tooeleite platy structure and its corresponding image (b) displaying 0.26 nm.

other elements, such as Pb, Sb and Zn, rose as particle size dropped. The smallest particle size fractions (0.053 mm) included the greatest concentrations of As (3.05%), Pb (3.23%) and Zn (0.1%). In contrast, the 0.25–0.125 mm fraction had the highest Sb content (85 mg/kg). Nevertheless, trace element concentrations were low in coarse fractions.

As determined by the trace element concentration and the mass percentage of each fraction, the proportion of trace elements loading in different size fractions compared to the bulk soil is shown in Figure 3.7. The results demonstrated that there were differences in particle size fractions in the % mass loading of trace elements. Similar to this, for soil samples taken from Webbs Consols and Halls Peak, the fine particle percentage reflected a larger mass loading to the overall concentration of trace elements. The coarse particle fraction (4 mm) had the lowest mass loading of trace elements, whereas the fine particle fraction (0.053 mm) had the greatest mass loading (23.04% for As, 21.3% for Pb, 17.66% for Sb and 19.71% for Zn).

The findings revealed that the % mass loading of trace elements varied among particle size fractions. Similarly, for soils collected from Webbs Consols and Halls Peak, the fine particle fraction reflected a larger mass loading to the overall concentration of trace elements. The fine particle fraction (0.053 mm) had the greatest mass loading of trace elements (23.04% for As, 21.3% for Pb, 17.66% for Sb and 19.71% for Zn), whereas the coarse particle fraction (42 mm) had the lowest mass loading. Liao et al. (2016) discovered the maximum trace element mass loading in the 0.25 mm particle size fraction, followed by the coarse fraction (> 2 mm).

The study by Bari et al. (2020) concluded that soils from derelict mine sites have elevated concentrations of potentially toxic elements, including As, Pb, Sb and Zn. Trace metal loadings were mostly accumulated in the smallest size fraction (<0.053 mm) in soils collected from Webbs Consols and Halls Peak, whereas the coarse size fraction (4.2 mm) from Mole River samples showed the highest mass loadings of trace elements. Interestingly, this study found tooeleite, an As^{III}-Fe^{III} sulphate mineral, for the first time in Australian mine soils (Figure 3.8). Besides tooeleite, authors have also reported the existence of scorodite and arsenopyrite, reflecting the complex nature of As geochemistry at derelict mine sites.

3.6.1.2 Case Study II

Derelict mine sites around the world deal with acid mine drainage (AMD) or acid rock drainage (ARD), and it is one of the most significant environmental challenges faced by the mining industry around the world (Hudson-Edwards, 2016; Simate and Ndlovu, 2014). Sulphide-rich metal ores oxidise when exposed to the atmosphere and release high quantities of sulfuric acid, which, in turn, dissolves the metals in the ores and induces acid mine drainage. In other words, any activity that disturbs mineralised sulfidic materials can lead to AMD generation, a strong acidic wastewater

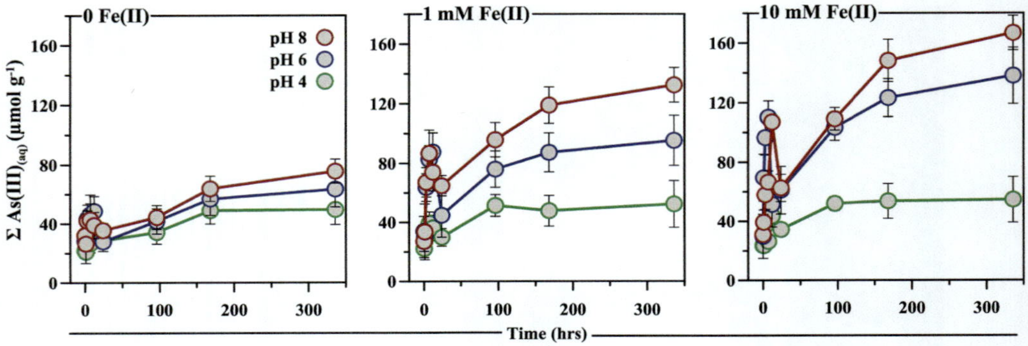

FIGURE 3.8 Temporal evolution of $As(III)_{(aq)}$ during the (a) 0 Fe(II), (b) 1 mM Fe(II), and (c) 10 mM Fe(II) induced reductive dissolution of tooeleite as a function of pH (4–8) over 14 days'.

FIGURE 3.9 Geographical location of Ottery mine and creek (a); few key locations along with AMD flow-path which was shown with solid-blue line, red and yellow zones represent slimes dam and waste rock, respectively; slimes dam wall with slimes dam seepage (b); mixing zone between discharge from the slimes dam and disused adit (c); downstream of sediment retention dam with V-notch weir (d).

with a high content of dissolved ferrous and non-ferrous metal sulphates (Georgopoulou et al., 1996; Skousen et al., 2019). This process occurs in open pits, waste rock piles and mine tailings (Sheoran and Sheoran, 2006). Although the oxidation of sulphide-rich mine sites controls AMD generation, the presence of microorganisms and also the availability of water play a significant role (Johnson and Hallberg, 2005).

Arsenic and several toxic contaminants have been reported in several derelict mine sites in NSW. Ottery Mine is one of the oldest derelict mine sites in northern NSW and its mineralogy was dominated with scorodite, arsenopyrite, cassiterite with abundant pyrite and arsenolite (Hebbard et al., 2017). Although several studies have been conducted to better understand the geochemical transformation of Fe(III) minerals, relatively few studies examined the As-imposed constraints on the formation and fate of As retained Fe(III) minerals, including jarosite and schwertmannite in As rich AMD conditions. These are critical to development of accurate risk assessment models to formulate suitable remediation practices. To address this knowledge gap, Burton et al. (2021) conducted an extensive investigation to understand the influence of As on Fe(III) mineralogy in strongly acidic AMD conditions (Figure 3.9). Simultaneously, they investigated the associated impact of jarosite and schwertmannite on As mobility in an AMD environment.

Mineralogy investigation at Ottery mine AMD flow path (Figure 3.10) suggested diversity in accumulation of various minerals along the flow path. Zone A (0–20 m from slimes waste) suggested that 2-line ferrihydrite and amorphous ferric arsenate were the dominant minerals and >95% of As in solid-phase samples was present as As(V). AMD released from the adit has a pH of 2.6-2.8. Zone B (20–40 m from slimes waste) predominantly contained abundant quantities of jarosite, schwertmannite with some minor quantities of scorodite.

Burton et al. (2021) highlighted the development of reducing conditions on the remobilisation of As via the reductive dissolution of As(V)-bearing Fe(III) minerals and subsequent reduction of As(V) to highly toxic and mobile As(III). In addition, this research has identified a crucial and previously unknown function for As(V) in jarosite production. The findings suggest that increasing levels of As(V) incorporation in schwertmannite enhance transformation to jarosite. This remarkable revelation reconsiders the widely held belief that As(V) retards schwertmannite transformation.

3.7 REMEDIATION APPROACHES ACROSS THE GLOBE

3.7.1 Phytoremediation

The phytoremediation technique has been trialled in several areas in Australia. A few researchers surveyed the potential of native species suitable for the remediation of derelict mine sites. The potential of phytoremediation technology depends on the concentration of heavy metals and biomass turnover and the rate of tolerance of plant species in derelict mine tailings (Goessler and Francesconi, 2002). Grant et al. (2002) investigated the rehabilitation of three different derelict mine sites in NSW through glasshouse pot trials. They tested 13 native species, including trees, shrubs and grasses, in four treatments, specifically (i) control, (ii) tailings, (iii) mulch and (iv) tailings/mulch treatment. The grass species *Cynodon dactylon* and *Dicanthium sericeum* had the greatest emergence rate (40%), although 11 out of 13 species had more than 12% emergence rates.

Grant et al. (2002) further reported that, compared to the control and tailings treatments, the mean annual height growth and cover of all species were generally more significant in the mulch and tailings/mulch treatments. The tree species *Eucalyptus camaldulensis* and *Casuarina cunninghamiana* also showed quite strong growth, with *Cynodon dactylon* recording the greatest mean annual height gain among the species. The mean biomass estimates for all species were considerably greater in the mulch and tailings/mulch treatments compared to the control and tailings treatments, except *Dicanthium sericeum*, *Aristida ramosa* and *Casuarina cunninghamiana*.

In a similar study, Marrugo-Negrete et al. (2016) screened 24 native plant species for the phytoremediation of Hg-contaminated mine sites in Colombia, and accumulation, translocation and

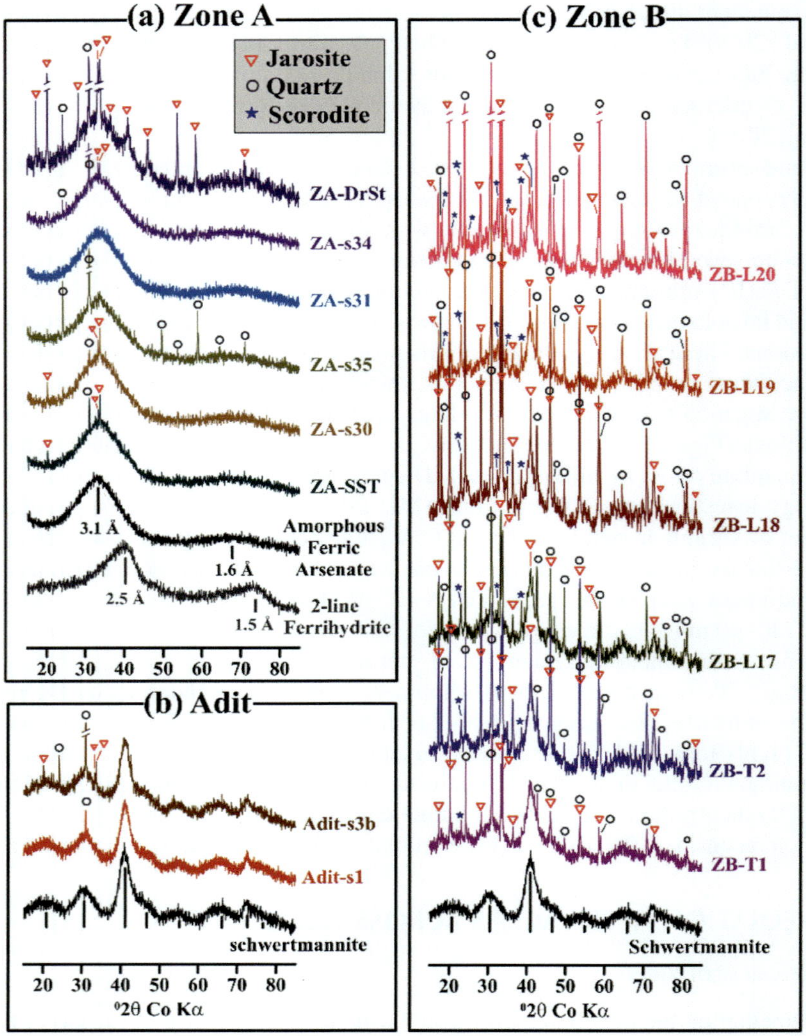

FIGURE 3.10 XRD patterns for samples collected from zone A (a), adit entrance (b), and zone B (c). XRD patterns for synthetic AFA, 2-line ferrihydrite, and schwertmannite are also presented to assist interpretation. The labels to the right of each XRD pattern denote the sample identification code (or the identity of synthetic phases).

bioconcentration factors were determined. Different tissues from all plant species were categorised in decreasing Hg accumulation sequence: roots > leaves > stems. Among 24 plants tested, herbs such as *Piper marginathum* and *Stecherus bifidus*, and the sub-shrubs *Jatropha curcas* and *Capsicum annuum,* were identified as promising native plants with the potential to be used in the phytoremediation of Hg-contaminated derelict mine tailings.

The utilisation of energy crops and vegetable species to remediate heavy metal-contaminated derelict mine tailings can be a win-win solution for bioenergy production and phytostabilisation. Ruiz Olivares et al. (2013) investigated the potential for using *Ricinus Communis* L. (castor oil plant) as an energy crop to remediate heavy metal-polluted derelict mine tailings. They reported that seed production in mine tailings with high concentrations of Zn, P and Cd yielded high oil content and

R. Communis L plant accumulated low metal content in the shoot but high metal content in the root, with low metal translocation factors involved in the phytostabilisation.

Potentially toxic elements such as As and U contamination of surface waters in derelict mine sites pose a serious threat to the environment and human health. *Lemana gibba* (gibbous duckweed) was reported to be growing naturally in tailing ponds of former uranium mines in Lengenfeld and Neuensalz-Mechelgrün (Germany). Mkandawire et al. (2004) examined *L. Gibba* for its efficiency in removing U and As from surface waters of derelict mine sites and found that it immobilised 84.5 and 88.2% of U and As from 1 L of nutrient solution over 21 days. The quantity of uranium and arsenic removed and the yield and quick turnover reported in this research imply that *L. gibba* might be used as a phytoremediation species for uranium, arsenic and other heavy metals of concern in surface waters of abandoned uranium mines.

Phytoextraction is also an alternative option to extract toxic metals and metalloids from derelict mine sites. When cultivated in a metal-rich environment, hyperaccumulators are plants with leaves that can accumulate at least 100 mg/kg of Cd, 1000 mg/kg of As, Cu, Pb, Ni, Co, Se or Cr, or 10,000 mg/kg of Mn and Zn (dry weight) (Brooks, 1994; Watanabe, 1997). Hyperaccumulating plant species can tolerate, uptake and accumulate high concentrations of metals. Aquatic macrophyte *Eleocharis acicularis* was tested for its efficiency in extracting multiple heavy metals from derelict mine sites in Hokkaido, Japan (Hoang Ha et al., 2009). They reported that *E. acicularis* is a hyperaccumulator of Pb and also reported high Si content (2.08%) in the same plant.

3.7.2 Impact of Organic/Inorganic Amendments

Due to poor soil qualities, such as acidity, low nutrient content, high metal concentrations, poor substrate structure and poor water retention, it is often challenging to develop plants on abandoned mine tailings (Whitbread-Abrutat, 1997). The application of amendments will perform essential primary functions, including (i) supplying nutrients to plants and microbes, and (ii) improving physico-chemical characteristics such as pH and organic matter content (Lebrun et al., 2022).

The utilisation of organic and inorganic residues as soil amendments helps to reduce the bioavailability of toxic heavy metals and enhances soil fertility, hence promoting the survival and development of plants in derelict mine sites. The decrease in plant- and bioavailable metal fractions with the application of amendments may be through adsorption, complexation and precipitation processes. The adsorption processes depend on the type of metals, soil type and the amount of organic matter in the soil.

Biosolids from sewage treatment facilities are a significant source of organic matter for improving the macro- and micronutrients and physical properties of the mine tailings and soils to support the growth of a vegetative cover (Gardner et al., 2010). Biosolids treatments may influence the carbon sequestration capacity of tailings and other mining wastes at the outset of reclamation by initiating soil development processes and enhancing primary production, ultimately leading to an improved accumulation of soil organic carbon in soils (Antonelli et al., 2018). Whitbread-Abrutat (1997) examined the role of ammonium nitrate, mineral phosphate and sulphate of potash fertilisers, cake sewage sludge and diatomite. Among these amendments, only cake sludge and diatomite positively affected plant growth.

Organic amendments, including manures, biochar, manure compost and sewage sludge application to soils, can improve the fertility of derelict mine tailings and mine soils. A greenhouse study conducted by Chiu et al. (2006) also evaluated the impact of manures and sewage sludge on remediation of Pb, Zn and Cu enriched mine tailings. They tested the feasibility of using *Vetiveria zizanioides* (vetiver) and *Phragmites australis* (common reed) for revegetation and reclamation. Plant tissue analysis revealed that manure compost and sewage sludge drastically decreased Pb uptake and accumulation in vetiver and common reed but no observed impact on Cu translocation. They further reported that these amendments decreased DTPA-extractable Pb, Zn and Cu in tailings. Sewage sludge increased the biomass of both plants in Pb-Zn mine tailings but not manure compost.

Similarly, Harris et al. (2021) investigated the application of biosolids on improving carbon and available nutrients and metals in derelict copper mine tailing facilities in British Columbia, Canada. Their research found that 17 years after a one-time application of biosolids, carbon, nitrogen, phosphorus and biomass levels in tailings were higher than in tailings that did not receive biosolids. There was no change in carbon pool content over 17 years, suggesting the long-term effect of a one-time application of biosolids in enhancing tailings for vegetation establishment.

3.7.3 Bioremediation of Derelict Mine Sites

Because heavy metals are toxic and difficult to remove from the environment, industrial and agricultural operations that contaminate natural ecosystems risk affecting the environment and public health (Boularbah et al., 2006; Zhong et al., 2020). Compared to physicochemical approaches for remediation of heavy metals–induced contamination, including chemical precipitation, ion exchange and sorption, bioremediation technologies are particularly appealing since they may be more affordable and practical for low metal concentrations (Newsome and Falagán, 2021; Rani et al., 2018).

Multiple strains of bacteria separated from derelict mine sites have been reported to precipitate, reduce or transform toxic contaminants into their less toxic forms. Natarajan (2008) isolated sulphate-reducing bacteria from abandoned mine tailings and used them to precipitate dissolved metals, including Cu, Zn, Fe and As. In particular, *Thiomonas* spp. has been shown to oxidise highly toxic and mobile As(III) to less toxic As(V).

Microbially induced carbonate precipitation of toxic metals and subsequent removal are alternatives for mitigating pollution in derelict mine sites. Dissolved carbonate and ammonium ions have the potential to bind with toxic metals and form carbonate precipitates (Kumari et al., 2016). Biomineralisation of toxic metals using ureolytic bacteria can mitigate Pb contamination in mine sites. Kang et al. (2015) isolated *Enterobacter cloacae* KJ-46 from abandoned mine sites, which showed a high removal rate of Pb (68%) by forming $PbCO_3$ crystals following incubation for two days, confirmed through SEM-EDX analyses. They attributed this precipitation to the microbial hydrolysis of urea resulting in the rapid generation of carbonate ions, followed by the precipitation of $PbCO_3$.

Bacteria, yeast, fungi and algae have the potential to *passively uptake* toxic metal ions, which can be entrapped in the cellular structure and subsequently biosorbed onto the binding sites of the cellular structure. Metal ions can also pass through the cellular membrane through the cell metabolic cycle (*active uptake*) and both these uptakes are collectively considered as bioaccumulation. An indigenous yeast strain, *Rhodosporidium toruloides,* isolated from flood water of a former uranium mining site, removed uranium via bioaccumulation and inactive biosorption mechanisms; the rate of removal was highly dependent on the temperature of the solution (Gerber et al., 2018).

REFERENCES

ABARES (2016) Land use of Australia 2010–11. Australian Bureau of Agricultural and Resource Economics and Sciences.

ABARES (2019) Australian Bureau of Agricultural and Resource Economics and Sciences. https://www.agriculture.gov.au./abares.

Abraham J., Dowling K. and Florentine S. (2018) Assessment of potentially toxic metal contamination in the soils of a legacy mine site in Central Victoria, Australia. *Chemosphere* **192**, 122–132.

Akcil A. and Koldas S. (2006) Acid Mine Drainage (AMD): Causes, treatment and case studies. *Journal of Cleaner Production* **14**, 1139–1145.

Antonelli P. M., Fraser L. H., Gardner W. C., Broersma K., Karakatsoulis J. and Phillips M. E. (2018) Long term carbon sequestration potential of biosolids-amended copper and molybdenum mine tailings following mine site reclamation. *Ecological Engineering* **117**, 38–49.

ANZECC (2000) Australian and New Zealand guidelines for fresh and marine water quality. *Australian and New Zealand Environment and Conservation Council and Agriculture and Resource Management Council of Australia and New Zealand, Canberra* **1**, 1–314.

Ashelford M. and Gore D. B. (2020) Elemental and mineralogical constraints on environmental contamination from slag at Gulf Creek copper mine. *Minerals Engineering* **154**, 106407.

Ashley P., Lottermoser B., Collins A. J. and Grant C. (2004) Environmental geochemistry of the derelict Webbs Consols mine, New South Wales, Australia. *Environmental Geology* **46**, 591–604.

Ashley P. M. and Lottermoser B. G. (1999) Arsenic contamination at the Mole River mine, northern New South Wales. *Australian Journal of Earth Sciences* **46**, 861–874.

Bari A. F., Lamb D., Choppala G., Bolan N., Seshadri B., Rahman M. A. and Rahman M. M. (2020) Geochemical fractionation and mineralogy of metal(loid)s in abandoned mine soils: Insights into arsenic behaviour and implications to remediation. *Journal of Hazardous Materials* **399**, 123029.

Batty L. C. (2005) The potential importance of mine sites for biodiversity. *Mine Water and the Environment* **24**, 101–103.

Bennett K. E. and Lacy H. W. B. (2012) Closure planning and decommissioning of tailings storage facilities. In *Mine Closure 2012: Proceedings of the Seventh International Conference on Mine Closure* (eds. A. B. Fourie and M. Tibbett). Australian Centre for Geomechanics, Brisbane, pp. 139–148.

Berger A. C., Bethke C. M. and Krumhansl J. L. (2000) A process model of natural attenuation in drainage from a historic mining district. *Applied Geochemistry* **15**, 655–666.

Boularbah A., Schwartz C., Bitton G. and Morel J. L. (2006) Heavy metal contamination from mining sites in South Morocco: 1. Use of a biotest to assess metal toxicity of tailings and soils. *Chemosphere* **63**, 802–810.

Britt A., Summerfield D., Senior A., Kay P., Huston D., Hitchman A., Hughes A., Champion D., Simpson R. and Sexton M. (2017) Geoscience Australia (2017) Australia's identified Mineral Resources 2017. *Geoscience Australia, Canberra.* https://doi.org/10.11636/1327-1466.2017.

Brooks R. R. (1994) Plants that hyperaccumulate heavy metals. *Plants and the Chemical Elements: Biochemistry, Uptake, Tolerance and Toxicity*, 87–105.

Burton E. D., Karimian N., Johnston S. G., Schoepfer V. A., Choppala G. and Lamb D. (2021) Arsenic-imposed effects on Schwertmannite and Jarosite formation in acid mine drainage and coupled impacts on arsenic mobility. *ACS Earth and Space Chemistry* **5**, 1418–1435.

Butler B. A. (2009) Effect of pH, ionic strength, dissolved organic carbon, time, and particle size on metals release from mine drainage impacted streambed sediments. *Water Research* **43**, 1392–1402.

Bycroft B. M., Coller B. A. W., Deacon G. B., Coleman D. J. and Lake P. S. (1982) Mercury contamination of the Lerderberg River, Victoria, Australia, from an abandoned gold field. *Environmental Pollution Series A, Ecological and Biological* **28**, 135–147.

Carbone C., Dinelli E., Marescotti P., Gasparotto G. and Lucchetti G. (2013) The role of AMD secondary minerals in controlling environmental pollution: Indications from bulk leaching tests. *Journal of Geochemical Exploration* **132**, 188–200.

Chiu K. K., Ye Z. H. and Wong M. H. (2006) Growth of Vetiveria zizanioides and Phragmities australis on Pb/Zn and Cu mine tailings amended with manure compost and sewage sludge: A greenhouse study. *Bioresource Technology* **97**, 158–170.

Coelho P. C. S., Teixeira J. P. F. and Gonçalves O. N. B. S. M. (2011) Mining activities: Health impacts. In *Encyclopedia of Environmental Health* (ed. J. O. Nriagu). Elsevier, Burlington, pp. 788–802.

DMP (Department of Mines and Petroleum) (2015) Abandoned Mines Policy, Government of Western Australia, Draft for Consultation, July 2015.

Dwiki S. (2018) Development of environmental policy in Indonesia regarding mining industry in comparison with the United States and Australia: The lesson that can be learned. *Evergreen*, **5**(2), 50–57.

Ericsson M. and Löf O. (2019) Mining's contribution to national economies between 1996 and 2016. *Mineral Economics* **32**, 223–250.

Fergusson L. (2017) Anthrosols and technosols: The anthropogenic signature of contaminated soils and sediments in Australia. *Water, Air, & Soil Pollution* **228**, 269.

Ferreira R. A., Pereira M. F., Magalhães J. P., Maurício A. M., Caçador I. and Martins-Dias S. (2021) Assessing local acid mine drainage impacts on natural regeneration-revegetation of São Domingos mine (Portugal) using a mineralogical, biochemical and textural approach. *Science of The Total Environment* **755**, 142825.

Gardner W. C., Broersma K., Naeth A., Chanasyk D. and Jobson A. (2010) Influence of biosolids and fertilizer amendments on physical, chemical and microbiological properties of copper mine tailings. *Canadian Journal of Soil Science* **90**, 571–583.

Georgopoulou Z. J., Fytas K., Soto H. and Evangelou B. (1996) Feasibility and cost of creating an iron-phosphate coating on pyrrhotite to prevent oxidation. *Environmental Geology* **28**, 61–69.

Gerber U., Hübner R., Rossberg A., Krawczyk-Bärsch E. and Merroun M. L. (2018) Metabolism-dependent bioaccumulation of uranium by Rhodosporidium toruloides isolated from the flooding water of a former uranium mine. *Plos One* **13**, e0201903.

Goessler W. and Francesconi K. (2002) Arsenic species in an arsenic gyperaccumulation fern, pityrogramma calomelanos. *The Science of the Total Environment* **284**, 27–35.

Goudie A. S. (2018) *Human Impact on the Natural Environment.* John Wiley & Sons, Hoboken, NJ, USA.

Grant C. D., Campbell C. J. and Charnock N. R. (2002) Selection of species suitable for derelict mine site rehabilitation in New South Wales, Australia. *Water, Air, and Soil Pollution* **139**, 215–235.

Guerin T. F. (2006) A survey of sustainable development initiatives in the Australian Mining and Minerals Industry. *Minerals & Energy – Raw Materials Report* **20**, 11–44.

Haddaway N. R., Cooke S. J., Lesser P., Macura B., Nilsson A. E., Taylor J. J. and Raito K. (2019) Evidence of the impacts of metal mining and the effectiveness of mining mitigation measures on social–ecological systems in Arctic and boreal regions: A systematic map protocol. *Environmental Evidence* **8**, 9.

Harris M. E., Gardner W. C. and Pypker T. (2021) Influence of a one-time biosolids application on elemental and nutrient concentrations on mine tailings. *Canadian Journal of Soil Science* **101**, 703–716.

Hebbard E. R., Wilson S. A., Jowitt S. M., Tait A. W., Turvey C. C. and Wilson H. L. (2017) Regrowth of arsenate–sulfate efflorescences on processing plant walls at the Ottery arsenic–tin mine, New South Wales, Australia: Implications for arsenic mobility and remediation of mineral processing sites. *Applied Geochemistry* **79**, 91–106.

Hilson G. (2003) Defining "cleaner production" and "pollution prevention" in the mining context. *Minerals Engineering* **16**, 305–321.

Hintjens H. (2000) Environmental direct action in Australia: The case of Jabiluka Mine. *Community Development Journal* **35**, 377–390.

Hoang Ha N. T., Sakakibara M., Sano S., Hori R. S. and Sera K. (2009) The potential of Eleocharis acicularis for phytoremediation: Case study at an abandoned mine site. *Clean–Soil, Air, Water* **37**, 203–208.

Hudson-Edwards K. (2016) Tackling mine wastes. *Science* **352**, 288–290.

Issaka S. and Ashraf M. A. (2021) Chapter 9 - Phytorestoration of mine spoiled: "Evaluation of natural phytoremediation process occurring at ex-tin mining catchment". In *Phytorestoration of Abandoned Mining and Oil Drilling Sites* (eds. K. Bauddh, J. Korstad and P. Sharma). Elsevier, pp. 219–248.

Jafari Y., Jones B. G., Pacheco J. C. and Umoru S. (2020) Trace element soil contamination from smelters in the Illawarra region, New South Wales, Australia. *Environmental Earth Sciences* **79**, 372.

Jarvis A. P., Johnston D., Mayes W. M. and Potter H. ;(2008) Application of a new methodology for assessing the priorities for abandoned mine water pollution remediation at a national scale: Results and implications from a study across England and Wales. *Proceedings of the 10th International Mine Water Association Congress: Mine Water and the Environment.* VSB-Tech. Univ. Ostrava, Ostrava, Czech Republic.

Johnson D. B. (2003) Chemical and microbiological characteristics of mineral spoils and drainage waters at abandoned coal and metal mines. *Water, Air and Soil Pollution: Focus* **3**, 47–66.

Johnson D. B. and Hallberg K. B. (2005) Acid mine drainage remediation options: A review. *Science of The Total Environment* **338**, 3–14.

Johnson R. H., Blowes D. W., Robertson W. D. and Jambor J. L. (2000) The hydrogeochemistry of the Nickel Rim mine tailings impoundment, Sudbury, Ontario. *Journal of Contaminant Hydrology* **41**, 49–80.

Jung M. C. (2001) Heavy metal contamination of soils and waters in and around the Imcheon Au–Ag mine, Korea. *Applied Geochemistry* **16**, 1369–1375.

Kang C.-H., Oh S. J., Shin Y., Han S.-H., Nam I.-H. and So J.-S. (2015) Bioremediation of lead by ureolytic bacteria isolated from soil at abandoned metal mines in South Korea. *Ecological Engineering* **74**, 402–407.

Kapoor D. and Singh M. P. (2021) 10 - Heavy metal contamination in water and its possible sources. In *Heavy Metals in the Environment* (eds. V. Kumar, A. Sharma and A. Cerdà). Elsevier, pp. 179–189.

Kavehei A., Gore D. B., Wilson S. P., Hosseini M. and Hose G. C. (2021) Assessment of legacy mine metal contamination using ants as indicators of contamination. *Environmental Pollution* **274**, 116537.

Koehnken L., Clarke N., Dineen R. and Jones W. (2003) Present status of remediation of Mt Lyell acid drainage. Sixth International Acid Rock Drainage Conference, Cairns, Australia, 65–70.

Kumari D., Qian X.-Y., Pan X., Achal V., Li Q. and Gadd G. M. (2016) Chapter two - Microbially-induced carbonate precipitation for immobilization of toxic metals. In *Advances in Applied Microbiology* (eds. S. Sariaslani and G. M. Gadd). Academic Press, pp. 79–108.

Laurence D. (2006) Optimisation of the mine closure process. *Journal of Cleaner Production* **14**, 285–298.

Lebrun M., Bourgerie S. and Morabito D. (2022) Restoration of old mining sites polluted by metal (loid) s by using various amendments. In *Bioremediation of Toxic Metal (loid) s.* CRC Press, pp. 351–383.

Lottermoser B. (2003) Mine water. In *Mine Wastes: Characterization, Treatment and Environmental Impacts* (ed. B. Lottermoser). Springer Berlin Heidelberg, Berlin, Heidelberg, pp. 83–141.

Lottermoser B. G. and Ashley P. M. (2011) Trace element uptake by Eleocharis equisetina (spike rush) in an abandoned acid mine tailings pond, northeastern Australia: Implications for land and water reclamation in tropical regions. *Environmental Pollution* **159**, 3028–3035.

Luo C., Routh J., Dario M., Sarkar S., Wei L., Luo D. and Liu Y. (2020) Distribution and mobilization of heavy metals at an acid mine drainage affected region in South China, a post-remediation study. *Science of The Total Environment* **724**, 138122.

Marrugo-Negrete J., Marrugo-Madrid S., Pinedo-Hernández J., Durango-Hernández J. and Díez S. (2016) Screening of native plant species for phytoremediation potential at a Hg-contaminated mining site. *Science of The Total Environment* **542**, 809–816.

Mhlongo S. E. and Amponsah-Dacosta F. (2016) A review of problems and solutions of abandoned mines in South Africa. *International Journal of Mining, Reclamation and Environment* **30**, 279–294.

Miao Z. and Marrs R. (2000) Ecological restoration and land reclamation in open-cast mines in Shanxi Province, China. *Journal of Environmental Management* **59**, 205–215.

Mkandawire M., Taubert B. and Dudel E. G. (2004) Capacity of Lemna gibba L. (Duckweed) for uranium and arsenic phytoremediation in mine tailing waters. *International Journal of Phytoremediation* **6**, 347–362.

Mohapatra D. P. and Kirpalani D. M. (2016) Process effluents and mine tailings: Sources, effects and management and role of nanotechnology. *Nanotechnology for Environmental Engineering* **2**, 1.

Moncur M. C., Ptacek C. J., Blowes D. W. and Jambor J. L. (2005) Release, transport and attenuation of metals from an old tailings impoundment. *Applied Geochemistry* **20**, 639–659.

Naidu G., Ryu S., Thiruvenkatachari R., Choi Y., Jeong S. and Vigneswaran S. (2019) A critical review on remediation, reuse, and resource recovery from acid mine drainage. *Environmental Pollution* **247**, 1110–1124.

Natarajan K. A. (2008) Microbial aspects of acid mine drainage and its bioremediation. *Transactions of Nonferrous Metals Society of China* **18**, 1352–1360.

Navarro M. C., Pérez-Sirvent C., Martínez-Sánchez M. J., Vidal J., Tovar P. J. and Bech J. (2008) Abandoned mine sites as a source of contamination by heavy metals: A case study in a semi-arid zone. *Journal of Geochemical Exploration* **96**, 183–193.

Newsome L. and Falagán C. (2021) The microbiology of metal mine waste: Bioremediation applications and implications for planetary health. *GeoHealth* **5**, e2020GH000380.

Newton G., Reynolds S., Newton-Reed S., Tuffly M., Miller E., Reeves S., Mistchenko J. and Bailey J. (2000) California's abandoned mines. *A Report on the Magnitude and Scope of the Issue in the State* **1**.

Nirola R., Megharaj M., Aryal R. and Naidu R. (2016) Screening of metal uptake by plant colonizers growing on abandoned copper mine in Kapunda, South Australia. *International Journal of Phytoremediation* **18**, 399–405.

Parbhakar-Fox A. K., Edraki M., Hardie K., Kadletz O. and Hall T. (2014) Identification of acid rock drainage sources through mesotextural classification at abandoned mines of Croydon, Australia: Implications for the rehabilitation of waste rock repositories. *Journal of Geochemical Exploration* **137**, 11–28.

Park D.-U. K., Kim D.-S., Yu S.-D., Lee K.-M., Ryu S.-H., Kim S.-G., Yang W.-H., Park D.-Y., Hong Y.-S., Park J.-D., Lee B.-K., Moon J.-D., Sakong J., Ahn S.-C., Ryu J.-M. and Jung S.-W. (2014) Blood levels of cadmium and lead in residents near abandoned metal mine areas in Korea. *Environmental Monitoring and Assessment* **186**, 5209–5220.

Passariello B., Giuliano V., Quaresima S., Barbaro M., Caroli S., Forte G., Carelli G. and Iavicoli I. (2002) Evaluation of the environmental contamination at an abandoned mining site. *Microchemical Journal* **73**, 245–250.

Pepper M., Roche C. and Mudd G. (2014) Mining legacies—Understanding life-of-mine across time and space. *Proceedings of the Life-of-Mine Conference.* Australasian Institute of Mining and Metallurgy Brisbane, Australia, pp. 449–465.

Pyatt F. and Pyatt A. (2004) The bioaccumulation of tungsten and copper by organisms inhabiting metalliferous areas in North Queensland: An evaluation of potential health effects. *Journal of Environmental Health Research* **3**, 13–18.

QLD E. (1999) *State of the Environment Queensland 1999.* Environmental Protection Agency, QLD, Brisbane.

Rani N., Sharma H. R., Kaushik A. and Sagar A. (2018) Bioremediation of mined waste land. In *Handbook of Environmental Materials Management* (ed. C. M. Hussain). Springer International Publishing, Cham, pp. 1–25.

Roche C. and Judd S. (2016) Ground truths: Taking responsibility for Australia's mining legacies. Mineral Policy Institute Report. http://www.mpi.org.au.

Saunders D., Beattie A., Eliott S., Fox M., Hill B., Pressey B., Veal D., Venning J., Maliel M. and Zammit C. (1996) Chapter 4 biodiversity. *State of the Environment Australia 1996,* Australian Government: Canberra, Australian Capital Territory.

Shamsoddini A., Raval S. and Taplin R. (2014) Spectroscopic analysis of soil metal contamination around a derelict mine site in the Blue Mountains, Australia. *ISPRS Annals of Photogrammetry, Remote Sensing & Spatial Information Sciences* **2**.

Sheoran A. S. and Sheoran V. (2006) Heavy metal removal mechanism of acid mine drainage in wetlands: A critical review. *Minerals Engineering* **19**, 105–116.

Simate G. S. and Ndlovu S. (2014) Acid mine drainage: Challenges and opportunities. *Journal of Environmental Chemical Engineering* **2**, 1785–1803.

Siegel, F. R. (2002) Environmental geochemistry of potentially toxic metals ;(Vol. 32). Berlin: Springer.

Skousen J. G., Ziemkiewicz P. F. and McDonald L. M. (2019) Acid mine drainage formation, control and treatment: Approaches and strategies. *The Extractive Industries and Society* **6**, 241–249.

Struhsacker D. W. and Todd J. (1998) *Reclaiming Inactive and Abandoned Hardrock Mine Lands in the West: What Really is Happening.* The Association.

Taylor M. P., Mackay A. K., Hudson-Edwards K. A. and Holz E. (2010) Soil Cd, Cu, Pb and Zn contaminants around Mount Isa city, Queensland, Australia: Potential sources and risks to human health. *Applied Geochemistry* **25**, 841–855.

Tomiyama S., Igarashi T., Tabelin C. B., Tangviroon P. and Ii H. (2019) Acid mine drainage sources and hydrogeochemistry at the Yatani mine, Yamagata, Japan: A geochemical and isotopic study. *Journal of Contaminant Hydrology* **225**, 103502.

Unger C., Laurencont T., Keliher L., McCombe C. and Bartley G. (2003) Rehabilitation plan for the Mount Morgan minesite central Queensland. Department of Natural Resources and Mines, Queensland Government, Brisbane.

Unger C., Lechner A., Glenn V., Edraki M. and Mulligan D. (2012) Mapping and prioritising rehabilitation of abandoned mines in Australia. *Proceedings Life-of-Mine*, 259–266.

Venkateswarlu K., Nirola R., Kuppusamy S., Thavamani P., Naidu R. and Megharaj M. (2016) Abandoned metalliferous mines: Ecological impacts and potential approaches for reclamation. *Reviews in Environmental Science and Bio/Technology* **15**, 327–354.

Wager R. and Jackson P. D. (1993) *The Action Plan for Australian Freshwater Fishes.* Australian Nature Conservation Agency.

Watanabe M. E. (1997) Phytoremediation on the brink of commericialization. *Environmental Science & Technology* **31**, 182A–186A.

Weng L., Boedhihartono A. K., Dirks P. H., Dixon J., Lubis M. I. and Sayer J. A. (2013) Mineral industries, growth corridors and agricultural development in Africa. *Global Food Security* **2**, 195–202.

Whitbread-Abrutat P. H. (1997) The potential of some soil amendments to improve tree growth on metalliferous mine wastes. *Plant and Soil* **192**, 199–217.

WorldBank (2002) It's not over when it's over: Mine closure around the world. World Bank Group Mining Department Washington, DC, p. 19.

Woodside, D. and O'Neill, D. (1995) Mining in relation to habitat loss and reconstruction. In Conserving Biodiversity: Threats and Solutions (eds. R.A. Bradstock, T.D. Auld, D.A. Keith, R.T. Kingsford, D.Lunney and D.P. Sivertsen), Surrey Beatty and Sons, Chipping Norton, pp. 76–93.

Worrall R., Neil D., Brereton D. and Mulligan D. (2009) Towards a sustainability criteria and indicators framework for legacy mine land. *Journal of Cleaner Production* **17**, 1426–1434.

Wright I. A., Paciuszkiewicz K. and Belmer N. (2018) Increased water pollution after closure of Australia's longest operating underground coal mine: A 13- month study of mine drainage, water chemistry and river ecology. *Water, Air, & Soil Pollution* **229**, 55.

Young K. P., Murray B. R., Martin L. J. and Murray M. L. (2021) Lost but not forgotten: Identifying unmapped and unlisted environmental hazards including abandoned mines. *Sustainability* **13**, 11011.

Zhong X., Chen Z., Li Y., Ding K., Liu W., Liu Y., Yuan Y., Zhang M., Baker A. J. M., Yang W., Fei Y., Wang Y., Chao Y. and Qiu R. (2020) Factors influencing heavy metal availability and risk assessment of soils at typical metal mines in Eastern China. *Journal of Hazardous Materials* **400**, 123289.

Zhou J., Nyirenda M. T., Xie L., Li Y., Zhou B., Zhu Y. and Liu H. (2017) Mine waste acidic potential and distribution of antimony and arsenic in waters of the Xikuangshan mine, China. *Applied Geochemistry* **77**, 52–61.

Zou T., Li T., Zhang X., Yu H. and Huang H. (2012) Lead accumulation and phytostabilization potential of dominant plant species growing in a lead–zinc mine tailing. *Environmental Earth Sciences* **65**, 621–630.

4 Managing Orphaned and Abandoned Mines – A Canadian Perspective

G.A. Tremblay and C.M. Hogan

4.1 INTRODUCTION

The legacy of orphaned and abandoned mines, including environmental liability, human health concerns, and the costs of cleanup and long-term monitoring and maintenance, is a serious issue in many countries, and Canada is no exception. Orphaned or abandoned mines are those for which the owner cannot be found, or is financially unable or unwilling to remediate the site. The most serious environmental issues for orphaned and abandoned mines are acidic drainage and metal leaching from underground workings, open-pit mine faces and workings, waste rock piles and tailings impoundment areas. When these contaminants enter the receiving environment, they can cause much damage. Public health and safety hazards result from mine openings, mine wastes, abandoned infrastructure and subsidence. Policy and financial implications include fiscal responsibility, cost of reclamation and long-term management, funding and jurisdictional liabilities and possible reuse of the land (for future mining, recreational activities, etc.).

Mining has been central to the Canadian economy for nearly 150 years, and Canada is a supplier of mineral commodities worldwide. The long history of mining in Canada has resulted in a large number of abandoned sites; it is estimated that there are more than 10,000 abandoned sites (Mackasey, 2000). These sites vary widely with respect to size, features and requirements for rehabilitation. In the past, mines became "abandoned" because there were no legislative mechanisms in place to make mine owners accountable for selection and design of a mine site, through to its closure and decommissioning. Furthermore, there was little understanding of the physical and environmental hazards that could be involved in walking away from a site. Today, mining legislation in all Canadian jurisdictions requires the preparation of mine closure plans that describe how mine sites should be rehabilitated throughout their lifecycle, and how they will be decommissioned when mining activities cease.

There are significant differences between closure and management of orphaned and abandoned mines, operating mines and new mines. An important concept is that cost-effective prevention of acidic drainage/metal leaching and other potential environmental impacts is primarily accomplished in the assessment and design phases. As shown in Figure 4.1, over time the costs involved in implementing preventive technologies or strategies dramatically increase. Closure planning during the mine design stage provides an opportunity to design for closure more effectively. Since many legacy sites did not include closure planning (it was not a regulatory requirement), dealing with the remediation of these sites currently can be very expensive.

The National Orphaned/Abandoned Mines Initiative (NOAMI) was formally established in 2002 at the request of the federal, provincial and territorial mines ministers in Canada. The groundwork for the initiative was laid out in 1999 when a number of stakeholders requested the Canadian Mines

DOI: 10.1201/9781315141855-4

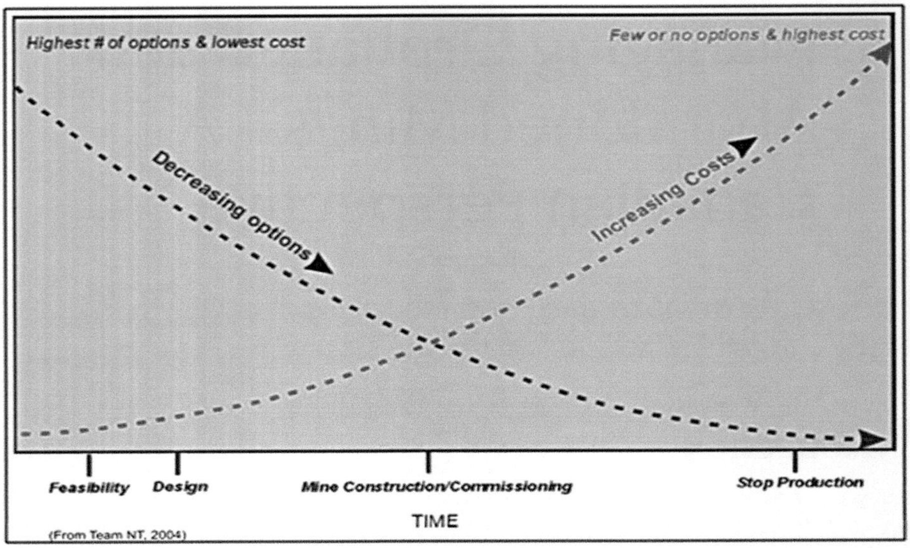

FIGURE 4.1 Options and costs over time – from the GARD Guide, INAP, 2009).

Ministers to establish a joint industry-government working group, assisted by other stakeholders, to review the issue of orphaned and abandoned mines. The Ministers supported this initiative and requested that a multi-stakeholder workshop be organized to identify key issues and discuss priorities for action.

This workshop – held in Winnipeg in 2001 – determined the key issues for remediation of orphaned and abandoned mine sites, identified common ground among various communities of interest and identified strategies for making progress. Operating principles and a series of guidelines were laid down at the workshop, which still apply to NOAMI as it exists today.

NOAMI is guided by a multi-stakeholder advisory committee that brings together representatives from the Canadian mining industry; federal, provincial and territorial governments; non-governmental organizations; and Aboriginal Canadians. Together, they assess issues and make recommendations for the collaborative implementation of remediation programs and policies for orphaned and abandoned mines across Canada. The NOAMI Advisory Committee takes direction from the federal, provincial and territorial mines ministers and, in turn, reports progress annually to the Energy and Mines Ministers Conference. NOAMI's activities are, first, jointly funded by the federal, provincial and territorial governments; the Mining Association of Canada; and the Prospectors and Developers Association of Canada; and, second, administered by a secretariat at Natural Resources Canada.

Over the past 15 years, NOAMI has made significant contributions to the state of knowledge of orphaned and abandoned mines in Canada and approaches for addressing them. These efforts are outlined in the *NOAMI 2002–2008 Performance Report* (NOAMI, 2009) and the *NOAMI 2009–2015 Performance Update* (NOAMI, 2015). NOAMI activities, publications and reports are available on the NOAMI website at www.abandoned-mines.org.

The NOAMI program examines two aspects of orphaned and abandoned mines – issues surrounding remediation and long-term management of existing sites and best practices to prevent the occurrence of future abandonments. In the initial years of the program from 2002 to 2008, the work focused primarily on the remediation and policies for existing abandoned sites. The later years (2009 to the present day) examined strategies to prevent future abandonments and built on aspects covered in the early years.

4.2 WORK OF NOAMI AND PRODUCTS – THE EARLY YEARS

NOAMI, by itself, does not directly fund or clean up orphaned and abandoned mine sites. NOAMI has no budget for this and nor does it have a mandate for this type of work. Instead, NOAMI examines the legislative, policy and program framework in Canada for addressing the issue of orphaned and abandoned mines and makes recommendations for improvements. NOAMI's multi-stakeholder model of collaboration is an ideal structure for this undertaking. NOAMI provides a forum for partners to discuss barriers and issues for the remediation of orphaned and abandoned mine sites.

When NOAMI was established in 2002, task groups were created to work on the following key issues identified at the Winnipeg 2001 workshop:

1. Web-based Inventory..
2. Community Engagement;.
3. Legislative and Institutional Barriers to Collaboration.
4. Funding Approaches.
5. Jurisdictional Legislative Review.

Communication, the sharing of information and knowledge, underpins all the activities carried out by the NOAMI program.

Program summaries for the task groups are provided below.

4.2.1 Web-Based NOAMI Inventory

One of NOAMI's objectives has been to devise an inventory of orphaned and abandoned mines based on compatible inventories from each province and territory. All Canadian provinces, territories and several federal agencies with a history of mining maintain their own inventory of mining and exploration sites that pose a risk to human health and safety or the environment. However, there is a large discrepancy in the level of detail and coverage of these inventories across Canada. An added challenge relates to variations in the way that hazards or features are defined or categorized, and which hazards are included in any particular inventory.

Research and compilation of information on abandoned mines is necessary to assist in sound decision-making, cost-efficient planning and sustainable rehabilitation. Such information is also necessary to ensure sound decision-making and access to information by governments, civil society, industry and other stakeholders. With this in mind, the first step to building the inventory was to reach consensus on the definitions and terminology applied to orphaned and abandoned mine sites. The definition of "orphaned" and "abandoned" mines or sites varies among jurisdictions in Canada. A primary criterion for the inventory was to develop a set of definitions under which information from all jurisdictions could be defined and compiled. In addition, many of the existing inventory definitions, which could be in conflict, had to be avoided. A comprehensive review of Canadian and international efforts to inventory orphaned and abandoned mine sites, along with high-level definitions, was documented in *Capacity Building for a National Inventory of Orphaned/Abandoned Mines in Canada* (Cal Data Ltd., 2005). The proposed definitions were suggested to be independent of most existing definitions and would provide a framework in which the existing definitions could be linked.

Following this work, in 2009 NOAMI decided to classify each site according to the environmental hazards each presented to the public and the environment. The feature-based classification system used for the NOAMI inventory was adapted from the four-class ranking of hazards developed for the province of Ontario's Abandoned Mines Information System (AMIS). Each site is classified into one of four categories (Class A, B, C or D) based on the degree of potential risk each poses to the public and the environment. Class A sites represent the highest concern, while Class D sites pose little risk.

To obtain a Canada-wide perspective, the NOAMI inventory was designed to bring together and build on the strengths of the individual inventories but allow for discrepancies in coverage, detail and standardization. A web-based approach was created to provide users with a tool that offers a single-window access to the data from multiple jurisdictions. A web portal is used, whenever possible, to provide users with a link to the original data source. The site includes an interactive user-friendly map to display the orphaned and abandoned sites, along with their feature-based or status classifications. A key consideration for the jurisdictions was to ensure that the system does not affect their current operational status and would accommodate the inclusion of more detailed information or updates in the future. The inventory is available to the public at www.noami.org.

4.2.2 Community Involvement

An objective of NOAMI is to develop the means to foster community involvement in making decisions about closure and reclamation standards and to ensure that targeted end-use and reclamation standards are acceptable to local communities. In 2003, NOAMI published a report titled *Lessons Learned on Community Involvement in the Remediation of Orphaned and Abandoned Mines – Case Studies and Analysis* (NOAMI, 2003a). Case studies related to community involvement were completed for three abandoned Canadian mine sites (Giant Mine, NT; Deloro Mine, ON; and Mount Washington Mine, BC), along with experiences in community involvement at abandoned mines in the United States. To reflect a diversity of perspectives, the report findings were based on personal interviews with key contacts in the community. These mines entail more complicated issues than most orphaned and abandoned mines. Case studies illustrate the importance of effective community involvement in decision-making. The lessons learned from these studies were developed into guidelines published in the brochure titled *Best Practices in Community Involvement: Planning for and Rehabilitating Abandoned and Orphaned Mines in Canada* (NOAMI, 2003b).

The brochure presents eleven guiding principles developed to support community involvement in policy development, site management and process administration:

- Communication.
- Inclusiveness.
- Representation.
- Fostering Confidence in Decision-Making.
- Information Dissemination & Communication.
- Participation and Representation.
- Resources and Assistance;
- Facilitation.
- Integration.
- Consistency of Involvement.
- Respecting Local Cultures.

These principles can serve as a template for use by governments, industry, local communities and other parties for the development of policy and citizen engagement plans prior to, during and after the rehabilitation of mine sites.

As part of other studies, NOAMI will continue to examine ways to foster meaningful community involvement and engagement in abandoned mine remediation and closure.

4.2.3 Legislative and Institutional Barriers to Collaboration

NOAMI wanted to understand better the legal and institutional problems that prevent third parties from collaborating on clean-up activities and addressing associated liabilities related to orphaned and abandoned mines. NOAMI examined the efficacy of various approaches in a background study titled *Barriers to Collaboration: Orphaned and Abandoned Mines in Canada* (Castrilli, 2002). This report examined regulatory or institutional barriers, liability disincentives and collaborative opportunities for voluntary abatement, remediation and reclamation of orphaned and abandoned mines in Canada and selected international jurisdictions. Particular emphasis was put on four approaches: "Good Samaritan" legislation; permit blocking; allocative versus joint and several responsibility; and non-compliance registries.

The report findings provided background for a multi-stakeholder workshop held in 2003 *Legal and Institutional Barriers to Collaboration Relating to Orphaned and Abandoned Mines* (Versteeg, 2003). Workshop objectives were to identify the key barriers that are inhibiting governments, the private sector and voluntary organizations from effectively financing and administering the remediation of problematic abandoned and orphaned mine sites in Canada and begin developing approaches to overcome these barriers. The presentations detailed experiences, in both Canada and the United States, on existing and alternative legal and administrative mechanisms that have been implemented to address the issue. Approaches to overcome these key barriers were developed and subsequently integrated into a review on jurisdictional legislations (Castrilli, 2005).

4.2.4 Funding Approaches

Another objective of NOAMI was to identify funding approaches and preferred options for the remediation of orphaned and abandoned mine sites across Canada that could be adapted to the needs of each jurisdiction. Typically, orphaned and abandoned mines revert to the Crown with their remaining environmental liabilities and inadequate or non-existent securities to finance remediation. A report titled *Potential Funding Approaches for Orphaned/Abandoned Mines in Canada* (Castrilli and Watson & Associates, 2003) was prepared and it outlined a variety of funding approaches to be considered for the cleanup or management of liabilities related to orphaned and abandoned mine sites. Advantages and disadvantages of each approach were evaluated, and preferred options were recommended for governments to consider. This report concluded that no single funding approach was a complete solution; a combination of a number of approaches would likely be required.

A multi-stakeholder workshop on *Assessing Liabilities and Funding Options* was held in 2005 that further developed funding approaches and related issues for orphaned and abandoned mine sites (Stratos Inc., 2006). A roll-up discussion identified gaps and future priorities for NOAMI. One recommendation was to produce a plain language toolkit of funding options, outlining a series of options and illustrated with case studies. This would be a resource document for use by jurisdictions across Canada to help guide the establishment of potential funding options for the remediation of orphaned and abandoned mine sites. This funding "toolkit" was completed and released in 2006, titled *Rehabilitating Abandoned Mines in Canada: A Toolkit of Funding Options* (Cowan Minerals Ltd., 2006).

4.2.5 Jurisdictional Legislative Review

In 2003, the Mines Ministers requested NOAMI to complete a set of guidelines for a jurisdictional legislative review that could be used by jurisdictions to evaluate their own policies with respect to collaboration, liability and funding. The work was conducted by NOAMI with a view to ensuring that approaches across jurisdictions are consistent, fully disclosed, coordinated and efficient. A series of guidelines and a checklist with associated questions were developed to facilitate a focused review of the existing legislation and related policies and practices that relate to orphaned and abandoned mine sites as well as contaminated and operating sites (where relevant to legacy issues). A study was undertaken using the guidelines laid out in the report *Guidelines for Legislative Review* (NOAMI, 2004).

Consultation with all the jurisdictions was done to ensure that existing legislation, policies and practices were identified and verified. A synthesis of the jurisdictional analysis was completed that included an assessment of gaps, limitations, barriers and opportunities, along with a summary of observations. Recommendations to address remediation of orphaned and abandoned mines were drawn from previous NOAMI reports on legislative barriers to collaboration and potential funding approaches.

In 2005, the publication titled *Report on the Legislative, Regulatory, and Policy Framework Respecting Collaboration, Liability, and Funding Measures in relation to Orphaned and Abandoned, Contaminated, and Operating Mines in Canada* (Castrilli, 2005) was released. This report is often referred to as the *Jurisdictional Legislative Review,* and it serves as an invaluable reference material on Canadian legislation, regulations and programs in relation to orphaned and abandoned mines, contaminated sites and operating mines in Canada.

4.3 RETURN OF MINE LANDS – A MULTI-PHASED PROJECT

One of the NOAMI's guiding principles is that "work toward eliminating future abandonment must continue, including the tightening of regulatory approaches". In 2009, NOAMI began to examine the legislative tools and policy approaches across Canada to ensure that current operating mines can be closed properly so that they do not become abandoned mines in the future. The committee members have long recognized the need for a clear policy framework for mine closure, long-term liabilities and return of mining lands to the Crown. Such a framework would address the closure and post-closure of mine sites and their long-term management and related liabilities in a manner that details the responsibilities of the mine owner/operator and government regulatory agencies. The policy framework would examine the questions surrounding the transfer of mining lands back to the Crown through the issuance of a release, including questions of when and under what conditions should such a return be possible (Holmes and Stewart, 2011; Gardiner et al., 2015).

It is recognized that once a mine is closed, the owner/operator may wish to return (relinquish, give up, surrender, abandon, exit) the site back to the Crown for various reasons. The owner may not wish to list the property on their books forever, and the ongoing liabilities need to be reported in corporate financial reports. Ultimately, the owner is going to vanish through business closure or some other mechanism, and the mineral right leases or permits will expire. In the long term, the lands will revert to the Crown, either on managed terms or through unplanned abandonments. Jurisdictions not wishing to receive abandoned lands and their associated liabilities need to have some agreed-upon procedure in place to accept lands (Cowan Minerals Ltd., 2006).

4.3.1 Mine Closure and Management of Long-Term Liabilities

NOAMI commissioned a study on mine closure and management of long-term liabilities titled *The Policy Framework in Canada for Mine Closure and Management of Long-Term Liabilities: A Guidance Document* (Cowan et al., 2010). The document examined major components related to mine closure and post-closure site management, which can include long-term maintenance and monitoring, financial assurance, relinquishment[1] and institutional care. A valuable reference tool, the report presents a policy framework, together with recommendations for preventing further accrual of abandoned mine hazards.

A relevant finding was that while processes of closure planning and provision of financial assurance are well-developed and consistently applied across Canada, policy around long-term management of sites beyond closure, including methods to return mining lands to the Crown, is almost non-existent.

A key message of the report was:

> *Jurisdictions should have a managed relinquishment process, which is clear and unfettered. Hitherto closure plans have been prepared on a "design for closure" basis. This should specifically include when and why relinquishment is not acceptable to the jurisdiction. It is suggested that a more*

forward-looking approach be embraced and that a "design for relinquishment" approach be adopted (Cowan et al., 2010).

Building on the findings of the guidance document, a multi-stakeholder workshop titled *Exploring the Management of Long-term Liabilities and the Return of Mining Lands to the Crown* was held in 2011. Advice and guidance were obtained from the workshop delegates who assisted NOAMI in developing a strategic roadmap for managing long-term liabilities and issues relating to the return (relinquishment) of mining lands to the Crown. The themes of risk management, funding, legislation and policy and regulation were explored during the workshop and presented in the proceedings (Tunis and Associates, 2011). A key recommendation generated by the workshop participants was for NOAMI to create a decision tree that would assist in determining the return of mining lands to the Crown.

The Advisory Committee agreed that the creation of a decision tree for assessing the return of mine lands to the Crown would be a useful tool and a logical next step for this project. The decision tree would provide information and options for Canadian jurisdictions to consider in the development of new legislation or to revise current legislation. In the past, NOAMI has used case studies to examine processes, and as tools to study factors involved in dealing with orphaned and abandoned mines. Examination of several mine sites in Canada that were successfully returned to the Crown or closed sites that had elements related to their potential return would provide information on processes involved and some lessons learned. The information collected from the case studies, along with other relevant documents, could be used to establish a decision tree.

4.3.2 Decision-Making Process for the Relinquishment of Closed Mine Sites

NOAMI commissioned a two-part study, *Case Studies and Decision-making Process for the Relinquishment of Closed Mine Sites* (Cowan et al., 2013). The first part examined case studies from different Canadian jurisdictions that provided information on relinquishment. The second part, a decision process, identified key issues and questions that had to be addressed to determine if a site should be brought under government jurisdiction, or remain the responsibility of the operator. The decision process was not to be prescriptive; instead, it was to focus on providing guidelines for how to make good decisions. Ultimately, it will be up to each jurisdiction to decide whether they will develop a return of mine lands legislation and, if so, what elements they may wish to adopt.

Case studies were undertaken for sites in different Canadian jurisdictions: Quebec Lithium Mine, Quebec; Renabie Gold Mine, north-central Ontario; Gregg River Coal Mine, Alberta; Contact Lake Gold Mine, northern Saskatchewan; Polaris Mine, Nunavut, and Farley East Tailings Management Area, northern Manitoba. There was much variation in the information garnered from the sites due to the age and nature of the projects: changes in ownership; management and operational personnel; regulatory regimes; and reporting requirements (Cowan et al., 2013).

The six case studies gave rise to a series of 24 lessons learned and dozen key ones are cited below (Cowan et al., 2013):

1. Jurisdictions must have clear and well-written legislation and policies in place to facilitate the return of lands. It is necessary to have a system to receive and manage funds to ensure the viability needed to address long-term issues.
2. Where planning and procedures for relinquishment are not in place, defaulting will eventually occur at taxpayers' expense.
3. Few jurisdictions have a satisfactory funding regimen in place to deal with surrendered, or about to be surrendered lands.
4. Most jurisdictions providing for rehabilitation releases and/or relinquishment of lands do not have complete releases for environmental responsibility or liability.
5. It appears that only some jurisdictions have institutional control plans for relinquished lands, i.e. there is no administrative unit with direct responsibility for institutional issues following surrender, e.g. records maintenance, land use planning inputs.

6. Public consultation requirements/efforts are generally considered inadequate relative to current expectations, especially with regard to Aboriginal communities. Consultation requires good information on the issues and needs of impacted communities.
7. The length of post-closure monitoring periods needs greater evaluation, especially where documentation/identification of features is lacking.
8. Well-defined risk assessment procedures for property returns are not available for many situations. More information on quantitative risk assessment is required to support chemical and physical stability determinations.
9. Where cutting-edge technology is involved, longer-term monitoring and scientific assessment may be expected, e.g. new schemes related to permafrost or climate change issues.
10. Effective cost-estimation procedures for long-term care and maintenance need development.
11. Storage and maintenance of mine plans and records is essential for technical assessment. Several case studies were hindered by the loss of or unavailability of (historic) documentation.
12. Provision of a final closure report detailing all completed decommissioning and reclamation work is an excellent best practice.

As noted at the NOAMI 2011 workshop, and illustrated in the report by Cowan et al. (2013), there are a vast array of factors to consider when developing a process for relinquishment. In addition, each jurisdiction would have its own set of parameters, as well as legislative and policy framework. Many of the parameters could be considered at different stages of the process, and would be amenable to a step-wise approach for planning. As part of a managed relinquishment process, a decision process should be generated that provides guidance and clarity, and which outlines key factors that need to be assessed or refined. These steps would determine whether responsibility for the site should remain with the operator or could be transferred to the Crown.

The five-step decision-making process (Cowan et al., 2013) provides a starting point for jurisdictions developing or revising a program for relinquishment. Although national consistency would be beneficial, each jurisdiction would need to tailor the decision-making process to meet its own regulatory regime and policies.

As shown in Figure 4.2, the five-step decision-making process follows the natural progression from "submitting the application" through to "implementation". For each step, consensus must be reached by the review committee before proceeding to the succeeding step, i.e., all "yes" components must be met. It is recommended that an appeal process be available through every step.

The step-wise approach provides guidance as to whether a mining property has been closed as set-out in the closure requirements and closure plan, and whether technical and financial long-term monitoring and maintenance needs have been addressed sufficiently (Cowan et al., 2013).

The steps are briefly summarized here:

1. *Submitting the Application*: Ensures that the proponent qualifies for relinquishment with respect to closure and post-closure plans and objectives. Specifically, the site is physically and chemically stable, and existing permits are fulfilled or transferable to the jurisdiction. All necessary parties, affected stakeholders and Aboriginal peoples must be notified of the application.
2. *Site Assessment*: Locational and technical issues are evaluated to determine if the site is sufficiently reclaimed to meet legal requirements under the closure plan, and will meet proposed future land use requirements. Any interim monitoring requirements must be identified for evaluation.
3. *The Long-Term:* Requirements for long-term monitoring, maintenance or capital replacement of rehabilitation works and associated cost estimating are evaluated. A peer-reviewed risk assessment by qualified persons is required for unforeseen events. This is a critical

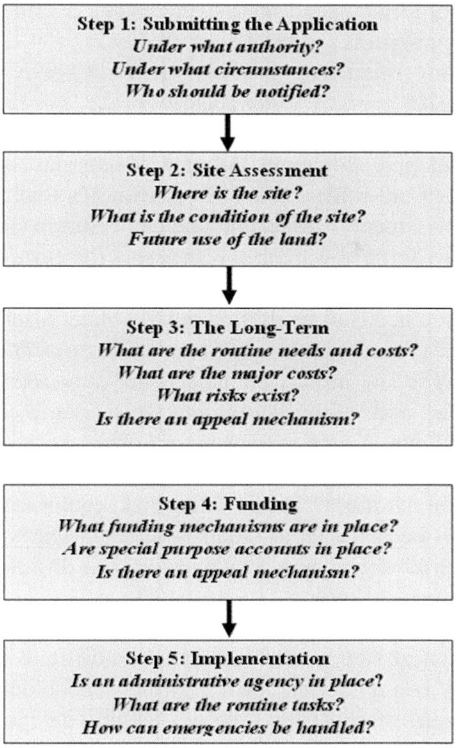

FIGURE 4.2 Five-step decision-making process (Cowan et al., 2013).

point in the process, as the proponent may decide to retain the properties due to prohibitive funding requirements for relinquishment.

4. *Funding:* Considers what future activities will need to be funded, their costs and the method of fund management. All funds are provided by the proponent.
5. *Implementation:* Once a site is approved for relinquishment, the regulator must have a management system in place to manage funds, ensure the site is monitored and maintained and manage data securely. Emergency protocols should be in place.

The report recommends that mining projects should be designed with the objective of reclaiming the site for relinquishment and future beneficial use; however, it is noted that relinquishment may not be possible where the environmental, social, political or financial risks/costs are too high (Cowan et al., 2013). The relinquishment process would be straightforward for fully reclaimed sites that are physically and chemically stable and require no or minimal monitoring, care or maintenance. A more complex form of relinquishment will be needed for reclaimed sites that require ongoing monitoring, care and maintenance. Long-term care of these sites would need to be fully funded by the proponent and managed by the jurisdictions through some form of institutional care. Cowan et al. (2013) asserted that a well-designed and well-managed relinquishment policy and program can lead to a win/win situation in many instances.

4.3.3 Key Criteria for Effective Long-term Stewardship

The policies and procedures for relinquishment need to be linked to the regulatory environment of the jurisdiction. A strategy needs to be in place to address the various aspects involved in the management or stewardship of relinquished sites, such as site assessment and an effective long-term monitoring and maintenance program.

NOAMI undertook a study titled *Key Criteria for the Effective Long-term Stewardship of Closed, Orphaned/Abandoned Mine and Mineral Exploration Sites* (Kingsmere Resource Services Inc., 2015) wherein criteria were devised to assess these sites in order to evaluate their condition. A list of site attributes was developed, along with a review of national and international best practices of the management of post-closure mine sites. The report notes that effective stakeholder engagement at every stage of the assessment is important and can result in significant benefits.

The initial step toward long-term stewardship is to assess the current state of the site. The report provides a summary of features that could exist at the site, with a focus on identifying the hazards. Before the inspection, research needs to be carried out to collect information on the site, and a site visit safety plan should be developed, containing potential hazards and mitigation controls, logistical details and safety rules. The site inspection aspects are numerous, and include site access and condition, mine type, facilities and infrastructure and physical and chemical hazards. To assist in this, a field report was created that can be used to record observations during the site visit.

The next step is an assessment of the level of risks posed by the identified hazards. Various types of risk are considered in the report: public safety risk, ecological risk and human health risk. Subsequently, a decision can be made as to whether remediation is required, and to what level. Reaching consensus on the level of risk posed by a hazard is a difficult process and must take into account the concerns of all the stakeholders.

An effective stakeholder engagement plan must be in place to address site risks and possible remediation of that risk. The next step is site remediation; a difficult stage since a balance must be achieved between the level of risk reduction desired by the stakeholders and the cost to achieve that level. The last step for long-term stewardship involves keeping data and/or records of the activities on the site and regular inspections made by qualified personnel. To ensure the long-term stewardship of closed, orphaned and abandoned mine sites, an institutional control program is recommended.

This report builds on several of the lessons learned from the Cowan et al. report (2013). In addition, it addresses the first two steps laid out in the Five-Step Decision-making Process, specifically, *Step 1: Submitting the Application* and *Step 2: Site Assessment.*

4.3.4 Cost and Risk Estimation Study: Orphaned and Abandoned Mines

The logical next study for NOAMI to consider was *Step 3: The Long Term.* A report was commissioned, known as *Orphaned and Abandoned Mines: Risk Identification, Cost Estimation, and Long-term Management* (Kingsmere Resource Services Inc. (KRS), 2016) to address elements of long-term or perpetual care. The report documented a risk identification/cost estimation framework for use as a tool so that a wide variety of multi-stakeholder users could:

- Identify and characterize risks and liabilities at orphaned and abandoned mine sites.
- Review methodologies to assess the significance of identified risks and to estimate the costs to mitigate the risk.
- Summarize and evaluate long-term management methods.
- Identify types of costs associated with long-term management of a site, and explore funding methods.
- Define major areas of potential cost underestimation/overruns, conflict and/or uncertainty and discuss ways to address these issues.

Long-term management of orphaned and abandoned mine sites, as well as closed mine sites, can involve a wide range of activities. These activities will depend on the site conditions and/or the residual hazards after remediation. Key components of a program for the long-term stewardship of a site would include:

- Information management.
- Site monitoring and maintenance.
- Unforeseen events (i.e. responses when remedies or controls fail).
- Application/enforcement of legal or other mechanisms to restrict future use.

Few jurisdictions in Canada or the world have developed a formal institutional control management framework that provides for long-term stewardship of mine sites once the operator has fulfilled their decommissioning and reclamation obligations and proven that the site is chemically and physically stable. The KRS (2016) report identifies five options for funding long-term stewardship costs and summarizes the Institutional Control Program developed in the province of Saskatchewan (Canada) as an example of a long-term management framework and funding mechanism.

4.3.5 Institutional Control Program

One of the few jurisdictions with a clear process and policy in place is the "Institutional Control Program (ICP)" in Saskatchewan (SK). The ICP has garnered international attention and NOAMI has identified the program as the most advanced Canadian regulatory regime that addresses all aspects of site relinquishment.

Relinquishment of a mine site back to the Crown is the final step in the mining sequence. Once a mine site has obtained a closed status, it can be put under institutional care of a custodial authority using a controlled process. This is very different from regulatory authorities having to assume ownership in an unmanaged process as for orphaned and abandoned mines. Under legislation of several Canadian mining jurisdictions, the government is able to issue a release, or a similar instrument, that would legally allow a mining operator to terminate its operations and responsibility for a site, with no further obligations. However, few jurisdictions other than Saskatchewan have issued such instruments because of the lack of clarity around the process.

The ICP outlines conditions for which it will accept closed mine sites into the program and provide for their long-term management. Similarly, SK will accept orphaned or abandoned mines into the program once the reclamation is completed, and monitoring has shown the site is chemically and physically stable (KRS, 2016).

In 2007, Saskatchewan enacted legislation to implement the ICP through the Reclaimed Industrial Sites Act (RISA) and Regulations (RISR). The ICP has two components: the Registry and the Institutional Control Funds, which includes the Monitoring and Maintenance Fund Unforeseen Events Fund. Records and information for accepted sites are archived in the Registry, and access is provided to both regulatory authorities and the public. The Registry provides the regulatory authorities with information to ensure that the prescribed monitoring and maintenance are executed and that the site remains compliant with land use restrictions.

As a condition of acceptance into the ICP, the operator must submit a monitoring and maintenance plan that lays out the long-term requirements for the site and the future costs associated with those requirements. The value of the future costs determines the amount to be deposited in the Monitoring and Maintenance Fund. The deposit is site-specific and must be calculated in order to provide sufficient revenue to pay for future costs incurred in perpetuity. In addition, the operator must contribute to the Unforeseen Events Fund. This contribution must attain a sufficient value to cover costs associated with unforeseen events, and eventually, once it builds up in value, the site would be released from this financial requirement. These funding measures ensure that future generations do not have to bear the cost of long-term management or future remediation of these sites. It is important to note that the funds are managed by the province but "stand alone" from provincial revenue.

The Institutional Control (IC) Program accepted six sites (one gold and five uranium) into the IC Registry in 2009 and conducted the scheduled inspections on the sites in 2014. A number of sites are under assessment for acceptance in 2017 and the regulatory regime is being reviewed to further enhance the effectiveness and sustainability of the program.

Several excellent documents on this subject have been produced by the team who developed the IC framework. These include papers by Hovdebo et al. (2015), and KRS (2016). Detailed information is available on Saskatchewan's Ministry of the Economy website: http://economy.gov.sk.ca?Institutional_Control-Decommissioned_Mines/Mills

4.3.6 Return of Mine Lands Project – Conclusion

The return of mine lands project has produced several important tools and guidance documents that will contribute to the prevention of future abandoned mines. Taken together, these projects constitute important tools that will make a major contribution toward the development of a pan-Canadian policy framework that would address all aspects of managing orphaned mine liabilities in the long term, and preventing future abandonments. NOAMI will continue to explore development of tools to assist in mine closure and relinquishment.

4.4 JURISDICTIONAL UPDATES

Provinces, territories and Indigenous and Northern Affairs Canada (for sites north of the 60th parallel) in Canada have made significant progress in the remediation of abandoned mines in their jurisdictions. Since 2002, the jurisdictions have spent more than C$1 billion addressing orphaned and abandoned mines. Various funding partnerships and collaborative approaches have been used, and this information is invaluable and can be applied on a national and international basis (Cowan Minerals, 2006). A number of partnerships have been formed to remediate orphaned and abandoned mines in Canada. Several of these approaches are of particular relevance to NOAMI's mandate for the development of collaborative partnerships in the implementation of remedial programs.

Highlights of their work on orphaned and abandoned mines were provided by several jurisdictions for the *Mining Sector Performance Report 2006–2015* (IGWG, 2016) and excerpted below. Additional information on jurisdictional activities and partnerships is provided in *NOAMI's 2002–2008 Performance Report* (NOAMI, 2009) and the *NOAMI 2009–2015 Performance Update* (NOAMI, 2015).

4.4.1 Federal Government – Indigenous and Northern Affairs Canada

The Northern Contaminated Sites Program (NCSP), within Indigenous and Northern Affairs Canada, was created in 1991 to manage remediation of contaminated sites across the North. In 2005, the Federal Contaminated Sites Action Plan was established by the federal government. This program committed C$3.5 billion over a 15-year period for the assessment and remediation of contaminated sites under the federal government's responsibility, which includes abandoned mines in the Yukon, the Northwest Territories (NWT) and Nunavut.

As of April 1, 2016, over 1,000 contaminated sites in the North have been assessed by the NCSP, of which 97 were classified as high priority for action. Remediation has been completed at 48 sites across the three territories. Work is ongoing at 76 sites, including two of the highest priority northern sites, the Giant Mine in the NWT and the Faro Mine in the Yukon. In the case of the Faro Mine, INAC works very closely with the Government of Yukon.

The Program continues to promote social and economic opportunities in the North by engaging First Nations, Inuit and other Northerners in all aspects of the site management and remediation process. It will continue to ensure that contaminated sites are properly managed to ensure the protection of human health, safety and the environment for all Northerners.

4.4.2 Province of British Columbia

The province of British Columbia established the Crown Contaminated Sites Program (CCSP) in 2003 based on a report by the Office of the Auditor General seeking improvements in the management of contaminated sites. The mandate of CCSP is to identify and remediate high-risk contaminated sites that are located on Crown land where no responsible person can be identified and the responsibility for remediation falls to the province. Remediation undertaken complies with the Environmental Management Act, Contaminated Sites Regulation and the Hazardous Waste Regulation. The 2016 Crown Contaminated Sites Program Biennial Report, which describes the province's effort and the results achieved, is available at: http://www2.gov.bc.ca/assets/gov/environment/air-land-water/site-remediation/docs/reports-and-presentations/biennial_report.pdf

To date, 82 sites have been investigated, of which 48 have been determined to be low risk where no immediate action is required, 18 have been fully remediated and 16 are under investigation or undergoing remediation. Orphaned and abandoned mine sites comprise about 95% of the contaminated sites within CCSP portfolio. As of March 31, 2016, the province of British Columbia has recognized contaminated sites liabilities totaling C$508 million. Of this, over C$192 million has been spent on site remediation.

4.4.3 Province of Manitoba

In 2000, Manitoba established the Orphaned/Abandoned Mine Site Rehabilitation Program to address the environmental, health and public safety concerns of orphaned and abandoned mines in the province. Within this program, 149 former mine sites were identified, including five high-priority sites and 31 high-hazard sites. To date, 30 of the 31 high-hazard sites have been remediated and work is underway to complete the one remaining site. Manitoba is now addressing lower priority sites, largely in response to communities and government concerns, and has remediated 10 low- to moderate-risk sites. As of March 31, 2016, the province had spent C$214.8 million on orphaned and abandoned mine site rehabilitation.

Manitoba supports the rehabilitation of orphaned and abandoned mines through its programs, partnerships, community involvement and funding initiatives. Further information may be found at http://www.gov.mb.ca/iem/mines/oa_rehabilitation.html.

4.4.4 Province of Newfoundland and Labrador

Orphaned and abandoned mines in Newfoundland and Labrador are mostly historic and predate the province joining the Confederation of Canada in 1949; all of the sites predate the Mining Act of 2000. These properties, ranging from exploration sites to large-scale former producing mines, can pose safety risks to the public and some have environmental issues.

Newfoundland and Labrador has spent over C$30 million on orphaned and abandoned mines in recent years. In addition, it has implemented a program of dam safety reviews and repairs of tailings dams at orphaned and abandoned mines with the goal of bringing the dams to Canadian Dam Association standards.

4.4.5 Province of Nova Scotia

There are approximately 7,500 abandoned mine openings (AMOs) in Nova Scotia, of which approximately 2,200 are located on Crown land. The Abandoned Mine Opening Remediation Program was created in 2001, and as of December 2015, has invested about C$760,000 to remediate the most hazardous of these openings. To date, 689 AMOs have been remediated through the program, including all the most hazardous (inescapable) mine openings. The work was conducted on 40 different mine sites.

4.4.6 Province of Ontario

The Ontario Abandoned Mines Rehabilitation Program helps to ensure that abandoned mine sites that once contributed to the province's economic growth do not become a lasting environmental and health concern. Ontario established its Abandoned Mine Rehabilitation Fund (AMRF) in 1999.

From September 1999 to March 2016, C$142.4 million was spent on rehabilitating Crown-held mine sites. Rehabilitation has been conducted on more than 80 of the highest-priority abandoned mine sites. The largest of these sites is the Kam Kotia Mine, where more than C$75 million has been spent to date.

The Abandoned Mines Information System (AMIS) is a database containing information on all known abandoned and inactive mine sites and mine hazard features located on both Crown and privately held lands within the province of Ontario. There are currently 5,762 known abandoned mine sites under Mining Act jurisdiction throughout the province, which contain approximately 17,456 mine features. Additional information on the AMIS system can be obtained at:

http://www.mndm.gov.on.ca/en/mines-and-minerals/applications/ogsearth/abandoned-mines

4.4.7 Province of Saskatchewan

In 2007, the Government of Saskatchewan enacted legislation to implement an Institutional Control Program for the post-closure management of decommissioned mine and mill sites on provincial Crown land. Information on the IC Program was presented in Section 4.3.6.

Project CLEANS (Clean-up of Abandoned Northern Sites) is a multi-year, multi-million-dollar project aimed at assessing and reclaiming the Gunnar Mine, Lorado mill and 36 satellite sites in northern Saskatchewan. In 2006, the governments of Saskatchewan and Canada signed a Memorandum of Agreement to share equally in the costs to clean up the Gunnar and satellite legacy uranium sites in northern Saskatchewan.

4.4.8 Province of Québec

Projects to rehabilitate mining sites have significantly increased since the early 1990s due to efforts made by the industry and the provincial government. During the early 1990s, the Government of Québec injected more than C$30 million into research, closure projects and financial assistance to rehabilitate mining waste areas. In April 2006, the Québec government undertook a major project to restore contaminated sites, mainly mining sites.

Quebec has explored novel approaches for site rehabilitation. Research was directed at the development of efficient, economical closure methods. In several cases, an exhaustive characterization of the site has led to innovative closure approaches using various waste materials, such as forest residue, sludge from sewage treatment plants, septic tanks or paper mills and ash from co-generation power plants. These technologies have reduced costs and offered a win/win solution to the problem of stockpiling, for at least some of the waste materials.

Different types of partnership approaches were created to address some of the contaminated sites (Tremblay and Hogan, 2010). A diverse partnership base is essential to the success of rehabilitating closed mines, and it is critical to involve partners early in the process. The innovative open approach used by the Québec government for mine rehabilitation is envied by many other Canadian jurisdictions.

In November of 2016, the Government of Quebec announced that it had allocated C$620 million for the rehabilitation of contaminated sites; 80% of this amount is for abandoned mines.

4.5 COMMUNICATIONS – SHARING INFORMATION AND KNOWLEDGE

Sharing information and knowledge with its partners and the public is an important function of NOAMI. The collaborative activities and work of NOAMI has been recognized internationally as an excellent model of partnership in the remediation of abandoned mines.

Workshops are the preferred vehicle to share information and knowledge and obtain feedback from the multi-sectoral mining community. Efforts are being made to include impacted communities in these workshops, and to ensure a sectoral balance of industry, government, environmental non-government organizations, academia, Aboriginal peoples and consultancies. The workshops maintain a focus on facilitated multi-stakeholder dialogue and expert discussion panels to develop key recommendations and a "toolbox" of options to help NOAMI ensure that issues are solved and progress is made. NOAMI workshops concentrate on a key issue or are based on a central theme and have included the following:

- Exploring the Management of Long-term Liabilities and the Return of Mining Lands to the Crown in Canada, 2011.
- Workshop to Explore Perspectives on Risk Assessment for Orphaned and Abandoned Mines, 2008.
- Orphaned and Abandoned Mines: A Workshop to Explore Best Practices, 2006.
- Assessing Liabilities and Funding Options Workshop, 2005.
- Legal and Institutional Barriers to Collaboration Workshop, 2003.
- Orphaned and Abandoned Mines in Canada Workshop, 2001.

In addition to workshops, NOAMI facilitates information sharing through documents such as the NOAMI Nugget (newsletter) and other bulletins, which are distributed electronically to a large national and international mailing list. NOAMI's website is bilingual and is regularly updated with workshop proceedings, pamphlets, announcements and newsletters. Links to national jurisdictional programs on abandoned mines are provided. NOAMI reports are also available on the website and can be downloaded for free (www.abandoned-mines.org).

4.5.1 NOAMI's 2002–2008 Performance Report and 2009–2015 Performance Update

NOAMI's first Performance Report, covering the years 2002–2008, was a major undertaking and described the activities and infrastructure of the initiative in considerable detail. It was widely distributed in Canada and abroad, in English and French, and effectively promoted NOAMI, while showcasing the activities of Canadian jurisdictions in the remediation of orphaned and abandoned mine sites.

NOAMI released a second performance report in 2015 covering the period 2009–2015. While the format is more focused than that of the first report, it once again highlighted both NOAMI's major achievements, and the efforts of Canadian jurisdictions to address the potential legacy issues associated with orphaned and abandoned mines throughout the country.

4.5.2 International Linkages

NOAMI looks to develop links with other initiatives dealing with orphaned and abandoned mines, with the objective of sharing information and resources. Organizations such as the International Council on Mining & Metals (ICMM), National Association of Abandoned Mine Land Programs (NAAMLP-U.S.), United Nations Environment Programme (UNEP), Sustainable Mining Institute (Australia) and Sernageomin (Chile) and Fundacion Chile have developed projects or approaches for more effective management of abandoned sites.

4.5.3 NOAMI – A Model of Collaboration

Mining with environmental stewardship involves not only improving the technical performance of mining and extraction processes so as to maximize the profitability of an operation but also entails leaving a positive environmental and social legacy. Up until the 1970s, the focus was primarily on generating profits. Since that time, there has been a growing awareness of the need to minimize the damage that mining has done to the natural environment. In the 1980s, a collective approach to problem-solving emerged in Canada. The Mine Environment Neutral Drainage (MEND) program, created in 1989, was the first international multi-stakeholder program to develop scientifically based technologies to reduce the effect of acidic drainage and metal leaching, a major issue for orphaned and abandoned mines. This approach allows policy decisions to be made based on sound science. Since then, this model of collaboration has been used by both Canadian and international programs to address issues of national importance. For example, NOAMI adopted the MEND model to develop a policy-based program for the remediation of orphaned and abandoned mine sites in Canada. The success of these programs can be attributed to several factors:

- The partnerships developed among the two levels of government, the mining industry, environmental groups and Aboriginal Canadians working together to develop solutions to a major environmental problem.
- An extensive peer-review process, both formal and informal, that results in enhanced credibility of the information base.
- A small dedicated secretariat that coordinates activities, manages the accounting, reporting and technology transfer, and serves as the "glue" that holds the program together. The secretariat for both MEND and NOAMI is located at Natural Resources Canada, which provides much synergy between the two programs.
- The dynamic approach taken for transferring the knowledge gained during these multi-stakeholder initiatives.

NOAMI's and MEND's collaborative activities and accomplishments are recognized internationally as excellent models of partnership in addressing issues of national importance, such as acidic drainage and metal leaching and the remediation of orphaned and abandoned mines. The secretariat and committee members are invited to speak at national and international events to share Canadian expertise, experience, knowledge and approaches to these issues. Through these efforts, significant advances in environmental management have been achieved, which, in turn, contributed to the long-term sustainability of the mining industry and the environment.

4.5.4 Abandoned Mines, MEND and Acidic Drainage

Serious environmental issues for orphaned and abandoned mines are related to acidic drainage and metal leaching from underground workings, open-pit mine faces and workings, waste rock piles and tailings impoundment areas. Since 1989, the multi-stakeholder Mine Environment Neutral Drainage (MEND) program has worked to develop technologies to prevent and control acidic drainage and metal leaching.

Through MEND, tremendous progress has been made in the areas of prediction, prevention and control, disposal technologies, active and passive treatment, monitoring and verification of closure technologies. A toolbox of technologies was developed to plan for, operate and decommission mines in an environmentally acceptable manner.

MEND report 3.50.1 *Study to Identify BATEA for the Management and Control of Effluent Quality from Mines* (Hatch, 2014) identified the best available technologies economically achievable (BATEA) to manage and control effluent from metal, diamond and coal mines in Canada. The study provided reference information for forthcoming changes to the Canadian mining effluent regulations. The study describes the effluent management and treatment technologies and techniques currently employed at metal (base metal, precious metal, uranium, iron ore), diamond and coal mine operations in Canada. The study provides an overview of each (sub)sector's water management and effluent treatment practices and establishes a model effluent treatment process and treated effluent quality for future use in BATEA selection.

MEND is currently undertaking a state of practice assessment of tailings management technologies. This study focuses on tailings dewatering processes (i.e., thickened, paste and filtered tailings) for surface disposal (including alternative disposal methods) and assessing their strengths and weaknesses, both physically and geochemically for Canadian projects and in environments similar to Canada. This report was finalised in 2017. (https://mend-nedem.org/mend-report/study-of-tailings-management-technologies-2017/).

The transfer of information on developed technologies to partners and the public has always been an important part of MEND. A bilingual website http://mend-nedem.org was created in the 1990s and underwent several major updates as technologies advanced. MEND has published more than 200 technical reports, guidance documents and manuals; these are all available electronically, for free, on the MEND website. Workshops are considered the best route for the timely and efficient transfer of technologies and case studies, and MEND has sponsored or co-sponsored over 40 workshops, including the hosting of three International Conferences on Acid Rock Drainage (ICARD). As a result of the efforts of MEND and other partners, our understanding of acidic drainage has greatly improved, and significant advances have been made in environmental stewardship.

Strong relationships are forged among MEND and international industry and government programs. MEND is the Canadian partner in the Global Alliance, an international partnership among organizations involved in acidic drainage research led by the International Network for Acid Prevention (INAP). A collaborative effort for the Global Alliance was the Global Acid Rock Drainage (GARD) Guide (www.gardguide.com). However, many challenges remain and new challenges are emerging. The MEND program has a proven track record for addressing these challenges effectively and to help ensure the long-term sustainability of both our mineral resources and the environment.

4.6 CONCLUSIONS

Over the last 15 years, NOAMI and its jurisdictional partners have made tremendous progress in dealing with issues surrounding orphaned and abandoned mines in Canada. In the initial years, from 2002 to 2008, the work was centered on remediation and long-term management of existing sites. From 2009 onward, the program examined mine closure and relinquishment and identified best practices to prevent the occurrence of future abandonments. The guidance documents and workshops produced by NOAMI on orphaned and abandoned mines have been valuable tools for a diverse array of stakeholders, both nationally and internationally.

ACKNOWLEDGMENTS

Thank you to the past and present members of the NOAMI Advisory Committee and to the contractors who committed their time and expertise to make the program a success.

Note 1. In NOAMI reports the phrase "return of mining lands" is interchangeably used with relinquishment.

REFERENCES

Cal Data Ltd. (2005) Capacity building for a national inventory of orphaned and abandoned mines in Canada. Prepared for NOAMI, viewed 16 March 2017, http://www.abandoned-ines.org/en/document/publication/

Castrilli, J.F. (2002) Barriers to collaboration: Orphaned and abandoned mines in Canada. Prepared for NOAMI, viewed 16 March 2017, http://www.abandoned-mines.org/en/document/publication/

Castrilli, J.F. (2005) Jurisdictional legislative review. Report on the legislative, regulatory, and policy framework respecting collaboration, liability, and funding measures in relation to orphaned and abandoned, contaminated, and operating mines in Canada. Prepared for NOAMI, viewed 16 March 2017, http://www.abandoned-mines.org/en/document/publication/

Castrilli, J.F. and C.N. Watson and Associates. (2003) Potential funding approaches for orphaned and abandoned mines in Canada. Prepared for NOAMI, viewed 16 March 2017, http://www.abandoned-mines.org/en/document/publication/

Cowan Minerals Ltd. (2006) Rehabilitating abandoned mines in Canada: A toolkit of funding options. Prepared for NOAMI, viewed 16 March 2017, http://www.abandoned-mines.org/en/document/publication/

Cowan, W.R., Mackasey, W.O. and Robertson, J.G.A. (2013) Case studies and decision making process for the relinquishment of closed mine sites. Prepared for NOAMI, viewed 16 March 2017, http://www.abandoned-mines.org/en/document/publication/

Cowan, W.R., Mackasey, W.O. and Robertson, J.G.A. (2010) The policy framework in Canada for mine closure and management of long-term liabilities: A guidance document. Prepared for NOAMI, viewed 16 March, 2017, http://www.abandoned-mines.org/en/document/publication/

Gardiner, E.J., Cowan, W.R., Tremblay, G.T. and Hogan, C.M. (2015) Managing long-term orphaned mine liabilities: Exploring the relinquishment of closed sites. In: Proceedings Tenth International Conference on Mine Closure, 1–3 June 2015, Vancouver, Canada, pp. 14–22.

Hatch. (2014) Study to identify BATEA for the management and control of effluent quality from mines. MEND Report 3.50.1. Viewed 16 March 2017, http://mend-nedem.org/wp-content/uploads/MEND3.50.1BATEAAppAD.pdf

Holmes, R. and Stewart, G. (2011) A guidance document for mine closure and management of long-term liabilities: Examining a policy framework in Canada. In: Proceedings Sixth International Conference on Mine Closure, 18–21 September 2011, Alberta, Canada, pp. 21–28.

Hovdebo, DG, Cunningham, KE, Kristoff, DM and Webster, MS 2015, 'Post-closure stewardship of mine sites: institutional control in Saskatchewan – a case history', Proceedings of the 10th International Conference on Mine Closure, Vancouver, Canada, 2015, InfoMine, pp. 23–36.

IGWG (Intergovernmental Workshop Group on the Mineral Industry). (2016) Mining sector performance report 2006–2015. Prepared for the energy and mines ministers' conference. Winnipeg, Manitoba. August 2016. Viewed 16 March 2017, http://www.nrcan.gc.ca/mining-materials/publications/18912

The International Network for Acid Prevention (INAP). (2009) Global Acid Rock Drainage Guide (GARD Guide). Viewed 16 March 2017, http://www.gardguide.com

Kingsmere Resource Services Inc. (2015) Key criteria for the effective long-term stewardship of closed, orphaned/abandoned mine and mineral exploration sites. Prepared for NOAMI, viewed 16 March 2017, http://www.abandoned-mines.org/en/document/publication/

Kingsmere Resource Services Inc. (2016) Orphaned and abandoned mines: Risk identification, cost estimation and long-term management. Prepared for NOAMI, viewed 16 March 2017, http://www.abandoned-mines.org/en/document/publication/

Mackasey, W.O. (2000) Abandoned mines in Canada. Prepared for MiningWatch Canada. Viewed 16 March 2017, http://miningwatch.ca/publications/2000/2/17/abandoned-mines-canada

NOAMI. (2003a) Lessons learned on community involvement in the remediation of orphaned and abandoned mines – Case studies and analysis. Viewed 16 March 2017, http://www.abandoned-mines.org/en/document/publication/

NOAMI. (2003b) Best practices in community involvement: Planning for and rehabilitating abandoned and orphaned mines in Canada. Viewed 16 March 2017, http://www.abandoned-mines.org/en/document/publication/

NOAMI. (2004) Guidelines for legislative review. Viewed 16 March 2017, http://www.abandoned-mines.org/en/document/publication/

NOAMI. (2009) National orphaned/abandoned mines initiative 2002–2008 performance report. Viewed 16 March 2017, http://www.abandoned-mines.org/en/document/publication/

NOAMI. (2015) National orphaned/abandoned mines initiative performance update 2009–2015. Viewed 16 March 2017, http://www.abandoned-mines.org/en/document/publication/

Stratos Inc. (2006) NOAMI Workshop Proceedings. Assessing liabilities and funding options. 2–3 November, 2005. Ottawa, ON. Prepared for NOAMI, viewed 20 March 2017, http://www.abandoned-mines.org/en/workshops/

Stratos Inc. (2006) NOAMI Workshop Proceedings. Orphaned and abandoned mines: A workshop to explore best practices. 26–27 October, 2006. Winnipeg, MB. Prepared for NOAMI, viewed 20 March 2017, http://www.abandoned-mines.org/en/workshops/

Stratos Inc. (2008) NOAMI Workshop Proceedings. A workshop to explore perspectives on risk assessment for orphaned and abandoned mines. 13–14 November, 2008. Vancouver, BC. Prepared for NOAMI, viewed 20 March 2017, http://www.abandoned-mines.org/en/workshops/

Tremblay, G.A. and Hogan, C.M. (2010) Mine closure case studies of orphaned and abandoned mines in Canada. In: Proceedings Fifth International Conference on Mine Closure, 23–26 November 2010, Vina del Mar, Chile, pp. 623–633.

Tunis, E.S. and Associates. (2011) NOAMI Workshop Proceedings – Exploring the management of long-term liabilities and the return of mining lands to the Crown in Canada. 31 May – 1 June, 2011. Ottawa, ON, Prepared for NOAMI, viewed 20 March 2017, http://www.abandoned-mines.org/en/workshops/exploring-the-management-of-long-term-liabilities-and-the-return-of-mining-lands-to-the-crown-in-canada/

Versteeg, H. (2003) NOAMI Workshop Proceedings – Legal and institutional barriers to collaboration relating to orphaned and abandoned mines. 24–25 February 2003. Ottawa, ON. Prepared for NOAMI, viewed 20 March 2017, http://www.abandoned-mines.org/en/workshops/

5 Mining Restoration and the Contribution of the Mining Industry to China's Economic Development over the Past 70 Years

Xiaodong Yang and Yongchao Gao

5.1 INTRODUCTION

According to the "China Mineral Resources Report 2019", China has now found 173 kinds of minerals, and proven reserves for 162 kinds of minerals. China's mineral resources accounted for about 12% of the world's total, ranking 3rd for mineral deposits (China's MOLR 2019). China is one of the few countries with complete types of mineral resources and abundant deposits and it is the world's largest producer and consumer of minerals (China's MOLR 2019). For example, China is the world's largest producer of steel, accounting for 50% of global output. China is a major producer of aluminum, magnesium, antimony, salt, talc, barite, cement, coal, fluorspar, gold, graphite, iron, lead, mercury, molybdenum, phosphate rock, rare earths, tin, tungsten, bismuth and zinc. China exports antimony, barite, rare earths, fluorspar, graphite, indium and tungsten, and it leads the world in the domestic mining of gold, zinc, lead, molybdenum, iron ore and coal (Blaettler 2018). However, China's per capita mineral resources are still at a relatively low level, accounting for just over half (58%) of the world average, and the country ranks 53rd globally (China's MOLR 2019). It is only 70 years since the mining industry in China got going, and it has played an important role in the national economy and improvement of people's lives (Zhou et al. 2010). For example, it has supplied nearly 46 billion tons of coal, 18 billion tons of cement, 9 billion tons of iron ore, 5 billion tons of crude oil, and it has provided nearly 4.6 million jobs. However, over the past 70 years, China's mining industry has long relied on the traditional extensive model of production, which takes up a large amount of cultivated land and causes several environmental problems, such as surface subsidence, pollution of water, air and soil, and irreversible damage (Xi 2019; Guang and Qiang 2000). Since the 1980s, the Chinese people and the government began to realize that environmental problems caused by the traditional mining industry required relevant government departments to urgently devise a series of laws and regulations. These included the "Mineral Resources Law of the People's Republic of China", "Land Reclamation Regulations", "Compiling Regulation on the Project of Mining Geo-environmental Protection and Integrated Renovation", to decrease mine pollution and environmental deterioration, in an effort to reclaim land in mining areas and restore mines (NEA 1996; Zhao 2007).

In 2017, "Ecological and environmental protection" and "Clear waters and green mountains are as valuable as mountains of gold and silver / Guided by the conviction that lucid waters and lush mountains are invaluable assets, the country advocates harmonious coexistence between humans

 DOI: 10.1201/9781315141855-5

and nature, and sticks to the path of green and sustainable development" were established as a guideline and policy of national governance in China. Their objective was to ensure sustainable development of the economy and save the environment. Mine restoration and a sustainable mining industry have been an important aspect of governance and protection of the environment in China, and must be maintained in the future (Xinhuanet 2015).

5.2 DEVELOPMENT OF CHINA'S MINING INDUSTRY

Before 1949, China's mining industry was very inefficient and small-scale, with only about 300 mines, and the annual outputs of crude oil, coal, ferrous metals, nonferrous metals and nonmetals were less than 120,000 tons, 32 million tons, 170,000 tons, 13,000 tons and 1 million tons, respectively (China's NSBC 2018). After the founding of the People's Republic of China, the industry underwent rapid development. China successively established many large bases in producing mine resources, such as: the large oil bases of Daqing, Shengli, Liaohe and Karamay; the large coal bases of Datong, Yanzhou, Pingdingshan, Huaibei and Junggar; the large steel bases of Anshan, Wuhan and Panzhihua; and the large bases of non-ferrous metal of Baiying, Jinchuan, Tongling and Dexing (China's SCIO 2004; Xinhuanet 2019). To this day, China is a top producer of the world's mine resources. In this paper, based on the research results of Zhu (2010), a former minister of China's Ministry of Geology and Mineral Resources in 2009, the progress of China's mining industry was divided into five stages over the past 70 years: 1949–1978, 1979–1994, 1995–2003, 2004–2013 and 2014–. Respectively, these were named as slow, moderate, depressed, fast and relatively stable development stages. These phases are explained in more detail below.

5.2.1 Slow Development Stage of China's Mining Industry

At the start of the People's Republic of China, the mining industry was very undeveloped and accounted for a small proportion of the economy (China's NSBC 2018). In 1949, the production amounts of coal, oil, iron ore and copper ore were only 32.43 million, 118,000, 560,000 and 5,000 tons, accounting for, respectively, 1.80, 0.20, 0.30 and 0.20% of world production (Figure 5.1). The productions of natural gas, crude steel and cement at this time were virtually non-existent. In contrast, China had 21% of the world's population (China's MOLR 2019). The per capita quantity of mineral resources was much less than the international average (Zhu 2010). In this case, China's Communist Party established a centrally planned economy as the key to economic development, taking mining as an important national pillar industry and investing a lot of financial and human resources to develop it. However, due to the backwardness of the national economy, the weak foundations of the mining industry and the shortages of talent, technology and funds, the mining industry made slow progress. Until 1978, the outputs of coal, oil, natural gas, iron ore, crude steel, copper ore and cement were 620.00 million tons, 110.00 million tons, 13.73 billion cubic meters, 34.79 million tons, 31.78 million tons, 160,000 tons and 65.24 million tons, respectively (Figure 5.1) (China's NSBC 2018). Coal, oil, iron ore and copper ore only accounted for 17.00, 3.00, 6.50 and 2.00% of the total global output, respectively (Figure 5.1) (China's NSBC 2018).

5.2.2 Moderate Development Stage of China's Mining Industry

From 1979 to 1994, the economic market-oriented reforms and opening-up policy attracted a large amount of capital and technology into China, which in turn enabled the mining industry to enter a new phase. In 1994, China produced 885.72 million tons of raw coal, 146.08 million tons of crude oil, 17.559 billion cubic meters of natural gas, 297.41 million tons of iron ore, 92.61 million tons of crude steel, 350,000 tons of copper ore and 421.18 million tons of cement, respectively (Figure 5.1) (China's

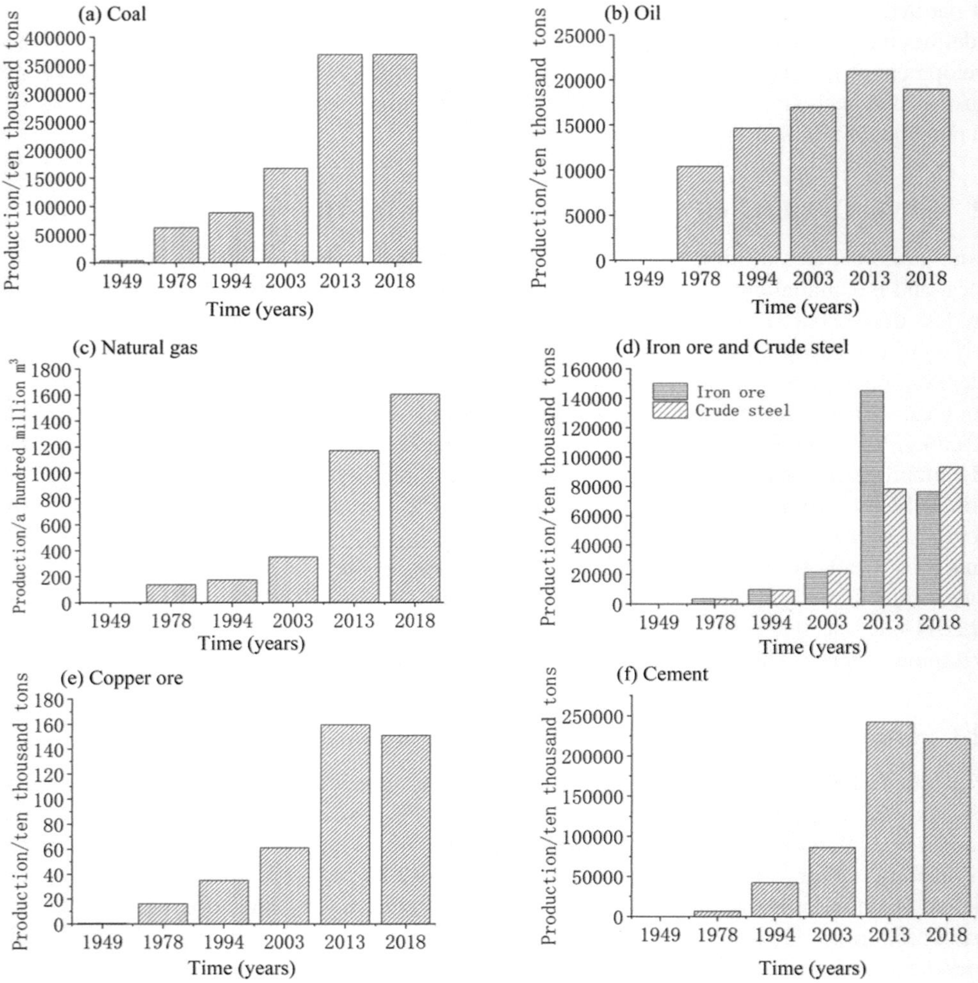

FIGURE 5.1 Changes in production of major minerals in China from 1949 to 2018

NSBC 2018). Compared with 1978, the outputs of coal, oil, natural gas, iron ore, crude steel, copper ore and cement increased by 43, 41, 28, 180, 191, 11 and 546%, respectively (Figure 5.1).

5.2.3 Depressed Development Stage of China's Mining Industry

1995–2003 marked the depressed development stage of China's mining industry. In 2003, the production of coal, crude oil, natural gas, iron ore, crude steel, copper ore and cement amounted to 16.67 million tons, 16.96 million tons, 35.015 billion cubic meters, 21.367 million tons, 22.34 million tons, 610,000 tons and 862.08 million tons in China, respectively (Figure 5.1) (China's NSBC 2018). These outputs were all lower than the previous stage because China's mining industry was affected by the global economic slowdown. The period 1995–2003 was the end of Western developed countries' domination of industrialization, and the shortages in new economic growth slowed down the global economy. As a result, China's economic growth rate also declined. The Chinese government invested less capital and human resources in the mining industry, resulting in it entering a depression. For example, the output dropped by 3.3% compared with the previous stage. From

1995 to 2003, the mining industry's personnel declined from 7.93 million to 5.65 million, and the number of employees in mineral exploration fell from 995,000 to 198,000 (China's NSBC 2018).

5.2.4 Fast Development Stage of China's Mining Industry

The rapid expansion of the domestic market drove the fast growth of China's economy from 2004 to 2013. Within this process, due to the continuous increase of infrastructure construction investments and better quality of people's lives, the demand for mineral resources rose simultaneously, leading to the fast development of China's mining industry. In 2013, the outputs of coal, oil, natural gas, iron ore, crude steel, copper ore and cement were, respectively, 368 million tons, 29 million tons, 117.05 billion cubic meters, 145 million tons, 780 million tons, 1.596 million tons and 242 million tons. These increased by 120.8, 23.2, 234.3, 578.6, 250.8, 161.6 and 180.7% compared with 2004, respectively (Figure 5.1) (China's NSBC 2018). It was also during this period that China began to import mineral resources from other countries (such as Australia, South Africa and Brazil) because the Chinese domestic demand for mineral resources exceeded its production. The dependence of the mining industry on imported mineral resources continued to grow in China. The mining industry had undergone a transformation from a domestically to internationally oriented one.

5.2.5 Relatively Stable Development Stage of China's Mining Industry

With the improvement in people's standard of living, environmental protection and environmental pollution caused by mining attracted more attention of the Chinese government and the general public (Sun et al. 2020). Improvements in the mining industry enabled China's mining industry to enjoy a period of stability since 2014. The production of major minerals has remained unaltered since 2014. For example, until 2018, the outputs of raw coal, oil, natural gas, iron ore, crude steel, copper ore and cement were 3.68 billion tons, 18.90 million tons, 16.03 billion cubic meters, 76 million tons, 93 million tons, 1.51 million tons and 221 million tons, respectively (Figure 5.1) (China's MOLR 2019). With the exception of natural gas and iron ore, the growth of output in other major minerals was less than 10% from 2014 to 2018.

5.3 CHINA'S MINERAL TYPES AND RESERVES

More mineral types are being found and mined since the start of the People's Republic of China due to sustained exploration and advances in technology (China's MOLR 2019). By the end of 2018, China had discovered 173 mineral types (Table 5.1). More specifically, there were 13 types of energy minerals (i.e., oil, natural gas, coal, uranium and geothermal resources), 59 types of metallic minerals (i.e., iron, manganese, copper, aluminum, lead, zinc and gold), 95 types of non-metallic minerals (graphite, phosphorus, sulfur and sylvite) and 6 types of hydro-gas minerals (groundwater, mineral water, etc.). According to the market value of proven reserves at 2018, China's proven minerals accounted for about 12% of the world total, ranking 3rd in the world. Moreover, compared with other countries, China has 45 dominant minerals, and 24 minerals, making the country home to the world's third-largest reserves. Reserves of rare earth, combustible ice, titanium, tungsten, antimony and bentonite are all ranked as first in the world (China's MOLR 2019).

5.4 PROBLEMS IN CHINA'S MINING INDUSTRY

Over the past 70 years, consistent problems have led to discrepancies between China and the Western world in terms of mining technology and equipment. China mainly used the traditional mining model, i.e., the extensive mining model which paid little attention to environmental protection, to produce mineral resources (Xi 2019). Additionally, China's mining management system was very centralized, and the front-line coal miners were mostly migrant workers of low quality.

TABLE 5.1
Changes in Types and Reserves of Major Minerals over the Past 30 Years

Number	Types	Mineral Name	Unit	2006	2010	2018
1	Energy and Minerals	Oil	100 million tons	27.6	31.7	35.73
2		Natural gas	100 million cubic meters	30009	37793	579368
3		Coalbed methane	100 million cubic meters			3046
4		Shale gas	100 million cubic meters			2160
5		Coal	100 million tons	11598	13408	170863
6	Black metals	Iron ore	100 million tons	607.3	727.0	852
7		Manganese ore	100 million tons	7.67	8.86	18.16
8		Chromite ore	10,000 tons	1008	1490	1193
9		Vanadium ore	V_2O_5 10,000 tons			6561
10		Titanium ore	TiO_2 100 million tons	7.0	7.2	8.26
11	Non- ferrous metals	Copper ore	metal 10,000 tons	7048	8041	11443
12		Lead ore	metal 10,000 tons	4141	5509	9216
13		Zinc ore	metal 10,000 tons	9710.9	11596	18756
14		Bauxite ore	100 million tons	27.8	37.5	51.7
15		Nickel ore	metal 10,000 tons	801.4	938.0	1188
16		Cobalt ore	metal 10,000 tons	66.1	68.2	69.65
17		Tungsten	WO_3 10,000 tons	558.4	591.0	1072
18		Tin ore	metal 10,000 tons	476.9	431.9	453.1
19		Molybdenum	metal 10,000 tons	1094	1402	3029
20		Antimony ore	metal 10,000 tons	225.1	255.0	327.7
21		Gold mine	metal ton	4997	6865	13638
22		Silver mine	metal 10,000 tons	14.4	17.2	32.91
23		Platinum group metals	metal tons	339.6	334.6	401
24		Strontium	Lapis lazuli 10,000 tons	4652	4375	5642
25		Lithium	ore Oxide 10,000 tons			1092
26		Magnesite	ore 100 million tons	35.9	36.4	31.03
27	Non- metals	Fluorite	mineral 100 million tons	1.72	1.80	2.57
28		Refractory clay	ore 100 million tons	23.4	24.6	26.38
29		Pyrite	ore 100 million tons	54.1	56.9	63
30		Phosphate rock	ore 100 million tons	169.8	186.3	252.8
31		Potassium salt	KCl 100 million tons	8.81	9.30	10.16
32		Boron ore	B_2O_3 10,000 tons	7276	7309	7837
33		Sodium salt	NaCl 100 million tons	13126	13338	14240
34		Glauber's salt	Na_2SO_4 100 million tons	207.7	934.2	1173
35		Barite	ore 100 million tons	3.83	10.71	3.73
36		Limestone for cement	ore 100 million tons	789.5	1021	1432
37		Glass siliceous material	ore 100 million tons	53.4	64.7	96.13
38		Plaster	ore 100 million tons	682.5	769.1	824.9
39		Kaolin	ore 100 million tons	18.3	21.0	34.96
40		Bentonite	ore 100 million tons	29.4	28.0	29.96
41		Diatomite	ore 100 million tons	4.7	4.3	5.11
42		Facing granite	100 million cubic meters	21.5	23.2	53.8
43		Facing marble	100 million cubic meters	14.1	15.3	17.78
44		Diamond	mineral kg	3645	3702	3127
45		Crystalline graphite	mineral 100 million tons	1.64	1.85	4.37
46		Asbestos	mineral 10,000 tons	9552	8975	9259
47		Talc	ore 100 million tons	2.58	2.67	2.88
48		Wollastonite	ore 100 million tons	1.65	1.55	2.29

These caused a series of problems such as poor mining efficiency, resources waste, unsafe working conditions and accidents, surface subsidence and environmental pollution (Guang and Qiang 2000).

According to statistics, the annual loss of ground subsidence caused by mining was more than 400 million yuan (RMB). For example, based on the survey data of Shanxi Provincial Development and Reform Commission, more than 2,146 episodes of ground subsidence, ground cracks, landslides and collapses were caused by coal mining in Shanxi from 1978 to 2008, destroying 3,309 villages and seriously affecting 660,000 people's lives (Xinhuanet 2019). Besides, mining also caused serious environmental problems, such as water and soil pollution, in China. In the process of mining, a large amount of groundwater was extracted and drained, leading to a substantial drop in the groundwater depth, and surface vegetation recession and soil erosion (Guang and Qiang 2000). More seriously, it also caused some toxic substances to enter the groundwater and then accumulate in the human body through the food chain. In addition, the gangue produced by mining occupied a large amount of surface land, which led to serious shortages of land resources in China (Li et al. 2014). For example, coal mining shrank the boundary of Xinjiang Kalamely Mountains Ungulate Wildlife Reserve by tens of kilometers and the area declined by more than 20% from 2005 to 2015.

5.5 THE CONTRIBUTION OF THE MINING INDUSTRY TO CHINA'S ECONOMY

Over the past 70 years, investment in the mining industry has shown a tendency of first increasing and then decreasing, which peaked in 2013, with total fixed investment reaching 1.48 trillion yuan (RMB) (Figure 5.2). The fixed investment in the mining industry increased from 1.112 billion yuan (RMB) in 1953 to 958.7 billion yuan (RMB) in 2018 and has risen almost 861-fold. Fixed investment in coal, oil and natural gas, ferrous and non-ferrous metals and non-metals increased by 783, 3,756, 284, 397 and 1,471 times, respectively.

Since primary energy consumption was not documented in China before 1985, the change in energy consumption has been reported from 1985 to 2018. China's primary energy consumption continued to increase from 1985 to 2018. Compared with 1985, the consumption of primary energy increased 6-fold in 2018 (Figure 5.2). Comparing the energy structure between 1985 and 2018, it is found that China's energy structure has to some extent improved from dirty to clean energy. The proportion of coal in energy consumption declined, whereas the proportions of oil, natural gas, hydropower, nuclear power and wind power increased (Figure 5.2). In 2018, the proportion of coal

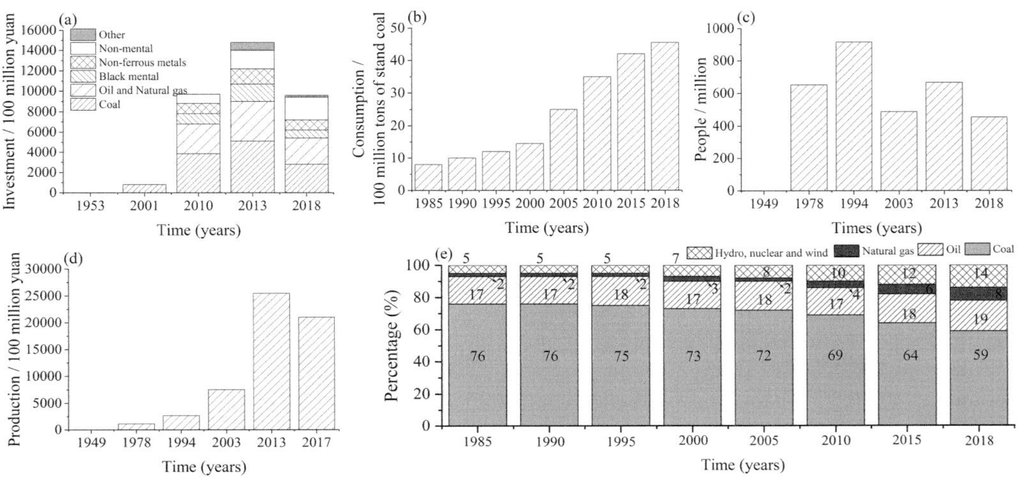

FIGURE 5.2 Changes in the fixed investment of the mining industry (a), primary energy consumption (b), mining personnel (c), output value of the mining industry (d), energy consumption structure (e) over the past 70 years.

TABLE 5.2
Change in Mineral Resource Taxes in China from 2008 to 2018

Years	Mineral Resource Taxes (billion yuan/RMB)	Proportion of Mining Resource Taxes in Total National Tax Revenue (%)
2008	301.8	0.56
2010	417.6	0.57
2012	904	0.90
2014	1083.6	0.91
2016	951	0.82
2018	1353	0.94

in energy consumption decreased by 16.8% compared with 1985. In contrast, the proportions of oil, natural gas and new energy (hydropower, nuclear and wind, etc.) in energy consumption rose by 1.8, 5.6 and 9.4%, respectively (Figure 5.2). Mineral products accounted for more than 70% of the total supply of goods and materials, providing important support for China's economic development. For example, since the proclamation of the People's Republic of China, the mining industry has supplied nearly 80 billion tons of coal and 6 billion tons, which made economic progress possible (Figure 5.2).

As an important sector of the national economy, the mining industry has contributed more than 100 million yuan (RMB) to China's economic development over the past 70 years. In 2018, the contribution of the mining industry to GDP exceeded 30%. In some western provinces, the mining industry contributed over 80% of local GDP (Yu et al. 2017). As well, the mining industry solved a large number of employment problems over the past 70 years. According to the report of the China National Bureau of Statistics, the industry generated 6.52 million, 9.15 million, 4.883 million, 6.665 million and 4.54 million jobs in China in 1978, 1994, 2003, 2013 and 2017, respectively. In 2018, miners accounted for 2.4% of the country's workforce (China's NSBC 2018).

China's government began to collect mineral resource taxes of coal, oil and natural gas in 1984, and taxes were extended to non-metallic ores, salt, ferrous and non-ferrous metal ores in 2011 (China's MOLR 2019). Based on the report of the China National Bureau of Statistics, the total amount of mineral resources tax steadily increased since 1984 (China's NSBC 2018). In 2018, China collected 163 billion yuan (RMB) of mineral resource taxes, accounting for 1.04% of the country's total revenue (Table 5.2).

5.6. STATUS AND METHODS OF MINE REHABILITATION IN CHINA

5.6.1 Distribution and Number of China's Mines

Mines were widely distributed in various provinces of China. More specifically, Anhui, Guizhou, Hebei, Guangxi, Hubei, Hunan and Liaoning provinces have more than 1,000 mines (Table 5.3). The provinces of Guangdong, Heilongjiang and Jilin have the most types of minerals, despite having fewer mines. Among all the provinces of China, Xinjiang, Inner Mongolia, Sichuan, Liaoning and Shanxi have the richest mineral reserves. The main mineral resources are unevenly distributed among Chinese provinces. Coal mines were mainly distributed in Xinjiang, Inner Mongolia, Ningxia and Shanxi provinces. Crude oil and natural gas were mainly located in Xinjiang, Liaoning, Shaanxi, Shandong, Sichuan and Qinghai provinces. Ferrous metals were mainly distributed in Hebei, Liaoning and Sichuan provinces. Nonferrous metals were mainly found in Gansu, Guangxi, Yunnan, Guangdong and Hunan provinces (Table 5.3).

TABLE 5.3
Distribution and Number of China's Mines in 2018

Province / Municipality	Number	Major Mineral Types
Beijing	327	Coal, iron, molybdenum, zinc, gold, cement raw materials, granite
Anhui	>2400	Coal, oil, natural gas, iron, manganese, titanium, copper, lead, zinc, gold, pyrite, cement limestone
Fujian	646	Coal and metal ore
Guangdong	360	Coal, oil shale, petroleum, natural gas, iron, manganese, titanium, copper, lead, zinc, aluminum, magnesium, cobalt, lithium, gold, silver, platinum, graphite, pyrite
Guangxi	1003	Coal, oil, natural gas, iron, manganese, titanium, copper, lead, zinc, aluminum, magnesium, gold
Guizhou	1268	Mercury, bauxite, phosphate, barite, rare earth
Hebei	1379	iron ore
Henan	854	Coal, oil, natural gas, aluminum, gold, thorium, molybdenum, dolomite
Heilongjiang	594	Coal, oil, natural gas, uranium, oil shale, iron, titanium, manganese, chromium, vanadium, copper, lead, zinc, nickel, cobalt, tungsten, tin, pyrite, phosphate, boron
Hubei	6942	Phosphate ore, rector clay, garnet, olivine for fertilizer, light rare earth ore
Hunan	1438	Coal, stone coal, iron, gold, silver, manganese, vanadium, ilmenite, rutile
Jilin	464	Coal, oil shale, petroleum, natural gas, uranium, iron, manganese, chromium, titanium, vanadium nickel, molybdenum, copper, lead, zinc, cobalt, tungsten, bismuth, gold, silver, boron, pyrite, potassium feldspar
Jiangsu	433	Coal, petroleum, iron, copper, lead, zinc, silver, strontium
Jiangxi	937	Copper, tungsten and tantalum niobium
Liaoning	>1600	Coal, oil, natural gas, oil shale, uranium, thorium, iron, manganese
Inner Mongolia	786	Niobium, rare earth, zirconium, glacial stone, coal, chromium, tantalum, fluorite, gypsum, thenardite and trona
Ningxia	91	Coal, gypsum, limestone, silica, clay, dolomite
Qinghai	272	Natural gas, lithium, bromine, boron, cobalt, silver, chromium, lead, nickel, arsenic
Shanxi	765	Coal, bauxite, iron, refractory clay, thenardite, pyrite, gypsum, limestone and gold
Shaanxi	563	Oil, gas, geothermal
Sichuan	174	Coal, copper, lead, zinc, phosphorus, sulfur, salt, calcium mirabilite, lapis lazuli
Tibet	164	Coal, peat, oil shale, geothermal, iron, chromite, boron
Xinjiang	207	Coal, peat, petroleum, natural gas, oil shale, iron, magnesium, chromium, copper, lead, zinc, tungsten, gold, silver
Yunnan	286	Tin, germanium, pyrite
Zhejiang	535	Coal, chemical limestone, building materials and minerals, lead, zinc, molybdenum

Note: Data cited from http://www.sohu.com/a/327190349_100176237.

5.6.2 Major Methods of Mine Rehabilitation

5.6.2.1 *China's Policy of Mine Management and Restoration*

In order to improve the ecology of mining areas and solve the geological and environmental problems caused by mining, some mining enterprises and local governments have spontaneously carried out pollution control and land reclamation strategies in mining regions since the 1950s (Yan 2019). However, due to the shortages in technology, capital and policy support, the area and number of mine restoration were both limited. At the country level, the first policy of mine restoration and management, known as the "Mineral Resources Law of the People's Republic of China", was introduced by the State Council in 1986. After that, many of China's national functional departments, such as the National People's Congress Standing Committee, the State Development Planning Commission, the State Economic Commission, the Ministry of Water Resources, Ministry of Finance, Environmental Protection Bureau and Land Resources Bureau, developed 28 laws and regulations dealing withf mine management and restoration (China's MOLR 2016, 2018). At the provincial level, the first local policy of mine management and restoration was introduced by the authorities of Shanxi province in 2013. Subsequently, many provinces, such as Anhui, Zhejiang, Shaanxi and Inner Mongolia, introduced a series of laws and regulations to control pollution and repair the contaminated mining areas from 2015 to 2018 (China's MOLR 2016; Johnston and Sibly 2018). A list of China's main policies of mine management and restoration is shown in Table 5.4.

TABLE 5.4
China's Main Policies of Mine Management and Restoration

			Level	
Years	**Department**	**Policy Name**	**Country**	**Province**
1986	The State Council	Mineral resources law of the People's Republic of China	√	
1988	The State Council	Land reclamation regulations	√	
1994	The State Council	Detailed rules for the implementation of the law of the People's Republic of China	√	
1998	Ministry of Land and Resources	Measures for registration management of Mineral Resources Exploration	√	
1998	Ministry of Land and Resources	Measures for registration management of mineral resources exploitation	√	
2001	Ministry of Land and Resources	Tailings safety management regulations	√	
2002	The State Council	White paper on China's policy on mineral resources	√	
2005	The State Council	Notice of the State Council on comprehensively rectifying and regulating the mining order of mineral resources	√	
2008	Office of Central Institutional Organization	Notice of further clarifying the division of responsibilities of supervision for mine closure	√	
2008	Ministry of Land and Resources	Notice on strengthening the treatment of abandoned mines	√	
2009	Ministry of Land and Resources	Provisions on the protection of the geologic environment of mine	√	
2009	Ministry of Land and Resources	National land & resources information (2008-2015)	√	
2010	Ministry of Land and Resources	The guideline on implementing the national mineral resources plan and developing green mining and building green mines	√	
2011	The State Council	Land Reclamation Regulations	√	
2011	Ministry of Land and Resources	Specifications for compilation of geological environment protection and restoration management in mines	√	
2013	National Development and Reform Commission	Mine water utilization development plan	√	
2013	National Energy Board	Guiding opinions on filling and exploiting coal mines	√	
2013	The National People's Congress.	Coal Law of the People's Republic of China (Revised in 2013)	√	
2013	The State Council	Air pollution prevention and control action plan	√	

(Continued)

TABLE 5.4 (CONTINUED)
China's Main Policies of Mine Management and Restoration

Years	Department	Policy Name	Level	
			Country	Province
2013	Department of Coal Industry of Shanxi Province	Measures for the assessment and rating of coal mine safety in Shanxi province		√
2013	People's Government of Henan Province	Implementation plan of mine restoration in Henan province		√
2013	Bureau of land resources and housing management of Chongqing	Measures for margin management of environmental protection and mine restoration in Chongqing		√
2014	Ministry of Land and Resources	Regulations for economical and intensive use of land resource	√	
2014	National Development and Reform Commission	Environmental protection law of the People's Republic of China (revised in 2015)	√	
2015	Ministry of Land and Resources	Measures for the administration of reclamation and utilization of abandoned lands of historic mining industry	√	
2015	People's Government of Shanxi Province	Opinion of the Shanxi Provincial People's Government on deepening the reform of the coal management system		√
2015	People's Government of Anhui Province	Regulations of Anhui Province on Non-coal Mine Management		√
2015	People's Government of Shaanxi Province	Requirements of inspection, standards and technology of mine remediation projects in Shaanxi Province (Trial)		√
2015	People's Government of Zhejiang Province	The 13th five-year mine ecological environmental protection and management plan		√
2016	Ministry of Land and Resources	Guidance on strengthening mine geological environment recovery and comprehensive treatment	√	
2016	Ministry of Land and Resources	The national plan of mineral resources (2016-2020)	√	
2017	National six ministries	Implementation opinions on accelerating the construction of green mines	√	
2017	People's Government of Inner Mongolia Autonomous Region	Implementation plan for the development and construction of green mining in Inner Mongolia Autonomous Region		√
2018	Ministry of Land and Resources	Rules and regulations of safety production in metallurgical and non-ferrous metal enterprises	√	
2018	People's Government of Zhejiang Province	Master plan of mineral resources of Zhejiang Province (2016-2020)	√	
2018	People's Government of Zhejiang Province	Master plan of ecological environmental protection and management of mine in Zhejiang Province (2016-2020)	√	
2019	Ministry of Land and Resources	Regulations on protection of mine geological environment (revised in 2019)	√	

5.6.2.2 *Methods of Mine Restoration in China*

5.6.2.2.1 *Improvement in Soil Quality and Fertility*

The need to improve or remediate soil quality and fertility was the basis of environmental restoration in mining areas, because soil was the foundation of vegetation growth. In China, the improvement of soil quality and fertility mainly focused on the surrounding areas of mines (that is, relatively flat areas without significant topographic changes caused by mining) (Yan 2019). Specifically, the contaminated soil was covered using uncontaminated soil, and if soil attained the high or heavy pollution level, it would be removed first, and then backfill employed for uncontaminated soil. After that, soil microorganisms, nitrogen fertilizers, organic fertilizers and other soil amendments were put into the soil to adjust and improve soil pH, organic matter content and soil structure. The aim of enhancing soil quality and fertility was to ensure plant growth (Yuan 2000). At present, as the main strategy of mine restoration, improving soil quality and fertility has been applied to tens of thousands of mine restoration projects. For example, in an abandoned silica mining area of Yinchuan city, Ningxia province, China, the use of this method repaired about 100,000 km^2 of polluted areas (Feng 2019). In the polluted area of a mercury mine in Xunyang county of Shaanxi province, more than 6.60 km^2 of seriously polluted soil was remediated with this method. The repaired soil reached the requirements for crop planting (Lan 2019).

5.6.2.2.2 *Wastewater Treatment*

In the process of mining, the waste can dissolve into the surface and underground runoffs via water circulation and as a result cause environmental pollution and large-scale health problems (Wang et al. 2017). Hence, the treatment of wastewater caused by mining was also a focus of mine restoration. In China, to reduce the dangers posed by water pollution, it was required to build water storage ponds in mining areas before mining commenced. In addition, flood drainage and slope protection were required to construct to reduce the sewage overflow from storage ponds caused by natural disasters such as rainstorms. For the stored wastewater, constructed wetlands and sewage plants were built to remove polluted substances found in wastewater (Zhang 2018). Up to now, under the strict supervision of the government, the vast majority of discharge water from mining enterprises is now considered clean in China.

5.6.2.2.3 *Land Reclamation in Mining Areas*

Reclaimed land in mining areas can be mainly divided into four types: agricultural reclamation, forestry reclamation, fishery reclamation and nature reserve reclamation (Yan 2019). Due to the limited area of cultivated land in China, agricultural reclamation and forestry reclamation were considered the most common type of land reclamation in mining areas (Mckenna et al. 2020). Previous studies showed that the method of land reclamation in mining areas generally included engineering and biological reclamation (Yan 2019; Zhou et al. 2010). Based on the geographic and geomorphic conditions, engineering reclamation referred to the use of mechanical equipment to backfill, level off and cover the mining damaged land by using backfill materials (i.e., gangue and coal ash) (Gu et al. 2010). The purpose of engineering reclamation was to produce soil and provide a foundation for subsequent biological reclamation.

Engineering reclamation was used in many regions to increase cultivated land in mining areas, especially in mining pits (Li 2013). For example, about 3,000 km^2 of land was repaired by using engineering reclamation in the coal mine area of Ordos, Inner Mongolia autonomous region of China (Guo 2017). The aim of biological reclamation was to plant suitable plants on reclaimed land and let them grow rapidly (Li 1995). The process of biological reclamation mainly included suitability assessment of plant growth, species selection, plantation, plant management and fertilization (Li 2016). Over the past 30 years, China has used these technologies and reclaimed about 1.01 million ha of land in mining areas.

5.6.2.2.4 *Treatment of Soil Heavy Metal Pollution in Mining Areas*

Soil heavy metal pollution caused by mining did the most environmental damage in China (Cui et al. 2004). Like other countries, China has undertaken a lot of work to treat heavy metal contaminated soil (Tahmasebi et al. 2020). The methods for remediating heavy metal contaminated soil were three main types: physical, chemical and bioremediation. Of these, the most commonly used physical remediation was replacement of contaminated soil by uncontaminated soil (Yang 2009). Chemical remediation mainly adopts the chemical improvement method in China. The most commonly used chemical remediation was amelioration. More specifically, some soil amendments, i.e., lime, biochar and organic matter, were added to heavy metal contaminated soil in order to diminish health and ecological risks via the soil physicochemical and biological processes, such as adsorption, ion exchange, organic complexation and antagonism (Huang et al. 2014). For example, chemical remediation was employed to control the serious soil plumbum and uranium pollutions in Yunnan and Jiangxi provinces, respectively (Zhu et al. 2013; Yu et al. 2002).

Compared with physical and chemical remediation, bioremediation was the most widely used method for treating soil heavy metal pollution in mining areas (Sun et al. 2006). It uses special plants, animals and microorganisms that can biologically enrich or convert heavy metals, reducing health and ecological risks. It has been reported that bioremediation has been used in all of China's mine areas to reduce soil heavy metal pollution. For example, the use of centipede grass to absorb arsenic from soil has achieved the desired results in Hunan, Yunnan and Guangxi provinces (Chen et al. 2010). Phytoremediation increased more than 180 km^2 of farmland in the coal mining area in Shenmu city, located in Shaanxi province.

5.6.2.2.5 *Tourism in Remediated Mine Areas*

Over the past few decades, China has remediated about 1,004,600 ha of polluted regions in mining areas, which included 884,600 ha polluted region in a mining area, and 120,000 ha of abandoned mines. China has remediated 7,298 mines which are mainly located in Inner Mongolia, Ningxia, Shandong, Anhui, Xinjiang, Shaanxi, Shanxi, Jiangsu and Qinghai provinces (China's MOLR 2019). Under the overall management and guidance of China's Ministry of Land and Resources, the local governments implemented mine park projects in 2004. In this project, based on the fact that mines were distributed in mountainous areas, mine restoration, environmental management and tourism were combined together to develop tourist attractions (China's MOLR 2006). By the end of 2017, China had built 88 national remediated mine parks, including coal, oil, gold, copper and non-metallic minerals (China's MOLR 2019). This method introduced new insights and made a contribution to China's economy. From 2015 to 2017, the country's remediated mine parks have hosted more than 1 billion tourists, and contributed more than 100 billion yuan (RMB) to local government authorities.

REFERENCES

Blaettler, K. G. 2018. A list of China's natural resources. https://sciencing.com/list-chinas-natural-resources-5375735.html.

Chen, T. B., H. X. Li, M. Lei. 2010. Accumulation of N, P and K in Pteris vittata L. during phytoremediation: A five-year field study. *Acta Sci. Circum.* 30: 402–408. (Chinese with English abstract)

China's Ministry of Land and Resources (MOLR). 2006. Notice of Ministry of Land and Resources on strengthening the construction of national mining parks. http://f.mnr.gov.cn/201702/t20170206_1435972.html.

China's Ministry of Land and Resources (MOLR). 2016. Documents on laws and regulations of the development and management of mineral resources (sixth edition). https://flk.npc.gov.cn/detail2.html?MmM5MDlmZGQ2NzhiZjE3OTAxNjc4YmY2YTFkNTA0ZDM.

China's Ministry of Land and Resources (China's MOLR). 2018. National green mine standard. https://www.mnr.gov.cn/dt/ywbb/201810/t20181030_2291165.html.

China's Ministry of Land and Resources (China's MOLR). 2019. China mineral resources report 2011–2019. http://zrzy.bynr.gov.cn/zwgk/zcjd/202008/P020220303417257831486.pdf.

China's National Bureau of Statistics (China's NSBC). 2018. China statistical yearbook. http://www.stats.gov.cn/sj/ndsj/2018/indexch.htm

China's State Council Information Office (China's SCIO). 2004. China's mineral resources policy white paper. *Land Res. Newsl.* 1: 39–44.

Cui, L. P., J. F. Bai, Y. H. Shi, et al. 2004. Pollution of heavy metals in coal mining soil by mining activities. *Acta Pedol. Sin.* 41: 896–904. (Chinese with English abstract)

Feng, K. 2019. Present situation of pollution prevention and control and soil remediation in mercury mining area of Xunyang County. *Environ. Dev.* 31: 43–44. (Chinese with English abstract)

Gu, B., J. Dong, Y. Dong, et al. 2010. Discussion on land engineering reclamation technology in mining area. *Eng. Technol.* 3: 117. (Chinese with English abstract)

Guang, Q. C., W. Qiang. 2000. Environmental issues and land reclamation in open pit mining. *Energy Environ. Prot.* 14: 28–32. (Chinese with English abstract)

Guo, W. Y. 2017. Study on pollution control in ordos coal mining area. MA. Disse., Inner Mongolia Univ.

Huang, K., X. F. Zhang, D. Li. 2014. Research progress of modifiers for remediation of heavy metal contaminated soil. *Jiangsu Agr. Sci.* 42: 304–308. (Chinese with English abstract)

Johnston, A. S. A., R. M. Sibly. 2018. The influence of soil communities on the temperature sensitivity of soil respiration. *Nat. Ecol. Evol.* 2: 1597–602.

Lan, X. Q. 2019. Practice and effect of ecological restoration in Suichang Gold Mine. *Zhejiang Economy* 12: 60–61. (Chinese with English abstract)

Li, J. 2016. Research on open-pit coal mine reclamation. *Coal* 197: 68–77. (Chinese with English abstract)

Li, S. Z. 1995. Biological reclamation technology. *Coal Environ. Prot.* 2: 18–22. (Chinese with English abstract)

Li, W. D. 2013. Application of filling reclamation technology in land reclamation project. *Sci. Technol. Innov.* 5: 255. (Chinese with English abstract)

Li, Z. Y., Z. W. Ma, T. J. Van Der Kuijp, et al. 2014. A review of soil heavy metal pollution from mines in China: Pollution and health risk assessment. *Sci. Total Environ.* 468: 843–853.

Mckenna, P. B., A. Lechner., S. Phinn., P. Erskine. 2020. Remote sensing of mine site rehabilitation for ecological outcomes: A global systematic review. *Remote Sens.* 12: 3535.

National Energy Administration (NEA). 1996. Mineral resources law of the People's Republic of China. http://www.nea.gov.cn/2017-11/02/c_136722865.htm.

Sun, Y. P., Y. Yang., N. Huang, X. Zou. 2020. The impacts of climate change risks on financial performance of mining industry: Evidence from listed companies in China. *Res. Policy* 69: 101828.

Sun, Y. Q., F. P. Li, S. D. Cui. 2006. Using bioremediation technology to control ecological environment in mining areas. *Express Inf. Min. Ind.* 25: 39–42. (Chinese with English abstract)

Tahmasebi, P., M. Taheri, M. H. M. Gharaie. 2020. Heavy metal pollution associated with mining activity in the Kouh-e Zar region, NE Iran. *B. Eng. Geol. Environ.* 79: 1113–1123.

Wang, Q., J. Zhen, L. Y. Zhang. 2017. Discussion on prevention and control of water pollution in Mining Area. *World Nonferr. Metal* 7: 310–312. (Chinese with English abstract)

Xi, L. 2019. Analysis of problems existing in mining engineering and countermeasures. *Coal* 86–87. (Chinese with English abstract)

Xinhuanet. 2015. Opinions of the CPC central committee and the state council on accelerating the construction of ecological civilization. http://www.xinhuanet.com/politics/2015-05/05/c_1115187518_2.htm.

Xinhuanet. 2019. Shanxi Province: Change from coal mining subsidence area into ecological landscape. https://baijiahao.baidu.com/s?id=1638023145768324068&wfr=spider&for=pc.

Yan, Z. 2019. Research on present situation and countermeasures of mine land reclamation in China. *Land Res.* 48–49. (Chinese with English abstract)

Yang, H. L. 2009. Remediation of heavy metal pollution soil. *Environ. Sci. Manag.* 34: 130–135. (Chinese with English abstract)

Yu, G. F., X. Jiang, W. X. He, et al. 2002. Effect of humic acids on species and activity of cadmium and lead in red soil. *Acta Sci. Circum.* 22: 508–513. (Chinese with English abstract)

Yu, L. H., Y. N. MA, Y. N. Su, et al. 2017. Analysis of changes in China's mining market and outlook between 2016 and 2017. *China Land Res. Econ.* 1: 28–34. (Chinese with English abstract)

Yuan, D. 2000. The management of polluted soils by heavy metal. *Environ. Exploit.* 25–28. (Chinese with English abstract)

Zhang, A. P. 2018. Study on ecological environment status and ecological restoration in Fuxin mining area. *Heilongjiang Environ. J.* 42: 95–98. (Chinese with English abstract)

Zhao, S. L. 2007. The foreign mine environmental protection system and the benefits for China. *China Min. Mag.* 43–46. (Chinese with English abstract)

Zhou, L. B., J. Wang, H. W. Dai. 2010. *Research and practice on ecological restoration of abandoned land in mining area*. Beijing: China Environmental Science Press.
Zhu, G. X., Q. J. Guo, J. X. Yang, et al. 2013. Research on the effect and technique of remediation for multi-metal contaminated tailing soils. *Environ. Sci.* 9: 3690–3697. (Chinese with English abstract)
Zhu, X. 2010. *Chinese mining history*. Beijing: Geology Press.

6 Mining in Colombia
An Opportunity for Better Management of Environmental Impacts

Alfonso R. Rodriguez Pinilla and Victor Andres Arias Espana

6.1 INTRODUCTION

The geological conditions and strategic location of Colombia create great potential for the extraction of minerals with more than 60 million hectares identified throughout the national territory as sources for these minerals (Acosta, 2014).

According to the Agencia Nacional de Mineria (2020), Colombia produced during the third trimester of 2019, 23,419,721 pounds of nickel, 288,418 ounces of gold, 102,187 carats of emeralds (includes raw, polished and set), 154,765 tons of steel, 86,738 tons of salt and 20,464,382 tons of coal. Mining is therefore and will continue to be a key driver of Colombia's economic progress. Traditionally, mining represents around 2% of the national Gross Domestic Product (GDP), 7% of the non-urban national GDP, 28% of the GDP of the mining municipalities, 28% of and 22% of direct foreign investment in the country (Asociacion Colombiana de Mineria, 2017; Martínez & Aguilar, 2013).

The main minerals exported by Colombia are coal, gold, nickel, emeralds and, more recently, copper. Since 2009, Colombia has exported on average US$12.474 billion per year in minerals. According to Asociacion Colombiana de Mineria (2017), exports of coal represent 51% of the external sales of minerals, gold (16%), ferronickel (5%) and emeralds (1%) (Figures 6.1 and 6.2).

Additionally, it is estimated that formal mining creates around 350,000 direct jobs and about 900,000 indirect jobs (Martínez & Aguilar, 2013). In a study carried out by Martínez and Aguilar (2013) it emerged that for every peso generated by mining, 2.44 pesos are created in other sectors, mainly the transportation, services and construction industries. Conversely, it is well known that mining operations, regardless of their scale, are essentially disruptive to the environment (Makweba & Ndonde, 1994), producing large quantities of waste that can generate serious problems for people, flora and fauna for several decades (UNEP, 2020). In particular, the use of mercury by (artisanal) gold mining for extraction purposes has caused extensive mercury and cyanide pollution, very high levels of river siltation, land degradation and many other environmental as well as social problems (Veiga, 2018).

Colombia is no exception and as indicated by the MASD (Ministerio de Ambiente y Desarrollo Sostenible) the mining sector is the largest contributor of brownfields with 42% (Ministerio de Ambiente y Desarrollo Sostenible, 2015a). The country has been ranked as the world's third most contaminated nation in terms of the quantity of mercury released to the environment, after China and Indonesia but is also the world's worst per capita mercury polluter (Cordy et al., 2011; Veiga et al., 2018). This is despite only being ranked 24th in terms of the amount of gold produced (Agencia Nacional de Mineria, 2020). More alarmingly, the highest concentration of mercury in urban air (1000 μg/m^3) ever reported was found in mining towns in the Department of Antioquia with an

 DOI: 10.1201/9781315141855-6

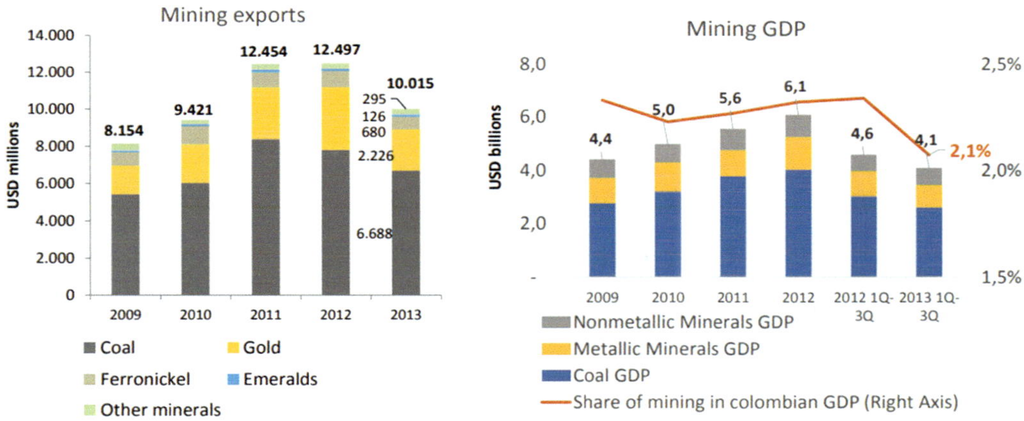

FIGURE 6.1 Mining Exports (left) and Mining GDP (right). Source: (Acosta, 2014)

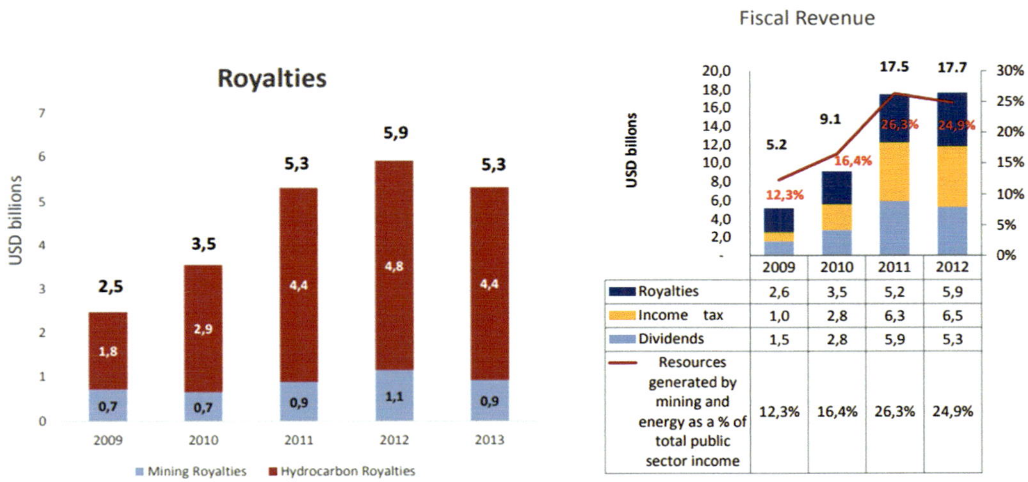

FIGURE 6.2 Mining Royalties (left) and Mining Fiscal Revenue (right). Source: (Acosta, 2014)

average level in residential areas amounting to 10 μg/m³ (Cordy et al., 2011). This chapter reviews past and existing mining in Colombia and the major pollution issues linked to artisanal gold mining, policy(ies) on mining and mine rehabilitation. The objective here is to identify deficiencies in policies and regulations. Finally, successful mine rehabilitation case studies are presented.

6.2 PAST AND EXISTING GOLD MINING

Mining in Colombia existed prior to Spanish colonization. During the pre-Columbian period, the indigenous communities exploited gold and precious gems for artisanal activities and rituals as well as non-metallic minerals such as coal, clay and salt, which were important elements of the economy prior to the arrival of the Spanish (Ministerio de Minas y Energia, 2010). During the Spanish colonial period, the arrival of slaves from Africa during the 16th century contributed to the expansion of the mining sector (mainly gold, silver and platinum) (Bonet, 2007; R. Mosquera, 1978; OECD, 2017). The mining sector slowed after Colombia became independent in 1810 and with the abolition of slavery in 1851. In the 19th century, the mining sector recovered with the arrival of

FIGURE 6.3 Artisanal mining photographs: a small jaw crusher operating in an artisanal mine in Colombia (left); Manual rock selection in small sluice box (right) in Colombia. Source: (Bardos, 2017)

foreign companies, such as US Granger Mining Company and the British Anglo Colombian Gold Development (Leal, 2009; J. Mosquera, 2013). In the early 1920s, mining concessions were granted for many regions throughout the country.

During most of the 20th century, the mining sector operated mainly as traditional ancestral artisanal mining and large-scale mechanized mining (OECD, 2017) (Figure 6.3). Currently, there are large explorations of minerals like ferronickel in the Department of Cordoba and large open coal pits in the Departments of Guajira and Cesar (Ministerio de Minas y Energia, 2010), to name just a few.

As stated earlier, gold extraction in Colombia can be traced back to the pre-colonial period. The areas of gold extraction today are the same areas exploited during Spanish rule and similarly the forms of exploitation have mostly remained unchanged (COLNODO – Red de Desarrolo Sostenible, 2018). Recent data indicated that Colombia produced 35.6 tons of gold in 2018, making it the sixth largest producer of gold in Latin America (Agencia Nacional de Mineria, 2020) (Figure 6.4). However, most of it is produced in an artisanal manner with 200,000 miners officially producing 30 tons of gold per annum (Cordy et al., 2011). This represents more than 80% of the gold production (COLNODO – Red de Desarrolo Sostenible, 2018). This is a consequence of poverty and lack of employment opportunities in rural areas.

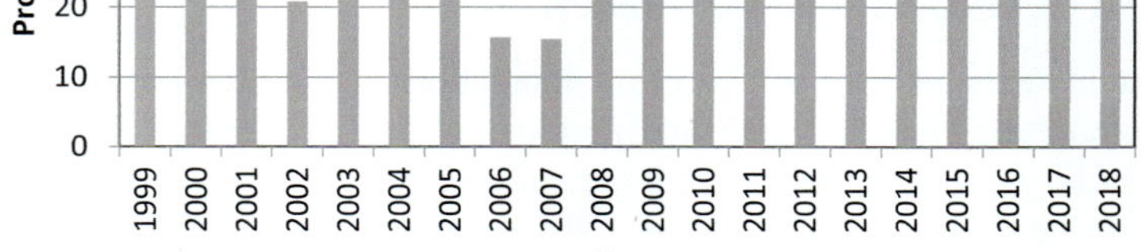

FIGURE 6.4 Gold production in Colombia 1996–2018. Data from Banco de la República, Ministerio de Minas y Energía, Minercol, Ingeominas (2004–2011), ANM (2012–2016), Agencia Nacional de Mineria (2017–2018).

FIGURE 6.5 Mercury discharges to the environment. Mercury is spread on the sluice box (left) Source: (Bardos, 2017) or on the ground (right). Source: (Pure Earth, 2019)

Another aspect to consider is the presence of illegal mining in the country. As indicated by the Contraloría General (Garavito & Ramirez, 2018), only 37% of the 14,357 surveyed mining units have a title and only 19.6% are actually legal. Another example is the high percentage of illegal mining (99.2%) reported in the Department of Chocó, which is considered one of the most important regions in terms of biodiversity (Pardo, 2013).

The use of mercury in artisanal gold mining has resulted in extensive mercury emissions and has become one of the most serious environmental problems the country faces (Arias, Rodriguez, Bardos, & Naidu, 2018) (Figure 6.5). The Colombian National Mercury Inventory reported that mercury emissions are approximately 77 tons each year (47 tons of mercury being released to the atmosphere, 15 tons to water and 15 tons to soil) due to gold mining and related activities (OECD/ECLAC, 2014), which falls within the 50–100 tons range as indicated in previous studies (Telmer & Veiga, 2008).

According to the United Nations Industrial Development Organization (UNIDO), Colombia was ranked as the world's third most contaminated country in terms of the quantity of mercury released to the environment after China and Indonesia. It is the number one country in terms of per capita mercury pollution (Cordy et al., 2011), despite being only ranked 24th in terms of the amount of gold produced (Agencia Nacional de Mineria, 2020).

Environmental and health problems caused by mercury are well known and include accumulation in sediments and ecosystems, and for people it can produce long-term neurological and renal degeneration (Howard, Trotz, Thomas, et al., 2011; Tschakert, 2010; Tschakert & Singha, 2007). In addition, mercury vapor can cause respiratory problems and symptoms, including chest pains, dyspnea, cough, hemoptysis, impairment of pulmonary function and interstitial pneumonitis (Levin, Jacobs, & Polos, 1988; Stopford, 1979).

Several studies have strongly emphasized that mercury pollution in Colombia has caused bioaccumulation of mercury in fish (Marrugo-Negrete, Benitez, & Olivero-Verbel, 2008; Olivero & Solano, 1998; Salazar-Camacho, Salas-Moreno, Marrugo-Madrida, Marrugo-Negrete, & Díez, 2017). The end result is incidental ingestion of fish by people living downriver from hotspots and consequently increasing the risk of mercury intoxication. Likewise, there is evidence of neurological complications in individuals and a greater prevalence of irregular menstruation in women (Rodríguez-Villamizar, Jaimes, Manquian-Tejos, & Sanchez, 2015) exposed to mercury vapor in Colombian mining towns, where "entables", which are processing centers where miners take their ores to be processed for a nominal fee, and gold shops process gold amalgams without any control measures (De Miguel, Clavijo, Ortega, & Gomez, 2014; Garcia et al., 2015).

A number of projects have set out to address contamination caused by mercury pollution. For instance, UNIDO in joint efforts with the Government of Antioquia, National University of

Colombia and University of British Columbia implemented "The Colombia Mercury Project" (Arias et al., 2018). This project intended to minimize mercury use and losses and included assessment of mercury losses, health monitoring and awareness campaigns. Results obtained were very encouraging and led to a significant reduction in mercury losses (63%) and, as a consequence, about 46 to 70 tons per year of mercury were prevented from entering the Colombian environment when compared to 2010 data (García et al., 2015).

In addition, the Ministry of Mines and Energy has made a remarkable effort to formalize mining – Directorate of Formalization (DF). The DF achieved, at the end of 2017, the basic stage of formalization of 1,012 Mining Production Units (UPM) of the total 14,357 UPMs (7%), which are fully or are on the way to becoming formalized. When compared to the rest of Latin America (where less than 1% of artisanal mining operations are formalized (Marshall & Veiga, 2017), it can be considered an outstanding success. Furthermore, all formalized units will be provided with permanent technical support and government support to obtain financial credit in the form of bank loans.

In a comprehensive study, Veiga (2018) identified the following key points related to mercury in Colombia:

- The whole ore amalgamation process is the main cause for mercury loss to the tailings.
- "Entables" and gold shops are the main sources of mercury pollution.
- Mercury emissions in urban air (1,000 $\mu g/m^3$) are a critical public health issue.
- The main environmental issue caused by mercury use in artisanal mining is the cyanidation of mercury-contaminated tailings.

To complement existing measures, the following aspects should be considered and implemented where possible as indicated by Veiga (2018):

- The strategy should focus on reducing mercury use before eliminating it completely.
- The measures to remove mercury must include all artisanal gold miners.
- Technologies to stop the use of mercury should be simple and easy to operate, with materials accessible to artisanal miners, low cost, and recover substantial gold. A successful example was the approach used by the CICAN (Colleges and Institutes Canada) Project in Colombia known as the MacGyver approach (Veiga et al., 2018). It used low-cost locally found materials to teach miners and their trainers how to make their own processing equipment.
- These technologies need to be complemented by training together with simple and easy-to-follow financial systems.

6.3 POLICY(IES) ON MINING AND MINE REHABILITATION

Law 685 of 2001 "By which the Code of Mines is issued and other provisions are dictated" comprehensively regulates everything linked to the legal agencies of the State with individuals and those of each other, and the work and activities of the mining industry in Colombia are no exception. This law and its regulations comprise a harmonious set of norms aimed at regulating everything concerning mining activities (MDS, 2014). Within this law it is also established that the person interested in realizing a mining exploitation project must present an Environmental Impact Study to the relevant environmental authority and obtain the Environmental License, which will be valid until the operations cease. Table 6.1 summarizes some the most relevant regulatory advances that have strengthened mining activities in Colombia.

The most current mining regulations in Colombia are compiled in "The National Mining Policy", which was adopted under Resolution 40391 of 2016 as set out by the Ministry of Mines and Energy of Colombia. This policy is based on six pillars (Agencia Nacional de Mineria, 2016) as follows:

TABLE 6.1
Regulations for Mining in Colombia

Resolution No. 18-1467 of 2011	National Mining Safety Policy. The policy seeks to establish the pillars for the improvement of mining safety and define technical and operational occupational health and safety guidelines to prevent work-related accidents. The policy is based on the following pillars: prevention, stricter technical requirements, active participation, commitment of all interested parties and consolidation of a public information system regarding mining safety.
Law 1658 of July 2013	Mercury Law. This law provides for the commercialization and use of mercury as established in the respective of industrial activities. It establishes requirements and incentives for mercury use reduction and disposal among others.
Resolution No. 90719 of 2014	National Policy for the Formalization of Mining in Colombia. The policy focuses on generating support tools for informal small and medium-scale miners, preferably with government support.
Resolution 0256 of 2014	National Mining Management Plan. The purpose of this plan is to achieve an orderly and responsible use of mining resources, thus promoting the conversion of mining capital into other forms of capital that translate into greater well-being and development for the producing regions and the country.
Resolution 40391 of 2016	The National Mining Policy. This policy is described below.

Source: (Ministerio de MInas y Energia, 2016)

- Legal security: clear and stable legal framework for the development of the sector.
- Competitive conditions for the mining industry so that:
  - the minerals can be extracted.
  - taxes and royalties can be paid.
  - employment is generated.
  - an efficient development of the industry is enabled.
- Legitimate trust: relationship of mutual trust between the government, the productive sector and the mining communities.
- Infrastructure: management for the expansion and modernization of the transport infrastructure in the country.
- Information: knowledge of Colombian territory from the geological-mining point of view, as well as the mining information systems that support mining activity.
- Strengthened and efficient mining institutions: improving the efficiency of the mining authorities in all aspects such as response times and speed of processing.

From an environmental point of view, the national mining plan, Horizon 2025, shows that efforts in improving regulation and removing loopholes are focused on three fundamental aspects: 1) processes of mine closure, 2) management of abandoned mining areas, and 3) In the case of gold mines, removing the use of mercury (UPME, 2017).

6.3.1.1 Mine Closure

Although mine closure is defined as the set of activities to be implemented throughout the life cycle of the mine, in order to meet the specific environmental criteria and achieve the desired social objectives after the exploitation stage (Ministerio de Minas y Energia, 2016), there is no specific regulation that determines the parameters of the closure process. In Colombia, the legislation tends to be scattered and, in practice, involves all environmental aspects such as soil, air, water discharges and emissions impacted during mine closure not being comprehensively dealt with. For this reason, it was recommended not to create a new regulation but rather prepare a fourth guide (of three existing ones, whose preparation was the joint responsibility of the Ministry of Mines and Energy and the Ministry of Environment and Sustainable Development) on the different legal provisions governing the closure of mines, and reinforce guarantees and monitoring programs on the long-term impacts (Ospina & Molina, 2013).

Currently the Ministry of Mines and Energy, the National Mining Agency and the Ministry of Environment and Sustainable Development are developing technical instruments to ensure rigor in the different stages of mine closure and seek its regulation, establishing it as an additional contractual obligation and demanding financial assurances (UPME, 2017).

6.3.1.2 Management of Abandoned Areas

Since 2015, the Ministry of Environment and Sustainable Development has been developing "The Strategy for the Management of Environmental Liabilities in Colombia", which includes the management of areas abandoned by mining interests (Ministerio de Ambiente y Desarrollo Sostenible, 2015a). This strategy includes the process of inventorying, characterizing and prioritizing abandoned mining areas, as well as managing the required resources for prompt intervention (UPME, 2017).

In 2016, the Ministry of Environment and Sustainable Development developed eight (8) technical instruments that are part of the strategy of environmental liabilities management. These instruments (Ministerio de Ambiente y Desarrollo Sostenible, 2016) are:

1. Technical guide with a list of potential activities generating environmental liabilities.
2. Methodological guide and implementation protocol for preliminary risk assessment at sites in areas with suspicion, considered as potential generators of environmental impacts not addressed in a timely manner.
3. Methodological guide and implementation protocol for the evaluation of risks to human health and the environment.
4. Guidance and protocol for soil and groundwater sampling, including methodologies and protocols for establishing concentrations, or acceptable limits in contaminated sites.
5. Permissible standards or limits of substances of interest in soils and groundwater according to their use and environmental importance.
6. Guide and implementation protocol for the establishment of general criteria in processes of monitoring and control of intervened areas.
7. Technical guide for implementation of remediation techniques.
8. Guide for the requirement of environmental concepts in land with changes in land use and merger or liquidation.

In 2017, a study was carried out to establish the required financial resources and funding sources for managing environmental liabilities and the necessary institutional arrangements for this purpose. One proposal is the creation of an independent agency to define the legal process for the management of environmental liabilities (Ministerio de Ambiente y Desarrollo Sostenible, 2017). Currently, a draft law for the management of environmental liabilities is underway in the Congress of the Republic, which is expected to have a satisfactory conclusion by 2020.

6.3.1.3. Elimination of Mercury Use

During the last 25 years, the Colombian government has advanced significantly in terms of developing regulations for the use of mercury in mining and other industries (Figure 6.6).

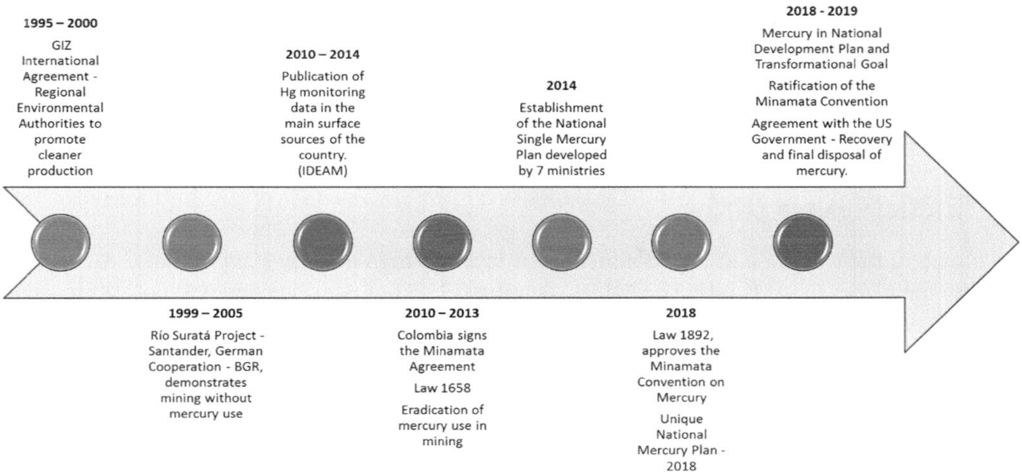

FIGURE 6.6 Timeline on the progress of mercury regulations in Colombia. Source: (Ministerio de Ambiente y Desarrollo Sostenible, 2019b)

From July 16, 2018, onward, the use of mercury in mining in Colombia has been prohibited, in accordance with the provisions of Law 1658 of 2013, with which the National Government assumed the challenge of eliminating this element in gold mining in five years, that is, between 2013 and 2018. During this period, the Ministry of Mines and Energy, with the support of the Ministry of Environment and Sustainable Development and regional environmental authorities, strengthened the mining sector by advocating alternative technologies that did not require mercury (Ministerio de Ambiente y Desarrollo Sostenible, 2019a). On the other hand, some doubts about the efficiency of this law have been raised as opponents argue that banning the use of mercury could create more illegality as miners will start working and burning mercury secretly and illegally at home, polluting his/her family members and neighbors as well as contributing to the illegal entry of mercury into the country (Veiga, 2018). The former phenomenon has been observed in Ecuador (Gonçalves, Marshall, Moreno-Chavez, & Veiga, 2017), and opponents argue that it will simply be replicated in Colombia.

There is another aspect relevant to the elimination of mercury use in gold mining, and that is the variability of emissions/discharges to the environment. These vary widely according not only to the technologies and techniques used for exploitation but also with the traditions of the miners and the region where this activity is carried out, since it has been done from generation to generation. The mineralogical composition of the material (e.g. sulfur content, quartz or other metallic elements) influences the amount of mercury required for extraction (UPME, 2017). It is expected that with the formal issuance of the technical instruments prepared by the Ministry of Environment and Sustainable Development, especially those corresponding to risk assessment of contaminated sites and the establishment of permissible limits of contaminants in the soil (Ministerio de Ambiente y Desarrollo Sostenible, 2016), these will cover part of the regulatory loopholes that currently exist.

Additional challenges have been identified that must be addressed as a priority in order to provide a comprehensive solution to the existing regulation regarding the issue of mercury use. As indicated by Veiga (2018), the Colombian government must first identify the suppliers and users of mercury, identify the main sources (e.g. "entables" and gold shops) and prioritize strategies/mechanisms to address pollution at the source that will put an end to the disrespect shown by the polluters to the general public. These strategies would replace the existing monitoring approach.

Furthermore, there is no specific regulation related to the management of mercury in mining activities such as storage, transportation and final disposal. Some efforts have been made by local governments to establish the technical criteria necessary to develop a strong, applicable and convenient regulation. Some examples are: (1) Assistance for the proper management of mercury waste in Colombia (UNIDO, 2019), (2) Guide for handling and storage of tailings contaminated with mercury (CNPML, 2019), and (3) Promoting Responsible Recovery and Handling of Mercury from Contaminated Artisanal Gold Mining Tailings in Colombia (Pure Earth, 2019).

6.4 ABANDONED MINES

In Colombia, the closure and abandonment of mines are regulated by Law 605 of 2001 as established in the Mines Code (Codigo de Minas) (Figure 6.7). They are considered to constitute a phase of mining activity, which is included within the parameters of exploitation and therefore a legal obligation for development/implementation by the entity carrying out the mining activity. This is presented in the plan for closing and abandonment of a mine and must be prepared before the exploitation stage since it is a requirement for submission in the Program/Plan of Works (Agencia Nacional de Mineria, 2016).

As well, mining activities must be complemented by feasibility and environmental permits, which are required to commence exploration works. These are issued by the appropriate environmental authorities and establish the environmental measures that any project owner (including mining title holder) must adhere to during the dismantlement phase and abandonment. The related environmental regulations are Law 99 of 1993 and Decree 1075 of 2015 (Agencia Nacional de Mineria, 2016).

Despite regulatory efforts to ensure that mining closures are carried out in an environmentally correct manner, it was necessary to develop a methodology for the characterization and intervention of Areas with Mining Activities in Abandoned States in Colombia (AAMEA), which was prepared by the Institute of Environmental Studies of the National University of Colombia (IDEA) (Ministerio de Minas y Energía, 2014). Further to making an inventory of areas abandoned in the country, the methodology's scope covers the following issues:

- Review of the legal status of the mine.
- Risk assessment of physical hazards.
- Risk assessment of technological hazards.
- Risk assessment of contamination.
- Social vulnerability.

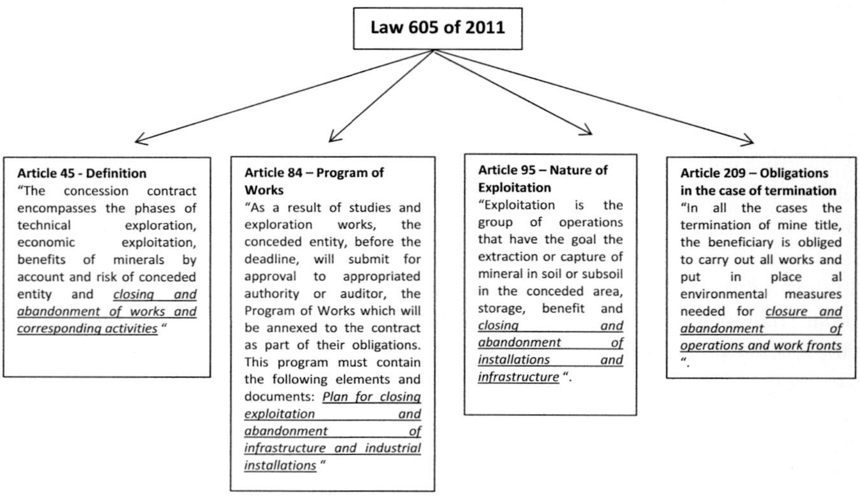

FIGURE 6.7 Relevant articles of Law 605.

The application of this methodology from 2014 to December 2018 has managed to identify 423 abandoned mining sites, located in Antioquia, Cauca, Guanía and Santander. Of these, 274 correspond to gold mining, 99 to coal mining and 50 to mining of construction materials (Portafolio, 2019). The IDEA Final Report also highlights that these abandoned mines present a risk to the community, the environment and the surrounding infrastructure, so the government has planned to intervene immediately (Portafolio, 2019).

Colombia does not have a standardized procedure for the remediation or restoration of abandoned mines. However, since 2015 some important parameters were established for the restoration of areas affected by closing and abandonment of mines under the framework of the "National Restoration Plan" (Ministerio de Ambiente y Desarrollo Sostenible, 2015b). This Plan incorporates three implementation approaches: first, ecological restoration; second, rehabilitation; and third, recovery, which depends on the type of intervention, level of degradation of the area and the objective(s) of restoration (Ministerio de Ambiente y Desarrollo Sostenible, 2012). In addition, the Plan considers intervention into areas identified as protected, for instance national parks and basins, as these are most likely to be affected by mining abandonment (Figure 6.8).

The Plan established a period of 20 years to achieve the restoration of these areas and specified timeframes to achieve specific goals (short-term goals achieved at three years, medium-term

FIGURE 6.8 Areas identified for intervention in the National Restoration Plan. Source: (Ministerio de Ambiente y Desarrollo Sostenible, 2015b)

at eight years and long-term at 20 years). The Plan also determines priority and essential actions for compliance, targets and indicators, responsible parties, potential sources of funding and a cost-effective approach to comprehensive and proper management (Ministerio de Ambiente y Desarrollo Sostenible, 2015b).

6.5 SUCCESSFUL REHABILITATION – CASE STUDIES

Despite not having a specific regulation for the restoration of abandoned mine areas, some efforts have been promoted by the central government, international agencies and private companies that followed the framework of the "National Restoration Plan". Successful rehabilitation case studies are presented in more detail below.

6.5.1 Reforestation of Areas Damaged by Mining – Bajo Cauca (Antioquia)

This pilot project was developed under the ENV / Mining program funded by the United States Agency for International Development (USAID) and executed by Reforestadora Integral de Antioquia S.A. (RIA) (USAID, 2014) (Figure 6.9).

Pilot tests identified *Acacia Mangium* as the best species for the rehabilitation of abandoned alluvial mine lands located in the Bajo Cauca (Antioquia).. The recovery program will consist of planting (90%) *Acacia Mangium* and (10%) native species such as Abarco, Tambor, Caracolí, Roble and Samán (Figure 6.10).

The Program plans to introduce beekeeping as an ancillary economic activity to commercial plantation schemes. *A. Mangium* produces excellent quality nectar that attracts bees, which makes the two activities compatible, generating income for landowners while plantations mature and reach the age of their first commercial thinning.

6.5.2 Protective Forest Plantations – Nechí River (Antioquia)

This project was developed by the private company Mineros SA during the early 1980s for the lower basin of the Nechí River and includes the planting of protective and productive native species. These displayed rapid growth and resistance to harsh conditions found in areas degraded by mining and low fertility soils (Mineros SA, 2020).

To date, the program has planted about 240,000 trees of different native species on 120 hectares of land affected by alluvial mining and the mass production of 200,000 seedlings (Mineros SA, 2020). It included the planting of 250,000 trees of Cativo, which is a forest species that is in danger of extinction due to excessive consumption and deforestation, on 150 hectares of floodplains in the area of direct influence, i.e. an area not of interest for dredging in the alluvial operation. This

FIGURE 6.9 Pilot project recovery of areas damaged by mining in Bajo Cauca. Source: (USAID, 2014)

FIGURE 6.10 Plantation of *Acacia Mangium* in Bajo Cauca. Source: (USAID, 2014)

FIGURE 6.11 Protective forest plantations – Rio Nechí (Antioquia). Source: (Mineros SA, 2020)

reforestation project included the participation of 23 local communities in the planting of seeds and seedlings (Figure 6.11).

The program brings several benefits to the local community, including employment for the rural workforce, whose annual average income for the next seven years is estimated to be 3,816 salaries. Plant materials produced are donated to the municipalities of the local government of "Bajo Cauca Antioqueño" and the organized community (Mineros SA, 2020).

6.5.3 Restoration and Remediation of Gualcolda Mine – Andes (Nariño)

Miners in the municipality of the Andes, Nariño stopped using mercury to process gold ore in 2017, but after decades of mining with mercury, contaminated soils remained. To address the issue, the miners' association ("La Fortaleza" with 13 members, with technical support from the Alliance for Responsible Mining (ARM), decided to restore and remediate the impacted area (Ethicalmetalsmiths, 2019).

The restoration and remediation project focused on cleaning 400 tons of sand contaminated with mercury located on the banks of the Honda River, municipality of the Andes, Nariño. The project was funded by a fund-raising campaign (Better without Mercury) and included strategies for revegetation, reforestation (Figure 6.12), adaptation of trails, maintenance and care of local water

FIGURE 6.12 Better without Mercury Project. One of several gardens at the Gualconda mine (Left), a Gualconda miner preparing seedlings for later planting at the remediated site (Right). Source: (Ethicalmetalsmiths, 2019)

resources and implementation of traditional mining pedagogical models. In this way, the mining community was better educated about what to do (Fortaleza, 2019).

More than 400 tons of contaminated land were safely remediated, and this helped to prevent the spread of pollutants into the environment. To date, 1,000 meters of nature trails have been completed as part of the restoration project (Fortaleza, 2019).

6.5.4 Achiote Production – Puerto Claver (Choco)

Under the "Oro Legal" (translates Legal Gold) program, also financed by USAID, it is intended to rehabilitate 11,500 hectares of land degraded by mining from 2015 to 2020 (USAID, 2016). One of the areas intervened so far corresponds to the municipality of Puerto Claver in the Department of Chocó, where the program has two fundamental objectives: first, to reforest degraded land; and, second, to formalize the economy of families that subsist on mining in that municipality (J, Angel et al., 2019). To date, the program has successfully resulted in the planting of more than 201 hectares of achiote (Figure 6.13) and the rehabilitation of 4,640 hectares, including neighboring municipalities such as Peine Mono, Abarco and Mataratón (Pais MInero, 2019).

FIGURE 6.13 Achiote plant from degraded area in Chocó. Source: (Enamorate del Chocó, 2019).

Moreover, the physical and chemical characteristics of the achiote or *Bixa orellana* seed have attracted the attention of a private sector company dedicated to the dye industry in Colombia. The "Bija", as it is known in the region, has characteristics that give competitive advantages over other crops of the same species (Enamorate del Chocó, 2019). The program aims to rehabilitate 1,200 hectares by planting achiote to immobilize mercury by 2020 and benefit the livelihoods of 600 families (Pais Minero, 2019).

6.6 CONCLUDING REMARKS

Mining makes a significant contribution to the economic development of Colombia, representing around 2% of the national gross domestic product (GDP), 7% of the non-urban national GDP, 28% of the GDP of the mining municipalities, 28% of exports and 22% of the direct foreign investment in the country (Asociacion Colombiana de Mineria, 2017; Martínez & Aguilar, 2013). Yet, it is well known that mining operations, regardless of their scale, can damage the environment and human health. In Colombia, the main environmental issue is closely linked to the use of mercury by artisanal gold miners. There is evidence of environmental and health problems caused by mercury, including bioaccumulation in fish downriver from hot spots and neurological complications in individuals and higher prevalence of irregular menstruation in women exposed to mercury vapor in Colombian mining towns.

A number of projects to address contamination caused by mercury pollution have been implemented, such as “The Colombia Mercury Project”, which aimed to minimize mercury use and losses and included assessment of these losses, health monitoring and awareness campaigns. The Ministry of Mines and Energy has made a remarkable effort to formalize mining – Directorate of Formalization (DF) achieving 7% of Mining Production Units (UPM) to be or on the way to becoming formalized and provided all formalized operations with permanent technical support and government support to obtain financial credit for bank loans.

In Colombia, 423 abandoned mining sites have been identified, of which 274 correspond to gold mining, 99 to coal mining and 50 to mining of construction materials. Finally, a number of successful case studies that followed the framework of the "National Restoration Plan" were presented. The positive results included:

- Pilot tests identified *Acacia Mangium* as the best species for the rehabilitation of abandoned alluvial mine lands located in the Bajo Cauca (Antioquia).. The recovery program will consist of planting (90%) *Acacia Mangium* and (10%) native species such as Abarco, Tambor, Caracolí, Roble and Samán.
- The plantation of about 240,000 trees of different native species on 120 hectares of land affected by alluvial mining, the mass production of 200,000 seedlings and planting of 250,000 trees of Cativo in the Nechí River (Antioquia) region. The program brings several benefits to the local community, including jobs for the rural workforce and plant materials being donated to the municipalities of Bajo Cauca antioqueño and elsewhere.
- Cleaning 400 tons of sand contaminated with mercury located on the banks of the Honda River, municipality of the Andes, Nariño.
- Restoration of 1,000 meters of nature trails.
- The planting of more than 201 hectares of achiote and the rehabilitation of 4,640 hectares in Puerto Claver (Choco), including neighboring municipalities like Peine Mono, Abarco and Mataratón.

REFERENCES

Acosta, M. A. (2014). Evolución reciente de la minería en Colombia y desarrollo sostenible. Ministerio de Minas y Energía. PowerPoint Presentation (worldbank.org).

Agencia Nacional de Mineria. (2016). Consulta-Plan de cierre de minas. Radicado No 20161200423111. Bogota D.C.

Agencia Nacional de Mineria. (2020). Producción minerales. Retrieved January 13, 2020, from BOLETIN2019-160222 (anm.gov.co)

J. Ángel, J., Ordoñez, M., Olivero, J., Echavarría, C., Ayala, H., & Cabrera, M. (2019). Consideraciones sobre la minería en el departamento del Chocó y recomendaciones para mejorar la gestión. Geopatrimonio – Universidad de Cartagena - IIAP - WWF. Páginas 58. Cali – Colombia. Retrieved fom consideraciones _sobre_la_mineria_en_el_departamento_del_choco_doble_pagina.pdf (panda.org)

Arias, E. V. A., Rodriguez, P. A. R., Bardos, P., & Naidu, R. (2018). Review: Contaminated land in Colombia: A critical review of current status and future approach for the management of contaminated sites. *Science of The Total Environment, 618*, 199–209.

Asociacion Colombiana de Mineria. (2017). Colombian mining sector development 2018–2022. Bogotá D.C.

Bardos, P., Cundy, A., Maco, B., Kovalick, W., Rodríguez, A., Hutchings, T., Hall, E., & Rodríguez, A. (2017). Strategies for rehabilitating mercury- contaminated mining lands for renewable energy and other self-sustaining re-use strategies. Output 2. https://doi.org/10.13140/RG.2.2.29241.47205. Bogotá: r3 Environmental Technology.

Bonet, J. (2007). *¿Por qué es pobre el Chocó?* Bogotá: Banco de la República.

CNPML. (2019). *Guía para el manejo y almacenamiento de relaves contaminados con mercurio.* Medellin: Centro Nacional de Producción Más Limpia.

COLNODO Red de Desarrolo Sostenible - IPEN. (2018). Mercury trade and supply in artisanal and small-scale gold mining in Colombia. Public_ver_Reporte ASGM Colombia EN INGLES (ipen.org)

Cordy, P., Veiga, M. M., Salih, I., Al-Saadi, S., Console, S., Garcia, O., . . . Roeser, M. (2011). Mercury contamination from artisanal gold mining in Antioquia, Colombia: The world's highest per capita mercury pollution. *Science of the Total Environment, 410–411*, 154–160.

De Miguel, E., Clavijo, D., Ortega, M. F., & Gomez, A. (2014). Probabilistic meta-analysis of risk from the exposure to Hg in artisanal gold mining communities in Colombia. *Chemosphere, 108*, 183–189.

Enamorate del Chocó. (2019). Achioteo bija del Choco en el mercado internacional de colorantes naturales. Retrieved from https://enamoratedelchoco.co/achiote-o-bija-del-choco-en-el-mercado-internacional-de-colorantes-naturales/

Ethicalmetalsmiths. (2019). Project. Better without mercury. Retrieved from https://ethicalmetalsmiths.org/better-without-mercury

Fortaleza. (2019). Mejor sin mercurio. Proyecto de elimación de areas contaminadas y restauración. Asociación Fortaleza – Alliance for Responsible Mining.

Garavito, M. A., & Ramirez, W. (2018). Ecological restoration based on the Colombian Legal System. In E. Ceccon & D. R. Pérez (Eds.), *Beyond restoration ecology: Social perspectives in Latin America and the Caribbean*, Chapter 9. Vazquez Mazzini Editorial.

García, O., Veiga, M. M., Cordy, P., Suescún, O. E., Molina, J. M., & Roeser, M. (2015). Artisanal gold mining in Antioquia, Colombia: A successful case of mercury reduction. *Journal of Cleaner Production, 90*, 244–252. https://dx.doi.org/10.1016/j.jclepro.2014.11.032

Gonçalves, A. O., Marshall, B. G., Moreno-Chavez, J., & Veiga, M. M. (2017). Mercury and cyanide use at artisanal and small-scale gold processing centers in Zaruma-Portovelo, Southern Ecuador. *Cleaner Production, 165*, 836–845.

Howard, J., Trotz, M., Thomas, K., et al. (2011). Total mercury loadings in sediment from gold mining and conservation areas in Guyana. *Environmetnal Monitoring and Assessment, 179*, 555–573.

Leal, C. (2009). La compañía Minera Chocó Pacífico y el auge del platino en Colombia. *Historia Crítica*, 150–164.

Levin, M., Jacobs, J., & Polos, P. (1988). Acute mercury poisoning and mercurial pneumonitis from gold ore purification. *Chest, 94*, 554–556.

Makweba, M., & Ndonde, P. (1994). *The mineral sector and the national environmental policy.* Paper presented at the Proceedings of the Workshop on the National Environmental Policy for Tanzania, Dar es Salaam, Tanzania.

Marrugo-Negrete, J., Benitez, L. N., & Olivero-Verbel, J. (2008). Distribution of mercury in several environmental compartments in an aquatic ecosystem impacted by gold mining in Northern Colombia. *Archives of Environmental Contamination and Toxicology, 55*(2), 305–316.

Marshall, B. G., & Veiga, M. M. (2017). Formalization of artisanal miners: Stop the train, we need to get off! *The Extractive Industries and Society Journal, 4*, 300–303.

Martínez, A., & Aguilar, T. (2013). Study on socio-economic impacts of the mining sector in Colombia: Sectorial chains. Fedesarrollo.

MDS. (2014). *Análisis de acciones técnicas ambientales para el cierre y abandono minero que contemple la recuperación de servicios ecosistémicos*. Bogotá: INTESEG.
Mineros SA. (2020). Sostenibilidad - Gestion Ambiental, 2020. Retrieved from https://mineros.com.co/es/Sostenibilidad/Memoria-de-Sostenibilidad
Ministerio de Ambiente y Desarrollo Sostenible. (2012). Guías técnicas para la restauración ecológica de los ecosistemas de Colombia: Ministerio de Ambiente y Desarrollo Sostenible – Academia de Ciencias Exactas, Físicas y Naturales – Grupo de Restauración Ecológica Universidad Nacional de Colombia.
Ministerio de Ambiente y Desarrollo Sostenible. (2015a). Diseño de una estrategia comprensiva para el manejo de pasivos ambientales en Colombia. Consultoria Innova Ambiental. Bogotá.
Ministerio de Ambiente y Desarrollo Sostenible. (2015b). Plan Nacional de Restauración. Restauración ecológica, rehabilitación y recuperación de áreas disturbadas.
Ministerio de Ambiente y Desarrollo Sostenible. (2016). Diseño de instrumentos técnicos para el manejo de pasivos ambientales en Colombia. Consultoria Aquaviva.
Ministerio de Ambiente y Desarrollo Sostenible. (2017). Diseño de Instrumentos Financieros y arreglo Institucional para el manejo de pasivos ambientales en Colombia.
Ministerio de Ambiente y Desarrollo Sostenible. (2019a). Abc del Mercurio. Retrieved from abc mercurio (minambiente.gov.co)
Ministerio de Ambiente y Desarrollo Sostenible. (2019b). Presentacion. Avances en la MIneria en Colombia. Vision desde el Ministerio. *Proyecto. Promoción de la elimib¿nación del Mercurio por Relaves Contaminados en MIneria Artesanal de Oro. Departamento de Estdo. Estados Unidos.*
Ministerio de Minas y Energia. (2010). *Diseño y validación del marco conceptual y metodológico para caracterizar, priorizar y valorar económicamente los pasivos ambientales mineros en Colombia*. Bogotá D.C.
Ministerio de MInas y Energia. (2016). *Politica Nacional Minera. Bases para la MIneria del Futuro*. Bogotá – Colombia: Ministerio de Minas y Energia.
Ministerio de Minas y Energía. (2014). Metodología para la intervención de áreas afectadas por actividades mineras en estado de abandono en Colombia: Ministerio de Minas y Energía – Universidad Nacional de Colombia, instituto de estudios ambientales (IDEA).
Mosquera, J. (2013, September 18). El oscuro expediente de una compañía. *Semana.com.*
Mosquera, R. (1978). Pasado, Presente y futuro de la minería en el Chocó. Bogotá: SOGECOL, Boletín 112.
OECD. (2017). Due diligence in Colombia's gold supply chain gold mining in Chocó.
OECD/ECLAC. (2014). OECD environmental performance reviews: Colombia 2014. Retrieved from https://www.scribd.com/document/232430949/Colombia-environmental-performance-review-Highlights
Olivero, J., & Solano, B. (1998). Mercury in environmental samples from a waterbody contaminated by gold mining in Colombia, South America. *Science of the Total Environment, 217*, 83–89.
Ospina, E., & Molina, J. (2013). Colombian legislation on mine closure. Is it really necessary? *Boletín Ciencias de la Tierra, 34*, 51–62.
Pais MInero. (2019). Potencial de Formalización del Programa de Oro Legal. Retrieved from https://paisminero.co/mineria/mineria-colombiana/18393-portafolio-potencial-de-formalizacion-del-programa-oro-legal
Pardo, L. (2013). La conflictividad por el territorio, el control de los RNNR y la renta minera. El choque de las locomotoras mineras en Colombia. In L. J. Garay (Ed.), Minería en Colombia: institucionalidad y territorio, paradojas y conflicto. Contraloría General de la República. Bogotá, Colombia.
Portafolio. (2019). Gobierno Intervendria 423 minas que se encuentran en abandono. Retrieved November 25, 2019, from https://www.portafolio.co/economia/gobierno/gobierno-intervendria-423-minas-que-se-encuentran-en-abandono-523600
Pure Earth. (2019). Promoting responsible recovery and handling of mercury from contaminated artisanal gold mining tailings in Colombia. Technical Report on Mercury Recovery from Tailings, Pure Earth. New York. Retrieved from Recovery_Mercury_Tecnologias_Colombia_Workshop_Final_Draft_AUTHORS_Disclaimer.pdf (pureearth.org)
Rodríguez-Villamizar, L. A., Jaimes, D. C., Manquian-Tejos, A., &, & Sanchez, L. H. (2015). Human mercury exposure and irregular menstrual cycles in relation to artisanal gold mining in Colombia. *Biomedica, 35*, 38–45.
Salazar-Camacho, C., Salas-Moreno, M., Marrugo-Madrida, S., Marrugo-Negrete, J., & Díez, S. (2017). Dietary human exposure to mercury in two artisanal small-scale gold mining communities of northwestern Colombia. *Environment International, 107*, 47–54.
SIMCO & UPME. (2021). Mapas:Áreas con potencial mineral para definir Áreas de Reserva Estratégica del Estado. Retrieved January 13, 2020, from https://www1.upme.gov.co/simco/Paginas/Mapas.aspx

Stopford, W. (1979). Industrial exposure to mercury. In J. Nriagu (Ed.), *The biogeochemistry of mercury in the environment* (pp. 367–397). Amsterdam: Elsevier/North-Holland Biomedical Press.

Telmer, K., & Veiga, M. (2008). World emissions of mercury from small scale artisanal gold mining and the knowledge gaps about them. In N. Pirrone & R. Mason (Eds.), *Mercury fate and transport in the global atmosphere: Measurements models and policy implications* (pp. 96–129). UNEP United Nations Environmental Program.

Tschakert, P. (2010). Mercury in fish: A critical examination of gold mining and human contamination in Ghana. *International Journal of Environment and Pollution, 41*, 214–228.

Tschakert, P., & Singha, K. (2007). Contaminated identities: Mercury and marginalization in Ghana's artisanal mining sector. *Geoforum, 38*, 1304–1321.

UNEP. (2020). Sustainability reporting in the mining sector. Retrieved from Sustainability Reporting in the Mining Sector | UNEP – UN Environment Programme.

UNIDO. (2019). Asistencia para la gestión adecuada de los residuos de mercurio en Colombia: United Nations Industrial Development Organization.

UPME. (2017). Plan Nacional de Desarrollo Minero con Horizonte 2025.

USAID. (2014). Environmental assessment of the ENV/mining program: United States Agency for International Development – Ambiental Consultores y Cía Ltda - Chemonics International.

USAID. (2016). Oro legal project. Fact Sheet.

Veiga, M. M. (2018). Report on characterization of artisanal gold processing in Colombia and measures to reduce/eliminate mercury use. Vancouver, BC: The University of British Columbia.

Veiga, M. M., Masson, P., Perron, D., Laflamme, A.-C., Gagnon, R., Jimenez, G., & Marshal, B. G. (2018). An affordable solution for micro-miners in Colombia to process gold ores without mercury. *Journal of Cleaner Production, 205*, 995–1005.

7 Mine Land Reclamation in India

Policies on Mine Closure and Rehabilitation

R. E. Masto, Siddharth Singh, S. Mukhopadhyay, N. K. Srivastava, Siddharth Bari, M. C. Manna, V. A. Selvi, A. K. Singh and P. K. Singh

7.1 INTRODUCTION

Mining contributes significantly to the economy of India. About 2.38% of the country's gross domestic product (GDP) is contributed by mining industries. India ranks 4th in world in mine production after China, the United States and Russia. About 95 minerals are produced in India, of which coal, lignite, petroleum and natural gas are the four fuel minerals. Though the Ministry of Mines, the government of India is targeting a growth rate of 10-–2% in the mining industry to increase its GDP contribution to 5%; however, the current growth rate is only 1.2–1.4%, and this is mainly due to the strict environmental, regulatory and land acquisition policies. There are about 1,405 mines in India, of which coal, limestone, iron ore are the major ones, and they cover 50% of the country's mines. The total land area under mining lease covers approximately 0.366 million hectares. About 0.1% of the total land area of India is occupied by the mining industry.

Though mining imposes long-lasting impacts on the land, mining is by its nature a temporary activity. After exhausting the economically exploitable mineral, the mine will be closed and the land is available for other uses. A thorough understanding of the mine's lifecycle will be helpful in planning and implementing effective land rehabilitation. Dumping of overburden, especially by opencast mining, damages – often irreparably – topography, vegetation and soil quality. The volume of overburden is too high and generally it creates artificial mountains consisting of a mixture of rock fragments, soil and soil-parent materials. Sometimes the overburden materials contain toxic substances due to the admixture of ore materials. These toxic substances may have secondary effects on the nearby environment through leaching and fugitive emissions. Covering such overburden with vegetation could stabilize the dump, reduce emissions of particles, erosion, leaching of potentially toxic substances, etc. Natural succession of the mine spoil takes longer to achieve, so planned revegetation after land treatment is highly recommended.

According to the data produced by the Indian Council of Agricultural Research (ICAR), the total land degradation due to mining now amounts to 0.26 million ha. Although the share of mining to the overall degradation of land is less, there are many associated and ongoing problems like erosion, blocking of drainage channels, acid drainage, leaching of chemicals (e.g. sodium aluminates and carbonate, from bauxite mines), deforestation, disturbed biodiversity and ecosystems, degraded quality of water resources and damage done to livestock and human health. Added to these, the pollution of air, water and soil is widespread (MoM, 2016). The Indian government's Ministry of Mines is playing a pivotal role in guiding the rehabilitation of mine land. It has to be achieved by enforcing

DOI: 10.1201/9781315141855-7

various policies, promoting research and new technologies and involving many stakeholders. This chapter reviews the mine land reclamation practices in India covering mine lifecycle, land area affected by mining, government policies and guidelines and some case studies on mine reclamation.

7.2 MINERAL PRODUCTION AND MINE LANDS IN INDIA

Total coal production in India was 606.89 million tons, of which 94.4% is non-coking coal and 5.6% is coking coal. According to the 2016–17 data (IBM, 2019a), there are 476 coal mines in India, most of which are located in the states of Jharkhand (132), West Bengal (76), Madhya Pradesh (64), Maharashtra (57), Chhattisgarh (55), Telangana (47) and Odisha (29). There are 19 lignite mines operating in Gujarat (10), Rajasthan (6) and Tamil Nadu (3). In the current financial year (2020–21) to date, Coal India Limited has produced 240.93 million tons of coal, where 94% has originated from opencast mines. Most of the coal and lignite are produced from opencast mines, which creates a lot of land disturbance. Regarding coal mines, in 2018–19, the total excavated area under Coal India Limited was 4162.63 hectares. Out of the total mined-out land area, 12.29% has been reclaimed with plants, 47.17% is under backfilling for reclamation and 40.54% is under active mining (CMPDI, 2020).

Like coal, steel production is an important indicator of a nation's economic progress. Indian steel production rose from 1.1 million tons in 1951 to 104.98 million tons in 2017–18, and the corresponding iron ore production is 200.955 million tons. Total limestone production in the country amounts to 3 million tons. There were about 711 mines and Rajasthan is the leading limestone-producing state, contributing 22% of the total production. Other major states are Madhya Pradesh (13%), Andhra Pradesh (11%), Chhattisgarh (11%), Karnataka (9%), Telangana (8%), Gujarat (7%) and Tamil Nadu (6%) (IBM, 2019b). Production of coal, limestone, iron ore or any other minerals leads to the generation of huge amounts of mine spoil and land degradation. In India, about 0.14% of the total land area is under mining lease arrangements (excluding fuel, atomic and minor minerals). The year-wise distribution of mines in India is presented in Figure 7.1. There are 1,405 mines in India out of which 780 are non-metallic and 625 are metallic operations (MoM, 2019). Data on the mining lease and area under different minerals are presented in Figure 7.2. Mining f limestone (0.165 million ha), iron ore (0.073 million ha), bauxite (0.028 million ha) and manganese (0.014 million ha) are the major sectors that occupy huge tracts of land. About 50% of the mines are categorized as > 500 ha/mine, while 20% of the mines can be classified in the 200–500 ha/mine group (MoM, 2019).

According to the Indian Bureau of Mines (IBM), there are 297 abandoned mines which have not yet been reclaimed. However, a few mines (24) continue to be operational, so the remaining mines

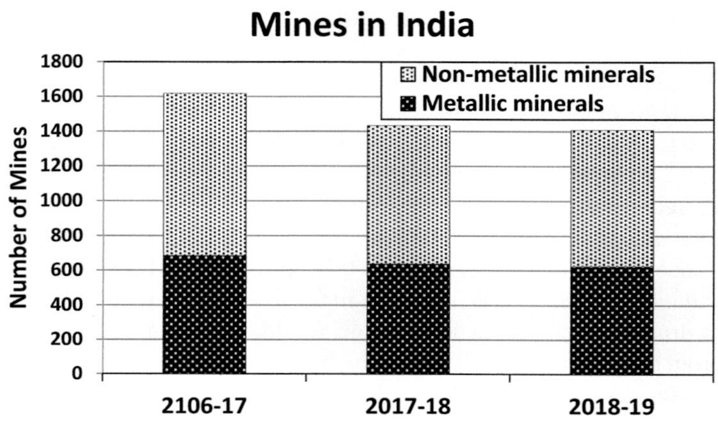

FIGURE 7.1 Metallic and non-metallic mines in India.

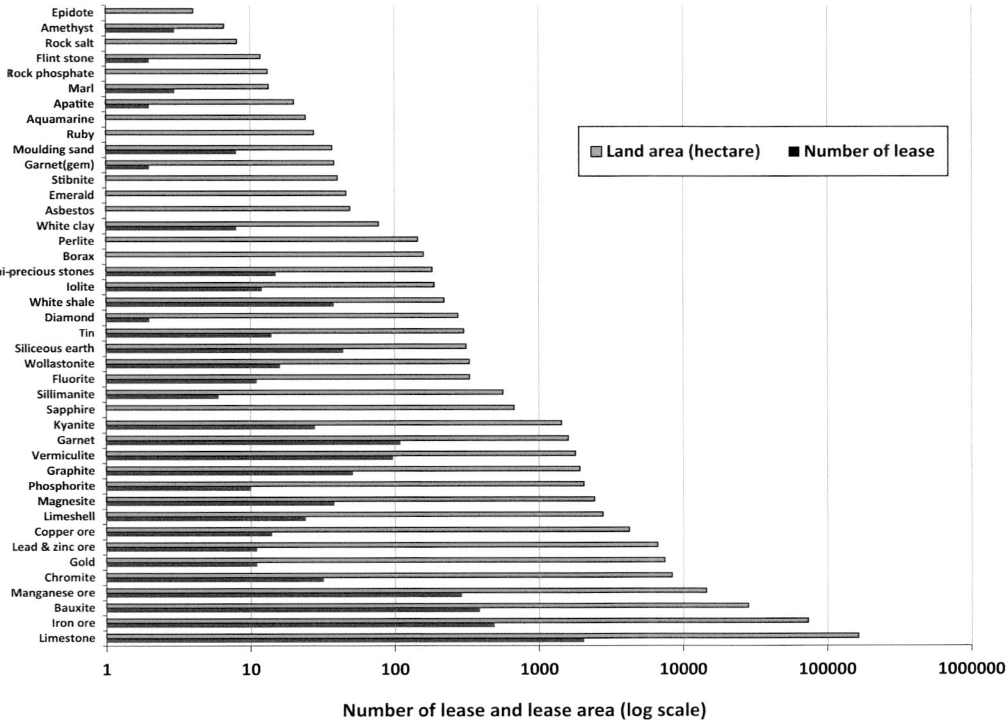

FIGURE 7.2 Data on mine leases for various minerals (excluding atomic, coal, lignite, petroleum, naturals gas and minor minerals) [source: MoM (2019)].

are yet to be reclaimed. These mines are located mostly in Madhya Pradesh, Andhra Pradesh, Tamil Nadu and Jharkhand (IBM, 2020). As stated in the Wastelands Atlas of India, which is produced by the Department of Land Resources, mining wasteland includes mine dumps consisting of rocks and stone quarries, sand and gravel, soil, etc. The total land area categorized as mining wasteland amounts to 2256.3 km^2, which is equivalent to 0.07% of the total geographical area of India (DoLR, 2019).

7.3 MINING CYCLE

Extraction of minerals through mining requires proper planning and execution. The various stages of mining combine to form a cycle (Fig 3) having four stages: exploration, development, extraction, closure and reclamation. Mining is a temporary usage of land and, after mining operations have ceased, the land is theoretically available for other purposes. However, massive rehabilitation of the disturbed land is essential for converting the mined-out land into usable land. In the first stage, exploration includes identification and mapping of suitable land sites for the proposed mining operations. This can take a few to several years of research. Modern advances in technology have helped to cut short the time consumed in this process. Modern techniques like geological and satellite mapping have made it easier to identify the best sites for carrying out mining. After successful exploration of the mining site, a proper set of data is made in order to plan the expenditure and expected profits.

The development stage includes the planning of mine construction and extraction processes (Figure 7.3). When the blueprints are made, the mine is to be developed, which requires a huge amount of manpower and investment. Scientists, geologists and experts from various fields develop and construct the mine for proper and efficient extraction of minerals or ores. It must be kept in mind that

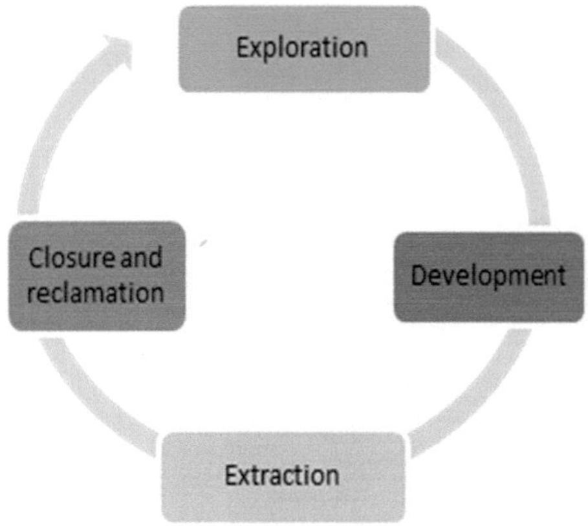

FIGURE 7.3 Flow chart of the mining cycle.

necessary legal obligations like proper licensing and environmental care must be taken into account before the actual mining. The extraction or production stage is where the actual mining of the valuable materials beneath the surface is carried out. The duration and cost of extraction depends on the assumptions made during the development stage, such as how far and to what extent the mining is to last. In some mines, the exploration and development stages go hand in hand with the extraction stage.

Once the extraction from the mine is complete, the next and last stage is to restore the mining site. As the depleted mines are of no use and the global aim is to ensure sustainable development of an economy, it is very necessary that after the mine is deemed completely depleted the site must be remediated for natural vegetation to occur. There should be a plan and sufficient resources in place to carry out reclamation. Reclamation should be planned for at the beginning of the mining project. This includes backfilling the mine, contouring slopes, revegetating and allowing for natural flooding of the open pit.

7.4 MINE CLOSURE AND REHABILITATION POLICIES

Post-independent India realized the need to expand and modernize its mines for more efficient production of minerals. Coal and metal mining industries were the key to economic growth, to boost people's livelihoods and supply them with fuel and raw materials so that the railways and manufacturing industries remained viable. Extensive mineral exploration, opening of new mines and mechanization of production and transport facilities were initiated to meet the rising production targets. Mine leases were largely owned by private sector interests and often operated to simply pursue profits while compromising on safety and protecting the environment. This has resulted in several serious – if not catastrophic - accidents and loss of lives and capital.

The government enacted legislation and regulations to reform the mining industry's woeful safety practices and workers' welfare and employment conditions. Central and state governments jointly regulate the Indian mining sector through several policies, acts, rules and regulations. The very first was the Mines Act, 1952, enacted by the Ministry of Labor and Employment, on March 15, 1952, and enforced on July 1, 1952. This statute mainly focused on safety and labor regulations. The Act defines the duties and responsibilities of mines owner as well as Agents or Managers, Inspectors, Surgeons and other statutory bodies. The Act also makes laws about health and safety issues in mines and illustrates various matters related to laborers like minimum working hours and wages, etc. The enforcement of this Act occurred under the auspices of the Directorate General of Mines Safety (DGMS).

The Mines Act led to secondary legislation which was administered by DGMS: Mines Rules, 1955; Coal Mines Regulations, 1957; Metalliferous Mines Regulations, 1961; Mines Vocational Training Rules, 1966; Mines Creche Rules, 1966; Oil Mines Regulations, 1984; and Mines Rescue Rules, 1985. The Mines Act thus covers issues concerning mines regulations, workers' safety and welfare, child labor, etc. However, the very important aspect of environmental issues in and around mining areas and rehabilitation of the project affected persons was ignored. A major landmark development by the central government came in the form of legislation titled Mines and Minerals (Development and Regulation) Act, 1957 (MMDR Act), which set out to provide necessary legislative support for the mining sector (MoM, 1957). This act recognized that environmental degradation was being perpetrated by mining companies and the state authorities had to do something about reclaiming and remediating mined-out areas.

Under the aegis of MMDR, 1957, the government of India, on October 24, 1988, enforced the rule for conservation and development of minerals, namely; Mineral Conservation and Development Rule, 1988. This is the first rule which incorporated the concerns about the environment and pollution of it, arising from the Environment (Protection) Act, 1986. This remarkable change in mining regulations was very much linked to the landmark 1987 report "Our Common Future" produced by the Brundtland Commission. The report successfully linked the environment to the increasingly recognized concept of "sustainable development". The report strongly advocated the implementation of institutional and legal changes to protect the environment. In subsequent action, the government of India amended the Mineral Concession Rule 1960 and Mineral Conservation and Development Rule 1988, and provided guidelines for when mines were to be closed. These amendments led to the submission of the "Progressive Mine Closure Plan" and "Final Closure Plan", which became mandatory as required by the Indian Bureau of Mines (IBM). The "Progressive Mine Closure Plan" has to be submitted within 180 days from the date of notification, along with prescribed financial securities. Further, the mining lessee must submit its "Final Mines Closure Plan" two years prior (Rule 24 of MCDR) to the actual closure of a mining site.

Another milestone achieved by the "World Summit on Sustainable Development 2002" (WSSD, 2002) took the form of the Johannesburg Plan of Implementation (JPOI). This focused on the environmental issues of mining and workers' health and safety, and advocated to establish partnerships between governments, mining industries and all other stakeholders such as indigenous communities throughout the lifecycle of mine operations. Included here were mine closure and rehabilitation, value added processing, reclamation and rehabilitation of degraded sites. In the lifecycle the main phases are: site exploration; development; active mining; disposal of overburden and waste rock; ore extraction and transportation; beneficiation (i.e. transformation of a mineral (or a combination of minerals) to a higher value product); tailings disposal; and reclamation and closure of the site and where the specific environmental footprint for each phase is documented. Although it is the last phase of a mine lifecycle the mine closure plan process should commence before the extraction of minerals and should continue through different stages of the mining cycle till the final closure (Asr et al., 2019; Srikanth and Nathan, 2018).

7.4.1 Progressive Mine Closure Plan (PMCP)

MCDR 2017 defines the "progressive mine closure plan" as a plan aiming to introduce various steps to protect, reclaim and rehabilitate mine lands. This plan should be prepared in tandem with the guidelines issued by the Indian Bureau of Mines or the Director, Atomic Minerals Directorate. The plan covers various land use activities to be done continuously throughout the mining cycle. The IBM guidelines describe that PMCP will be reviewed every five years during the mining cycle. Generally, the PMCP may include environmental baseline information, impact assessment, progressive reclamation plan, disaster management and risk assessment, emergency management plan and financial assurance matters.

7.4.2 Final Mine Closure Plan (FMCP)

MCDR 2017, through Rule 21(4), states that before the closure of a mining project, the FMCP is to be implemented, and all the protective, reclamation and rehabilitation work carried out should be certified by the appropriate authority. FMCP shall be submitted two years prior to the lease expiry along with all the letters/formats for approval/rejection/bank guarantee, etc., and relevant procedures *shall be consistent with the* Mining Plan (IBM, 2018). The proposals documented in the FMCP shall follow the directions of the IBM, 2018 guidelines as stated below:

- The FMCP document shall begin with information about reasons for closure, statutory obligation and closure plan preparation.
- It shall contain descriptions about the mine, local and regional geology, reserves, mining method and mineral beneficiation.
- Appropriate methodology for review of implementation of mining plan / scheme of mining, including five years PMCP to final closure of the mine.
- Closure plan as stated in the document shall contain details about land area affected by mining, water and top soil management, waste management, tailing dam management, infrastructure, disposal of mining machinery, safety and security, disaster management, risk assessment and care and maintenance during temporary discontinuance.
- Economic implications of mine closure and manpower retrenchments shall be dealt with in order to take care of post-closure economic, social and safety issues. Information concerning local residents, their occupations, compensation given, satellite occupation, rehabilitation status and any anticipated impact on nearby communities.
- The document shall contain the schedule for the site's intended abandonment.
- The cost of abandoning the mine is to be estimated based on activities required for implementing protective and rehabilitation measures; this involves the post-closure maintenance and monitoring for three years, etc. Financial assurance by the bank is mandatory and the mining company shall open an Escrow account in any nationalized bank.
- A certification is issued under rule 21(4) of MCDR, 2017, if the protective, reclamation and rehabilitation work by mine lessee has been found, after verification, to be in accordance with the FMCP.

Mine closure must aim at rehabilitation and leave the land in a state that should not be a burden for or pose a danger to local communities. Srikanth and Nathan (2018) emphasized the need for scientific studies and enforced Indian Standards for implementing mine closures and reclamation so that the environment/ecology was protected as much as possible.

7.4.3 Rehabilitation and Resettlement

Land is a finite resource and owners' rights are protected under the constitution of India. The strong dissatisfaction and resentment amongst the stakeholders with the Land Acquisition Act, 1894, forced policymakers to formulate and implement a new land acquisition law called the Right to Fair Compensation and Transparency in Land Acquisition, Rehabilitation and Resettlement (RFCTLARR) Act, 2013. This statute is one dealing with land acquisition for industries and infrastructure development. Any land acquisition must be done with fair compensation in mind and minimal disturbance to the interests of land owners and families in consultation with local grass-roots- level institutions. There is provision for resettlement and rehabilitation of affected persons who are partners in such development, and it is intended to improve and protect their social and economic interests after land has been acquired. In order to make the rehabilitation more effective and improve the basic amenities and infrastructure in the affected area, often for underdeveloped areas, the Indian government amended the existing MMDR, 1957 through the creation of District Mineral

Foundation (DMF). This is a statutory trust, with separate allocations of funds and frameworks that operates through *Pradhan Mantri Khanij Kshetra Kalyan Yojana (PMKKKY)* (MoM, 2015). The government of India announced the launch of the PMKKKY on September 17, 2015. This program is to provide for the welfare of areas and people affected by mining-related operations, using the funds generated by District Mineral Foundations (DMFs).

In 2015, the government of India established a District Mineral Foundation (DMF) in every district throughout the nation that was affected by mining-related operations. Often, these operations largely affect less developed and remote areas and vulnerable sections of the population especially scheduled tribes. There is a need for an organized and well-structured strategy to extend the benefits of mineral wealth of a given region to the vulnerable sections of society to improve their standard of living. The DMF has been given the status of a statutory trust and has the authority to distribute funds received from lessees/licensees/permit holders/auctioneers of major and minor minerals. These funds are utilized according to the provisions of PMKKKY. The major thrusts of the PMKKKY scheme are to:

(a) Implement developmental and welfare projects/programs in mining-affected areas.
(b) minimize/mitigate environmental, health and socio-economic impacts during and after mining.
(c) ensure long-term sustainable livelihoods for affected communities in and around mining areas.

7.5 REVEGETATION OF MINE SPOIL

Open cast coal mining degrades valuable land and replaces the ecosystem with unwanted overburden rocks and soil (coaly matter/shale) in the form of mine spoil (Chaulya, 2004; Singh and Singh, 2006). Mine spoils are poor in nutrients, organic matter, water retention, microbial activity, and soil aggregation (Gonzalez-Sangregorio et al., 1991; King, 1988), resulting in soil degradation and loss of valuable topsoil (Dutta and Agrawal, 2002; Singh et al., 1996). As such *in situ* revegetation of mine spoil is required for the establishment of a self-sustaining ecosystem through environmentally friendly technologies. The successful reclamation of mining waste land is greatly shaped by appropriate plant species, apart from other edaphic and climatic parameters (Roberts et al., 1988). Revegetation of mine land helps with the biological reclamation of land by modifying the quality of mine soil, restoring soil fertility and preventing soil erosion (Bradshaw, 1997; Dutta and Agrawal, 2002). The extensive root systems of tree species penetrate the deeper layers of compacted soil and affect the water and nutrient mobility in the soil profile, making possible the accumulation of organic matter in different layers of mine spoil and establishing soil biological activity, etc. (Andrews et al., 1998; Palumbo et al., 2004). Planting mixed tree species provides support for growing new tree seedlings and it means reestablishing canopy cover (Rai et al., 2011). Hazarika et al. (2022) evaluated 42 native plant species for revegetation of backfill area of a coal mine overburden dump in Assam and found that 17 species showed 60–70 % survival after two years of planting. Inoculation of seedlings with arbuscular mycorrhizal fungi, plant growth promoting rhizobacteria, and application of lime and farm yard manure was found to be the best treatment for reclamation of overburden dump. Kumari et al. (2022) studied the steel and iron ore solid waste dump revegetated with grass (*Pennisetum pedicellatum*) and legume (*Stylosanthes hamata*) plants in central India; and found that biomass (32.5–45.8 Mg/ha) and mulch density (11.4–28.0 Mg/ha) increased significantly due to revegetation.

For revegetation of mine spoil, proper levelling, compaction and formation of suitable terraces are required depending on the existing slope. In a case study conducted in the Jharia area of Dhanbad, a total of 6500 pits (60 x 60 x 60 cm^3) at 2 m intervals were dug on the leveled and stepped dumps. The excavated pits were first filled with pesticide to inhibit the growth of soil-borne pathogens and then subsequently filled with soil, fly ash and farmyard manure (4:2:1). Other

TABLE 7.1
List of Different Plant Species Planted on Mine Spoil

S. No.	Botanical Name	Local Name	Family
1.	*Albizzia lebbek*	Kala Siris	Mimosaceae
2.	*Dalbergia sissoo*	Shisham	Papilionaceae
3.	*Cassia siamea*	Chakundi	Caesalpiniaceae
4.	*Acacia auriculaeformis*	Acacia	Mimosaceae
5.	*Delonix regia*	Gulmohar	Caesalpiniaceae
6.	*Ficus religiosa*	Pipal	raceae
7.	*Mangifera indica*	Aam	Anacardiaceae
8.	*Psidium gujava*	Amrud	Myrtaceae
9.	Ziziphus jujube	Ber	Rhamnaceae
10.	*Melia sp.*	Bakain	Meliaceae
11.	*Terminalia arjuna*	Arjun	Combretaceae
12.	*Casuarina equisitifolia*	Casuarina	Casuarinaceae
13.	*Bougainvillea sp.*	Bogainvillea	Nyctaginaceae
14.	*Callistemon lanceolatus*	Bottle Brush	Myrtaceae
15.	*Swietenia macrophyla*	Mahogany	Meliaceae
16.	*Thuja occidentalis*	Thuja	Cupressaceae
17.	*Areca sp.*	Palm	Palmae
18.	*Vetiveria zizanioides*	Khus grass	Poaceae
19.	*Cymbopogon flexuosus*	Lemon grass	Poaceae
20.	*Dendrocalamus strictus*	Bamboo	Poaceae

amendments such as coco peat (250 g/pit), *Rhizobium* + *Phosphobacterium* mixture (@ 5g each/pit) and recommended dose of fertilizers: 50:50 g/pit as urea and DAP were thoroughly mixed in order to maintain the pits' uniform fertility. Healthy saplings (1 year old) of different plant species (Table 7.1) were planted. The survival rate of plant species was about 90–95%. The increase in different morphometric parameters of planted species is attributed to steady improvement in the fertility status and biological activity in the mine refuse together with the beneficial effects of the leguminous plant species. The leaf area of the plant species plays a major role in regulating the photosynthetic/physiological activities. The number of stomata depends on the leaf area and subsequently regulates photosynthetic activities. Leaf area was maximum for *Acacia* followed by *Albizia* and *Dalbergia*. The leaf area of *Acacia* varied from 11.2 cm^2 to 29.3 cm^2; *Albizia* from 5.9 to 11.4 cm^2 and Dalbergia from 3.8 to 9.3 cm^2. Similar trends were observed for two other species, whereby *C. flexuosus* grass had a relatively higher leaf area than *V. zizaniodes*. High leaf area promotes better canopy cover and enhanced photosynthesis.

The photosynthetic rate was the highest during summer followed by the monsoon season and the least during winter. Among these plant species, in winter *Dalbergia* had the highest photosynthetic rate (6.5-9.9 mol/m^2/sec) followed by *Acacia* (5.7-9.45 μmol/m^2/sec) and Albizia (4.9-9.4 mol/m^2/sec). In summer, *Acacia* indicated the maximum photosynthetic rate (12.3-14.1 μmol/m^2/sec) followed by *Dalbergia* (11.2-13.4 μmol/m^2/sec) and *Albizia* (6.9-9.5 μmol/m^2/sec). A similar trend is also observed for the *monsoon season*. The incremental photosynthetic rate among the planted species clearly highlights the fact that CO_2 sequestration took place gradually, made possible by carbon assimilation in the plants as photosynthates. Thereby CO_2 flux, a major greenhouse gas in the surrounding environment, was reduced. It is important to note here that *Dalbergia* – being a leguminous plant (Papilionaceae) with relatively higher phytomass particularly in winter as compared to the biomass of other species – played a significant role in higher photosynthesis whereas

Acacia (Mimosaceae) showed relatively higher photosynthesis during summer and the rainy season. The reduced photosynthetic activity, in turn, discourages cambial activity and consequently curtails biomass production (Pozgaj et al., 1996). As such, the rate of photosynthesis varied from species to species and also from one growth stage to the other due to different canopy structure, leaf area and biomass.

Like photosynthesis, transpiration rate reached its maximum in summer followed by the rainy and winter seasons. As such, the transpiration rate of *Acacia* was the best followed by *Dalbergia* and *Albizia* in all seasons and age of the planted species together with the accompanied linear increase in the transpiration rate at different growth stages. The variation in the transpiration rate of these plant species was caused by the leaf area and canopy characteristics of the individual plant species. Leaves, which are the primary route of gaseous uptake, are controlled by stomatal aperture (Kangasjärvi et al., 1994). The stomatal behavior is more important in regulating CO_2 availability to the leaves than O_2 penetration. Stomata close when there is a heavy increase in CO_2 concentration (Norby and O'Neill, 1991). The opening and closing of stomata have an important effect on the regulation of photosynthesis. The stomatal conductance of *Dalbergia* is at its maximum in all seasons followed by *Acacia* and *Albizia*, which is probably due to the fact that *Dalbergia*, having greater nitrogen-fixing capacity, revealed a higher rate of biomass production. This resulted in higher stomatal conductance when compared to other species.

Singh and Kumar (2022a) evaluated the performance of exotic (*Acacia auriculiformis, Casuarina equisetifolia, Cassia siamea and Grevillea pteridifolia*) and native species (*Albizia lebbeck, Albizia procera, Dendrocalamus strictus and Tectona grandis*) on a coal mine overburden dump and found that the native species are more effective in mine spoil rehabilitation. Native species have a significantly higher survival rate, stem diameter, biomass and net primary production than exotic woody plantations (Singh and Kumar, 2022b).

Despite all the reclamation efforts by the forest department and mining industry, the satellite images showed there is a decrease in the area under dense and sparse vegetation in the coal-mining areas of Anuppur district, Madhya Pradesh (Thakur et al., 2022). In another mining site (Sohagpur and Bishrampur coal mines, central India), during the last 20 years (2001 to 2020), the forest cover decreased (5.25 and 6.02 km^2), the mine/overburden dumps have increased by 10.30 km^2 in Sohagpur and 5.38 km^2 in Bishrampur. In Jharia coalfield also Siddiqui and Jain (2022) reported significant changes in vegetation cover due to mining. Thus, the reclamation activities in the mining areas need to be intensified.

7.6 RESTORATION OF THE QUALITY OF MINE SOIL

With increases in age and growth of tree species, the plant root-soil and litter-soil interaction also increases. This interaction triggers the pedogenic process and improves physical, chemical and biological properties of the minespoil. During revegetation, some amendments like topsoil, organic manures, biosolids, biochar, nitrogen-fixing bacteria, phosphorus-solubilizing bacteria, inorganic fertilizers, etc., were also added to assist plant establishment and growth. Reclaimed chronosequence sites are good repositories for understanding the sequence of soil development in mine areas. A brief compilation of some key studies on the reclamation of mine spoil is presented in Table 7.2.

Soil quality indices (SQIs) provide integrated information on the health of the mine spoil. SQI was 0.220 in a 2-year old reclaimed site and the corresponding SQI for a 17-year old site was 0.670. SQI > 0.500 is considered to be a sign of successful reclamation (Mukhopadhyay et al., 2014). Based on the SQI, Seybold et al. (2004) concluded that lack of soil aggregation, lower soil moisture and water retention and high bulk density are the major limiting factors when a mine land is converted to agricultural use. Soil organic carbon, C:N ratio, soil microbial activity and diversity are the sensitive indicators for a restored mine land in semi-arid areas. Soil functions based on these indicators are too low when compared to the undisturbed reference site (Muñoz-Rojas et al.,

TABLE 7.2
Studies on the Reclamation of Mine Spoil

	Location of Mine	Plants Grown	Brief Improvement in Mine Soil Parameters	Reference
1.	Vishwakarma opencast mines, Dhanbad, India.	*Cassia siamea, Acaccia auriculiformis, Dalbergia sissoo, Gmelina arborea, Leucaenea leucocephala, Terminalia arjuna.*	Maximum overall improvement in soil nutrient status and biological properties was reported with *C. siamea* and *D. sissoo.*	Mukhopadhyay et al. (2013)
2.	Muraidih opencast mines, Dhanbad, India	*Acacia auriculiformis, Cassia siamea, Dalbergia sissoo, Leucaena leucocephala, Albizia lebbeck, Azadirachta indica, Delonix regia, Prosopis juliflora, Zizyphus jujube.*	Soil bulk density declined whereas WHC, organic carbon and CEC increased. Mine spoil N (+60.6%), P (+113%), K (+19.9%), Ca (+49.6%), Mg (+12.2%), S (+46.7%), dehydrogenase activity (+197%), and microbial biomass carbon (+115%) increased over the unreclaimed control site.	Mukhopadhyay et al. (2016b)
3.	Chandan opencast mines Dhanbad, India.	*Acacia auriculiformis, Albizia lebbeck, Cassia siamea, Delonix regia,* and *Dalbergia sissoo.*	Labile C, microbial biomass carbon and dehydrogenase activity were significantly higher in the rhizosphere than the non-rhizosphere soil.	Mukhopadhyay et al. (2016a)
4.	Chandan opencast mines Dhanbad, India.	Top soil addition on mine spoil for growing *Stylosanthes hamata*, and *Cenchrus ciliaris*	Legume cover contained more soil organic matter, C, and N, and the corresponding accumulation of soil C (1.57 Mg SOC ha^{-1} $year^{-1}$) and N (0.47 Mg N ha^{-1} $year^{-1}$).	Kumari and Maiti (2019)
5.	North Karanpura Area, India.	*Acacia auriculiformis, A. mangium, A. catechu, C. siamea, D. sissoo, G. arborea, L. leucocephala, Terminalia arjuna, Dendrocalamus strictus, Ziziphus jujube, Bombax malabaricum, Melia azedarach, Pongamia pinnata, etc.*	The soil quality of the 17-year old site is comparable to a nearby forest site.	Mukhopadhyay et al. (2014)
6.	Rohini opencast project, Ranchi, India.	Dominant tree species: *Cassia siamea, Dalbergia sissoo, Heterophragma adenophyllum, Acacia auriculiformis* and *Acacia mangium.*	Soil C and N content increased progressively and maximum content observed for an 11-year old site. Performance of *D. sissoo* and *C. siamea* was comparatively better.	Ahirwal et al. (2017b)
7.	Rohini OCP located south of the Damodar River in the Ranchi district of Jharkhand, India.	*Acacia auriculiformis, Cassia siamea, Acacia catechu, Azadirachta indica* and *Dalbergia sissoo.*	Soil properties and C accumulation increased progressively and maximum was observed for a 15-year old site.	Ahirwal et al. (2018)

(*Continued*)

TABLE 7.2 (CONTINUED)
Studies on the Reclamation of Mine Spoil

	Location of Mine	Plants Grown	Brief Improvement in Mine Soil Parameters	Reference
8.	Sonepur-Bazari opencast project, Raniganj, India.	*Acacia auriculiformis, A. nilotica, Albizzia lebbeck, Alstonia scholaris, Butea monosperma, Cassia siamea, Dalbergia sissoo,* and *Tectona grandis.*	Improvements in mine spoil quality parameters (organic C, N, texture, nutrients, enzyme activity, respiration, biomass C) were observed in the 2 – 21-year old reclaimed sites.	Kumar et al. (2015)
9.	Jayant Block, Singrauli, India	*A. lebbeck, A. procera* and *T. grandis.*	Improvement in soil physical properties, nutrient content, C and N accumulation was better for *A. lebbeck.*	Singh et al. (2004)
10.	Moher basin, Singrauli, India.	Grass species: *Stylohemata,* Dinanath, Sawai grass. Tree species: *Eucalyptus, Dalbergia sissoo, Pongamia* pinnata, *Prosopis juliflora, Azadirachta indica, Emblica officinalis, Cassia* siamea, *Leucena leucocephala, Acacia catechu, Shorea robusta,* Bamboo, *Madhuca* indica, *Ficus religiosa, Ficus bengalensis, Delonix regia, Cassia fistula, Gmelina arborea,* etc	After 19 years of reclamation, soil acidity (13%), sand content (6%), and bulk density (8%) decreased, whereas clay (8%), silt (39%), N (214%), P (62%), C (4 times), microbial biomass C (6.6 times) increased significantly.	Tripathi et al. (2016)
11.	Ramagundam project, Singareni, India.	*Prosopis juliflora, A. arabica, A. auriculiformis, A. indica, C. fistula, D. sissoo, D. regia, Eucalyptus sp., L. leucocephala, P. pinnata, P. juliflora* and *T. arjuna.*	C, N, and P stocks in the mine spoil increased progressively and maximum was observed for the 8 years old site.	Ahirwal et al. (2017a)

2016). SQI of the reclaimed soils varied from 68 to 87 on a scale from 0 to 100. Total N and cation exchange capacity are the main indicators of SQI for a reclaimed coal mine land. SQI was high for an 18-year-old reclaimed site (0.651) when compared to a natural forest (0.575) (Noviyanto et al., 2017). SQI and plant biodiversity were the highest in a 15-year-old reclaimed site (Ahirwal et al., 2018). In another study, SQI was highest in the forest (0.4-1.0); intermediate values were observed for 16-20-year-old reclaimed mine sites (0.4-0.7), while the lowest SQI was observed for less than 1-year-old site (0.1 to 0.3) (Domínguez-Haydar et al., 2019).

Evaluation of key soil indicators after 16 years of reclamation strongly suggested that the addition of organic matter during revegetation has only a short-term benefit in restoring the quality of mine land. In the long run, organic matter derived from revegetated plant species is a crucial factor in restoring the biological quality of mine spoil (Bendfeldt et al., 2001). Consequently, the selection of appropriate plant species for revegetation is very important and will complement the amendments used. The effects of *Acacia auriculiformis, Cassia siamea, Dalbergia sissoo* and *Leucaena* on the quality of a mine land were evaluated. Some key soil quality indicators like soil coarse fraction, pH, EC, organic C, phosphorus, calcium, sulfur and dehydrogenase activity were integrated into the soil quality index (SQI). When compared with the reference site, SQI was 93.1% higher for *D. sissoo* and 86.4% higher for *C. siamea*. It is recommended that these two tree species are planted for future reclamation activities of the mine land (Mukhopadhyay et al., 2016b). In a 25-year-old reclaimed mine site of Gevra, Chhattisgarh, Singh et al. (2022) reported that the microbial biomass carbon (1468.45 mg C/kg) and soil organic carbon stock (334.72 Mg C/ha) were highest under *A. indica*. Spectroscopic analysis showed the recalcitrant nature of the soil organic carbon in the 25-year-old site. In a chronosequence (3–16 years) site, Mukhopadhyay and Masto (2022) found that *C. siamea* performed comparatively better than *A. lebbeck,* wherein the age of the plant played a major role in increasing the soil C and nutrient stocks.

7.7 SUMMARY

Mining disturbs many soil functions, compromises the quality of surrounding land and has a huge impact on adjacent communities. Mine reclamation and rehabilitation plans must be devised and agreed to during the early stages of the mining cycle. The government of India has introduced several guidelines for the mining industry for proper mine closure and land rehabilitation and resettlement strategies. Revegetation of the disturbed land with viable plant species and the use of organic and inorganic amendments are required to save the ecological systems of disturbed areas. As the human demand for fuels and minerals rises, more and more mining wasteland will be generated in the coming years, so novel technological interventions and stringent enforcement of government policies are required if degraded land is to be remediated and has a viable future.

REFERENCES

Ahirwal, J., Kumar, A., Pietrzykowski, M., Maiti, S.K., 2018. Reclamation of coal mine spoil and its effect on Technosol quality and carbon sequestration: A case study from India. *Environmental Science and Pollution Research* 25, 27992–28003.

Ahirwal, J., Maiti, S.K., Satyanarayana Reddy, M., 2017a. Development of carbon, nitrogen and phosphate stocks of reclaimed coal mine soil within 8 years after forestation with Prosopis juliflora (Sw.) Dc. *Catena* 156, 42–50.

Ahirwal, J., Maiti, S.K., Singh, A.K., 2017b. Changes in ecosystem carbon pool and soil CO2 flux following post-mine reclamation in dry tropical environment, India. *Science of The Total Environment* 583, 153–162.

Andrews, J.A., Johnson, J.E., Torbert, J.L., Burger, J.A., Kelting, D.L., 1998. Minesoil and site properties associated with early height growth of eastern white pine. *Journal of Environmental Quality* 27, 192–199.

Asr, E.T., Kakaie, R., Ataei, M., Mohammadi, M.R.T., 2019. A review of studies on sustainable development in mining life cycle. *Journal of Cleaner Production* 229, 213–231.

Bendfeldt, E.S., Burger, J.A., Daniels, W.L., 2001. Quality of amended mine soils after sixteen years. *Soil Science Society of America Journal* 65, 1736–1744.

Bradshaw, A., 1997. Restoration of mined lands—using natural processes. *Ecological Engineering* 8, 255–269.

Chaulya, S.K., 2004. Assessment and management of air quality for an opencast coal mining area. *Journal of Environmental Management* 70, 1–14.

CMPDI, 2020. Land restoration/reclamation monitoring of 30 opencast coal mines and 08 clusters of CIL producing less than 5 mcm (Coal+OB) based on satellite data for the Year 2018–19. (Available at https://www.coalindia.in/DesktopModules/DocumentList/documents/Sat.%20Monitoring%20of%20Reclamation%20less%20than%205%20MCM%20for%202018-19.pdf).

DoLR, 2019. Wastelands Atlas of India. Department of Land Resources. Ministry of Rural Development, Government of India. (Available at https://dolr.gov.in/documents/wasteland-atlas-of-india).

Domínguez-Haydar, Y., Velásquez, E., Carmona, J., Lavelle, P., Chavez, L.F., Jiménez, J.J., 2019. Evaluation of reclamation success in an open-pit coal mine using integrated soil physical, chemical and biological quality indicators. *Ecological Indicators* 103, 182–193.

Dutta, R.K., Agrawal, M., 2002. Effect of tree plantations on the soil characteristics and microbial activity of coal mine spoil land. *Tropical Ecology* 43, 315–324.

Gonzalez-Sangregorio, M.V., Trasar-Cepeda, M.C., Leiros, M.C., Gil-Sotres, F., Guitian-Ojea, F., 1991. Early stages of lignite mine soil genesis: Changes in biochemical properties. *Soil Biology and Biochemistry* 23, 589–595.

Hazarika, P., Dutta, D., Hazarika, P., 2022. Reclamation of coal mined land with integrated biological approach. *Indian Forester* 148, 313–321.

IBM, 2018. Final Mine Closure Plan (Exclusively for leases wherein mineral exists and closure is as per the provisions of Section 8A of MMDR Amendment Act, 2015). IBM, Ministry of Mines, Govt. of India, Nagpur.

IBM, 2019a. Indian minerals yearbook 2018 (Part- III: Mineral Reviews) 57th edition. Coal and Lignite. Government of India, Ministry of Mines, Indian Bureau of Mines.

IBM, 2019b. Indian minerals yearbook 2018 (Part- III: Mineral Reviews). 57th edition. Limestone & Other Calcerous Materials. Government of India Ministry of Mines, Indian Bureau of Mines.

IBM, 2020. Abandoned mine sites. Indian Bureau of Mines, Ministry of Mines, Government of India (Available at https://ibm.gov.in/index.php?c=pages&m=index&id=90&mid=23897).

Kangasjärvi, J., Talvinen, J., Utriainen, M., Karjalainen, R., 1994. Plant defence systems induced by ozone. *Plant, Cell & Environment* 17, 783–794.

King, J.A., 1988. Some physical features of soil after opencast mining. *Soil Use and Management* 4, 23–30.

Kumar, S., Maiti, S.K., Chaudhuri, S., 2015. Soil development in 2–21 years old coalmine reclaimed spoil with trees: A case study from Sonepur-Bazari opencast project, Raniganj Coalfield, India. *Ecological Engineering* 84, 311–324.

Kumari, S., Maiti, S.K., 2019. Reclamation of coalmine spoils with topsoil, grass, and legume: A case study from India. *Environmental Earth Sciences* 78, 429.

Kumari, S., Ahirwal, J., Maiti, S.K., 2022. Reclamation of industrial waste dump using grass-legume mixture: An experimental approach to combat land degradation. *Ecological Engineering* 174, 106443.

MoM, 1957. Mines and minerals (Development and regulation) Act, 1957 (As amended up to 11th July, 2015) (Available at https://mines.gov.in/writereaddata/UploadFile/ MMDR %%20 Act,1957.pdf).

MoM, 2015. Pradhan Mantri Khanij Kshetra Kalyan Yojana (PMKKKY). No./16/7/2015-M.VI (Part), Govt. of India, Ministry of Mines, September 16, 2015. .

MoM, 2016. Proceedings of the Workshop on "Remediation off Mined–Out and Abandoned Mines–Status and Strategies" held at India Habitat Centre on 19th Dec 2016. Ministry of Mines, Government of India (prepared by TERI, New Delhi) (Available at https://mines.gov.in/UserView/index?mid=1570).

MoM, 2019. Annual report 2018–19. Government of India, Ministry of Mines. (Available at https://mines.gov .in/writereaddata/UploadFile/English637094270360450099.pdf).

Mukhopadhyay, S., Maiti, S.K., Masto, R.E., 2013. Use of Reclaimed Mine Soil Index (RMSI) for screening of tree species for reclamation of coal mine degraded land. *Ecological Engineering* 57, 133–142.

Mukhopadhyay, S., Maiti, S.K., Masto, R.E., 2014. Development of mine soil quality index (MSQI) for evaluation of reclamation success: A chronosequence study. *Ecological Engineering* 71, 10–20.

Mukhopadhyay, S., Masto, R.E., Cerdà, A., Ram, L.C., 2016a. Rhizosphere soil indicators for carbon sequestration in a reclaimed coal mine spoil. *Catena* 141, 100–108.

Mukhopadhyay, S., Masto, R.E., Yadav, A., George, J., Ram, L.C., Shukla, S.P., 2016b. Soil quality index for evaluation of reclaimed coal mine spoil. *Science of The Total Environment* 542, 540–550.

Mukhopadhyay, S., Masto, R.E., 2022. Comparative evaluation of Cassia siamea and Albizia lebbeck for their potential to recover carbon and nutrient stocks in a chronosequence post-mining site. *Catena* 208, 105726. https://doi.org/10.1016/j.catena.2021.105726

Muñoz-Rojas, M., Erickson, T.E., Dixon, K.W., Merritt, D.J., 2016. Soil quality indicators to assess functionality of restored soils in degraded semiarid ecosystems. *Restoration Ecology* 24, S43–S52.

Norby, R.J., O'neill, E., 1991. Leaf area compensation and nutrient interactions in CO2-enriched seedlings of yellow-poplar (Liriodendron tulipifera L.). *New Phytologist* 117, 515–528.

Noviyanto, A., Purwanto, P., Minardi, S., Supriyadi, S., 2017. The assessment of soil quality of various age of land reclamation after coal mining: A chronosequence study. *Journal of Degraded and Mining Lands Management* 5, 1009–1018.

Palumbo, A., Mccarthy, J.F., Amonette, J.E., Fisher, L., Wullschleger, S.D., Daniels, W.L., 2004. Prospects for enhancing carbon sequestration and reclamation of degraded lands with fossil-fuel combustion by-products. *Advances in Environmental Research* 8, 425–438.

Pozgaj, A., Iqbal, M., Kucera, L., 1996. Development, structure and properties of wood from trees affected by air pollution. In M. Yunus and M. Iqbal (Eds.), *Plant Response to Air Pollution*. John Wiley & Sons, Chichester, 395–423.

Rai, A.K., Paul, B., Singh, G., 2011. A study on physico chemical properties of overburden dump materials from selected coal mining areas of Jharia coalfields, Jharkhand, India. *International Journal of Environmental Sciences* 1, 1350–1360.

Roberts, J., Daniels, W., Bell, J., Martens, D., 1988. Tall fescue production and nutrient status on southwest Virginia mine soils. *Journal of Environmental Quality* 17, 55–62.

Seybold, C., Grossman, R., Sinclair, H., McWilliams, K., Struben, G., Wade, S., 2004. Evaluating soil quality on reclaimed coal mine soils in Indiana, National Meeting of the American Society of Mining and Reclamation and The 25th West Virginia Surface Mine Drainage Task Force. Anais do XXII National Meeting of the American Society of Mining and Reclamation, 1644–1663.

Siddiqui, A.U., Jain, M.K., 2022. Change analysis in land use land cover due to surface mining in Jharia coalfield through Landsat time series data. *Materials Today: Proceedings* 49, 3462–3468.

Singh, A.N., Raghubanshi, A.S., Singh, J.S., 2004. Impact of native tree plantations on mine spoil in a dry tropical environment. *Forest Ecology and Management* 187, 49–60.

Singh, A.N., Singh, J.S., 2006. Experiments on ecological restoration of coal mine spoil using native trees in a dry tropical environment, India: A synthesis. *New Forests* 31, 25–39.

Singh, R., Chaulya, S., Tewary, B., Dhar, B., 1996. Restoration of a coal-mine overburden dump-a case study. *International Journal of Rock Mechanics and Mining Sciences and Geomechanics Abstracts*, 331A.

Singh, P., Ghosh, A.K., Kumar, S., Jat, S.L., Seema, K., Pradhan, S.N., Kumar, M., 2022. A molecular and spectroscopic approach to reclamation of coal mine soil using tree species: A case study of Gevra mining area, Korba, India. *Journal of Soil Science and Plant Nutrition* 22, 2205–2220.

Singh, A.N., Kumar, A., 2022a. Comparative soil restoration potential of exotic and native woody plantations on coal mine spoil in a dry tropical environment of India: A case-study. *Land Degradation & Development* 33(12), 1971–1984.

Singh, A.N., Kumar, A., 2022b. Ecological performances of exotic and native woody species on coal mine spoil in Indian dry tropical region. *Ecological Engineering* 174, 106470. https://doi.org/10.1016/j.ecoleng.2021.106470.

Srikanth, R., Nathan, H.S.K., 2018. Towards sustainable development: Planning surface coal mine closures in India. *Contemporary Social Science* 13, 30–43.

Thakur, T.K., Dutta, J., Bijalwan, A., Swamy, S.L., 2022a. Evaluation of decadal land degradation dynamics in old coal-mines of Central India. *Land Degradation & Development*. Published on 06 June 2022. https://doi.org/10.1002/ldr.4383.

Thakur, T.K., Dutta, J., Upadhyay, P., Patel, D.K., Thakur, A., Kumar, M., Kumar, A., 2022b. Assessment of land degradation and restoration in coal mines of central India: A time series analysis. *Ecological Engineering* 175, 106493. https://doi.org/10.1016/j.ecoleng.2021.106493.

Tripathi, N., Singh, R.S., Hills, C.D., 2016. Soil carbon development in rejuvenated Indian coal mine spoil. *Ecological Engineering* 90, 482–490.

WSSD, 2002. A/CONF.199/20 – Chapter IV, Paragraph 46 of the Johannesburg Plan of Implementation (JPOI), WSSD 2002. https://sustainabledevelopment.un.org/index.php?page=view&type=2002&nr=101&menu=35.

8 Paringin Pit Lake at Coal Mine Site, Indonesia

A Sustainable Post-Mining Pit Lake Model

Didik Triwibowo, Muthia Elma, Eko Suhartono, Rony Riduan, Suhernomo, Tim Roberts and Ihsan Noor

8.1 BACKGROUND

8.1.1 Introduction

Indonesia is a country blessed with many mineral and coal resources. For coal, Indonesia has 143.7 billion tons and reserves of 38.84 billion tons, which are mainly found in Kalimantan and Sumatra (Ministry of ESDM, 2021). Coal is an integral part of the country's natural resources and mined for the people's welfare according to the mandate of the 1945 Constitution article 33, paragraph 3: Land, water and the natural resources contained therein are controlled by the state and used for the greatest prosperity of the people. According to Indonesian Law no. 4 of 2009 concerning Mineral and Coal Mining, mining is an activity within the framework of mineral or coal exploitation. It includes general investigation activities, exploration, feasibility studies, construction, mining, processing and refining, transportation and sales, as well as post-mining. Implicitly in the law of 2009 is the statement that mining not only takes the form of extracting minerals and coal for their economic benefits, but also involves post-mining activities to protect the environment and human communities. Law No. 4 of 2009 explains that post-mining is a planned, systematic and continuing activity to restore natural environmental and social functions to their previous state in all mining locations.

To provide guidelines for implementation, several regulations are relevant here as follows: Government Regulation No. 78 of 2010 concerning Reclamation and Post-mining; and Minister of Energy and Mineral Resources No. 26 of 2018 concerning the implementation of good mining principles and supervision of mineral and coal mining. This was later relegated to the Minister of Energy and Mineral Resources Decree No. 1827 of 2018 concerning Guidelines for the Implementation of Good Mining Practices contained in attachment VI.

Based on this description, it can be concluded that there is a link between the mandate of the 1945 Constitution and the technical guidelines for managing natural resources. All-natural wealth, in this case minerals and coal, is utilized for the greatest possible prosperity of the people and this needs to be translated into a more concrete explanation given that mineral and coal resources are finite and non-renewable. In 1987, the World Commission on Environment and Development published a groundbreaking report titled *Our Common Future*, which is also known as the Brundtland Report. It introduced the concept of sustainable development, namely "development that meets the needs of the present, without compromising the ability of future generations to meet their own needs" (Brundtland, 1987) as an answer to the many environmental problems resulting from industrialization and its destruction of flora, fauna and ecosystems. For this reason, the Indonesia government

DOI: 10.1201/9781315141855-8

seeks to strike a balance between harnessing economic development from its abundant coal and mineral resources and safeguarding the environment.

This chapter presents a compelling example of a successful and sustainable coal pit lake, serving as best practice within the context of a developing country like Indonesia, which has a tropical climate. This case serves as an integrative model for post-mining lakes. It demonstrates that maintaining a pit lake in good condition can yield environmental and social benefits for local communities, within not only the mining concession area but also downstream, extending beyond the concession area. This model offers valuable guidance to coal mining companies that are confronted with the task of establishing post-mining lakes at their sites. Ultimately, it contributes to addressing sustainable development goals within the mining sector.

8.1.2 Pit Lake and Sustainable Development in Indonesia

The concept of sustainability in the context of non-renewable mineral and coal mining is a challenge for mining companies to realize (Adomako and Tran, 2022). Adomako and Tran (2022) state that implementing sustainable development in the mining industry means taking into account environmental sustainability. The post-mining phase is the most crucial stage in the successful implementation of sustainable development. In areas that depend economically on mining, closures without any planning have resulted in areas experiencing pollution, destruction, toxicity, etc. (Syahrir et al., 2021). Syahrir et al. (2021) suggest strategies to encourage community/community-focused sustainable development, including sharing of mining facilities, use of mining facilities to increase workers' expertise and creation of a sense of community empowerment. What is very important is collaboration between companies and the government to get the best results. Mining activities are important for economic progress and the products of mining have been intrinsic to human life for millennia (de la Torre de Palacios and Espí Rodríguez, 2022). In Indonesia, this is evident in the tin mine that is located on Bangka Island (Sulista and Rosyid, 2022). However, the negative outcomes of open-pit mining are in the form of damaged landscapes wherein the bio-geophysical and social features in and around a community are compromised, so this needs to be mitigated through environmental and social recovery programs (Worlanyo and Jiangfeng, 2021; Hosseinpour et al., 2022). Without adequate restoration work in the form of post-mining planning, the damaging environmental, economic and social impacts after mine closure will continue to be felt by the community, as happened at the tin mining location on Singkep Island, Indonesia (Syahrir et al., 2020).

On the island of Kalimantan where there are many coal mining operations, a significant environmental outcome is shown by the abandoned coal mining pits on the eastern side, reaching 1,735 sites. These locations resulted in people drowning in ex-mining pits (Toumbourou et al., 2020). This fact attracted the attention of all levels of society, non-governmental organizations (NGOs), government officials and academics to find solutions to resolve the dangers posed by undetected post-mining lakes. It can be concluded that post-mining pits and post-mining lakes from an environmental, safety and social perspective are real problems but at the same time have the potential to provide benefits in the future if managed properly. To do this, it is necessary to plan for the restoration of environmental and social functions on a landscape scale (Kodir et al., 2017).

The mining activities responsible for extracting these resources create serious problems, affecting both the biogeophysical aspects of post-mined areas and social dimensions (Hosseinpour et al., 2022; Worlanyo and Jiangfeng, 2021). Particularly, open-pit mining operations may lead to the formation of expansive and deep voids, which, when inundated with water, can transform into pit lakes (Lund et al., 2020; McCullough and Vandenberg, 2020; McCullough, 2015; Sakellari et al., 2021). Pit lakes raise not only environmental concerns regarding water quality but also pose safety risks to the public (Marove et al., 2022; Olías et al., 2019; Toumbourou et al., 2020; Vandenberg et al., 2022). In the biodiverse hotspot of East Kalimantan, an area spanning 5,000,000 hectares was allocated for surface coal mine concessions (Woodbury et al., 2020). Abandoned pit lakes resulting from open-pit coal mining have become ecological challenges confronting the government

(Toumbourou et al., 2020). Open-pit coal mining operations have given rise to Acid Mine Drainage (AMD) as a water quality issue, which can endanger both human life and the environment (Elma et al., 2020). The source of acidic water stems from the disturbance of original rock layers during coal mining, exposing sulfidic materials that elevate metal content in waste dumps or overburden piles (Blodau, 2006; Moodley et al., 2018; Santofimia et al., 2013). Erosion of these sulfidic waste dump materials can trigger metal leaching and increased metal concentrations in water and sediments, raising concerns about their potential effects on ecosystems and human health (Marove et al., 2022; Rybnikova and Rybnikov, 2020).

Nevertheless, numerous examples exist of mine rehabilitation, including the creation of pit lakes as artificial bodies of water, which can bring about benefits for the environment, economy and society (Bernasconi et al., 2022; Blanchette and Lund, 2016; Keenan and Holcombe, 2021; Rosa et al., 2020; Seifollahi-Aghmiuni et al., 2019; Sjöberg et al., 2022). Natural lakes hold vital roles as primary components in river networks within landscapes (Huang et al., 2019). Similarly, pit lakes function as repositories for runoff water and groundwater from the surrounding area, making the land cover condition in the catchment area a critical factor influencing the quantity and quality of pit lake water (Grzybowski et al., 2023; Schmidt et al., 2022). Pit lakes can primarily be classified based on pit lake water balance as evaporative sinks, sources (surcharged) perched above local groundwater levels, or flow-through pit lakes ((McCullough and Schultze, 2018). Given that runoff from the catchment area. into the pit lake serves as a crucial source of pit lake water, the hydrological connectivity approach should be utilized to assess the interaction (Bracken et al., 2013).

In Indonesia, it is important to create a more sustainable post-mining future as thousands of former coal mining voids are on track to become pit lakes. To ensure their sustainable utilization, each prospective pit lake should meet specific criteria to be considered successful.

8.1.3 Post-Mining Success Criteria

According to Minister of Energy and Mineral Resources Decree 1827 of 2018 Appendix VI, the criterion for post-mining success is restoring not only environmental functions but also sociocultural circumstances. For this reason, it is important for mine companies to have a Mine Closure Plan and its comprehensive implementation in place. The restoration of environmental functions of all ex-mining land, both land and water, and the social aspect, in the form of the readiness of relevant stakeholders, especially the community that exists around mining operations, can benefit viably in the long term. Mining companies need to devise integrated strategies to restore the natural environment in tandem with socioeconomic requirements.

One approach that can be used to integrate post-mining objectives in a comprehensive and sustainable manner is a landscape restoration approach that focuses on improving ecological functions, and conserving biodiversity without neglecting aspects of human welfare (Chen et al., 2021). This is also in line with the post-mining objectives which aim to restore environmental and social functions according to local conditions. Ideally, this should be adopted by mining companies in their Mine Closure Plan (MCP) documents. In addition to these documents, mining companies must have a five-year Reclamation Plan document. This Reclamation Plan document forms the basis of annual reclamation implementation which is included in the company's Yearly Work Plan and Budget document. It is important to get approval from the Ministry of Energy and Mineral Resources. A mining company during its operational activities must carry out reclamation and post-mining program. Restoration of ex-mining land begins with reclamation so that the environment can start to be looked after.

Reclamation is an activity carried out throughout the stages of mining to organize, restore and improve the quality of the environment and ecosystem (Ministry of ESDM, 2018). In general, reclamation success indicators are divided into three parts, namely land management, revegetation and final settlement with respective weights of 60%, 20% and 20%. Assessing the success of reclamation is carried out by the Ministry of Energy and Mineral Resources at least one year after the

FIGURE 8.1 Conceptual coal mine operation and reclamation options in Indonesia.

reclamation activities are carried out as outlined in the Assessment Report. Success is stated as a percentage according to the indicators of land management, revegetation and final completion. Mining companies can generally obtain a 100% success rate over a period of 6–10 years, evidence for this being thorough revegetation of the canopy (Figure 8.1).

However, in areas where revegetation is not feasible, the Indonesian government offers alternative forms of reclamation for post-mining area utilization, including freshwater sources, tourism, productive land and settlements. The success criteria for these alternative reclamation methods are determined by mining companies, taking into account local conditions, and must receive government approval. Regarding pit lakes, the success criteria pertain to both water quantity and quality, ensuring their suitability for use as freshwater sources, as well as for eco-tourism and productive freshwater fishery (Figure 8.1).

8.2 PARINGIN PIT LAKE

8.2.1 Site Location

Paringin pit lake is geographically located at Latitude 2^0 18' 48.9" S and Longitude 115^0 31' 12.4" E in South Kalimantan Province, Indonesia, covering an area of 151.82 Ha. The area comprises 134.3 Ha of rehabilitated post-mining area as the pit lake catchment area., 17.52 Ha for the pit lake itself, and a downstream area of 1.31 km in length of Nungkaran River, which receives water from the pit lake's outflow. The lake is located within PT Adaro Indonesia's (Adaro) coal concession area and it has been mined out since 2009. Adaro remains responsible for maintaining and monitoring the water quantity and quality (Figure 8.2).

8.2.2 Site Characteristics

South Kalimantan Province encompasses a total wetland area of approximately 1,194,471.98 hectares, accounting for 32.39% of the total land area. Generally, Adaro's mining operations are situated in the upstream region, characterized by hilly terrain with limited wetlands, including swamps, rivers and their banks. Pit lakes resulting from open-pit coal mining operations serve as prime examples of man-made wetlands. These areas experience temporary or permanent water inundation due

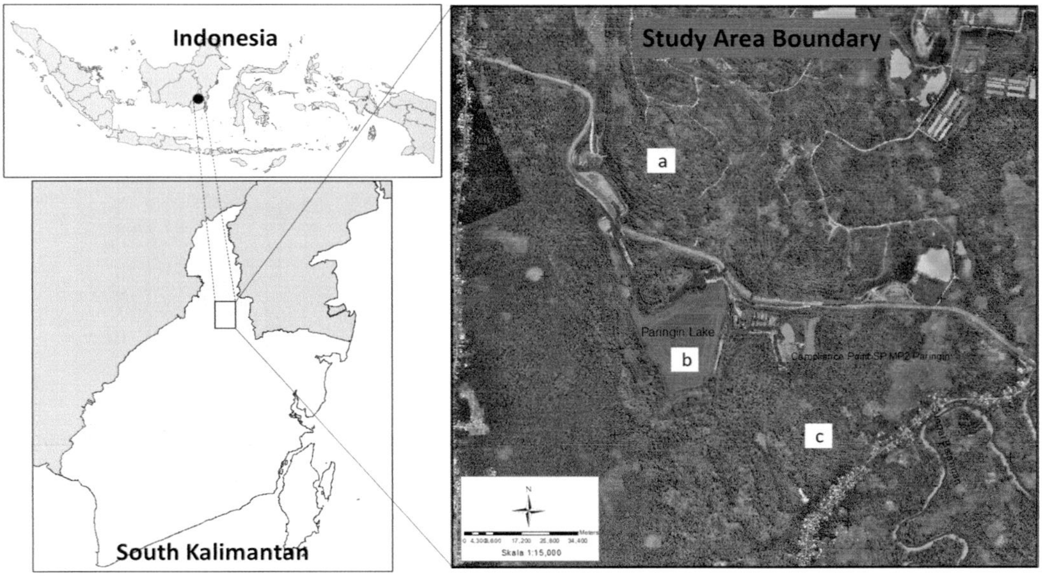

FIGURE 8.2 Site Locations of Adaro in South Kalimantan, Indonesia (left) and Paringin pit lake (b), its catchment area (a) and Nungkaran River (c) as receiving water body of Paringin pit lake (right).

to human activities, such as paddy fields, freshwater ponds, weirs, dams (reservoirs), lakes, ponds, ex-mining pits, settling ponds for mine wastewater, artificial swamps and more. Both natural and artificial wetlands can host a variety of viable ecosystems if effectively managed.

South Kalimantan experiences a tropical climate with an average annual rainfall of 2,647 mm/year. It exhibits two distinct seasons: the rainy season, influenced by monsoonal rains and characterized by peak rainfall (December, January, February), and the peak dry season (June, July, August), affected by monsoon winds from the Asian and Australian continents. The annual mean temperature in the study area, as monitored at Paringin pit lake, is 26.10°C, with recorded maximum and minimum temperatures of 26.96°C and 25.42°C, respectively.

Meanwhile, Adaro's active area covers a total of 16,000 Ha, comprising a 4,500 Ha of pit area (Tutupan, Wara and Paringin) and an 11,500 Ha of pit area that can be divided into 19 major catchment areas (Figure 8.3). It is planned to create six final voids at the end of mining operations, with the intention of repurposing them as valuable post-mining lakes or pit lakes. Presently, one of these voids in Paringin has already been turned into a pit lake, called Paringin pit lake.

The Paringin pit lake covers an area of 20 hectares and is part of a larger 150-hectare catchment area.. This pit lake has significant potential as an artificial wetland capable of providing a thriving ecosystem. Its benefits include the provision of raw water sources and fisheries, as well as serving as a habitat for flora and fauna, supporting ecotourism and offering various other activities. It plays a crucial role in hydrological functioning, essentially acting as a reservoir. Reservoirs are integral components of rural landscapes, contributing to ecosystem preservation, water retention and biodiversity conservation ((Fu et al., 2018).

As a source of freshwater, this pit lake provides a vital service, especially considering the escalating global challenges associated with water scarcity, particularly the difficulties households face in accessing clean water (Adams et al., 2022). However, it is essential to ensure that the quality of pit lake water meets the required standards. Water quality serves as an indicator of environmental health, encompassing both physical and chemical aspects, including key parameters that support aquatic organisms, such as Dissolved Oxygen (DO), temperature and pH levels ((Ullah et al., 2017).

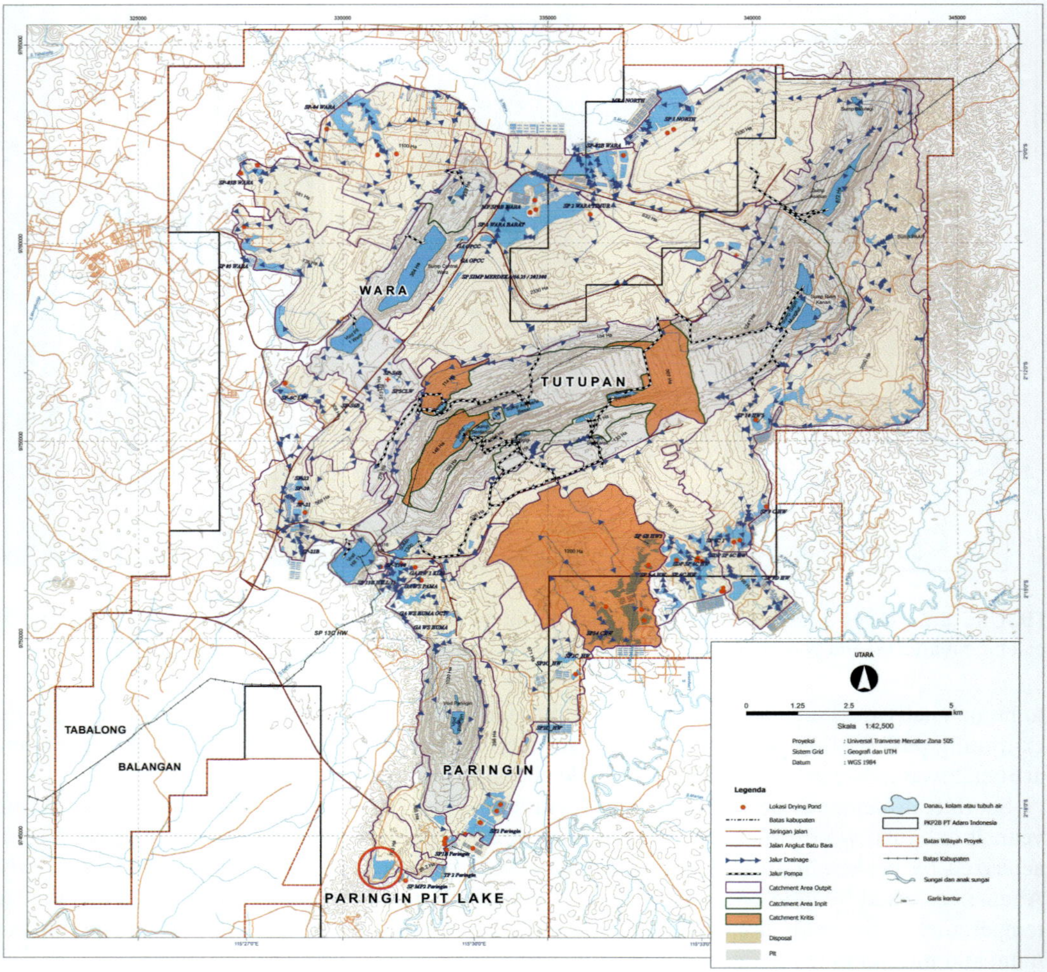

FIGURE 8.3 Adaro site features and location of Paringin pit lake (red circle).

8.3 STRATEGY FOR CREATING A SUCCESSFUL PIT LAKE

There are various strategies for creating the Paringin pit lake that can yield environmental and socioeconomic benefits for the community. Figure 8.4 illustrates the aspects that must be considered when establishing a post-mining lake to ensure sustainable outcomes for communities. These aspects include:

a. Stable Overburden/Waste Dump of Pit Lake Catchments: Ensuring the geochemical and geotechnical stability of waste dump material.
b. Catchment Management: Addressing the water system (water balance) and erosion control, as well as ecosystem reconstruction.
c. Stable Final Void Shape: Ensuring the geotechnical and geochemical stability of the final void pit wall.
d. Monitoring: Conducting daily and quarterly monitoring of water quality (pH, TDS, TSS, DO, dissolved metals, etc.) and key species of fauna.
e. Determining the Future Utilization of the Pit Lake: Engaging stakeholders in consultation, involving technical, social and economic considerations for all relevant parties.

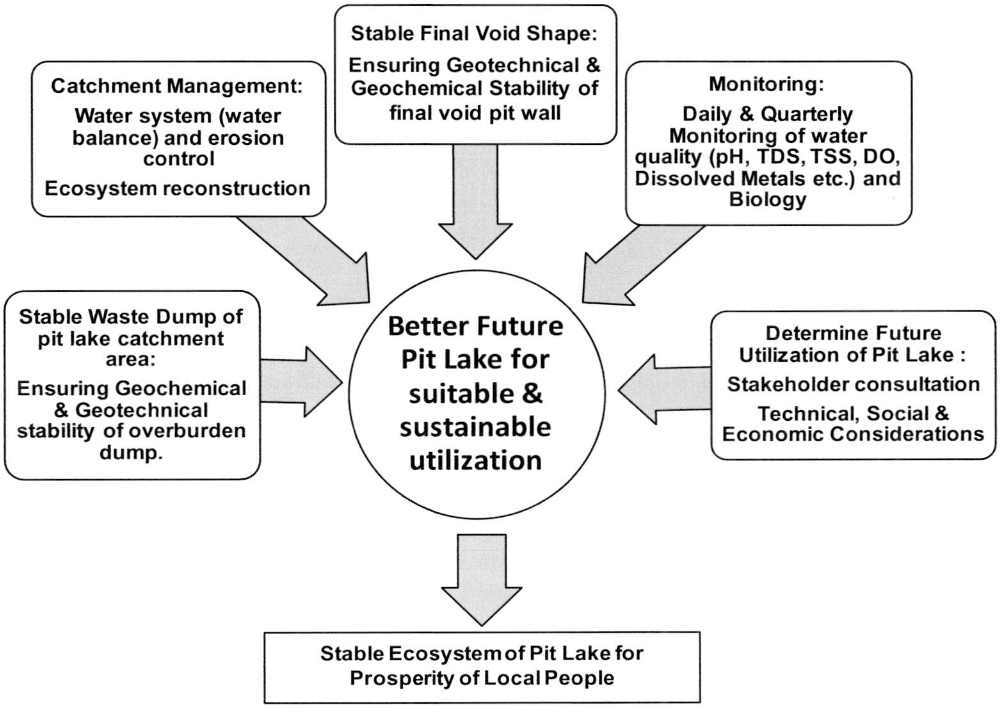

FIGURE 8.4 Schematic diagram for creating a better pit lake.

8.3.1 Stable Overburden/Waste Dump of Pit Lake Catchment Area: Ensuring Geochemical and Geotechnical Stability of Waste Dump Material

Prevention and mitigation of the formation of acid mine drainage (AMD) is done by identifying sulfide materials, i.e., iron sulfides like pyrite contained within overburden rock. The primary reaction in AMD formation is the oxidation of sulfide minerals when they are exposed to oxygen (O_2) and water (H_2O):

$$FeS_2\,(\text{pyrite}) + 7/2\ O_2\,(\text{oxygen}) + H_2O\,(\text{water}) \rightarrow Fe^{2+} + 2SO_4^{2-} + 2H^+$$

In this reaction, pyrite is oxidized to form ferrous ions (Fe^{2+}), sulfate ions (SO_4^{2-}) and hydrogen ions (H^+) that generate acidity. The products of the oxidation of sulfides include acidity, Total Dissolved Solids and metals. Metals derive from oxidized sulfides and dissolution of acid-consuming minerals. Factors that shape the rate of sulfide oxidation reactions are chemical and physical in character (type of mineral, surface area, crystal form, temperature, water source, pH), and biological factors (bacteria).

The successful prevention and mitigation of acid mine drainage (AMD) indicate the effectiveness of land reclamation and post-mining strategies. Rock geochemical analysis testing involved sampling through exploratory and geochemical drilling. From 2011 to 2019, Adaro conducted core drilling for AMD analysis at Paringin, Wara, and Tutupan, resulting in 1,646 samples, 1,591 samples and 2,729 samples, respectively, and a total of 60, 85 and 93 drill holes at each site. These samples form a database used to determine whether the rock is Potentially Acid Forming (PAF) or Non-Acid Forming (NAF) based on geochemical characterization analysis. Especially for Paringin, the number of samples from the Paringin pit drilling amounted to 1,646 samples from a total of 60 drill holes. Based on the classification of geochemical characteristics (see Table 8.1), the percentage of NAF is 80% while PAF Low Capacity, PAF Medium Capacity and PAF High Capacity are 10%, 9% and 1%, respectively.

TABLE 8.1
Geochemical Classification of Adaro Overburden Rocks Based on National Standard of Indonesia

No	Type	Geochemical Characteristics
1	NAF (Non-Acid Forming)	NAG pH (Net Acid Generation pH) ≥ 4.5 and NAPP (Net Acid Producing Potential) = negative
2	PAF LC (Potentially Acid Forming Low Capacity)	NAG pH (Net Acid Generation pH) < 4.5, NAG pH at pH 4.5 < 5 Kg H_2SO_4 /ton and NAPP (Net Acid Producing Potential) = 0 - 10 Kg H_2SO_4 /ton
3	PAF MC (Potentially Acid Forming Moderate Capacity)	NAG pH (Net Acid Generation pH) < 4.5, NAG pH at pH 4.5 > 5 Kg H_2SO_4 /ton and NAPP (Net Acid Producing Potential) ≤ 10 Kg H_2SO_4 /ton
4	PAF HC (Potentially Acid Forming High Capacity)	NAG pH (Net Acid Generation pH), and Paste pH < 4.5, NAG pH at pH 4.5 > 5 Kg H_2SO_4 /ton and NAPP (Net Acid Producing Potential) ≥ 10 Kg H_2SO_4 /ton

More detailed geochemical analysis has been conducted in 2023 by using 18 samples from the Paringin pit lake catchment area. These 18 samples were collected from two distinct locations within the Paringin Pit Lake catchment area. Remarkably, all these samples fell under the Non-Acid Forming (NAF) category. This classification signifies that the materials possess minimal potential for generating Acid Mine Water (AMD) due to their low sulfide content and a moderate Acid Neutralizing Capacity (ANC). Consequently, the likelihood of Acid Mine Water formation upon exposure to water and air/oxygen is notably reduced (Lund et al., 2013; Olías et al., 2019).

In terms of the waste dump's geotechnical stability, to ensure the geotechnical stability of the overburden dump, the dump design follows the geotechnical standard in accordance with Work Instruction Procedure WIN-AI-MSP-04-024. This is concerned with creating overburden dump design. Since the overburden dumping has met the geotechnical design requirements, there is no indication of dump failure in the field.

8.3.2 Catchment Management: Water System (Water Balance) and Erosion Control and Ecosystem Reconstruction

Numerous studies have concluded that land use types and morphological characteristics exert a significant impact on the water balance and water quality of receiving waters (Vrebos et al. (2017); Yu et al. (2019). However, concerning pit lakes, the catchment areas are generally small and characterized by sparse vegetation (Blanchette & Lund, 2016).

In general, Kalimantan experiences monsoonal rainfall characterized by a peak rainy season from December to February and a dry season from June to August. These seasonal patterns are influenced by monsoon winds originating from the continents of Asia and Australia (Arini et al., 2015). Arini et al. (2015) reported that the average peak rainfall ranges from 340 mm/month to the lowest rainfall of 60 mm/month. The annual rainfall recorded in the Paringin pit lake area for 2022 was 3,228.8 mm/year, which exceeded the average annual rainfall observed from 2011 to 2022, averaging 2,646.7 mm/year. Given the combination of high rainfall throughout the year and the significant total active area of Adaro's concession area, effective water management and erosion control measures are crucial aspects that need to be addressed.

In general, the water inflow rate into Paringin pit lake is dictated by the intensity of rainfall within the catchment area. The main and only source of water inflow into Paringin pit lake originates from rainfall within its catchment area, flowing either as surface runoff or as sub-surface flow.

The rehabilitation of the Paringin post-mined area, functioning as the Paringin pit lake catchment area, has been underway since 2014, utilizing a combination of slow-growing and fast-growing tree species for vegetation. Previously, this area served as a waste dump with relatively uniform overburden material from the coal mining operations of the Paringin pit. The flow-through Paringin pit lake receives water from two inflow drainages originating in its catchment area within the rehabilitated Paringin area. It then discharges water through a series of fishponds and settling ponds, ultimately leading to the Nungkaran River through a compliance point. By using F.J. Mock Method, the water inflow of Paringin pit lake catchment can be estimated based on rainfall data, hydrological flow area characteristics and evapotranspiration. The result of the average inflow into the Paringin pit lake derived from 10 years of rainfall data (2011–2022) is 0.68 m^3/second. Details regarding inflow estimation into Paringin pit lake 2011–2022 are presented in Table 8.2.

Post-mining landforms often exhibit higher levels of erosion and sedimentation compared to pre-mining conditions. This increase is primarily attributed to changes in morphology, topography, soil conditions and natural vegetation types (Lowry et al., 2019). Indonesian Minister of Energy and Mineral Resources Decree No. 1827 of 2018, Appendix VI, underscores the significance of implementing erosion control and water management measures in both active mining and reclamation areas. The water system (water balance) and erosion control are achieved through the ongoing process of revegetation, the aim being to reduce Total Suspended Solids (TSS). Consequently, water quality in post-mining lakes can serve as an indicator of environmental conditions in the catchment area.

Revegetation in the catchment part of the area could reduce the erosion rate from the catchment area's surface. Most of the revegetation at Paringin Pit Lake catchment occurred before 2014 while the latest revegetation was done in 2016. Figure 8.5 shows the Normalized Difference Vegetation Index (NDVI) Map 2023 based on satellite image of the new landform of the ex-coal mine and surrounding undisturbed landscape outside the concession area. The NDVI map shows a relatively similar condition of vegetation cover between the ex-coal mine and undisturbed area/natural vegetation, showing healthy vegetation cover (NDVI range 0.66 - 1). Therefore, Paringin Pit Lake has a healthy catchment area that can act as a hydrological function toward the pit lake.

The pH parameter can indicate the presence or absence of acid mine drainage. Without proper management of overburden materials with the potential to generate acidity (Potential Acid Forming/PAF), rain can trigger oxidation, leading to Acid Mine Drainage. The presence of a low/acidic pH is a key indicator of Acid Mine Drainage. Additionally, the Total Suspended Solids (TSS) parameter can reveal evidence of erosion and sedimentation. If the catchment area. is not adequately revegetated, the surface of the overburden area becomes prone to erosion, and the eroded sediment is transported to the former mining pit.

Beyond water quality, water quantity or water balance serves as another important indicator for post-mining lakes. Hydrologically, an analysis of the dimensions of the post-mining lake is conducted, including its holding capacity, inlet and outlet discharge. To achieve success in conducting revegetation in the Paringin Pit Lake catchment area, it should address the issue of soil fertility of post-mining land. The challenge is closely linked to soil conditions, which encompass characteristics such as low or high pH levels, compaction, limited water-holding capacity, erosion, high bulk density and insufficient micro- and macro-nutrient content. Soil plays a crucial role as a medium for plant growth and is vital for the success of revegetation and landscape restoration efforts (Stanturf and Callaham, 2021). To facilitate the best plant growth, the primary step in land restoration is the application of topsoil because it will serve as the growth medium. Approximately 15–30 cm of topsoil thickness is typically required for this purpose. The reclamation process, as outlined in the Indonesia Minister of Energy and Mineral Resources Decree, comprises three main activities: land contouring, revegetation and final completion (Table 8.3). Successful mine reclamation must fulfill all sub-activities, resulting in a total completion rate of 100%.

TABLE 8.2
Average Monthly Inflow into Paringin Pit Lake

Year	Average Monthly Inflow into Paringin Pit Lake (Q in m^3/second)											
	JAN	FEB	MAR	APR	MAY	JUN	JUL	AUG	SEP	OCT	NOV	DEC
2011	0.68	0.45	0.46	0.53	0.60	0.27	0.19	0.09	0.32	0.60	0.35	0.58
2012	0.53	0.28	1.25	0.74	0.13	0.10	0.10	0.09	0.08	0.06	0.06	0.80
2013	0.55	0.54	0.59	0.74	0.27	0.10	0.09	0.08	0.07	0.06	0.23	0.92
2014	0.54	0.51	0.55	0.34	0.31	0.38	0.09	0.08	0.07	0.06	0.21	0.72
2015	0.64	0.66	1.05	1.18	0.33	0.22	0.12	0.11	0.10	0.08	0.07	0.65
2016	0.09	1.12	1.08	0.65	0.35	0.39	0.32	0.11	0.10	0.69	0.57	0.43
2017	1.33	0.44	1.23	0.88	1.87	0.53	0.46	0.54	0.25	0.68	1.48	1.01
2018	0.51	0.95	0.64	0.85	0.13	0.11	0.10	0.09	0.08	0.06	0.32	0.73
2019	6.30	2.79	2.80	4.80	4.00	2.00	2.30	0.70	1.40	0.50	1.90	1.60
2020	1.26	1.20	1.11	1.26	0.77	0.42	0.20	0.19	0.58	0.52	1.64	0.80
2021	0.79	0.52	0.62	0.27	0.45	0.15	0.09	0.08	0.07	0.39	1.22	0.67
2022	0.78	0.82	0.48	0.97	0.85	0.26	0.15	0.11	0.67	1.09	1.01	0.27
Average	1.17	0.86	0.99	1.10	0.84	0.41	0.35	0.19	0.32	0.40	0.75	0.77

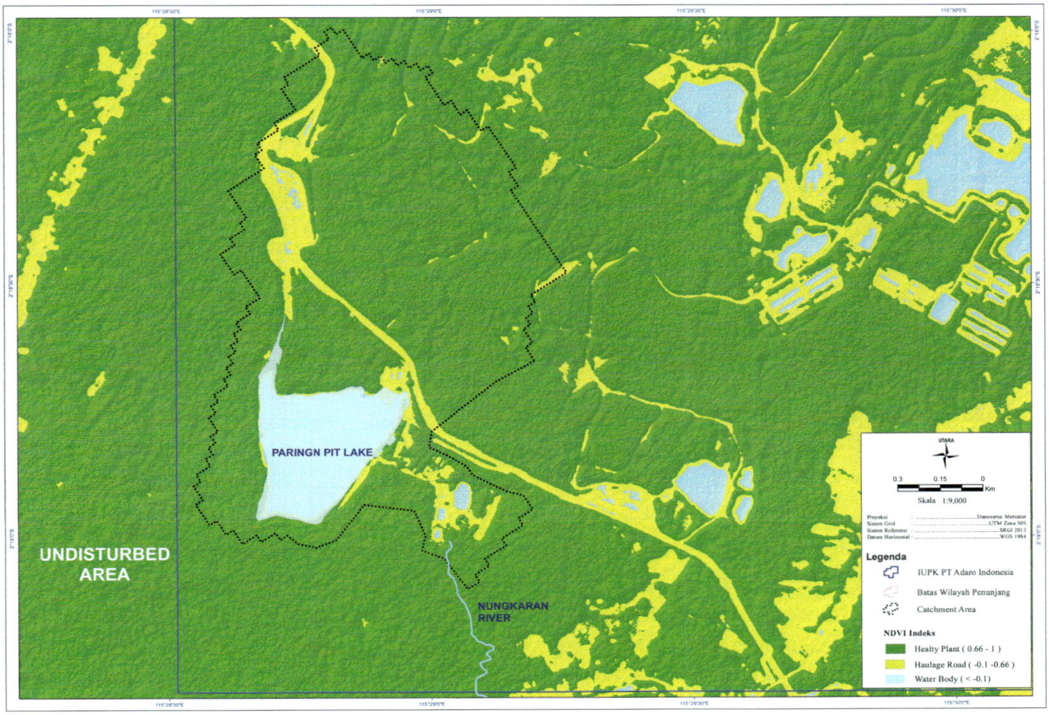

FIGURE 8.5 NDVI map year 2023 of Paringin pit lake, its catchment area, Nungkaran River and undisturbed area outside mining concession.

TABLE 8.3
Reclamation – Revegetation Success Criteria

Minister of Energy and Mineral Resources Decree No.1827.K/30/MEM/2018		
Activity	**Sub-Activity**	**Weight (%)**
Land contouring	Land contouring and inpit backfill	40
	Soil spreading	10
	Erosion and sedimentation control	10
Revegetation	Planting of cover crop	2.5
	Planting of fast-growing plants	7.5
	Planting of slow-growing plants	5
	Control of Acid Mine Drainage	5
Final completion	Canopy coverage	10
	Maintenance	10
Total		**100**

8.3.3 Stable Final Void Shape: Ensuring Geotechnical and Geochemical Stability of Final Void Pit Wall

Geotechnical stability of the final void refers to the structural integrity and stability of the final void wall, which is crucial for preventing wall collapses or failures when the final void becomes a pit lake. Creating a stable pit lake by ensuring geotechnical stability of the final void pit wall involves

FIGURE 8.6 Cover crop conditions at pit wall in Adaro mine site, one month (left) and one year (right) after hydroseeding application. Cover crop for preventing pit wall erosion.

taking measures to prevent wall collapses by implementing the recommendations of the geotechnical investigation. In the long term, the final void slope design must follow proper slope design to determine the safe angle and configuration of the pit wall, taking into account factors like soil type, rock type and the overall size of the final void.

Geochemical stability is especially needed to prevent the generation of acid mine drainage (AMD) from both the final void wall and waste dump area. To ensure the waste dump's geochemical stability, it requires overburden management with a focus on preventing the exposure of potentially acid-forming materials (PAF) from oxygen and water. In Adaro's operation, it can be done by covering PAF materials with non-acid-forming (NAF) materials to prevent oxidation and AMD.

In addition to ensuring the geotechnical and geochemical stability of the pit wall, Adaro employs a pit wall cover crop revegetation technique using the hydroseeding method. Cover crop vegetation plays a crucial role in stabilizing the pit wall's surface, preventing erosion and contributing to the geochemical stability of the final void. The hydroseeding technique is applied in combination with the installation of jute net/coco net on the pit wall's surface, as depicted in Figure 8.6. This approach has a relatively high success rate. However, during its implementation, various technical challenges are encountered, rendering this method ineffective and inefficient, with a very high level of risk, particularly during the net installation phase.

Implementing these geotechnical and geochemical stability measures makes it possible for mining companies to establish a secure pit lake, effectively reducing the risk of wall failures, water pollution and environmental damage. This approach not only enhances safety but also fosters the potential for sustainable land utilization and ensuring a viable ecosystem during the post-mining phase.

8.3.4 Monitoring: Monitoring of Water Quality (pH, TDS, TSS, DO, Dissolved Metals, etc.) and Key Species of Fauna

8.3.4.1 Paringin Pit Lake Water Quality

Daily and quarterly monitoring of the water quality at the pit lake was extensively conducted by Adaro, for such things as Total Suspended Solids (TSS), pH, Dissolved Iron (Fe), Dissolved Manganese (Mn) and Dissolved Oxygen (DO). Figure 8.7 shows quarterly Total Suspended Solids (TSS) data that resulted from the monitoring campaign from 2013 to 2022. It shows that the TSS concentration relatively fluctuated up to 2018 but, after that year, TSS concentration remained low, in fact below 25 mg/l, in which the regulatory requirement is 200 mg/l for mine wastewater and 50

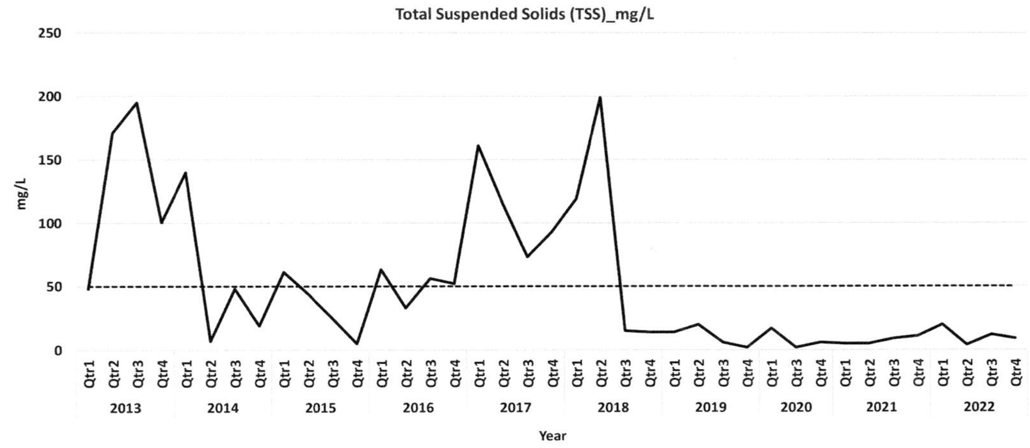

FIGURE 8.7 TSS data of Paringin pit lake, 2013–2022.

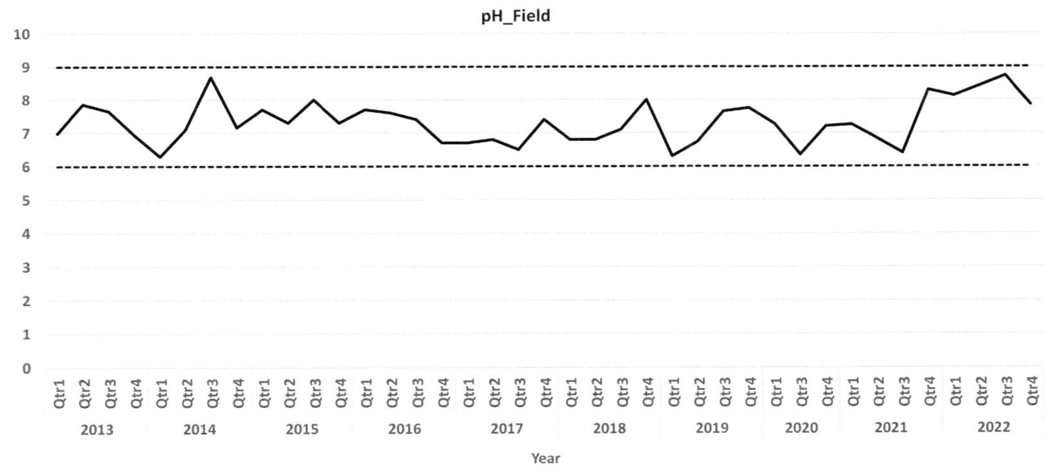

FIGURE 8.8 Historical pH data from 2013–2022 consistently meet the government standard_6 – 9.

mg/l for fishery purposes. The declining TSS level is due to the progressive revegetation that has been conducted. Vegetation cover will reduce erosion from the surface of the catchment area.

Meanwhile, Figure 8.8 shows the historical pH levels of the Paringin pit lake from 2013 to 2022, demonstrating a consistently near-neutral pH range between 6 and 9. This neutral pH range serves as a strong indicator that there is no evidence of acid mine drainage (AMD) generation within the Paringin pit lake catchment area., resulting in the absence of acidity in the inflow into the Paringin pit lake.

Figures 8.9, 8.10 and 8.11 show historical dissolved Iron (Fe), Manganese (Mn) and Cadmium (Cd) concentrations during 2013–2022 that should meet regulatory standards. The dissolved Iron (Fe) data show some fluctuation of Fe concentration, but overall, the concentration level always meets the government standard of <7 mg/l. The Manganese (Mn) concentration tends to become flat below <0.1 mg/l and always meets the government standard of <4 mg/l coal mine water. Meanwhile, the Cadmium concentration always met the standard of <0.01 mg/l for fishery purposes.

In terms of Dissolved Oxygen (DO) concentration, historically in the last 10 years, quarterly monitoring has shown that it is consistently above the minimum DO concentration of 4 mg/l set by the government for the freshwater fishery industry (Figure 8.12).

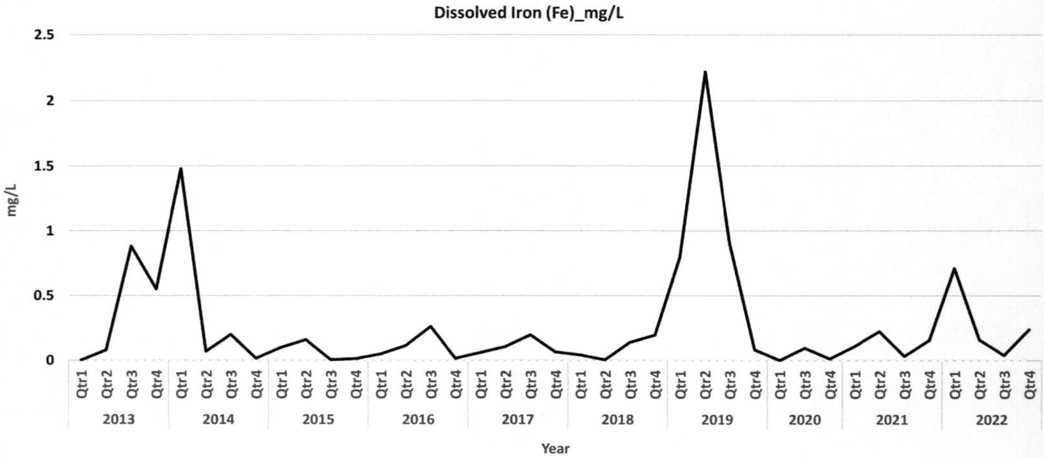

FIGURE 8.9 Dissolved Iron (Fe) 2013–2022 meets the government standard of below 7 mg/l.

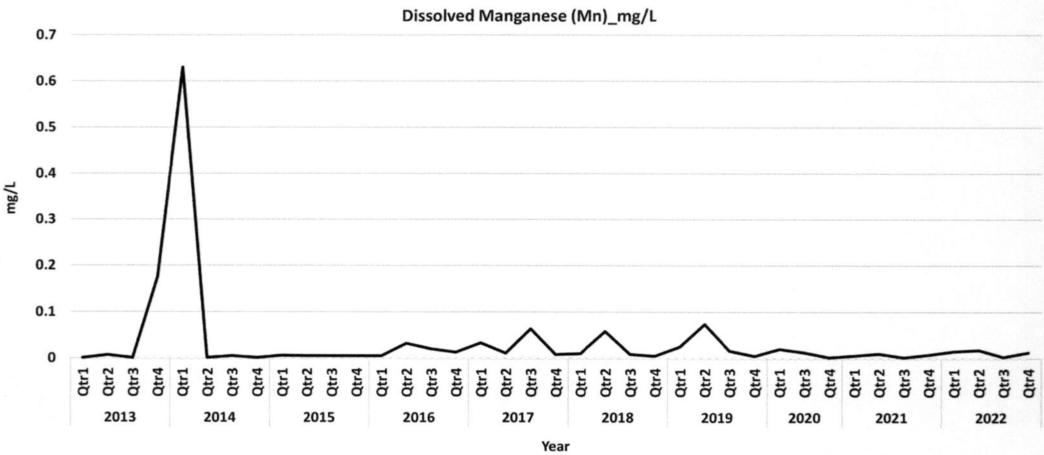

FIGURE 8.10 Dissolved Manganese (Mn) 2013–2022 meets the government standard below 4 mg/l.

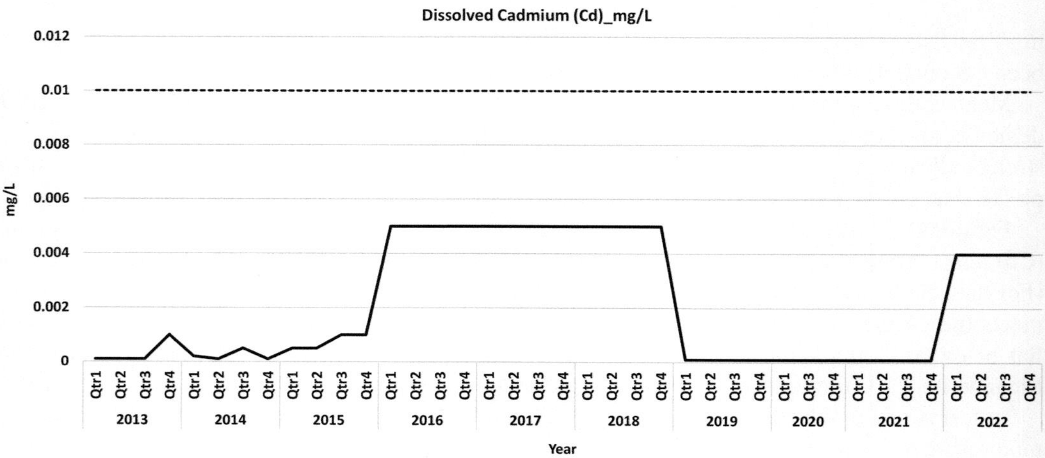

FIGURE 8.11 Dissolved Cadmium (Cd) 2013–2022 meets the government standard below 0.01 mg/l.

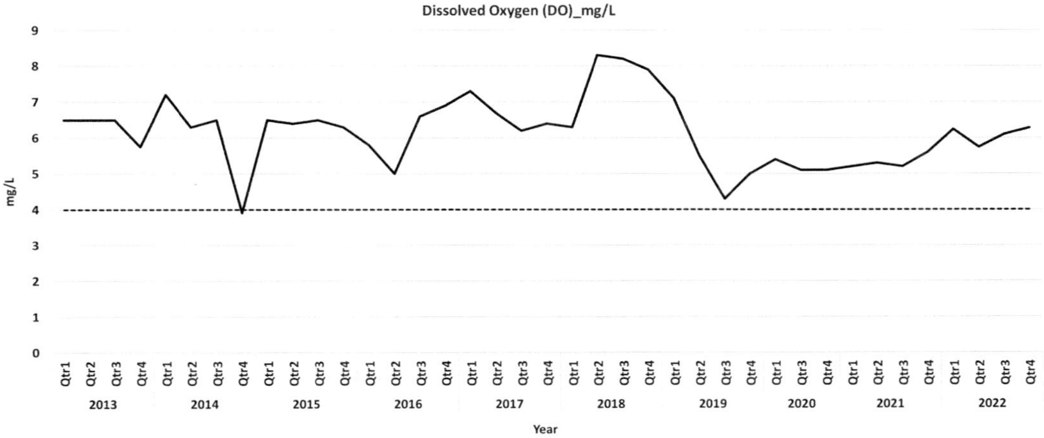

FIGURE 8.12 Dissolved Oxygen 2013–2022 meets the minimum standard of 4 mg/l.

FIGURE 8.13 The Proboscis monkey, now listed as a protected animal in Indonesia, is resting on the branch of *Albizia chinensis,* a vegetation planted in the Paringin pit lake catchment area.

8.3.4.2 Key Species Fauna Observation

Based on fauna observations conducted within Paringin pit lake in 2021 and 2022, several animal species were identified as protected animals under Minister of Environment and Forestry Regulation P.106 of 2018. These species include various bird species such as the Brontok eagle (*Nisaetus cirrhatus*), Tutut Takur (*Psilopogon rafflesia*), Sepahraja Honeybird (*Aethopyga siparaja*), Green pitta (*Pitta sordida*) and Striped fan (Rhipidura *javanica*). Additionally, protected primate species were identified, including Proboscis monkeys (*Nasalis larvatus*) and Gray monkeys (*Trachypithecus cristatus*). Figure 8.13 shows a male proboscis monkey with its distinctive long nose resting on the branch of *Albizia chinensis*, while a female proboscis monkey at the back is observed eating *Albizia chinensis* young leaves.

8.4 Determining the Future Utilization of Pit Lake: Stakeholders' Technical, Social and Economic Considerations

According to the Balangan Regency, South Kalimantan Spatial Planning for the period 2013–2032, the post-mining lake area has been officially designated a freshwater lake and wetland farming area

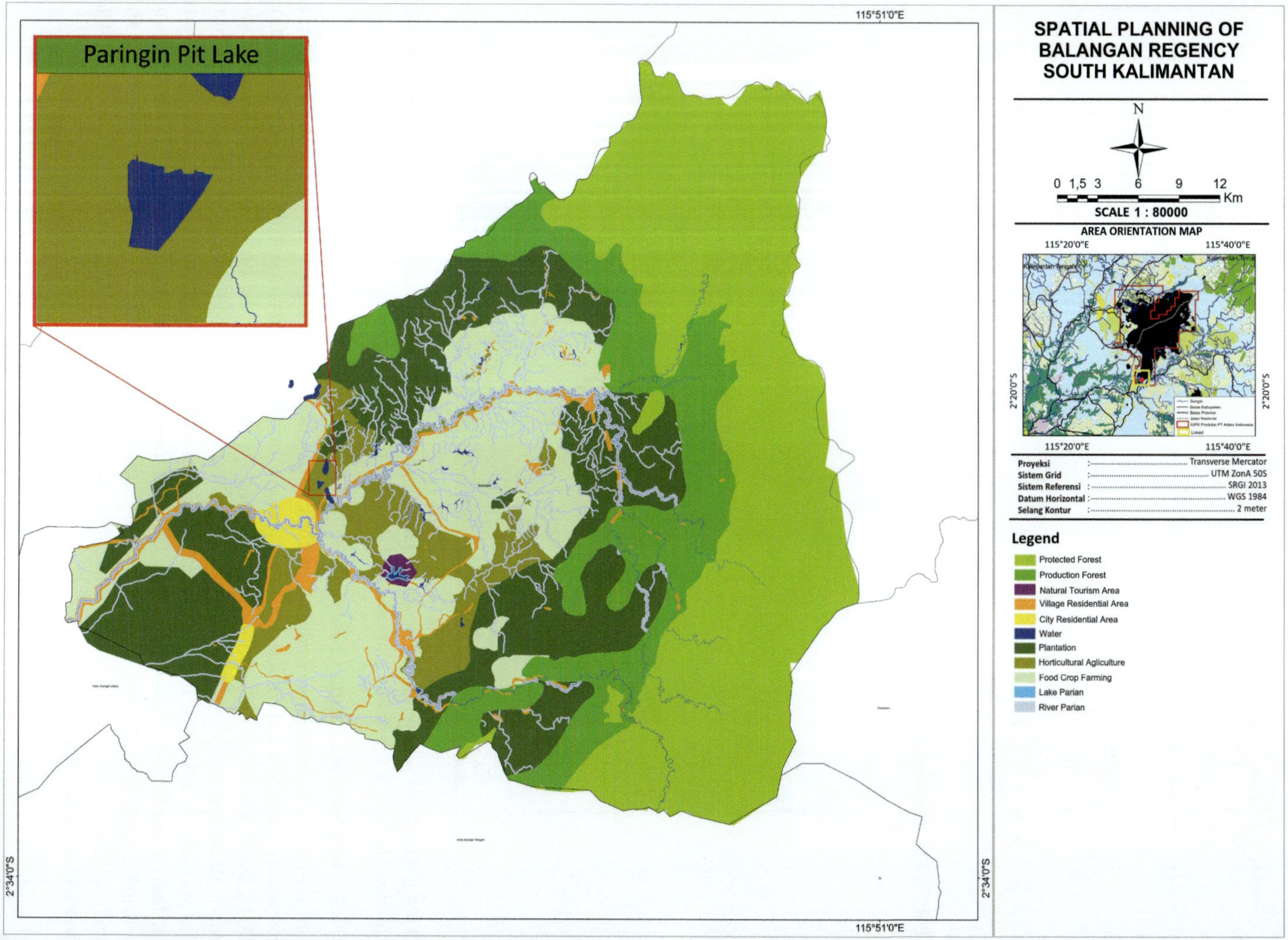

FIGURE 8.14 Spatial planning of Balangan Regency covers Paringin pit lake.

(see Figure 8.14). Consequently, it is imperative that this area functions as a pit lake that provides a sustainable ecosystem benefiting all stakeholders. The advantages of ensuring the quality of water flowing from the Paringin pit lake to downstream areas have been well documented. Notably, this directly impacts two villages in the downstream region, namely Balida and Murung Ilung Villages.

Regarding water availability of Paringin pit lake, Suhernomo conducted a study on the water balance simulation at the lake based on monthly rainfall data from 2021 (Suhernomo, 2022). However, this study lacked field measurement data for hydrological inflow and outflow from the lake and did not factor in water loss caused by evaporation. Table 8.4 provides a comprehensive overview of water balance for Paringin pit lake in 2022. Through bathymetric surveying, the pit lake's maximum capacity was determined to be 2.095 million cubic meters, with a critical surface water level set at 64.8 meters above sea level. Daily monitoring recorded a total inflow volume of 4.95 million cubic meters, with water losses from the pit lake amounting to 3.12 million cubic meters through outflow and 184,676 cubic meters through evaporation. In summary, given the pit lake's 1:7 proportion to the catchment area, coupled with a tropical climate featuring over 3,000 mm/year of rainfall and an evaporation rate of 1,000 mm/year, the lake functions as a flow-through system to receiving water of Nungkaran River.

8.5 PIT LAKE AND ITS CATCHMENT AREA UTILIZATION

8.5.1 Freshwater Source

Paringin pit lake is situated within a landscape where its catchment area, one that is already revegetated, encompasses the mine rehabilitation region, and the Nungkaran River serves as the recipient of water from its outflow (Figure 8.15). The hydrological connection commences with rainfall within the rehabilitation zone, which forms the catchment area responsible for supplying surface and sub-surface flows into Paringin pit lake. Excess water flows out of Paringin pit lake and subsequently enters the Nungkaran River. During this runoff or surface flow from the Paringin pit lake catchment area, sediments are transported along with the flow, as evidenced by the Total Suspended Solids (TSS) concentration in the inflow water. The majority of these sediments settle at the lake's bottom, resulting in the water flowing into the Nungkaran River having a lower TSS concentration. This highlights the sediment connectivity between Paringin pit lake and its catchment area, implying two important aspects: first, the relatively good water quality of the Nungkaran River's water flow; and, second, an increase in sedimentation, which reduces the effective capacity of Paringin pit lake. Maintaining high water quality in the river is essential for public health and socioeconomic sustainability (Castrillo and García, 2020). Paringin pit lake contains approximately 2.09 million cubic meters of freshwater that consistently generates an overflow to Nungkaran River with an average flow of 0.1 m^3/second.

Visual clarity (VC) is a critical parameter for assessing the quality of post-mining lakes' freshwater. VC is significantly influenced by suspended fine sediment, which is closely linked to the state of the catchment area., serving as its source. Evaluating VC necessitates a spatial analysis of the drainage patterns within the catchment area (Vale et al., 2022). The high visual clarity observed in the water flowing from Paringin pit lake into the Nungkaran River enhances the potential ecosystem services offered by the river, such as serving as a recreational water pool for children in the neighboring community (Figure 8.16).

8.5.2 Flora and Fauna Habitat of Key Species

The catchment area of the post-mining lake has become a good habitat for various protected species of birds and primates, for example, the endangered proboscis monkey. There are 105 species of birds found both in the revegetated area and in Paringin pit lake. Birds are using the revegetated

TABLE 8.4
Water Balance of Paringin Pit Lake 2022

Month	Monthly Rainfall	Total Inflow Volume from Catchment Area (I)	Total Evaporation from Pit Lake (E)	Total Outflow Volume (O)	Outflow Water Elevation (Level)	Water Balance [I - (E+O)]	Dynamic Water Capacity
	mm	**m3**	**m3**	**m3**	**m a.s.l**	**m**	**m3**
Jan	306.0	407,509.1	13,322.5	348,883.2	64.8	45,303.4	2,095,084.5
Feb	320.0	294,097.2	12,106.2	163,814.4	64.8	118,176.6	2,095,084.5
Mar	261.1	420,966.5	12,761.1	221,875.2	64.8	186,330.1	2,095,084.5
Apr	327.8	403,688.5	16,888.3	278,380.8	64.8	108,419.4	2,095,084.5
May	315.4	354,897.7	11,301.7	263,865.6	64.8	79,730.4	2,095,084.5
Jun	189.4	356,233.3	12,761.1	121,305.6	64.8	222,166.6	2,095,084.5
Jul	164.5	292,626.2	14,782.0	77,241.6	64.8	200,602.6	2,095,084.5
Aug	169.6	264,296.8	16,837.9	94,867.2	64.8	152,591.7	2,095,084.5
Sep	276.7	948,435.0	22,796.7	317,260.8	64.8	608,377.5	2,095,084.5
Oct	348.6	591,049.6	22,584.6	462,412.8	64.8	106,052.2	2,095,084.5
Nov	394.0	589,946.7	16,391.2	633,484.8	64.5	-59,929.3	2,035,155.2
Dec	155.7	24,108.4	12,143.7	134,784.0	64.1	-122,819.3	1,972,265.2
Total	**3,228.8**	**4,947,854.9**	**184,677.0**	**3,118,176.0**		**1,645,001.9**	

FIGURE 8.15 An aerial photo of Paringin pit lake, holding an average of 2 million cubic meters of freshwater, showcases the clear water surrounded by the successful reclamation area of a former waste dump.

FIGURE 8.16 Nungkaran River with good freshwater quality is now a playground for children.

area for shelter, finding food, breeding, resting and hiding from predators. Meanwhile, the proboscis monkeys use the revegetated area as a place to live and find food.

8.6 INTEGRATION WITH COMMUNITY DEVELOPMENT

A successful criteria model harmonizes post-mining requirements with sustainable development objectives. For instance, as outlined by Gastauer et al. (2018), the success of post-mining rehabilitation can be evaluated by assessing the functionality of the ecosystem following extensive rehabilitation efforts. The research conducted by Rosa et al. (2020) posits that meeting the success criteria for rehabilitation does not inherently guarantee the permanent realization of anticipated social benefits. Therefore, at Paringin pit lake, Adaro has endeavored to align its strategy with the development needs of the local community, particularly in the downstream areas of Balida and Murung Ilung villages.

The strategy for integrating the pit lake into the development plans of these two villages involves the implementation of the eco-riparian program along the Nungkaran River, which receives inflows from Paringin pit lake. The eco-riparian program is elaborated upon as follows. First, the program will be integrated into the post-mining requirements to improve the ecology and socioeconomic conditions of Balida and Murung Ilung villages. Second, the program aims to improve the ecological enrichment of hydrological plant species (e.g. ficus sp., bamboo) at the upstream area of Nungkaran River and Multi-Purpose Tree Species (MPTS) at the riparian zone of the middle and downstream areas of Nungkaran River. Third, the program to eliminate domestic waste pollution will improve villagers' awareness not to dump their domestic waste into the river. Fourth, integration with the Racah Mampulang Cultural Market Tourism and Balida Village Bamboo Bridge Tourism programs is envisaged (Figure 8.17).

8.7 CONCLUSION

Paringin pit lake serves as evidence that man-made landform alterations resulting from open-pit coal mining operations at the landscape level can harmoniously coexist with natural landform features downstream. This substantiates the conclusions of a previous study by Huang et al. (2021), which emphasized the potential impact of anthropogenic landscape changes on river water quality. In the case of Paringin pit lake, preserving the health of its catchment area remains crucial for sustaining its hydrological functions by maintaining a balance between surface and sub-surface flow into the lake.

Paringin pit lake stands as a noteworthy example of a healthy pit lake originating from a former coal mine open pit. This lake was meticulously planned and developed during the Paringin mining operations, seamlessly integrated into the rehabilitation and revegetation process. The key success factors for creating a more sustainable and suitable pit lake for future use encompass maintaining

FIGURE 8.17 Cultural tourism at Balida village, next to Nungkaran River.

stable waste dump areas within the lake's catchments, effective catchment management, ensuring a stable final void shape, determining the lake's future utilization and continuously monitoring the restoration of the Nungkaran Riverbanks. These measures are essential to ensure the enduring ecological and socioeconomic functions of the lake. Moreover, the sustainable integration of Paringin pit lake and the Nungkaran River eco-riparian program with the cultural tourism potential of Racah Mulang, Balida village, Paringin District, and Balangan Regency further bolsters community development efforts.

REFERENCES

Adams, E. A., Byrns, S., Kumwenda, S., Quilliam, R., Mkandawire, T., & Price, H. (2022). Water journeys: Household water insecurity, health risks, and embodiment in slums and informal settlements. *Social Science & Medicine*, *313*, 115394. https://doi.org/10.1016/j.socscimed.2022.115394

Adomako, S., & Tran, M. D. (2022). Sustainable environmental strategy, firm competitiveness, and financial performance: Evidence from the mining industry. *Resources Policy*, *75*, 102515. https://doi.org/10.1016/j.resourpol.2021.102515

Arini, E. Y., Hidayat, R., & Faqih, A. (2015). Rainfall simulation using RegCM4 model in Kalimantan during El Nino southern oscillation. *Procedia Environmental Sciences*, *24*, 70–86. https://doi.org/10.1016/j.proenv.2015.03.011

Bernasconi, R., Lund, M., & Blanchette, M. (2022). Non-charismatic waterbodies and ecosystem disservices: Mine pit lakes are underrepresented in the literature. *Frontiers in Microbiology*, *13*. https://doi.org/10.3389/fmicb.2022.1063594

Blanchette, M. L., & Lund, M. A. (2016). Pit lakes are a global legacy of mining: An integrated approach to achieving sustainable ecosystems and value for communities. *Current Opinion in Environmental Sustainability*, *23*, 28–34. https://doi.org/10.1016/j.cosust.2016.11.012

Blodau, C. (2006). A review of acidity generation and consumption in acidic coal mine lakes and their watersheds. *Science of the Total Environment*, *369*(1), 307–332. https://doi.org/10.1016/j.scitotenv.2006.05.004

Bracken, L. J., Wainwright, J., Ali, G. A., Tetzlaff, D., Smith, M. W., Reaney, S. M., & Roy, A. G. (2013). Concepts of hydrological connectivity: Research approaches, pathways and future agendas. *Earth-Science Reviews*, *119*, 17–34. https://doi.org/10.1016/j.earscirev.2013.02.001

Brundtland, G. H (Chairman). (1987). *Report of the world commission on environment and development: Our common future*. UN Documents: Gathering a Body of Global Agreements has been compiled by the NGO Committee on Education of the Conference of NGOs from United Nations web sites with the invaluable help of information & communications technology. https://sustainabledevelopment.un.org/content/documents/5987our-common-future.pdf

Castrillo, M., & García, Á. L. (2020). Estimation of high frequency nutrient concentrations from water quality surrogates using machine learning methods. *Water Research*, *172*, 115490. https://doi.org/10.1016/j.watres.2020.115490

Chen, M., Zeng, L., Huang, Z., Lei, L., Shen, Y., & Xiao, W. (2021). Evaluating suitability of land for forest landscape restoration: A case study of Three Gorges Reservoir, China. *Ecological Indicators*, *127*, 107765. https://doi.org/10.1016/j.ecolind.2021.107765

de la Torre de Palacios, L., & Espí Rodríguez, J. A. (2022). In mining, not everything is a circular economy: Case studies from recent mining projects in Iberia. *Resources Policy*, *78*, 102798. https://doi.org/10.1016/j.resourpol.2022.102798

Elma, M., Akhbar, Mustalifah, F. R., Suryani, L., Rampun, E. L. A., & Rahma, A. (2020). Wetland saline water and acid mine drainage desalination by Interlayefree silica pectin membrane from Banan peels. *Nusantara Science and Technology Proceedings*, 271–279. https://doi.org/10.11594/nstp.2020.0543

Fu, B., Xu, P., Wang, Y., Yan, K., & Chaudhary, S. (2018). Assessment of the ecosystem services provided by ponds in hilly areas. *Science of The Total Environment*, *642*, 979–987. https://doi.org/10.1016/j.scitotenv.2018.06.138

Gastauer, M., Silva, J. R., Caldeira Junior, C. F., Ramos, S. J., Souza Filho, P. W. M., Furtini Neto, A. E., & Siqueira, J. O. (2018). Mine land rehabilitation: Modern ecological approaches for more sustainable mining. *Journal of Cleaner Production*, *172*, 1409–1422. https://doi.org/10.1016/j.jclepro.2017.10.223

Grzybowski, M., Furgała-Selezniow, G., Koszałka, J., Kalinowska, J., & Jankun-Woźnicka, M. (2023). Correlation between catchment land use/cover and macrophyte assessment of lake ecological status. *Ecological Indicators*, *146*, 109857. https://doi.org/10.1016/j.ecolind.2022.109857

Hosseinpour, M., Osanloo, M., & Azimi, Y. (2022). Evaluation of positive and negative impacts of mining on sustainable development by a semi-quantitative method. *Journal of Cleaner Production*, *366*, 132955. https://doi.org/10.1016/j.jclepro.2022.132955

Huang, J., Zhang, Y., Bing, H., Peng, J., Dong, F., Gao, J., & Arhonditsis, G. B. (2021). Characterizing the river water quality in China: Recent progress and on-going challenges. *Water Research*, *201*, 117309. https://doi.org/10.1016/j.watres.2021.117309

Huang, W., Chen, X., Wang, K., Chen, J., Zheng, B., & Jiang, X. (2019). Comparison among the microbial communities in the lake, lake wetland, and estuary sediments of a plain river network. *MicrobiologyOpen*, *8*(2), e00644. https://doi.org/10.1002/mbo3.644

Keenan, J., & Holcombe, S. (2021). Mining as a temporary land use: A global stocktake of post-mining transitions and repurposing. *The Extractive Industries and Society*, *8*(3), 100924. https://doi.org/10.1016/j.exis.2021.100924

Kodir, A., Hartono, D. M., Haeruman, H., & Mansur, I. (2017). Integrated post mining landscape for sustainable land use: A case study in South Sumatera, Indonesia. *Sustainable Environment Research*, *27*(4), 203–213. https://doi.org/10.1016/j.serj.2017.03.003

Lowry, J. B. C., Narayan, M., Hancock, G. R., & Evans, K. G. (2019). Understanding post-mining landforms: Utilising pre-mine geomorphology to improve rehabilitation outcomes. *Geomorphology*, *328*, 93–107. https://doi.org/10.1016/j.geomorph.2018.11.027

Lund, M., Van Etten, E., & McCullough, C. (2013). *Importance of catchment vegetation and design to long-term rehabilitation of acidic pit lakes.* Conference: Proceedings of the International Mine Water Association (IMWA) Congress At: Bunbury, Australia. Editor: Wolkersdorfer, Brown (Figueroa (Editors). https://www.researchgate.net/publication/262224619_Importance_of_catchment_vegetation_and_design_to_long-term_rehabilitation_of_acidic_pit_lakes#fullTextFileContent

Lund, M., Van Etten, E., Polifka, J., Vasquez, M., Ramessur, R., Yangzom, D., & Blanchette, M. (2020). The importance of catchments to mine-pit lakes: Implications for closure. *Mine Water and the Environment*. https://doi.org/10.1007/s10230-020-00704-8

Marove, C. A., Sotozono, R., Tangviroon, P., Tabelin, C. B., & Igarashi, T. (2022). Assessment of soil, sediment and water contaminations around open-pit coal mines in Moatize, Tete province, Mozambique. *Environmental Advances*, *8*, 100215. https://doi.org/https://doi.org/10.1016/j.envadv.2022.100215

McCullough, C., & Vandenberg, J. A. (2020). Studying mine pit lake systems across multiple scales 多尺度规模的矿坑湖系统研究Untersuchung von Bergbaufolgeseen in unterschiedlichen ModellmaßstäbenEstudio de sistemas de lagos de hoyos de minas a través de múltiples escalas. *Mine Water and the Environment*, *39*. https://doi.org/10.1007/s10230-020-00678-7

McCullough, C. D. (2015). Consequences and opportunities from river breach and decant of an acidic mine pit lake. *Ecological Engineering*, *85*, 328–338. https://doi.org/10.1016/j.ecoleng.2015.10.001

McCullough, C. D., & Schultze, M. (2018). Engineered river flow-through to improve mine pit lake and river values. *Science of The Total Environment*, *640–641*, 217–231. https://doi.org/10.1016/j.scitotenv.2018.05.279

Ministry of ESDM (2018). Keputusan-Menteri-ESDM-Nomor-1827-K-30-MEM-2018. https://bdtbt.esdm.go.id/wp-content/uploads/2019/12/Keputusan-Menteri-ESDM-Nomor-1827-K-30-MEM-2018.pdf

MinistryofESDM. (2021). *Cadangan Batubara Masih 38,84 Miliar Ton, Teknologi Bersih Pengelolaannya Terus Didorong.* https://www.esdm.go.id/id/media-center/arsip-berita/cadangan-batubara-masih-3884-miliar-ton-teknologi-bersih-pengelolaannya-terus-didorong

Moodley, I., Sheridan, C. M., Kappelmeyer, U., & Akcil, A. (2018). Environmentally sustainable acid mine drainage remediation: Research developments with a focus on waste/by-products. *Minerals Engineering*, *126*, 207–220. https://doi.org/10.1016/j.mineng.2017.08.008

Olías, M., Cánovas, C. R., Basallote, M. D., Macías, F., Pérez-López, R., González, R. M., Millán-Becerro, R., & Nieto, J. M. (2019). Causes and impacts of a mine water spill from an acidic pit lake (Iberian Pyrite Belt). *Environmental Pollution*, *250*, 127–136. https://doi.org/10.1016/j.envpol.2019.04.011

Rosa, J. C. S., Geneletti, D., Morrison-Saunders, A., Sánchez, L. E., & Hughes, M. (2020). To what extent can mine rehabilitation restore recreational use of forest land? Learning from 50 years of practice in southwest Australia. *Land Use Policy*, *90*, 104290. https://doi.org/10.1016/j.landusepol.2019.104290

Rosa, J. C. S., Morrison-Saunders, A., Hughes, M., & Sánchez, L. E. (2020). Planning mine restoration through ecosystem services to enhance community engagement and deliver social benefits. *Restoration Ecology*, *28*(4), 937–946. https://doi.org/https://doi.org/10.1111/rec.13162

Rybnikova, L. S., & Rybnikov, P. A. (2020). Pit lake and drinking water intake example of coexistence (Middle Urals, Russia). *Mine Water and the Environment*, *39*(3), 464–472. https://doi.org/10.1007/s10230-020-00691-w

Sakellari, C., Roumpos, C., Louloudis, G., & Vasileiou, E. (2021). A review about the sustainability of pit lakes as a rehabilitation factor after mine closure. *Materials Proceedings*, *5*(1), 52. https://www.mdpi.com/2673-4605/5/1/52

Santofimia, E., López-Pamo, E., & Montero, E. (2013). Environmental management of aznalcollar mine and its influence in the hydrogeochemical of the pit lake. *Water Environment Research: A Research Publication of the Water Environment Federation*, *85*, 706–714. https://doi.org/10.2175/106143013X13698672323263

Schmidt, S. I., Hejzlar, J., Kopáček, J., Paule-Mercado, M. C., Porcal, P., Vystavna, Y., & Lanta, V. (2022). Forest damage and subsequent recovery alter the water composition in mountain lake catchments. *Science of the Total Environment*, *827*, 154293. https://doi.org/10.1016/j.scitotenv.2022.154293

Seifollahi-Aghmiuni, S., Nockrach, M., & Kalantari, Z. (2019). The potential of wetlands in achieving the sustainable development goals of the 2030 agenda. *Water*, *11*(3), 609. https://doi.org/10.3390/w11030609

Sjöberg, Y., Dessirier, B., Ghajarnia, N., Jaramillo, F., Jarsjö, J., Panahi, D. M., Xu, D., Zou, L., & Manzoni, S. (2022). Scaling relations reveal global and regional differences in morphometry of reservoirs and natural lakes. *Science of The Total Environment*, *822*, 153510. https://doi.org/10.1016/j.scitotenv.2022.153510

Stanturf, J. A., & Callaham, Mac A. (2021). *Soils and landscape restoration*. Elsevier Inc.

Suhernomo, F. L., Saidy, A. R., Priatmadi, B. J., Noor, I., & Triwibowo, D. (2022). Strategy water flow treatment of Paringin pit lake to meet waste water discharge compliance. *Journal of Degraded and Mining Lands Management, 10*(4), 4683–4695. https://doi.org/10.15243/jdmlm.2023.104.4683. https://jdmlm.ub.ac.id/index.php/jdmlm/article/view/1273

Sulista, S., & Rosyid, F. A. (2022). The economic impact of tin mining in Indonesia during an era of decentralisation, 2001–2015: A case study of Kepulauan Bangka Belitung Province. *The Extractive Industries and Society*, *10*, 101069. https://doi.org/10.1016/j.exis.2022.101069

Syahrir, R., Wall, F., & Diallo, P. (2020). Socio-economic impacts and sustainability of mining, a case study of the historical tin mining in Singkep Island-Indonesia. *The Extractive Industries and Society*, *7*(4), 1525–1533. https://doi.org/10.1016/j.exis.2020.07.023

Syahrir, R., Wall, F., & Diallo, P. (2021). Coping with sudden mine closure: The importance of resilient communities and good governance. *The Extractive Industries and Society*, *8*(4), 101009. https://doi.org/10.1016/j.exis.2021.101009

Toumbourou, T., Muhdar, M., Werner, T., & Bebbington, A. (2020). Political ecologies of the post-mining landscape: Activism, resistance, and legal struggles over Kalimantan's coal mines. *Energy Research & Social Science*, *65*, 101476. https://doi.org/10.1016/j.erss.2020.101476

Ullah, A. K. M. A., Maksud, M. A., Khan, S. R., Lutfa, L. N., & Quraishi, S. B. (2017). Dietary intake of heavy metals from eight highly consumed species of cultured fish and possible human health risk implications in Bangladesh. *Toxicology Reports*, *4*, 574–579. https://doi.org/10.1016/j.toxrep.2017.10.002

Vale, S. S., Smith, H. G.,Davies-Colley, R. J., Dymond, J. R., Hughes, A. O., Haddadchi, A., &Phillips, C. J. (2022). The influence of erosion sources on sediment-related water quality attributes. *Science of The Total Environment*, 160452. https://doi.org/https://doi.org/10.1016/j.scitotenv.2022.160452

Vandenberg, J. A., Schultze, M., McCullough, C., & Castendyk, D. (2022). The future direction of pit lakes: Part 2, corporate and regulatory closure needs to improve management. *Mine Water and the Environment*, *41*. https://doi.org/10.1007/s10230-022-00868-5

Vrebos, D., Beauchard, O., & Meire, P. (2017). The impact of land use and spatial mediated processes on the water quality in a river system. *Science of The Total Environment*, *601–602*, 365–373. https://doi.org/10.1016/j.scitotenv.2017.05.217

Woodbury, D. J., Yassir, I., Arbainsyah, Doroski, D. A., Queenborough, S. A., & Ashton, M. S. (2020). Filling a void: Analysis of early tropical soil and vegetative recovery under leguminous, post-coal mine reforestation plantations in East Kalimantan, Indonesia. *Land Degradation & Development*, *31*(4), 473–487. https://doi.org/https://doi.org/10.1002/ldr.3464

Worlanyo, A. S., & Jiangfeng, L. (2021). Evaluating the environmental and economic impact of mining for post-mined land restoration and land-use: A review. *Journal of Environmental Management*, *279*, 111623. https://doi.org/10.1016/j.jenvman.2020.111623

Yu, H., Liu, X., Kong, B., Li, R., & Wang, G. (2019). Landscape ecology development supported by geospatial technologies: A review. *Ecological Informatics*, *51*, 185–192. https://doi.org/10.1016/j.ecoinf.2019.03.006

9 Abandoned Mines in New Zealand

Jo Cavanagh and Robyn Simcock

9.1 INTRODUCTION

Mining is an important economic activity in New Zealand, and has a long history often linked with important cultural developments. Māori used and traded several types of minerals, either found on the surface or extracted by quarrying, for tools, weapons and ornaments. These included obsidian (matā), nephrite, commonly known as greenstone (pounamu), and argillite (pakohe) (Nathan 2006). From the 1840s early settlers prospected for metals (primarily gold) and coal. Small-scale mining of gold, silver, copper, lead, iron, chromium, zinc and antimony was underway by 1870. All the main goldfields and coalfields had been discovered, and coal was being mined in many areas around the country (Nathan 2006).

The initial gold rush period occurred from the 1860s to the early 1900s, starting with the mining of rich surface gravels and then moving to larger-scale underground mining, some of which continued until the 1950s. With the deregulation of the gold price in the 1970s, gold production increased (Christie 2019) large, open-cast mines were opened in the 1990s. In contrast to the fluctuations in gold-mining activity, coal mining steadily increased production to 2 million tonnes annually in 1910. It then stayed reasonably static for the next 70 years until the development of an export market for bituminous coal saw an increase in production from the 1990s (Nathan 2006). Then, in 2012 the international price of coking coal fell dramatically, leading to the closure of a number of coal mines. Annual production is currently slightly over 2 million tonnes (NZPAM 2023).

Coal and gold are the primary commodities currently mined in New Zealand. Coal is mined across New Zealand, as shown in Figure 9.1, with different coalfields having different characteristics. Rocks mined for coal on the West Coast of the South Island mainly belong to either the Brunner or the Paparoa coal measures and primarily yield bituminous coal, with much of the coal exported for steel production. Coal, and until recently lignite, were mined from eastern Southland lignite deposits, while smaller coal-mining areas occur in Canterbury and Otago. In the Waikato region, sub-bituminous coal is mined from several locations for use in steel production and power generation (Cavanagh et al. 2018a).

Hard-rock gold occurs in fissures and fracture zones, where it has been deposited in quartz veins, either near the Earth's surface (epithermal gold) or at greater depths (mesothermal/orogenic gold). Epithermal gold can be found, along with silver, in quartz veins on the Coromandel Peninsula. Other metals, such as copper and zinc, are primarily associated with epithermal deposits (Cavanagh et al. 2018b). Orogenic gold deposits are confined to the South Island. They occur in two main belts of metamorphosed sedimentary rocks that have been offset by the Alpine Fault. The western belt includes a range of geologically old (400 million years) metamorphosed sedimentary rocks in northwest Nelson, Buller, and Westland and in southwest Fiordland. The eastern belt includes Otago schist and its equivalent metamorphosed sedimentary rocks in Marlborough (Cavanagh et al. 2018c). Alluvial, or placer, gold deposits are derived from erosion and river transport of the gold from orogenic deposits.

As noted above, mining activity has been variable over the years, and there are multiple reasons for closure or abandonment of a mine, with economic viability a significant driver. Factors

DOI: 10.1201/9781315141855-9

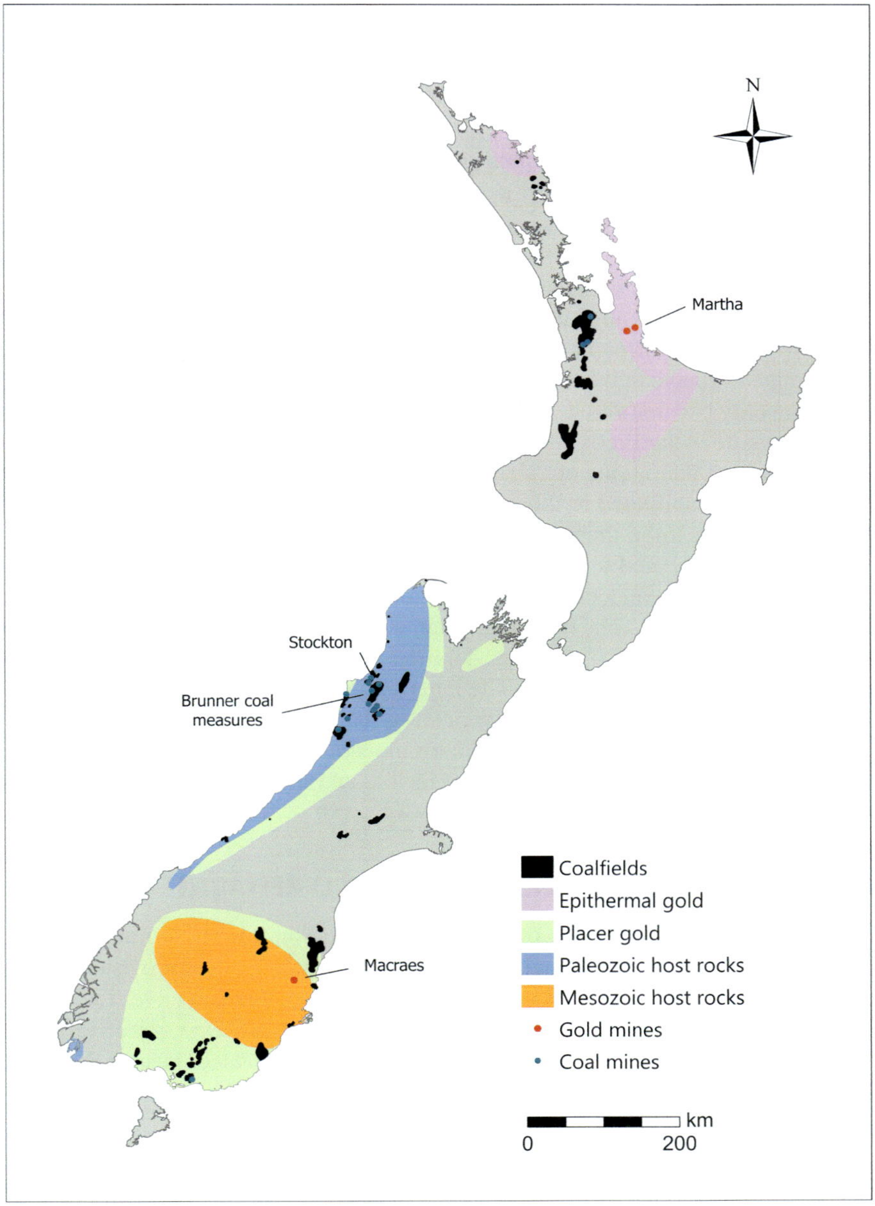

FIGURE 9.1 The approximate distribution of the main economic epithermal and mesothermal deposits and coal fields across New Zealand. Also shown are the locations of selected major active mines mentioned in the text, along with the generalised location of the Brunner Coal measures, an environmentally significant coal formation.

influencing viability include speculative operations not resulting in the anticipated production – such was the case for many of the minor minerals; downturns in commodity prices and/or increases in fuel prices resulting in previously economic operations no longer being viable. Examples of other factors that have influenced economic viability leading to abandonment include the presence of mercury impurities in metals (copper, lead, zinc) extracted from epithermal deposits at Tui mine in the Waikato region in 1973; the collapse of the Prohibition Shaft in 1957, which signalled the end for the Blackwater gold mine on the West Coast of the South Island, one of the largest producers of its time; prohibitive costs for rectifying instability in the tailings storage facility, arising from a large ancient landslide underlying most of the tailings dam and waste rock embankment (Jacobs 1999), and led to the abandonment of the Golden Cross mine in Waikato in the late 1990s; and an explosion at Strongman coal mine, resulting in 19 deaths, which led to the abandonment of the mine in 1967 (although Strongman 2 mine opened in 1994 and was abandoned in 2003). The most recent mining disaster in New Zealand, an explosion at the Pike River coal mine in 2010 resulting in 29 deaths, led to the abandonment of the mine as well as changes to Health and Safety legislation following a Royal Commission review. The Stockton coal mine, situated in the largest coal field in New Zealand, was at risk of being abandoned in 2015 when the owner, Solid Energy, entered into voluntary administration (https://solidenergy.co.nz/). All land, mines and other assets were offered for sale, and most were sold to different parties, with the Crown taking on additional liabilities associated with acid mine drainage in the sale of the mine (House of Representatives 2019). In the case of iron sand, mining only became viable when methods for smelting the iron sands to produce iron and steel became economic in the 1960s and extraction has remained profitable with no abandoned sites.

Under New Zealand definitions of “abandoned” mines, there is a variation in the condition of these mines, from unrehabilitated to those undergoing planned rehabilitation and closure. Derelict mines can be considered to be those unrehabilitated abandoned mines, where the mine owner or operator is no longer connected to the site. The varying definition of abandoned mines influences the environmental state of abandoned mines and more widely affected areas. Other factors include mine size and type (alluvial, underground, open-cast), geological setting, specific commodity (including any on-site processing), climate, duration of mining, and (arguably most importantly) the time of permitting and abandonment. The latter two are critical contributors to the state of the mine because they determine the extent to which legislative requirements for rehabilitation of the mine by the mining operator existed, and the environmental stringency of those requirements. This chapter explores the relationships between these factors and the environmental state of abandoned and derelict mines in the context of mining in New Zealand.

9.2 POLICY RELATING TO MINE CLOSURE AND REHABILITATION

There is no – and never has been – national policy or legislation relating to mine closure and rehabilitation in New Zealand. Some regional or district plans include specific provisions related to mining; for example, the Hauraki District Council Plan (2016) identifies objectives for specific mineral zones (e.g., the Martha mineral zone, Golden Cross mineral zone). One objective for the Martha mineral zone is to ensure that rehabilitation, once mining is completed, is such that the long-term uses of the area within this zone provide an appropriate and sustainable environment for Waihi (the township adjacent to the mine).

”Permission” to mine (including prospecting and exploration) has always been a legislative requirement, with “abandonment” often referred to in the legislation. Until 1991, coal-mining activities were legislated under different Coal Mine Acts, while mining for gold and other minerals was legislated under different Mining Acts. In addition to outlining legislative requirements for undertaking mining activities, these Acts included provisions relating to ”abandoned” mines and/or “abandoned shafts”. Specifically, from 1886, the abandonment of coal mines required that mine plans be provided to the Minister of Energy (or mining inspector from 1925) within one to three months, and that abandoned or disused shafts or mines be fenced off and secured. Similar requirements for fencing or securely “covering in” abandoned or disused shafts were specified in various Mining Acts from 1891. The provision of plans of abandoned mines to the mining inspector was not required until the Mining Act 1971, and then only for underground mines.

”Abandoned” is not defined in the legislation but appears to refer to when the mine ceases to be worked for a comparatively short time (one to three months). The wording of requirements for the protection of abandoned shafts in some early legislation suggests that the motivation for this was more to enable abandoned shafts to be reopened in the event of mining recommencing rather than to prevent people falling into the shafts. Nonetheless, the legislative requirements for the provision of mine plans for abandoned mines have probably been a significant contributor to enabling the identification of the location and nature of historically abandoned mines (see section 9.3).

9.2.1 Consideration of Damage to the Land from Mining: 1970s

The Mining Act 1971 and the Coal Mine Act 1979 were the first pieces of legislation to set conditions to "prevent, reduce, or make good, damage to the surface of the land". These conditions could be set by the Minister of Mines (Minerals Act) or Energy (Coal Mines Act) on granting a mining licence or "privilege", and they could be informed by reports from the Commissioner of Crown Lands, Catchment Board, Catchment Commission, Waikato Valley Authority or Soil Conservation and Rivers Control Council. A further consideration in setting conditions was to prevent, as far as reasonably practicable, any conflict with the purposes of the Soil Conservation and Rivers Control Act 1941, which focused on the conservation of soil resources, erosion prevention and protection from flood damage – and is still in force today.

Additional conditions could be imposed by the appropriate land-administrating department; for example, the New Zealand Forest Service or the Department of Lands and Survey before 1987, and the Department of Conservation (DOC) after 1987 (Simcock and Ross 2017). Under the Coal Mining Act 1979, a monetary deposit could also be required as security for compliance with the terms and conditions of a coal-mining right. These conditions, which often included a requirement to maintain land-use capability after mining (together with associated bonds), drove the process of land rehabilitation (Simcock and Ross 2017). The requirement for a similar bond is less clear under the Mining Act 1971, although the Act did allow for setting the amount of any monetary deposit or bond required with applications for mining privileges, including prescribing the circumstances under which deposits were to be wholly or partly refunded. Simcock and Ross (2014) commented on the inadequacy of the size of many bonds to cover the costs of rehabilitation of small alluvial and dredging operations, the issues relating to the bonding process (e.g. the council not being notified of mine closure until mining equipment is off site, or even when access to the mine was removed), and they provided advice on how the process could be improved to achieve better rehabilitation outcomes.

9.2.2 Environmental and Planning Legislation Reform: 1991

In 1991 there was an overhaul of environmental and planning legislation in New Zealand. Permitting of all resource extraction activities (including oil and gas) was captured under a single Act, the Crown Minerals Act 1991 (CMA), while environmental management requirements fell under the Resource Management Act 1991 (RMA). Under the CMA, mining operations include "other land improvements" and "wastewater produced from or consequent on the operations", suggesting that for as long as terrestrial rehabilitation and/or treatment of mine drainage is being undertaken, a mining permit is required. The duration of a mining permit is 40 years and may be extended to enable the permit holder to complete rehabilitation works. Rehabilitation is not defined in the Act, but it is generally taken to include terrestrial rehabilitation and/or treatment of mine drainage. Finally, when a permit is surrendered it "must be accepted by the chief executive of the relevant government department if everything is *in order*" [emphasis added], which could include meeting clauses in the permit, for example, around leaving a stable land form and/or whether conditions in consents obtained under the RMA have been met.

The purpose of the RMA is to provide for the sustainable management of natural and physical resources, which includes safeguarding the life-supporting capacity of air, water, soil and ecosystems as well as avoiding, remedying or mitigating any adverse effects of activities on the environment. Under the RMA, discharge and land-use consents required to undertake mining activities are granted by district councils (for noise, light and land use) or regional councils (for discharges to air and water, water takes and erosion control), and thus are regional- and/or site-specific. A new resource management system was to be phased in over the next 10 years, with the Natural and Built Environment Act 2023 the main replacement for the RMA, came into force in August 2023, however, after the completion of this chapter there was a change in government and this Act was repealed.

The final strand of current legislation relating to mining is the Health and Safety at Work Act 2015 (HSAW), which provides protection for mine workers. As a result of the Royal Commission review into the 2010 Pike River coal mine disaster (Royal Commission 2012), significant changes were made to health and safety legislation as it relates to mining. This included the first formal definition of abandoned mining operations under the Health and Safety at Work (Mining Operations and Quarrying Operations) Regulations 2016. Essentially, "abandoned" means that mining operations (which include extraction, processing, handling or maintenance of waste material, decommissioning of infrastructure and repair activities) are no longer being carried out and any hazards associated with the carrying out of those activities in that place have been eliminated or isolated. Under health and safety legislation, "hazard" refers to hazards to workers. "Suspended" operations means that no active resource extraction is occurring, and only maintenance of the mining operation may be being carried out, with a view to either enabling the mining operation to actively mine in the future or while the mining operation is winding down in preparation for being abandoned. Mining operations under HSAW include "repair" activities, which could include terrestrial rehabilitation and treatment of mine drainage.

Mining operations permitted before 1991 maintained conditions set under the old legislation. As a result, all or parts of the largest open-cast mines currently operating in New Zealand operate under pre-1991 conditions, including the largest coal mine (Stockton, which started opencast mining in 1944), and the two most productive gold mines, Martha mine (Waikato, started in 1988) and Macraes Flat (Otago, started in 1990) (Simcock and Ross 2017). The Golden Cross mine was permitted under the Mining Act 1971 but closed under the RMA; Jacobs (1999) noted the challenges associated with this change in regulatory regime, and that societal expectations with respect to long-term environmental stewardship during the life of the mine resulted in the requirements of closure being significantly different from the expectations at permitting. Rehabilitation and monitoring are ongoing in 2023, although the council expectation was that closure would occur within the life of the District Plan (Hauraki District Council 2014). Newer mining operations (e.g., extensions of the Waihi gold mine, the Cypress extension of the Stockton mine, the nearby Escarpment mine on the Denniston Plateau and the Globe-Progress mine at Reefton) have been consented fully under the RMA and have much more stringent environmental criteria, including more robust closure criteria for rehabilitation.

Martha mine is perhaps the only mine for which the general closure process is outlined; notably, that when the regulatory authorities consider the site has reached a "safe, stable, self-sustaining, rehabilitated state", a post-closure phase will commence with the formation of the Martha Trust (OceanaGold 2022). The trust will own, manage, monitor and maintain the land and facilities at the wetlands, the water treatment plant and the parklands adjacent to the lake. It will also be responsible for maintenance and monitoring of the lake. The company will hand over a sum of money to the trust, and this money and the interest it earns will be used to carry out the responsibilities of the trust in perpetuity.

A further strand relevant to derelict and some abandoned mines is that, under the RMA, regional councils have a responsibility to investigate land for the purposes of identifying and monitoring contaminated land, including on Crown land. This requirement had been further strengthened under the Natural and Built Environment Act 2023, which requires that regional councils identify all HAIL (Hazardous Activities and Industries List, Ministry for the Environment 2011b) land within its boundaries, and keep this information on a publicly available register. Mineral extraction, refining and reprocessing and storage and use comprise one such category in the HAIL; thus there was now a legislative requirement for councils to identify all locations where mining activities have occurred within their region, and also an obligation on the landowner to manage, investigate and monitor contamination. For many derelict and abandoned mines, this landowner is likely to be central government, such as the Department of Conservation (see also section 9.3).

9.2.3 Non-Legislative Guidance

Few tools are available to help mining companies and regulators assess and predict the environmental impacts of historical or planned mining operations on New Zealand's geology and environmental condition. An exception is guidance published in 2018 by the Centre for Minerals Environmental Research (CMER, https://www.cmer.nz/). This includes three mine environment life-cycle guides (MELGs), each covering the dominant New Zealand mine "types" based on geology: acid and non-acid coal mining, mesothermal (orogenic) gold mining and epithermal gold mining (Cavanagh et al. 2018 a,b,c). Specifically, the MELGs examine the geochemistry of mine waste rock and mine drainage, the freshwater ecology of mine-affected catchments and the effectiveness of water treatment and terrestrial rehabilitation techniques for different mine types and environments to achieve intended outcomes and consider this in the context of the life-cycle of a mine (as shown in Table 9.1). The MELGs built on more than 12 years of research into the environmental aspects of mining and are referenced in the *Minerals and Petroleum Resource Strategy for Aotearoa New Zealand: 2019–2029* (MBIE 2019). The MELGs include guidance on passive treatment systems and include many techniques and approaches for rehabilitation.

TABLE 9.1
Summary of a Conventional Mine Life-Cycle, the Mine Permitting/Consenting Regime Currently Operating in New Zealand and their Relationship to the Mine Environment Life-Cycle Stages as Used in the MELGS

Mine Life-Cycle Stage	Permitting/Consenting Regime	Mine Environment Life-Cycle Stage
Exploration	Prospecting (land access) Exploration (land access)	Exploration – prospecting and exploration activities
Mine planning	Mine permit – feasibility study Resource consent	Pre-operations – all activities leading up to the development of an operational mine; specifically, activities required to obtain land access and resource consent. Engagement with iwi (Māori tribal groups) and stakeholders should be initiated.
Operations (including commissioning and construction)	Annual reporting	Operations – all activities that are undertaken while the resource is being extracted (i.e. the mine permit is active).
Decommissioning	Annual reporting. A mine is considered to be closed when the mining permit has been surrendered.	Closure – when resource extraction ceases and rehabilitation activities are working towards the agreed post-mining environment. Under a staged mine plan, some areas of the mine may be in closure while others are still operational.
Post-closure	Agreement with regulators that appropriate conditions, including allowances for ongoing treatment, have been met to the extent that agreed post-closure outcomes will be met. The mine company is no longer involved on-site.	Post-closure – agreement with regulators that appropriate conditions, including allowances for ongoing treatment, have been met to the extent that agreed post-closure outcomes will be met. The mine company is no longer involved on-site.

9.3 TYPE AND NUMBER OF ABANDONED MINES

”Mine type” encompasses the mining approach, the commodity being extracted and the geological setting. Historically there has generally been a transition from underground to open-cast mining for both coal and other minerals. For example, early coal mining was underground, with miners using picks and shovels. In the early 1900s explosives and compressed-air machines were used to cut coal from the seam, and from the 1990s high-pressure water jets were used to cut the coal in some underground mines. By the 2000s the majority of mining was open cast (Sherwood and Philips 2021).

Most historical gold-mining production was from alluvial, or placer, gold deposits derived from eroded hard-rock sources in the South Island. Water was a critical feature for recovering gold, involving small dams and water races – channels that cut across contours, constructed at all goldfields to bring water from creeks to areas where gold was worked. Riffle or sluice boxes were the main method of recovering gold. These were long, terraced wooden boxes over which gold-bearing gravel was washed. Sluicing, where jets of water were aimed at gravel faces, helped to wash gold-bearing gravels down through the sluice boxes. From 1863 dredges were also used in rivers in the Otago and West Coast regions to dredge for gold: in the early 1900s many small dredges were used, but from the 1930s larger dredges took over, with dredging continuing in the Grey River on the West Coast until 2012. In the 1990s large-scale alluvial open-cast mining was carried out at Arahura and Ross on the West Coast, Nokomai in northern Southland and near Millers Flat and the Tokomairiro River in Otago (Walrond 2006).

Hard-rock gold was initially mined underground. The recovered quartz was crushed by stamper batteries, and cyanide was used to reclaim the gold – with the commercial use of cyanide for extracting gold pioneered internationally in the Karangahake gorge (Walrond 2006). By the 1950s hard-rock gold mining had largely ceased, but a rise in the gold price in the late 1970s saw a resurgence of hard-rock mining (Christie 2019). Large open-cast pits were dug at Macraes mine in Otago and Waihi mine in Coromandel, while both underground and open-cast methods operated at Golden Cross mine in Coromandel in the 1990s. Open-cast mining was also used at the Globe-Progress mine at Reefton, which reopened in 2007, with ore concentrate trucked and railed in containers 700 km to Macraes mine for final gold extraction. A mix of underground and open-cast mining is now being used at Waihi.

9.3.1 Minor Commodities

A small number of primarily underground mines operated briefly from the late 1800s to the early 1900s to extract various metals, including chromium, copper and tungsten. Of these, the underground and open-cast mercury mines at Puhipuhi, Northland, are probably the most significant, operating at various times from 1890 to 1945, with ore-processing facilities also on-site (Butcher 2010).

9.3.2 Number of Abandoned Mines

Some information on the number of abandoned mines is provided by NZ Mine Plans, a purpose-built database providing access to a catalogue of plans for recent and historical mines in New Zealand, linked to an interactive GIS webmap. NZ Mine Plans is available from the New Zealand Petroleum & Minerals website (https://www.nzpam.govt.nz/) and was developed jointly with WorkSafe New Zealand to provide regulatory oversight of mining activities in New Zealand and to support mining operators to ensure mine planning is informed and safe.

NZ Mine Plans identifies 450 abandoned mines, compared to 17 active (three gold, one iron sand, 13 coal) and five suspended operations (all coal); it includes opencast, alluvial and underground mines, as shown in Table 9.2. The number of mines is an underestimate, because it only includes the operations for which mine plans were available, which would also exclude many alluvial (placer)

TABLE 9.2
Summary of the Number and Type of Abandoned Mines Based on Information from NZ Mine Plans, as at July 2023

Commodity	Total Number	Underground	Open Cast	Alluvial
Coal	274	240	34	NA
Gold (and other commodities e.g. silver, copper, other)	150	140	0	10
Minor commodities (including copper, tungsten, mercury, antimony)	26	26	0	NA

NA – not applicable

operations. The number of gold-mine sites and placer/alluvial deposits identified in Christie 2019 is much greater than that shown in Table 9.2, while Christie and Barker (2013) identify a greater number of mines for the minor commodities than appear in the NZ Mine Plans database. Nonetheless, the location of abandoned as well as suspended and active (based on the definitions provided under Health and Safety Regulations, see section 9.2) mines shown in Figure 9.2 reflects the general spread and location of current and historically mined areas across New Zealand. A further point to note is that a significant number of abandoned mines fall on conservation estate; based on the point location provided in NZ Mine Plans, over 200 mines are located on conservation land, with another 127 mines within 500 m of conservation land boundaries.

9.4 ENVIRONMENTAL AND HUMAN HEALTH CONCERNS ASSOCIATED WITH ABANDONED AND DERELICT MINES

The environmental and human health concerns arising from abandoned and derelict mines are strongly influenced by the geological setting, including the commodity mined. For coal, environmental effects primarily depend on whether the resource is located in non-acid-forming or potentially acid-forming settings. Potentially acid-forming rocks release hydrogen and iron ions as pyrite oxidises. However, if carbonate is present in the coal measures, this may neutralise any acidity generated by pyrite oxidation, which means the release of acid mine drainage (AMD) depends on a balance between potentially acid-forming rocks and carbonate. Potentially acid-forming rocks are possible in all New Zealand coal-mining areas, but in practice mines in the Brunner coal measures on the West Coast (Figure 9.1) always produce AMD, while mines in other locations only occasionally produce AMD (Cavanagh et al. 2018a). AMD typically has pH values of 2–4 and elevated concentrations of dissolved iron, aluminium and sulphate, although elevated concentrations of other trace elements may also occur (Pope et al. 2010).

For gold and other metals, environmental concerns depend on whether the resource is located in epithermal or orogenic deposits. Epithermal deposits can give rise to mine drainage ranging from neutral to acidic. This can contain a wide variety of trace elements, including arsenic, lead, zinc and mercury, which are released from mining different deposits. In contrast, arsenic and, to a lesser extent, antimony are the primary contaminants of concern released from mining orogenic deposits. Additional sources for environmental concern at hard-rock gold-mining sites are large-scale tailings dams for processed overburden and ore materials, and ore processing (e.g. roasting plants, cyanidation tanks). Gold is harder to extract from orogenic deposits, with the release of arsenic and antimony also dependent on the sulphide content of the ore (Craw et al. 2015; Cavanagh et al. 2018c).

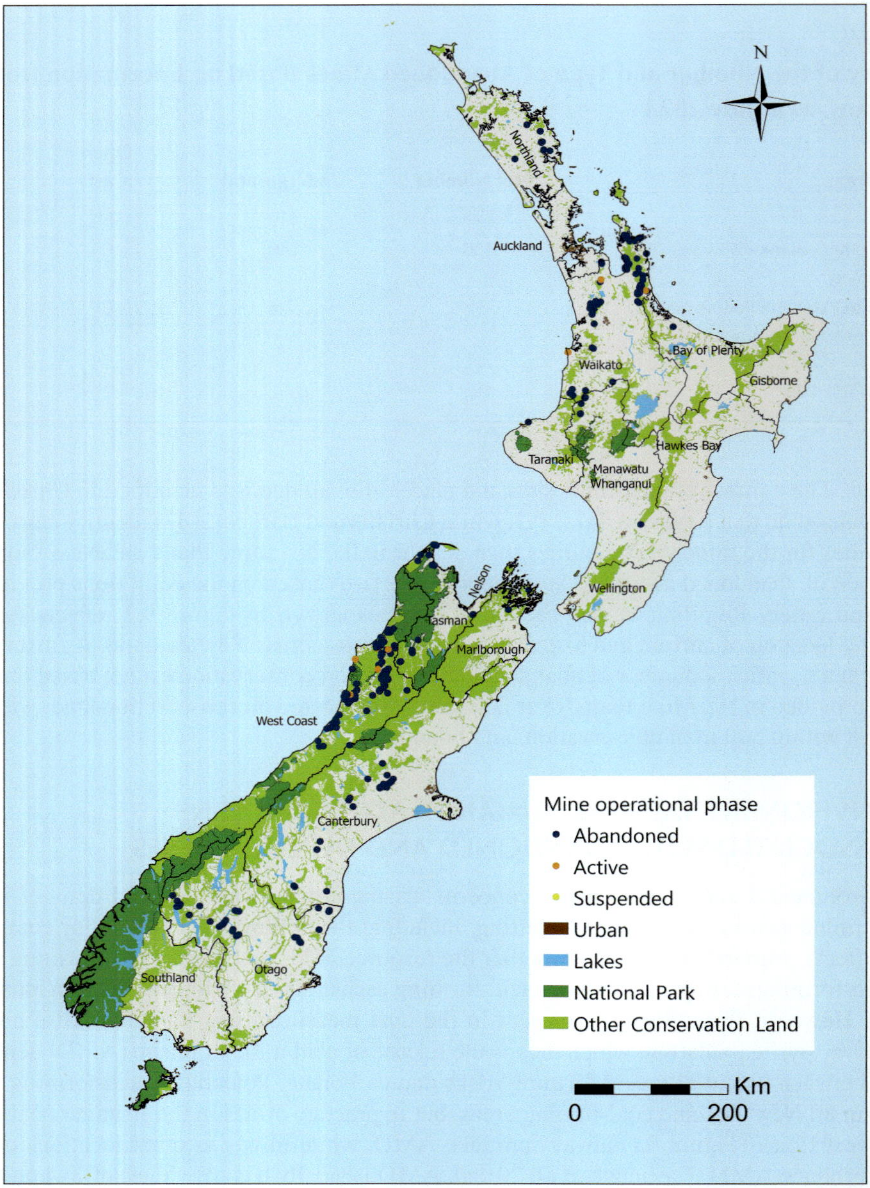

FIGURE 9.2 The approximate location of abandoned, suspended, and active mines across New Zealand, as captured in the NZ Mine Plans database, July 2023; these mines include only those with a mine plan.

The specific trace elements released determine the extent to which human health or environmental considerations. For example, arsenic, lead and mercury are of concern from a human health perspective, while copper and zinc have greater significance from an ecological perspective. A summary of the key mine types and the primary contaminants of concern is shown in Table 9.3.

9.4.1 Coal-Mining Areas

AMD's impacts on streams are the dominant negative environmental effects arising from derelict coal mines on the West Coast. These effects may arise through direct effects (e.g. toxicity associated

TABLE 9.3
Summary of Factors Influencing Environmental and Human Health Concerns Associated with Derelict Mines in New Zealand

Commodity	Mine Type	Geological Setting	Primary Contaminants of Concern	Contaminant-Related Impact
Coal	Opencast, underground	Acid-forming (pyritic rock)	pH, Al, Fe precipitates	Environmental
		Non-acid forming	Sediment	
Gold & other metals	Alluvial, hard-rock, underground, opencast	Mesothermal/orogenic deposits	As, Sb (sulphides)	Environmental, Human health
	Hard-rock – underground, opencast	Epithermal deposits	As, Sb, Pb, Zn, Hg, Cu	Environmental, Human health

Al = aluminium; Fe = iron; As = arsenic; Sb = antimony; Pb = lead; Zn = zinc; Hg = mercury; Cu = copper.

with low pH or high metal concentrations) or indirect effects associated with AMD (iron and aluminium oxyhydroxides), which can coat/cement the streambed and clog the areas around rocks where animals live, leading to poor habitat for all aquatic life.

Numerous studies investigating the biological impacts, particularly on macroinvertebrates, of AMD in streams on the West Coast have been undertaken since the 1990s (e.g. Winterbourn 1998; Harding et al. 2000). Impacts are often associated with marked changes in the taxonomic richness or composition of microbial (Lear et al. 2009), algal (Bray et al. 2008; Schowe et al. 2013; Schowe and Harding 2014), benthic invertebrate (e.g. Harding 2005, Grey and Harding 2012) and fish (Harding and Boothroyd 2004; Greig et al. 2010) communities. Hogsden and Harding (2012) found that mining streams receiving anthropogenic inputs of acidic and metal-rich drainage have much simpler food webs (fewer species, shorter food chains, fewer links) than those in naturally acidic, naturally high-metal and circum-neutral streams, which were also attributed to slowing leaf decomposition (Hogsden and Harding 2013). These studies have variously attributed pH or elevated metals, particularly aluminium (e.g. Waters and Webster-Brown 2013), as eliciting the primary contaminant of concern, although dilution was found to not be a practical solution to mitigating the effects of AMD on aquatic invertebrates (Waters and Webster-Brown 2016). Other studies found that metal hydroxides precipitating onto solid surfaces slowed the decomposition of leaves and wood, which affected associated biota (including microbes) more than water chemistry stressors (pH, metals) (Niyogi et al. 2013).

The West Coast also has numerous naturally acidic streams, which afford invertebrates sourced from these streams greater tolerance to AMD impacts than invertebrates from circum-neutral streams (Collier et al. 1990; O'Halloran et al. 2008). Native fish from West Coast streams also have a range of tolerance to pH (Greig et al. 2010; Jellyman and Harding 2014). Greig et al. (2010) recommended that to enable recolonisation of streams by fish, remediation of AMD-affected waters should aim to raise pH above $\approx$4.5 and reduce individual concentrations of dissolved aluminium and iron to less than 1.0 mg/L. A broader, catchment-scale study using data on algae, benthic invertebrates and fish communities from 186 stream sites, including both un-mined streams and those associated with 620 mines, found that mining was associated with a decline in stream biodiversity, irrespective of catchment size, and that recovery was not evident until at least 30 years after mining activities ceased (Clapcott et al. 2016).

Terrestrial impacts may also arise from pyritic rocks present at the surface. This was observed at Wangaloa, an open-cast coal mine in south Otago. Final rehabilitation was based on recontouring the surface, followed by planting over 100,000 nursery plants into 10 ha of unamended overburden. Large losses of nursery plants occurred on the overburden (60–65% mortality in the first three years), with some areas unsuccessfully replanted three times up to 2010 (Todd et al. 2009). Seedlings planted into coal-rich and pyrite-bearing overburdens died within onel year, repeatedly. These two hostile root zones had AMD (pH 1–3.5) from sulphate oxidation and high boron in the coal (400–500 mg/kg) and very dry conditions, which exacerbated plant stress, with some plants having up to 230 mg/kg boron (Rufaut et al. 2015). In contrast, overburden amended with imported topsoil and raw overburden rich in Cretaceous siltstone had the highest plant survival due to higher pH (4.5–5) and water retention compared to other types of overburden. Quartz-rich overburden also had relatively benign root zones, except when quartz pebble content exceeded 35% volume, at which point nutrient and water availability reduced to growth-limiting levels (Craw and Rufaut 2017). Nonetheless, ongoing rehabilitation at the site through a major planting programme and natural colonisation from nearby islands of established native species had the additional benefit of decreasing the severity of AMD at the site (Rufaut et al. 2015).

9.4.2 Gold Placer Mining

Environmental effects arising from placer mining are largely associated with the landscape effects of hydraulic sluicing, while dredging materials were often deposited on the land and returned to pastoral or forestry use (Simcock and Ross 2014). These materials were often highly nutrient deficient, with phosphorus being the primary limiting nutrient (Craw and Rufaut 2017). Another issue arising from unrehabilitated placer mines is unstable landforms that might require large-scale earthworks to create safe and stable sites (Simcock et al. 2004). Enduring environmental effects have arisen from poor-quality rehabilitation (Simcock and Ross 2014), although under some conditions historical mining activities have led to the creation of unique ecosystems.

9.4.2.1 Creation of Unique Ecosystems

In the Central Otago region, an arid climate has combined with the legacy effects of hydraulic sluicing from placer gold mining to support a distinctive and rare inland salt-tolerant or halophyte plant ecosystem (Druzbicka et al. 2015). Specifically, a rain shadow in the inland area of Central Otago, which has precipitation as low as ≈300 mm/year, combined with the deposition of marine aerosols dominated by salt-bearing sea spray, gives rise to some naturally saline areas (Craw et al. 2013). Historical sluicing at two placer gold mine sites has exposed the layer of clay-altered schist bedrock immediately below the gold-bearing rocks. Surface and ground water runoff are controlled by these impermeable surfaces, and the arid climate ensures that much of the mine water evaporates, leaving saline residues (Druzbicka et al. 2015; Law et al. 2016). Salination of the sites from mine water runoff has been going on for over 100 years and has limited the colonisation of tall native and exotic plants. Instead, the mine soils now support a distinctive and rare inland salt-tolerant or halophyte plant ecosystem with low-growing plants, and this natural rehabilitation has resulted in enhanced biodiversity for the area (Druzbicka et al. 2015; Law et al. 2016). These areas have recently been made into scientific reserves for that reason (Craw and Rufaut 2021). Further research is investigating ways in which these soil-free ecosystems can be enhanced and add to the collective knowledge of rehabilitation through enhancement of natural processes (Rufaut et al. 2023).

9.4.3 Orogenic Hard-Rock Mining

In orogenic gold deposits, arsenic, mainly in the form of arsenopyrite, can be significantly enriched compared to crustal background concentrations. Where natural weathering processes release these trace elements, concentrations remain relatively low, and dilution or attenuation in the downstream environment is rapid (Druzbicka and Craw 2015; Haffert et al. 2010). Where there are historical workings or wastes related to orogenic gold mining, the concentrations of trace elements (particularly arsenic) can be extreme (Haffert and Craw 2008a; Malloch et al. 2017).

A particular example is Blackwater mine, located in the Reefton goldfield, which was one of the largest producers of gold on the West Coast (Christie 2019). In 1951 there was a major collapse of the Prohibition shaft, resulting in the abandonment of the mine with no rehabilitation of the site. The nearby Waiuta historic town site is a popular destination for visitors, who also have unrestricted access to the mine site and associated wastes that were exposed at the surface. Some of the wastes associated with the processing plant contained arsenic at between 20 to 40 wt% (Haffert and Craw 2008a, 2008b), and those parts of the processing site were remediated in 2016 to limit the surface environmental exposure of arsenic and potential arsenic discharge to streams. Remediation included the removal of highly arsenic-contaminated material in barrels from the site for treatment and disposal at a specialised facility, with some less-contaminated material placed in a low-lying area lined with waste rock from the mine. Additional waste rock was placed on top of the contaminated material and compacted to create a solid surface and ensure people cannot be exposed to contamination.

McLachlan and Craw (2018) undertook an assessment of the residual contamination at the site, including a tailings area that had arsenic concentrations between 4,000 and 9,000 mg/kg, although leaching experiments showed the arsenic was largely immobile. While there were some bare areas, the tailings area has been recolonised naturally, predominantly with native shrubs and exotic grasses. Arsenic concentrations in the shrubs growing directly in the high-arsenic tailings material were generally low, although were up to 20 mg/kg dry weight in grass.

9.4.4 Epithermal Deposits

Epithermal gold mines may produce mine drainage ranging from neutral mine drainage, with neutral pH and low dissolved metal concentrations, to acid mine drainage (AMD), with acidic pH values and elevated concentrations of dissolved iron, copper, zinc, lead, manganese and/or other trace elements. There are few studies on the biological impact of mine drainage from epithermal systems in the North Island, although these studies found that mining impacts were clearly evident in streams (Livingstone 1987; Hickey and Clements 1998).

Perhaps one of the most extensively affected sites is Tui mine, which has also undergone an extensive remediation process. Mining of Te Aroha maunga (mountain) began in the 1880s, stopped about the turn of the century and resumed in the 1960s. A range of base metals, including copper, lead and zinc, were produced until 1973, when Tui mine, with five levels of adits, a processing plant, waste rock and tailings, was abandoned by Norpac Mining Co. The site is located within the catchments of the Tui and Tunakohoia Streams, both of which flow into the Waihou River. The Tui stream was affected by heavy metals arising from the tailings dam, while the Tunakohoia (meaning "plentiful eel" in Māori, the indigenous language) stream was affected by high dissolved metals concentrations and highly acid leachate from the adits, waste rock, and tailings dam. Technical reports found that tailings impoundment was at risk of collapse in a moderate seismic event or an extreme weather event, potentially resulting in liquefying over 90,000 m^3 of tailings containing $\approx$3,000 ppm lead through the town of Te Aroha.

A $NZ24 million remediation programme was undertaken from 2010 to 2016, involving treatment of adit discharges to reduce loads of metal contaminants entering the Tui and Tunakohoia Streams, and stabilisation of the tailings dam (Environment Waikato, undated). Specifically, the workings were partially plugged with bulkheads, and an alkaline solution (lime or cement) was injected into the ground. Chemical and physical stabilisation of tailings and waste rock with a layer of imported clean fill and topsoil was placed over the newly shaped land, which was grassed. The cap was designed to stop oxygen and water entering the stabilised tailings in the short term; the grass was to stabilise the surface against erosion. Tall trees must be prevented in perpetuity from establishing to protect the cap.

The remediation goal for water quality was to achieve an order of magnitude reduction, or preferably, water quality in the lower Tui and Tunakohoia Streams, that would not pose an unacceptable risk for contact recreation (URS 2014). For zinc, a 90% reduction in mass load was realised (URS 2014). Post-remediation monitoring of the site indicated an improvement in stream health, based on macroinvertebrate assessment, although key sites in the Tui and Tunakohoia Streams were still affected (PDP 2016). Additional studies assessing invertebrates and fish undertaken over 2014–2016 indicated that the streams remain affected, with elevated zinc likely to be exerting the dominant effect (Cavanagh et al. 2022). While concentrations of lead in the water were low, preliminary studies on eels collected from the Tui and Tunakohoia Streams showed elevated concentrations (Safran 2017). This poses a human health risk, although the magnitude of this risk depends on the extent to which eels from these streams are consumed, and the extent to which elevated lead is observed in fish in the surrounding waterways.

Poor historical mining practices resulted in ore and crushed rock containing elevated trace element concentrations being distributed widely and heterogeneously across the site. Zinc, copper and lead are the main contaminants with elevated concentrations, with concentrations in the order of 1,000 – 3,000 mg/kg. Zinc and copper are both essential elements for plant and animal growth (e.g., soil invertebrates), but at elevated concentrations they are toxic to plants and animals. Leaching or offsite movement of soil particles with elevated zinc or copper provides an additional contribution of these metals to the surrounding waterways. However, elevated lead in soil poses a greater human health risk rather than an environmental risk. The magnitude of this risk is dependent on the frequency of visits and the extent of potential incidental exposure to these soils (e.g., through ingestion of soil particles attached to hands or retained on footwear). Movement of soil particles with elevated lead into surrounding waterways contributes to elevated concentrations in stream sediments, which provide a potential exposure route for lead in eels. Arguably, the Tui Mine site remains the most contaminated derelict mine site in New Zealand, despite the significant rehabilitation efforts that have been undertaken to date.

A different example of effects associated with epithermal deposits is that of the Puhipuhi mercury mines, which were operational from around 1907 to 1945 via both underground and opencast methods (Butcher 2010). The mines are located within the Waikiore Conservation Area and a considerable amount of processing plant machinery is still present (Butcher 2010). Gionfriddo et al. (2015) found elevated mercury across the site and identified mine waste dumps, located upgradient from the Waikiore Stream, as a significant contamination risk. Fairgray (2019) undertook high-spatial-resolution monitoring to build an understanding of the distribution of key elements of concern, in particular mercury. High concentrations of sediment-bound arsenic and mercury were present in the upper reaches of sub-tributaries of the Waiariki Stream, and showed that sources were not limited to the ore-processing site but also arise from other areas of exposed rock, such as sinter outcrops and former quarry sites. Mercury concentrations were highest at the ore-processing site of the mine, while arsenic was low at the mine site but elevated in streams adjacent to the mine-affected Waikiore Stream.

Macroinvertebrate populations followed a similar pattern to mercury concentrations. Several sources of mercury in addition to the mine were identified, including the weathering of sinter rock, which may release mercury (cinnabar), iron (pyrite), arsenic (marcasite containing arsenic) and antimony (stibnite). Anthropogenic sources include those associated with the mining of rock, runoff from roads lined with quarried rock, machinery at the ore-processing site and vehicles on nearby roads. Ecological assessments of aquatic organisms indicate lower taxonomic richness and loss of sensitive taxa for sites with greater mercury and arsenic contamination (upper catchment), although opposite trends were observed in concurrent studies by PDP, reported in Fairgray (2019), who attributed greater macroinvertebrate richness in the upper catchment to habitat rather than contamination.

Mercury was also commercially extracted from Ngawha (Northland) and Mackaytown (Coromandel Peninsula); mining for mercury is now a prohibited activity under the Natural and Built Environment Act 2023.

9.5 MINE REMEDIATION AND REHABILITATION

A comprehensive discussion on mine rehabilitation in New Zealand is provided by Simcock and Ross (2017), who also discuss rehabilitation to pastoral land uses and native ecosystems. Further guidance on rehabilitation approaches (including passive treatment and terrestrial rehabilitation) – which are also relevant for derelict and abandoned mines – is provided in the Mine Environment Lifecycle Guides (Cavanagh 2018a, b, c). In the following sections we discuss and provide context specific to the rehabilitation of derelict mines, including opportunities provided by rehabilitation at active mine sites, and examples of alternative rehabilitation approaches, including the integration of cultural values into rehabilitation processes and approaches to enhancing natural regeneration at derelict mines.

9.5.1 Rehabilitation at Active Mines

Many current mining operations occur in areas that have been mined historically, which provides an opportunity to address environmental effects arising from derelict mines. One example is Stockton open-cast coal mine, the largest open-cast coal mine in New Zealand. The mine is located on the Stockton plateau. Many of these underground mines are a source of uncontrolled release of AMD. Current open-cast mining operations are digging out some underground tunnels, potentially enabling greater control of mining operations. The Mangatini Stream, a significant tributary of the Ngākawau River, was identified as the main source of poor water quality within the Ngākawau catchment, and in 2005 Solid Energy, then owner of Stockton, committed to a five-year programme to improve the water quality of the Ngākawau River. Water quality targets including pH greater than 4.7 (99% of the time); aluminium less than 1 mg/L (99% of the time) to be achieved >90% of the time during base flow conditions, were set (Elder et al. 2012). In 2009, the Mangatini Sump was commissioned for this purpose, with a total construction cost of approximately $20 million (Murray and Walker 2022). Marked improvements in water quality were subsequently observed. Fish surveys undertaken in 2005 and 2011 found an improvement in fish numbers in Ngākawau in 2011 (Grey et al. 2016), suggesting that the improved water quality is also leading to improved ecological health. The construction cost of the Mangatini Sump provides an indication of the magnitude of costs required to remediate high-load AMD impacted streams – and noting that operational costs will further add to the total costs.

9.5.2 Rehabilitation of Derelict Mines

There are an unknown number of derelict mines in New Zealand for which rehabilitation may be required, although the ability to achieve this is highly dependent on funding. The main mechanism for funding assistance currently available for investigating and rehabilitating most derelict mines is through the Contaminated Sites Remediation Fund provided by the Ministry for the Environment. This fund was established in 2003 for a three-year trial, following development of the Orphaned Sites Remediation Fund, which began in 1999 to clean up the Fruitgrowers Chemical Company site, considered to be New Zealand's most contaminated site at the time (Ministry for the Environment 2011a). The fund is available to support landowners who are unable to fund remediation or management themselves, and where site contamination has occurred before the RMA was enacted in 1991, or after 1991 but only if no enforcement could be undertaken. Regional councils and unitary authorities can apply for funding, on behalf of landowners, such as from Department of Conservation (DOC), for the remediation of contaminated sites that pose a high risk to human health and the environment (Ministry for the Environment 2013). In 2023, $NZ2.63 million is available annually, with a portion of this available on a contestable basis.

The Contaminated Sites Remediation Fund provided $21.7 million of the $24 million used for the remediation of Tui mine (Environment Waikato undated), and, jointly with the Department of

Conservation, \$2.6 million for the remediation of the Blackwater and Alexandra mines on the West Coast in 2016 (Source: Department of Conservation (NZ).

Three derelict mines or mining infrastructure are on the current list of contaminated sites prioritised for funding (Ministry for the Environment 2022). Of these, two are related to gold mining. The first, Victoria Battery, was the largest gold ore–crushing battery in New Zealand and operated from 1897 to 1952. This heritage-listed site is located on Public Conservation Land within the Karangahake goldfield area. The second gold mine is the Tokatea mine complex in Coromandel, which commenced in 1868 and operated in varying capacities until 1935. The final site being investigated is the Puhipuhi mercury mine in Northland, which operated from 1907 to 1945.

9.5.2.1 Rehabilitation Driven by Cultural Values

Te Aroha maunga, on which Tui mine is located, is one of the most significant landscapes for the tribes and people of the Hauraki region. Given this significance, an iwi (tribal) advisory group was established to help guide the outcomes of the remediation process, which also aimed to "address, as far as practicable, within the limitations of the project, the impacts of the Tui mine on the taonga (treasure) of the Te Aroha maunga mountain for iwi" (Waikato Regional Council, undated). Some outcomes from the remediation programme, such as the grassed, remediated tailings dam and ore stockpiles shown in Figure 9.3, were predetermined and reflected conventional engineered rehabilitation. An opportunity to give greater effect to the aspirations of local iwi arose from the discovery of additional contamination arising from dumping ore in a gully adjacent to the tailings dam, known as Area S. Rehabilitation of Area S was developed with the iwi advisory group and informed by a cultural monitoring plan (Anderson 2013) through an iterative process involving site visits before and during remediation works.

The rehabilitation approach for Area S was designed to promote natural succession by creating root zones that would allow seeds and transplants from the adjacent forest to establish (i.e. helping the maunga to heal itself). A more natural topography was created, avoiding linear and obviously "engineered" lines and vertical cuts or cleared areas; for example, the rip-rap-lined water course was gently sinuous, and tree ferns were placed to break up linear features and prevent long views down the stripped area, as shown in Figure 9.4. There was no importation of materials (no nursery

FIGURE 9.3 Conventional rehabilitation showing the grassed ore stockpile area with plantings at age 1.

FIGURE 9.4 Top: Upper third of Area S, Tui Mine, showing natural regeneration with transplanted tree ferns and kiekie cuttings at age 1 year, September 2017. Bottom: upper third of Area S showing regenerated wineberry, māhoe, and five-finger overtopping tree ferns and pampas, with koromiko, tī kōuka, and groundferns in the foreground, February 2022.

plants or imported soils were used); instead, suitable plants (especially tree ferns), soils and logs were obtained by salvaging them ahead of stripping and placed back into rehabilitation areas.

The approaches were supported by a cultural induction involving all contractors. Iwi representatives, an ecologist, engineers and earthworks contractors worked together to get options that were both technically and culturally feasible and to reduce risks to the contractors of "non-standard" works and outcomes. This included prioritising those plants to protect (a crane was used to avoid damage to plants screening the site, and high-value trees and ferns were physically identified to help protect them during earthworks), as well as understanding the cultural value of specific plants and how this affects their suitability for rehabilitating the site, which remains contaminated (e.g. using *Clematis paniculata* as a warning).

Area S had high concentrations of copper (480–870 mg/kg) and zinc (1,100–4,900 mg/kg), which might be expected to inhibit plant growth. Lead was present at high concentrations (3900–6400 mg/kg), although this is expected to have less effect on plants compared to the concentrations of copper and zinc. Similarly elevated trace element concentrations (zinc ≈3,000 mg/kg, copper ≈800 mg/kg) were found in an area known as the log stockpile area, where logs retrieved from the tailings impoundment area during stabilisation works had been placed, and where there was no natural regeneration. The reason for this difference in plant response is largely unknown, although it was observed that a much higher carbon:nitrogen ratio was present in the log stockpile, which may indicate nutrient limitation (Simcock and Cavanagh 2022; Cavanagh et al. 2022).

9.5.2.2 Enhancing Natural Regeneration

Enhancing natural regeneration can be an effective technique to rehabilitate abandoned mine sites where overburden is the dominant substrate. One example is the Wangaloa coal mine that operated for 45 years between 1945 and 1989. No topsoil was saved during mine development, so all vegetation has developed on substrates exposed by mining. As discussed in section 9.4.1 extensive plantings during intensive rehabilitation efforts commencing in 2000, resulted in significant failures on some of the more hostile substrates – although these plantings were important for establishing vegetation and ecosystems on the more benign loess substrates (Rufaut et al. 2015). Regeneration of native vegetation (mainly mānuka, *Leptospermum scoparium* and kanuka, *Kunzea ericoides*) had occurred during the mining period on parts of the unmined loess slopes, with additional seed sources for native shrub colonisation in semi-natural forest remnants bordering the mine. Gorse (*Ulex europeus*) developed alongside the manuka and kanuka, becoming more dominant with increasing distance from the forest remnants. In other areas, retained bush patches provided seed sources for natural colonisation by mānuka and kanuka onto waste rock. This natural colonisation had resulted in vegetation cover of over 30% of the waste rock area by 2015 (Rufaut et al. 2015). As exemplified in this example, critical to natural regeneration is the presence of proximal seed sources, this coupled with techniques such as fascining (see next section) and, where needed, control of browsing animals or competing plants can enhance natural regeneration. At Wangaloa, the more extensive revegetation coverage has contributed to a decrease in the severity of acid mine drainage (AMD), apparently becoming sustainable with-out further intervention (Rufaut et al. 2015). Natural regeneration – primarily by manuka and kanuka – has also contributed to revegetation on highly arsenic contaminated substrates (Craw et al. 2007, McLachlan and Craw 2018).

As noted in section 9.4.2.1, natural regeneration can lead to the creation of unique ecosystems under some conditions, also giving rise to opportunities for alternative rehabilitation approaches. A further example of enhancing natural regeneration is the Bellvue coal mine discussed in the next section.

9.5.2.3 Rehabilitation of the Bellvue Coal Mine

Bellvue mine is an underground coal mine located in the Brunner coal measures that was mined from c.1941 to 1965, with extracted coal transported to Rapahoe on a rope gravitational haulage system (West 2014). This derelict mine was a focus for research on rehabilitation for the Centre for Minerals Environmental Research (CMER) in partnership with local organisations (Grey District Council, the West Coast Coal Heritage Trust, Minerals West Coast, Tai Poutini Polytechnic, Go West Coast), and included treatment of AMD and rehabilitation of an extensive coal fines area.

This mine is the major source of AMD discharging into Cannel Creek (Trumm and Cavanagh 2006). The AMD emanates from an anoxic pool in the adit and flows down a steep, 55 m-long cascade. At the bottom is a large, flat area covered in coal tailings, small pebbles and gravels, and a hard, oxidised layer caused by AMD precipitates. These factors influenced the selection of passive treatment options, which were initially trialled by West (2014) and Forbes et al. (2018), with a full-scale sulphate-reducing bioreactor treatment system ultimately installed at the base of the cascade. The mine drainage was piped from the top of the cascade down to the treatment system, where a header system split the flow evenly between five 30,000 L plastic tanks filled with waste mussel shells (Fairgray 2019). AMD flowed up through the bioreactor treatment tanks and exited through pipes at the top of each tank. The treatment system outflow pipes discharged into a small channel, and the treatment tank effluent entered Cannel Creek via overland surface flow from the channel. The pH in Cannel Creek immediately downstream of the Bellvue mine AMD discharge point increased by 2.4 pH units.

After remediation, the concentration of dissolved iron, aluminium and nickel entering Cannel Creek from Bellvue mine was 80% lower than the average pre-remediation concentration at the site immediately downstream of the Bellvue mine AMD discharge point. The concentration of dissolved zinc in Cannel Creek was 66% lower at approximately 0.02 mg/l than before remediation with the bioreactor treatment system. There has been a significant decrease (over 90%) in the concentration of iron and aluminium in Cannel Creek immediately downstream of where the Bellvue mine AMD entered. However, the treatment system has an operating limit of 1 L/s AMD passing through each of the tanks to properly treat the AMD. At high flow, when this limit was exceeded, the excess AMD bypassed these tanks and entered Cannel Creek untreated (Fairgray 2019). While these studies demonstrated the potential for passive treatment to improve water quality in the stream, an absence of funding for continuing operation or further improvements has meant that AMD treatment has now been abandoned.

Aside from AMD discharge, there is a large area of coal fines remaining at the main coal storage area, which is defined by a sharp, dense edge of native vegetation (as shown in Figure 9.5). Scattered gorse seedlings (*Ulex europeus*, a woody, nitrogen-fixing shrub-weed with a long-lived seed bank) are present in drainage lines and along the western edge, probably from seed washed down from the road above. The coal stockpile has been slowly revegetating from the edges since the 1950s to early 1960s, but this is limited by high fluctuations in surface temperature (due to black coal) and very little protective micro-topography (a near-level surface), combined with low pH (restricting rooting depth and exacerbating drought stress).

Trials undertaken to passively revegetate about 40% of the exposed coal storage area used a range of techniques combined with targeted weed control. The overall post-remediation objective was to achieve resilient, low-maintenance, native plant cover along exposed road edges and revegetated areas. The methods focused on salvage and reuse of existing resources (such as live plants, seed-laden branches and wood), minimal use of imported materials, mitigation of existing weeds and prevention of new weeds. A revegetation technique known as *fascining*, where small- to medium-sized branches with ripe seed capsules or fruits cut from are laid in direct contact with the ground

FIGURE 9.5 Coal storage area at Bellvue mine in January 2021, showing the distinct edge of native vegetation surrounding the coal storage area along with a small 2015 trial in the foreground and a larger trial in the distance.

TABLE 9.4
Tallest Mānuka Seedling Height in Plots Amended with Biosolids, Compost, Lime + Compost or Lime Only Compared to Control Plots with Fascined Manuka (F) or No Fascining (N) 6 Yrs after Establishment

Height of Tallest Seedling (mm)	Biosolids	Compost	Lime + Compost	Lime	Control F	Control N
Mean (std dev)	2425 (377)	438 (360)	962 (672)	494 (381)	705 (689)	622.5 (663)
Annual growth rate (std dev)	485 (75)	88 (72	192(134)	99 (76)	141(138)	125 (133)

to maximise released seed fall, was the dominant method used to facilitate revegetation. Multiple branches can be woven on top of each other (brush layering) to also increase fallen seed density and shelter. The leaves fall, creating a protective mulch for the germinating seeds. Seedlings developed from fascining can seemingly tolerate fairly hostile "soil" conditions (see also Rufaut et al. 2015).

Amendments in the form of biosolids, compost and lime were used to mitigate the low pH and nutrient status of the coal fines. Plots (approximately 2 m × 2m) were treated with lime, compost, lime+compost, dried biosolids, or were a control, with mānuka branches with ripe seed capsules placed over the plot and wood used to hold down the branches. In addition, some flax (*Phormium tenax*) was salvaged from the road surface before it was upgraded; and some flax seedlings from a local nursery were planted. Flax was intended to attract nectar-feeding birds, which are not attracted by mānuka flowers/seeds, and which would assist with natural regeneration. The robust, 1 to 2 m long flax flower stalks are also very useful perches, and flax grows in open conditions.

Initial seed trials in autumn 2015 showed that mānuka seedlings could germinate on the surface and grow over the following 12 months if the surface was covered with mānuka branches and lime or compost added. No mānuka established in 'controls' (i.e. branches but no seed). In March 2017 most gorse, including all flowering gorse, was cut and the base immediately painted with herbicide to kill it while minimising herbicide damage to adjacent native plants and avoiding disturbing the surface. Seedling growth was measured in February 2023 and showed significant growth, particularly for plots that received biosolids, as shown in Table 9.4. In addition, non-native seedlings present in the plots were counted. Biosolids had five times the number of seedlings than any other treatment. Only two species were present across plots with more than two individuals and both are woody trees – native beech and broadleaf (*Griselinea littoralis*). This trial highlights the potential of low-cost, minimal intervention approaches to enhancing natural regeneration.

9.6 HERITAGE VALUE OF DERELICT MINES

In contrast to the often negative environmental and human health considerations associated with derelict mines, they can also provide significant heritage value. Several mining-related locations – the Denniston area (coal mining), Brunner coal mine historic area, Charming Creek (coal mining and timber), Karangahake Gorge (gold mining, Waikato) and the Bannockburn sluicings (alluvial gold mining, Central Otago) – are considered by DOC to be iconic sites when exploring New Zealand's cultural heritage (Source: Department of Conservation (NZ). DOC estimates it has as many as 1,000 gold-mining sites on lands it manages, with approximately 70 that are considered to represent most of New Zealand's significant gold-mining heritage actively managed and developed for visitors. Similarly, various coal mine, precious mineral and quarry sites are actively managed by DOC to capture cultural heritage, and many towns in mining areas across New Zealand leverage their mining heritage to attract tourists.

Where a derelict mine is of archaeological value (i.e. pre-1900), Heritage New Zealand has a statutory role to protect these sites under the Heritage New Zealand Pouhere Taonga Act 2014. DOC has an interest in more recent structures or remnants that reflect the heritage of a site and is a signatory to the International Council on Monuments and Sites charter. Thus, on conservation land, a heritage assessment of a derelict mine is required prior to initiating investigations, either for new mining operations or for site contamination. For new mining operations, management of historical mining artefacts can be a highly valued part of mine rehabilitation and compensation.

A specific example is the Globe-Progress open-cast mine, which is located in an area of historical underground coal and gold mining within the Victoria Conservation Park. The mining heritage includes walkways, abandoned mining equipment, buildings and other mining history stretching from Reefton to the Waiuta area (https://www.reefton.co.nz/discover).

Construction of the tailings dam affected parts of the Alborns coal mine, an underground mine that started in about 1880 and operated at a large scale from about 1935 before being abandoned in the 1950s. The area was not protected under Heritage New Zealand, but the mine is actively managed by DOC, and a mitigation package was agreed to preserve the legacy of the "old time" Alborns coal miners and enhance a resource for tourism. This included a full photographic and survey record of the affected areas, construction of a shed over a 1915 Leyland truck, interpretation signage, restoration of a coal wagon and relocation of recoverable artefacts in agreement with DOC (Hayton 2015).

However, there can be tensions between the desire to preserve mining heritage and ensuring the health and safety of visitors. For example, Prohibition mill and mine adjacent to the "Ghost-town" of Waiuta was promoted as a place to visit; however, "alarmingly high" arsenic concentrations (up to 40% arsenic) in the ruins of the condenser towers of the ore roasting plant – which at the time also provided a "nice spot to sit" – were discovered through geochemical investigations in the mid-2000s (Haffert and Craw 2008a,b, Haffert and Craw 2009). This site was subsequently closed and remediated before being reopened to the public (see section 9.). The Puhipuhi mercury mine is unique in New Zealand and hence has great cultural heritage value (Butcher 2010). However, mercury is a highly toxic metal, and contamination is known to exist in and around the mine with investigations currently underway to better understand site contamination. The Victoria Battery, Karangahake, is the site of the largest stamper plant in New Zealand and is heritage listed but is also currently undergoing investigations to assess site contamination issues.

9.7 CONCLUSIONS

Abandoned mines in New Zealand can span those that are unrehabilitated to, more recently, those undergoing planned rehabilitation and closure. The former are more easily considered as derelict mines, although rehabilitation activities that may have occurred at some other abandoned mines may be considered insufficient by today's standards to mitigate environmental and human health concerns. Thus, we have considered both unrehabilitated (derelict) and partially rehabilitated mines in this chapter. Following from this is that the environmental state of these mines is influenced by the time of the commencement of mining (and thus the legislation under which any rehabilitation activities might have been required) alongside mine type and geological setting.

Alluvial or placer mining has most often given rise to landscape effects, in particular those associated with hydraulic sluicing, while environmental concerns arising from underground and open-cast mining are more closely related to their geological setting. The most extensive environmental effects arise from acidic and metalliferous drainages and substrates associated with potentially acid-forming coal measures, such as the Brunner Coal Measures (e.g. Stockton Plateau), and epithermal deposits (e.g. Tui Mine). This particularly affects aquatic ecosystems but can also impact terrestrial rehabilitation. Human health concerns more often arise from derelict mines in orogenic or epithermal gold deposits where arsenic can be significantly elevated in processing ore or waste rock – the magnitude of this risk depends on the extent to which people are exposed to these materials.

Rehabilitation of derelict and partially rehabilitated mines depends on a number of factors, including the costs of rehabilitation and the availability of funding, with the costs of rehabilitation disproportionately affected by:

- presence of acidic or metalliferous drainage/substrates (e.g. Tui mine, Bellevue mine, Stockton mine),
- whether large scale earthworks are needed to create safe and stable sites, and/or specific hydrology (e.g. wetlands) (many of the alluvial gold mines, quarries, also some coal mines),
- location of the mine – many derelict mines are in remote locations (and on conservation estate) hence access is challenging; conversely, some are located in areas that are currently being mined, in which case concerted efforts should be made to also rehabilitate legacy impacts.
- potential for natural regeneration to achieve desired outcomes (requiring favourable substrates, proximal seed sources and sometimes control of browsing animals or competing plants). In some sites, the potential for natural regeneration can be released by exposing suitable substrates (Rufaut et al. 2023) or moderating surfaces by low-cost amendments alongside sheltered sites (Tui) and seed sources (Bellvue, Wangaloa).

Finally, there are some positive contributions of derelict mines that should be considered alongside the more commonly considered negative outcomes. The heritage value associated with legacy mines and mining activities should be recognised and enhanced, where appropriate, during rehabilitation processes – this rehabilitation also needs to assess and remediate, to the extent practicable, risks associated with any contamination on-site. Further, under some conditions, derelict mines can give rise to unique ecosystems, such as the halophyte communities of Central Otago, that, in turn, provide a positive environmental outcome.

REFERENCES

Anderson, A. 2013. Tui mine cultural monitoring plan 2012–2017. Tui Mine Remediation Project. Prepared for Tui Mine Iwi Advisory Group.

Bray, J. P., P. A. Broady, D. K. Niyogi, and J. S. Harding. 2008. Periphyton communities in New Zealand streams impacted by acid mine drainage. *Marine and Freshwater Research* 59:1084–91.

Butcher, M. 2010. The Puhipuhi mercury mine history and site description. Whangarei: Department of Conservation. https://www.doc.govt.nz/globalassets/documents/conservation/historic/by-region/northland/puhipuhi-mercury-mines-description-and-history.pdf

Cavanagh, J., J. Pope, J. Harding, et al. 2018a. *Mine environment life-cycle guide: Potential acid-forming and non-acid-forming coal mines*. Christchurch: CMER.

Cavanagh, J., J. Pope, R. Simcock, et al. 2018b. *Mine environment life-cycle guide: Epithermal gold mines*. Christchurch: CMER.

Cavanagh, J., J. Pope, R. Simcock, et al. 2018c. *Mine environment life-cycle guide: Mesothermal (orogenic) gold mines*. Christchurch: CMER.

Cavanagh, J., R. Simcock, and D. Trumm. 2022. A summary of research by the Centre for Minerals Environmental Research at Tui Mine, Te Aroha. LC Contract Report LC4186 prepared for the Iwi Advisory Group and Waikato Regional Council. Lincoln, NZ: Manaaki Whenua – Landcare Research.

Christie, A. B. 2019. Introduction to New Zealand hard rock gold deposits and their exploration models. *New Zealand Journal of Geology and Geophysics* 62(4):394–413.

Christie, A. B., and R. G. Barker. 2013. Mineral, coal and petroleum resources: Production, exploration and potential. In *Ecosystem services in New Zealand – conditions and trends*, ed. J. R. Dymond. Lincoln, NZ: Manaaki Whenua Press.

Clapcott, J., E. Goodwin, and J. S. Harding. 2016. Identifying catchment-scale predictors of coal mining impacts on New Zealand stream communities. *Environmental Management* 57:711–21.

Collier, K. J., O. J. Ball, A. K. Graesser, M. R. Main, and M. J. Winterbourn. 1990. Do organic and anthropogenic acidity have similar effects on aquatic fauna? *Oikos* 59:33–38.
Craw D., and Rufaut, C. 2017. Geochemical and mineralogical controls on mine tailings rehabilitation and vegetation, Otago Schist, New Zealand. *Journal of Geology and Geophysics*, 60, 176–187.
Craw, D., and C. Rufaut. 2021. Geoecological zonation of revegetation enhances biodiversity at historic mine sites, Southern New Zealand. *Minerals* 11:18. https://doi.org/10.3390/min11020181
Craw, D., J. Cavanagh, J. Druzbicka, et al. 2015. A geoenvironmental model for orogenic gold deposits to predict potential environmental effects. *Mine Water Research in New Zealand* 34(4):388–403.
Craw, D., J. Druzbicka, C. Rufaut, and J. Waters. 2013. Geological controls on paleo-environmental change in a tectonic rain shadow, southern New Zealand. *Palaeogeography, Palaeoclimatology, Palaeoecology* 370:103–16.
Craw, D., C. G. Rufaut, L. Haffert, and L. A. Paterson. 2007. Plant colonization and arsenic uptake on high arsenic mine wastes, New Zealand. *Water, Air and Soil Pollution* 179:351–64.
Craw, D., C. G. Rufaut, S. Hammitt, S. Clearwater, and C. M. Smith. 2007. Geological controls on natural ecosystem recovery on mine waste in southern New Zealand. *Environmental Geology* 51:1389–1400.
DoC. 2016. Prohibition mine arsenic remediation complete. Media release, 16 December 2016. Wellington, New Zealand: Department of Conservation. https://www.doc.govt.nz/news/media-releases/2016/prohibition-mine-arsenic-remediation-complete/
Druzbicka, J., and D. Craw. 2015. Metalloid attenuation from runoff waters at an historic orogenic gold mine, New Zealand. *Mine Water and Environment* 34:417–29.
Druzbicka, J., C. Rufaut, and D. Craw. 2015. Evaporative mine water controls on natural revegetation of placer gold mines, southern New Zealand. *Mine Water and Environment* 34:375–87.
Elder, D. M., M. H. Pizey, P. A. Weber, et al. 2012. Addressing the environmental effects of mining on the Ngakawau River. WaterNZ annual conference.
Environment Waikato. Undated. Tui Mine Clean up. https://www.waikatoregion.govt.nz/services/regional-services/waste-hazardous-substances-and-contaminated-sites/tui-mine/
Fairgray, M. E. 2019. Assessing the use of geochemical modelling in supporting mine site remediation. A thesis submitted in partial fulfilment of the requirements for the degree of PhD diss., University of Canterbury.
Forbes, E., D. Trumm, J. Pope, and D. Bell. 2018. Comparison of diversion well substrates for the treatment of acid mine drainage Bellvue Mine, West Coast, New Zealand. In *Risk to opportunity*, Vol 2, eds. C. Wolkersdorfer, L. Sartz, A., J. Burgess, and G. Tremblay, 116–120. 11th ICARD | IMWA | MWD Conference. Pretoria: ICARD.
Gionfriddo, G. M., J. M. Ogorek, M. Butcher, D. P. Krabbenhoft, and J. W. Moreau. 2015. Mercury distribution and mobility at the abandoned Puhipuhi mercury mine, Northland, New Zealand. *New Zealand Journal of Geology and Geophysics* 58(1):78–87.
Gray, P. D., and J. S. Harding. 2012. Acid Mine Drainage Index (AMDI): A benthic invertebrate biotic index for assessing coal mining impacts in New Zealand streams. *New Zealand Journal of Marine and Freshwater* 46:335–52.
Gray, P. D., J. S. Harding, and P. Lindsay. 2016. Remediation of a major acid mine drainage point source discharge restores headwater connectivity for a diadromous native fish. *New Zealand Journal of Marine and Freshwater Research* 50(4):566–80.
Greig, H. S., D. K. Niyogi, K. L. Hogsden, P. G. Jellyman, and J. S. Harding. 2010. Heavy metals: Confounding factors in the response of New Zealand fish assemblages to natural and anthropogenic acidity. *Science of the Total Environment* 408:3240–50.
Haffert, L., and D. Craw. 2008a. Mineralogical controls on environmental mobility of arsenic from historic mine processing residues, New Zealand. *Applied Geochemistry* 23:1467–83.
Haffert, L., and D. Craw. 2008b. Processes of attenuation of dissolved arsenic downstream from historic gold mine sites, New Zealand. *Science of the Total Environment* 405:286–300.
Harding, J. S., and I. K. G. Boothroyd. 2004. Impacts of mining. In *Freshwaters of New Zealand*, eds. J. S. Harding, M. P. Mosley, C. P. Pearson, and B. K. Sorrell, 36.1–36.10. Christchurch, NZ: New Zealand Hydrological Society Inc. and New Zealand Limnological Society Inc.
Haffert, L., and D. Craw. 2009. Field quantification and characterisation of extreme arsenic concentrations at a historic mine processing site, Waiuta, New Zealand. *New Zealand Journal of Geology and Geophysics* 52:261–72.

Haffert, L., D. Craw and J. Pope. 2010. Climatic and compositional controls on secondary arsenic mineral formation in high As mine wastes, South Island, New Zealand. *New Zealand Journal of Geology and Geophysics* 53:91–101.

Harding, J. S. 2005. Impacts of metals and mining on stream communities. In *Metal contaminants in New Zealand*, eds. T. A. Moore, A. Black, J. A. Centeno, J. S. Harding, and D. A. Trumm, 345–57. Christchurch: Caxton Press, 2005.

Harding, J. S., G. P. Quinn, and C. W. Hickey. 2000. Effects of mining and production forestry. In *New Zealand stream invertebrates: Ecology and implications for management*, eds. K. J. Collier and M. J. Winterbourn, 230–59. Christchurch, NZ: New Zealand Limnological Society.

Hauraki District Council. 2014. 5.18 Golden cross mineral zone. Hauraki District Plan.

Hauraki District Council. 2016. 5.17 Martha mineral zone. Hauraki District Plan.

Hayton S.2015. Mitigation of the Historic Alborn's Coal Mine and Remnants.OceanaGold Presentation to Minerals West Coast, Greymouth.

Hickey, C. W., and W. H. Clements. 1998. Effects of heavy metals on benthic macroinvertebrate communities in New Zealand streams. *Environmental Toxicology and Chemistry* 17:2338–46.

Hogsden, K. L., and J. S. Harding. 2012. Anthropogenic and natural sources of acidity and metals and their influence on the structure of stream food webs. *Environmental Pollution* 162:466–74.

Hogsden, K. L., and J. S. Harding. 2013. Leaf breakdown, detrital resources, and food webs in streams affected by mine drainage. *Hydrobiologia* 716: 59–73.

House of Representatives. 2019. The treasury and the financial statements of the Government of New Zealand for the year ended 30 June 2018. In: Appendix to the Journals of the House of Representatives, 2017–2020, Vol. 5, p. 14.

https://www.doc.govt.nz/our-work/heritage/icon-heritage-sites/

https://www.waikatoregion.govt.nz/services/regional-services/waste-hazardous-substances-and-contaminated-sites/tui-mine/

Jacobs, R. 1999. Modern mine closure in New Zealand – The Golden Cross case study. AusIMM New Zealand Branch Annual Conference.

Jellyman, P. G., and J. S. Harding. 2014. Variable survival across low pH gradients in freshwater fish species. *Journal of Fish Biology* 85:1746–52.

Law, S., C. Rufaut, K. Lilly, and D. Craw. 2016. Geology, evaporative salt accumulation, and geoecology at Springvale historic gold mine, Central Otago, New Zealand. *New Zealand Journal of Geology and Geophysics* 59(3): 382–95.

Lear, G., D. Niyogi, J. S. Harding, Y. Dong, and G. Lewis. 2009. Biofilm bacterial community structure in streams affected by acid mine drainage. *Applied and Environmental Microbiology* 75:3455–60.

Livingstone, M. E. (ed.). 1987. Preliminary studies on the effects of past mining on the aquatic environment, Coromandel Peninsula. Water & Soil miscellaneous publication number 104. Wellington, NZ: Ministry of Works and Development.

Malloch, K., D. Craw, and D. Trumm. 2017. Arsenic mineralogy and distribution at the historic Alexander gold mine, Reefton goldfield, New Zealand. *New Zealand Journal of Geology and Geophysics* 60(2):129–44.

MBIE. 2019. *Responsibly delivering value: Minerals and petroleum resource strategy for Aotearoa New Zealand: 2019–2029.* Wellington, NZ: Ministry of Business, Innovation and Employment.

McLachlan, C., and D. Craw. 2018. Environmental mineralogy and geochemistry of processing residues at Prohibition historic gold mine site, Waiuta, Westland. *New Zealand. New Zealand Journal of Geology and Geophysics* 61(2):180–94.

Ministry for the Environment. 2011a. Cleaning up Mapua: The story of the Fruitgrowers' Chemical Company site. Wellington, NZ: Ministry for the Environment.

Ministry for the Environment. 2011b. Hazardous activities and industry list. Wellington, NZ: Ministry for the Environment.

Ministry for the Environment. 2013. CSRF prioritisation tool user manual. Wellington: Ministry for the Environment.

Ministry for the Environment. 2022. Contaminated sites remediation fund priority list. https://environment.govt.nz/what-you-can-do/funding/contaminated-sites-remediation-fund/contaminated-sites-remediation-fund-priority/

Murray, J., and B. Walker. 2022. Cost effective management of AMD sludge at Stockton coal mine. In *Proceedings conference of the International Mine Water Association*, 6–10 November, Christchurch, New Zealand, eds. J. Pope, C. Wolkersdorfer, R. Rait, D. Trumm, H. Christenson, and K. Wolkersdorfer.

Nathan, S. 2006. Mining and underground resources: Metals known but not mined. In Te Ara – The Encyclopedia of New Zealand. http://www.TeAra.govt.nz/en/mining-and-underground-resources/page-3 (accessed August 30, 2023).

Niyogi, D., J. S. Harding, and K. Simon. 2013. Organic matter breakdown as a measure of stream health in New Zealand streams affected by acid mine drainage. *Ecological Indicators* 24:510–517.

NZPAM. 2023. Operating coal mine production figures. https://www.nzpam.govt.nz/nz-industry/nz-minerals/minerals-statistics/coal/operating-mines/, Wellington New Zealand Petroleum and Minerals.

OceanaGold. 2022. OceanaGold update, September 2022. Waihi, NZ: Oceana Gold. https://www.waihigold.co.nz/uploads/documents/updates/2022/OceanaGold-Waihi-Update-29-September-2022.pdf?_cchid=7a63a213108e0bcddbd87b0f4cabf55b

O'Halloran, K., J.-A. Cavanagh, and J. S. Harding. 2008. The response of a New Zealand mayfly to toxicity from acid mine drainage. *Environmental Toxicology and Chemistry* 27:1135–40.

PDP. 2016. Tui mine post remediation ecological monitoring 2015. Report prepared for Waikato Regional Council. https://www.waikatoregion.govt.nz/assets/WRC/Services/regional-services/Waste-hazardous-substances/TuiMine/Tui_Mine_Remediation_-_Ecological_Monitoring_Report_2015_3747598.pdf

Pope, J., N. Newman, D. Craw, D. Trumm, and R. Rait. 2010. Factors that influence coal mine drainage chemistry, West Coast, South Island, NZ. *New Zealand Journal of Geology and Geophysics* 53:115–28.

Royal Commission on the Pike River Coal Mine Tragedy. 2012. Volume 2. Wellington, NZ.

Rufaut, C., D. Craw, and A. Foley. 2015. Mitigation of acid mine drainage via a revegetation programme in a closed coal mine in southern New Zealand. *Mine Water and the Environment* 34:375–87.

Rufaut, C., D. Pillani, and D. Craw. 2023. Enhancement of alkaline saline soil-free bare substrates and specialist ecosystems, southern New Zealand. *Environmental Earth Sciences* 82:440. https://doi.org/10.1007/s12665-023-11145-1

Safran, A. 2017. Heavy metal bioaccumulation & toxicity in response to acid mine drainage remediation of the Tui Mine. Master's diss., University of Auckland.

Schowe, K. A., J. S. Harding, and P. A. Broady. 2013. Diatom community response to an acid mine drainage gradient. *Hydrobiologia* 705:147–158.

Schowe, K. A., and J. S. Harding. 2014. Development of two diatom-based indices: A biotic and a multimetric index for assessing AMD impacts in New Zealand streams. *New Zealand Journal of Marine and Freshwater Research* 48:16376.

Sherwood, A., and J. Phillips. 2021. Coal and coal mining. In Te Ara – The Encyclopedia of New Zealand. http://www.TeAra.govt.nz/en/coal-and-coal-mining (accessed August 29, 2023).

Simcock, R., and C. Ross. 2014. Guidelines for mine rehabilitation in Westland. Envirolink Advice Grant: 937-WCRC83. Report for the West Coast Regional Council.

Simcock, R. C., and C. W. Ross. 2017. Mine rehabilitation in New Zealand: Overview and case studies. In *Spoil to soil: Mine site rehabilitation and revegetation*, eds. N. S. Bolan, M. B. Kirkham, and Y. S. Ok. Boca Raton: CRC Press.

Simcock, R., C. Ross, and M. Pizey. 2004. Rehabilitation of alluvial gold and opencast coal mines from 1904 to 2004. In *Proceedings of the 37th Annual conference of the Australasian Institute of Mining and Metallurgy New Zealand Branch*, 29 August – 1 September, 77–82. Nelson, New Zealand.

Simcock, R., and J. Cavanagh. 2022. Healing Te Aroha maunga: Terrestrial rehabilitation at Tui Mine, 2013–2021. Manaaki Whenua – Landcare Research contract report LC 4187 for Iwi Advisory Group and Waikato Regional Council. Lincoln, NZ: Manaaki Whenua – Landcare Research.

Trumm, D., and J. Cavanagh. 2006. Investigation of acid-Mine-Impacted waters at Cannel Creek. Landcare Research Contract Report LC0506/169. CRL report 06-41101. http://www.wcrc.govt.nz/Documents/Environmental%20Management/BellvueMineRemediation.pdf

Todd, A. J., C. G. Rufaut, D. Craw, and M. A. Begbie. 2009. Indigenous plant species establishment during rehabilitation of an opencast coal mine, South-East Otago, New Zealand. *New Zealand Journal of Forest Science* 39: 81–98.

URS. 2014. Tui mine remediation: Site validation report. Technical report 42092212/R058/D Prepared for Waikato Regional Council by URS.

Walrond, C. 2006. Gold and gold mining. Te Ara – The Encyclopedia of New Zealand. http://www.TeAra.govt.nz/en/gold-and-gold-mining (accessed August 29, 2023).

Waters, A. S., and J. Webster-Brown. 2013. Assessing aluminium toxicity in streams affected by acid mine drainage. *Water Science and Technology* 67(8):1764–72.

Waters, A. S., and J. G. Webster-Brown. 2016. Is dilution a solution to aluminium toxicity in an acid mine drainage affected stream on the Stockton Plateau, New Zealand? *Mine Water and the Environment* 35:235–242.

West, R. 2014. Trialing small-scale passive systems for treatment of acid mine drainage: A case study from Bellvue Mine, West Coast, New Zealand. Master diss. University of Canterbury.

Winterbourn, M. J. 1998. Insect faunas of acidic coal mine drainages in Westland, New Zealand. *New Zealand Entomology* 21:65–72.

10 Mining and Mine Rehabilitation in Peru

Pedro Alberto Sifuentes Amez

10.1 INTRODUCTION

Peru is one of the world's 10 mega-diverse countries, it has 11 eco-regions, 84 of the planet's 117 life zones and 17 transitional zones of the 104 currently existing globally. Furthermore, it has 8 biogeographic provinces and 3 large hydrographic basins that contain 12,201 lakes and lagoons, 1,007 rivers, as well as 3,044 glaciers (SERNANP, 2016).

Geographically, Peru is characterized by the presence of the sea on the Pacific coast, the Andes mountain range, the Amazon rainforest and its latitudinal location. This geographical richness, due to the subduction of the Nazca plate and the formation of the Andes, has provided Peru not only with a high ecological diversity of climates and ecological floors but also with exceptional geological richness expressed in mineral deposits of different origins and varieties. Because of this wealth of mineral deposits, mining is an important economic activity for Peru. Thus, in 2022, Peru ranked second in world copper and zinc production, while in gold production, it ranks tenth in the world and second in the region, as seen in Table 10.1.

With the variety of mineral deposits that exist in the country, as well as the methods of mining exploitation, it is difficult to measure the different environmental impacts from which the great variety of ecosystems suffer.

Undoubtedly, the mineralogical potential and qualified personnel for the exploration and exploitation of these deposits are Peru's main strengths. However, competitive and sustainable mining under the principles of the Sustainable Development Goals presents a range of improvement opportunities

TABLE 10.1
Position of Peru in the World Ranking of Mining Production (Data from MINEM, 2023)

Product	Latin America	World
Gold	2	10
Copper	2	2
Silver	2	3
Zinc	1	2
Lead	2	5
Tin	1	4
Molybdenum	2	4
Cadmium	2	14
Phosphate rock	2	10
Diatomite	3	7
Andalusite	1	4
Selenium	1	10

DOI: 10.1201/9781315141855-10

and challenges that must be taken into account not only for current and future mining operations, but also to correct the mistakes made by mining in the past.

With regard to existing mining, one of the main issues is the environmental impacts that are left for future generations to deal with. In this sense, the most frequently asked question is: Is the environmental management of current mining adequate to avoid leaving mining environmental liabilities to future generations? Therefore, the improvement opportunities and challenges must first be addressed at the academic level, where more research and training for mine rehabilitation is required, followed by government with more realistic regulation regarding mine rehabilitation and post-rehabilitation as well as mining companies, which must show a greater commitment.

In relation to past mining, the remediation of mining environmental liabilities presents us with various improvement opportunities and challenges, such as mechanisms for finding financial sources, research into economically feasible technological alternatives for the remediation of mining environmental liabilities, technical guidelines and regulation that establishes mechanisms for improving community involvement in the remediation process in order to achieve sustainability.

This chapter provides an overview of past and existing mining in Peru, regulatory framework of environmental management in mining, mine rehabilitation, remediation of mining environmental liabilities as well as selected successful or emblematic cases of remediation of these liabilities, and some concluding remarks.

10.2 PAST AND CURRENT MINING

As it has previously stated, mining is an important economic activity for Peru, throughout its pre-colonial and colonial history. The book *Historia Ambiental del Perú, siglos XVIII y XIX* (MINAM, 2016b), collects information from different historians who have highlighted at length the predominant role of mining in the economic, administrative, political and social structure in the Viceroyalty. It also documents how mining caused a large number of damaging outcomes, due to the persistence in current times of mining environmental liabilities, whose origins date from the colonial era.

With respect to mining environmental liabilities, the reference book mentions that during the colonial era, certain mines were abandoned, either because all the ore had been extracted (exhausting the deposit), cost or because the required technology was lacking. At the end of that period, 1,131 mines were registered as abandoned, of which 1,124 were silver mines and 57 gold mines. Based on this, it can be deduced that during the years between 1750 and 1850 there was already a large amount of mining environmental liabilities.

In the publication *La Industria de la Minería en el Perú* (OSINERGMIN, 2017), the importance of regulatory and economic contributions, from the colonial period to the present day, is indicated. Important milestones referring to the introduction of technologies and methodologies of extraction through investments made by foreign capital mining companies, local mining companies and the state. Those activities led to an accumulation of serious environmental problems that today are being addressed through the management of mining environmental liabilities.

Likewise, it is important to highlight that the history of legislation on these matters during the Republican era began with the law promulgated in 1900, which came into force in 1901. However, the main milestone is the mining law of 1950, which constituted the guidelines for mining regulation, some of which are still in force in the mining institutional structure, such as the concession and its attributes. It should be noted that environmental regulations related to mining have only been in force since 1989, when a report was issued requiring a mining rights holder to accept environmental recommendations (MINAM, 2016a). This obligation was approved by a multi-sector commission of the Presidency of the Council of Ministers.

In 1990, the Environmental Code incorporated a mining regulation, and, in 1991, this was significantly modified by Legislative Decrees No. 708 and 757. Finally, in 1993, the Mining Environmental Regulation was issued, which gave meaning to the other sectorial environmental regulations and was modified in 2014. Based on the above, and on the information described in the book *Historia Ambiental del Perú, siglos XVIII y XIX* (MINAM, 2016), Table 10.2 summarizes the important milestones regarding the techniques used in mining and their associated environmental impacts.

TABLE 10.2
Important Milestones on Techniques Used in Mining in Different Periods (Author's Summary, 2020)

Period	Characteristics of Used Techniques	Environmental Impacts
Prior to 1537 (pre-colonial)	Opencast and small-scale mining. Metal processing (gold and silver), using metallurgical furnaces or *"huayras"* (MINAM, 2016b).	Generation of particulate matter.
1537-1582	The technique of the *"huayras"* existed until approximately 1582. Later it was gradually replaced by amalgamation with mercury (MINAM, 2016b).	Generation of particulate matter and mercury vapor and deforestation.
1556	Beginning of the exploitation of underground mines. In 1572, open-cast mining was prohibited. The method of "pots" was developed to obtain mercury in the Huancavelica region. Also, productivity decreased despite the evident excessive energy consumption (MINAM, 2016b).	Generation of tailings, waste rock, particulate matter and mercury vapor; deforestation and damage to health.
1572	Use of "cold" mercury amalgamation or "yard method" for silver production (MINAM, 2016b).	Dumping of tailings into sources of water containing suspended solids and mercury. Generation of mercury vapor, deforestation and damage to health.
1577	The use of "*buitrones*", which refers to amalgamation in ovens (MINAM, 2016b).	Increased generation of mercury vapors, deforestation and damage to health.
1596	The xabecca ovens of Almaden begin to be used (MINAM, 2016b).	Generation of mercury vapors, deforestation and damage to health
1633	Implementation of the mercury distillation system in the mines located in the Huancavelica region. The process consists of obtaining mercury through condensation of vapors (MINAM, 2016b).	Reduced generation of mercury vapors in quicksilver mines. Deforestation and damage to health.
1789	Technological improvements occur, including amalgamation in barrels (MINAM, 2016b).	Generation of mercury vapors, deforestation and damage to health.
1822-1878	Extraction of ore from the veins through underground mines. Transformation of ore into metal by the mercury amalgamation process, carried out in "*ingenios*" (small mining smelting centers) (J. Deustua, 1986).	Generation of waste rock, tailings and acid drainage.
1881-1898	Mainly silver ores were extracted; however, with the discovery of electricity and its applicability in a number of industrial uses, copper deposits began to be exploited (J. Deustua, 1986).	Generation of acid drainage, waste rock, tailings, particulate and mercury vapor
1903	Arrival of American capital, with the company Cerro de Pasco Mining Corporation. (OSINERGMIN, 2017). With an increase in exploitation volumes, the Andean lagoons served as tailings deposits. An example is Quiulacocha lagoon in 1920. Currently, in place of the lagoon, there is a tailings impoundment that represents a mining environmental liability requiring remediation (MINEM, 2022a).	Generation of acid mine drainage, waste rock, tailings, particulate matter, dumping of tailings into water resources.

(Continued)

TABLE 10.2 (CONTINUED)
Important Milestones on Techniques Used in Mining in Different Periods (Author's Summary, 2020)

Period	Characteristics of Used Techniques	Environmental Impacts
1922	La Oroya Metallurgical Complex began operations, processing three metals: lead, bismuth and copper. Subsequently (two decades later) the metallurgical complex processed 22 metals (A. Mendiola, 2017).	Generation of particulate matter, emissions to the atmosphere of lead, cadmium, arsenic and sulfur dioxide, generation of slag containing heavy metals and damage to health.
1956	Open-pit mining. In 1956, the open-pit mining of Cerro de Pasco began operations, being the first in the country. Before this mode of operation, mining was done underground (A. Mendiola, 2017).	Increase in the generation of waste rock and tailings, due to the open-pit stripping ratio is higher than underground mines. Generation of acid drainage, particulate matter, deforestation, loss of soil and damage to health.
1990	Open pit mining of disseminated oxide gold deposits began. Open pit mining method and gold processing were done using cyanide. In 1993 Minera Yanacocha began producing gold doré (A. Mendiola, 2017).	Waste rock, heap leaching pads, generation of acid drainage, cyanide discharge to the environment, particulate matter, deforestation and loss of soil.

10.3 REGULATORY FRAMEWORK FOR MINING AND MINE REHABILITATION

The current mining regulatory framework is based on the General Mining Law of 1981, from which subsequent laws and decrees were established to address mining from prospecting through construction and operation to mine closure and post-closure phases. Since its publication, the General Mining Law has been modified several times, and in 1992 it became necessary to establish the Orderly Unique Text of the General Mining Law through Supreme Decree No. 014-92-EM. With regard to mine rehabilitation, an important milestone in mining regulation was the approval of the Mine Closure Law, whose regulatory framework is applicable to, first, mining units that were in operation at the time the law came into force in October 2003; and, second, mining units that began or resumed operations subsequently.

It is important to note that, according to the Ministry of Energy and Mines (MINEM, by its Spanish acronym), a mining unit is an area where activities of exploration, exploitation, beneficiation, general labor, transportation and/or storage of minerals are carried out or planned, and where the facilities or components of the work are directly linked to each other. However, those mining units that had already completed their operations or had been abandoned by October 2003 are not covered by the Mine Closure Law. Thus, the facilities and waste from the aforementioned former mining units or orphan mines became environmental liabilities. For this reason, in 2005 it was necessary to approve a new regulatory framework to address the closure of former mining units, bringing to the fore the Law Regulating the Mining Environmental Liabilities.

This section addresses the regulatory framework for mining with a focus on environmental management, as well as mine closure and environmental liabilities.

10.3.1 Orderly Unique Text of General Mining Law

Approved by Supreme Decree No. 014-92-EM on June 3, 1992, with an updated publication in July 2022 (MINEM, 2022b).

The General Mining Law includes all matters related to the use of mineral substances from the soil and subsoil, except for oil and similar hydrocarbons, guano deposits, geothermal resources and medicinal mineral waters. In regard to environmental protection, the regulations for mining activities were approved within the regulatory framework of the General Mining Law. The regulations in force are the Environmental Protection and Management Regulations for Mining Exploitation, Benefit, General Labor, Transportation and Storage Activities approved in 2014, and the Environmental Protection Regulations for Mining Exploration Activities approved in 2017.

10.3.2 National System of Environmental Impact Assessment Law

Approved by Law No. 27446 on April 23, 2001, and amended by Legislative Decree No. 1078 and Legislative Decree No. 1394. The rules of the Law on National System of Environmental Impact Assessment were approved by Supreme Decree N° 019-2009-MINAM (MINAM, 2011).

Through this law, the National System of Environmental Impact Assessment (SEIA, by its Spanish acronym) was created as a unique and coordinated system of identification, prevention, supervision, control and early correction of problems caused by people who invested in mining projects.

In the same way, a uniform process is established, which covers requirements, stages and scope of the environmental impact evaluations of projects; and, additionally, mechanisms are established to allow community participation in the environmental impact evaluation process. The relevant assessment is developed through instruments that the projects need in order to obtain the required certification.

The environmental management instruments applicable to mining vary in requirement and complexity according to the type of activity to be developed and their dimensions. Among the instruments applicable to mining activities are the following: Environmental Impact Statement (Category I: minor impacts); Semi-detailed Environmental Impact Assessment (Category II: moderate impacts) and the Detailed Environmental Impact Assessment (Category III: significant impacts).

In that sense, the regulations of the SEIA Law establish that the Competent Authorities, such as MINEM, in the case of mining activities, may issue rules for the anticipated classification of projects and approve Terms of Reference for those with common or similar characteristics.

10.3.3 Environmental Protection and Management Regulations for Mining Exploration Activities

Approved by Supreme Decree No. 042-2017-EM of December 22, 2017. And amended by Supreme Decree No. 019-2020-EM (MINEM, 2017)

These regulations govern the environmental aspects of mining exploration activities in order to ensure compliance with legislation regarding prevention, minimization, mitigation, rehabilitation, remediation and, if applicable, compensation for damage done by mining, as well as corresponding closure and post-closure activities.

The environmental management instruments required for the environmental certification of mining exploration activities are Environmental Impact Statements and semi-detailed Environmental Impact Assessments.

Mine owner or operator is required to have environmental certification before starting activities, with the exception of prospecting and searches that do not require such certification. Once the exploration activities have been completed, the mine owner is required to adopt the necessary final closure and post-closure measures in order to restore physical and chemical stability, and long-term ecological processes, according to the terms and deadlines set forth in the approved environmental management instrument. The post-closure considers a maximum duration of 4 years.

Within this regulatory framework and consistent with the SEIA, the Common Terms of Reference were approved for the elaboration and evaluation of environmental studies for mineral exploration activities (MINEM, 2018).

10.3.4 Environmental Protection and Management Regulations for Exploitation, Benefit, General Labour, Transport and Mining Storage Activities

Approved by Supreme Decree No. 040-2014-EM on November 12, 2014 and amended by Supreme Decree No. 005-2020-EM (MINEM, 2014)

The regulations govern the protection and environmental management of mining activities, profits, general work, transport and storage in order to ensure that such activities are developed in a way that safeguards the constitutional rights of people to enjoy a functioning natural environment, within the framework of free private sector initiative and the sustainable use of natural resources.

The development of mining activities includes construction, operation, closure and post-closure. The mine owner or operator is required to obtain environmental certification from the relevant management instrument before starting the project. This instrument must also contain the respective Mine Closure Plan with a description of closure measures at the conceptual level, in accordance with the Mine Closure Law.

Within this regulatory framework and consistent with the SEIA, the Common Terms of Reference were approved for the elaboration and evaluation of detailed and semi-detailed environmental impact assessments for exploitation, processing, general work, transport and storage activities (MINEM, 2015).

10.3.5 Mine Closure Law (Mine Rehabilitation)

Approved by Law No. 28090 on October 14, 2003, and amended by Law 28234 and Law 28507. The Regulations to the Law on Mine Closure were approved by Supreme Decree No. 033-2005-EM, as amended. (MINEM, 2005a)

This law governs the obligations and procedures that mine holders are required to comply with for the elaboration, presentation and implementation of the Mine Closure Plan, as well as the constitution of environmental guarantees to mitigate any problems resulting in damaged health of people, the surrounding ecosystem and effects on property. It is subject to the principles of environmental protection, preservation and recovery. The mine closure plan is an environmental management instrument consisting of technical and legal actions, which must be executed by mine holders, in order to rehabilitate relevant areas. It is important that ecosystem characteristics compatible with a healthy and suitable environment for the development of life and the conservation of landscape are respected.

The aim of mine rehabilitation is that the areas used or impacted by mining attain:

- Chemical and physical stability.
- Recovery of local flora and fauna communities.
- Characteristics that represent minimal risk to human health.
- Conditions that allow for some subsequent use of the land, as far as possible.
- Other specific aspects related to the particular characteristics of these areas.

The mine closure plan should describe rehabilitation measures, their cost, timing and methods for control and verification of operation, final closure and post-closure stages. As well, the mine closure plan must indicate the required constitution of environmental guarantees including how much it will cost. The environmental guarantee entails the competent authority to cover the costs of rehabilitation measures for the final closure and post-closure periods. Lastly, the post-closure stage is the responsibility of the mining holder to manage for a period of not less than 5 years.

10.3.6 Law Regulating the Environmental Liabilities of Mining

Approved by Law No. 28271 on July 6, 2004, amended by Law No. 28526 and Legislative Decree No. 1042. The Regulations to the Law on Environmental Liabilities of Mining were approved by Supreme Decree No. 059-2005-EM, as amended (MINEM, 2005b).

The law regulates the identification and registration of mining environmental liabilities, as well as the responsibility and funding for the remediation of areas affected by them, aimed at their reduction and/or elimination, in order to mitigate any damage done to the health of the population, the surrounding ecosystem and property. Mining environmental liabilities are those facilities, effluents, emissions, tailings or waste rock deposits produced by abandoned or inactive mining operations and which constitute a permanent and potential risk to people, the surrounding ecosystem and property.

Compared to the regulatory framework for mine closure, which defines "rehabilitation", the mining environmental liabilities. regulatory framework introduces the term "remediation", which has similar definitions and objectives. Environmental remediation is implemented through the closure plans with the objective of ensuing the following:

- Long-term physical and chemical stability.
- Remediation of affected areas.
- Alternative use of such areas or facilities.
- Conditions of possible future use of such areas or facilities.

In that sense, the law and its regulations establish mechanisms for the Ministry of Energy and Mines to identify the mining environmental liabilities. and what or who has caused them. The results of the identification are published in the mining environmental liabilities inventory.

According to the regulatory framework, once identified, those responsible for mining environmental liabilities must submit the closure plan for these liabilities. This closure plan is an environmental management instrument and it is important to highlight that the implementation phase of the closure plan can take from 3 to 5 years, while the post-closure stage must be at least 5 years. Furthermore, when those responsible for mining environmental liabilities are not identified, the state assumes their remediation, under a criterion of prioritization and a mechanism for obtaining the funding. Finally, the regulatory framework sets out the alternatives of reuse and reclamation of mining environmental liabilities that are currently being applied.

10.3.7 Criteria for the Management of Contaminated Sites

Approved by Supreme Decree No. 012-2017-MINAM of December 2, 2017. These criteria cover aspects related to assessment and remediation, which must be regulated by the relevant authorities in order to protect human health and the environment (MINAM, 2017).

In accordance with the current regulatory framework, the contaminated site is the area in which the soil contains contaminants from anthropogenic activities in concentrations that may represent risks to health or the environment because they exceed the Environmental Quality Standards for Soil, international standards approved by the Ministry of Environment (MINAM, by its Spanish acronym), or background levels, as long as these levels present values that exceed those standards.

In the case of mining activities, the mining holders are responsible for the management of contaminated sites. It is divided into the following four phases: identification, characterization, preparation of the plan aimed at remediation and implementation of the remediation measures. Thereon, the plan aimed at remediation is an environmental management instrument, whose purpose is to save sites contaminated by past or current anthropogenic activities.

According to the mentioned criteria, remediation means the elimination or reduction, to acceptable levels, of risks to human health or the environment associated with the contaminated site.

Included here are actions that allow the subsequent use of the site or the reestablishment of it to a state similar before the damage or pollution occurred.

10.3.8 National System of Environmental Evaluation and Enforcement Law

Approved by Law No. 29325 on March 5, 2009, and amended by Law No. 30011 and Legislative Decree No. 1389. The Regulations to the Law of the National System of Environmental Evaluation and Enforcement (SINEFA, by its Spanish acronym) were approved by Supreme Decree N° 022-2009-MINAM (MINAM, 2009).

SINEFA is applicable to any natural or legal person, public or private, mainly for the entities of the national, regional and local government that exercise functions of evaluation, supervision, control and sanctioning power in environmental matters. This system has charge of the Agency for Environmental Evaluation and Enforcement (OEFA, using its Spanish acronym) as the governing body. OEFA evaluates, supervises, controls and sanctions compliance with environmental obligations established in legislation, as well as with the commitments derived from the environmental management instruments and the mandates or provisions issued by OEFA.

The environmental enforcement of mining in Peru is subject to the stratum to which the mining owner or operator belongs. The qualification of the stratum is decided through the variables referring to the extension of concessions and the installed production capacity. These are summarized in Table 10.3.

10.4 MANAGEMENT OF MINING ENVIRONMENTAL LIABILITIES IN PERU (REMEDIATION OF ABANDONED MINES)

This section concentrates on the management of environmental liabilities, and it is divided into the stages of identification of mining environmental liabilities, delimitation of those responsible, the entity in charge of the remediation of mining environmental liabilities assumed by the state, elaboration and approval of the environmental management instruments, environmental inspection and the certification of complete closure.

10.4.1 Identification: Mechanism for the Identification and Evolution of Inventories

In accordance with Article 3 of the Law regulating environmental liabilities of mining, the General Directorate of Mining of MINEM is responsible for elaboration and updating of the inventory of mining environmental liabilities. The initial inventory was published in 2006, where 850 mining environmental liabilities were identified; this inventory has been updated regularly. Figure 10.1 depicts the 15 updates of the inventory. The last one was published in September 2022, which presents 6,903 mining environmental liabilities identified (MINEM, 2022a).

TABLE 10.3
Strata Related to Mining Activities in Peru (Data from MINEM, 2022b)

Stratum	Extension of Concessions	Installed Production Capacity	Competent Environmental Enforcement Entity
Large-scale mining	Not established	Over 5000 MTD(1)	OEFA
Medium-scale Mining	Not established	350 to 5000 MTD	OEFA
Small-scale mining	Up to 2000 ha(2)	Up to 350 MTD	Regional Directorates of Energy and Mines
Artisanal Mining	Up to 1000 ha.	Up to 25 MTD	Regional Directorates of Energy and Mines

(1) MTD: Metric Tons per Day, (2) ha: Hectares

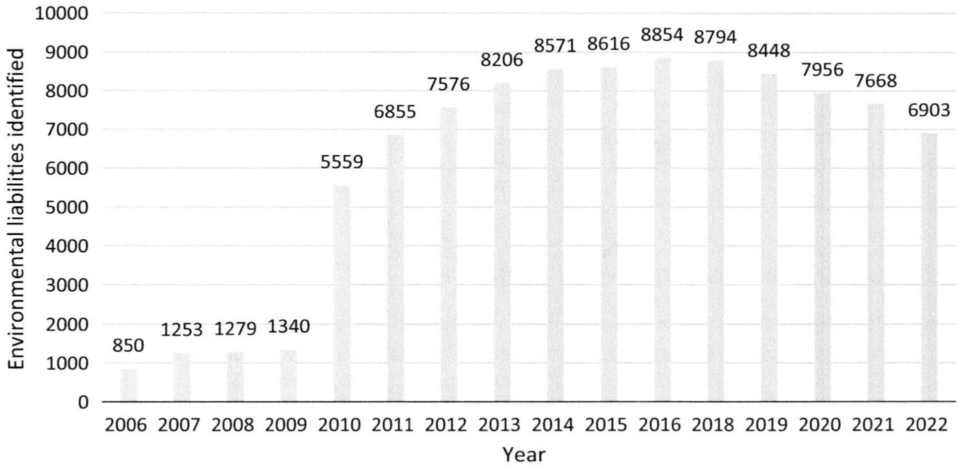

FIGURE 10.1 Evolution of the mining environmental liabilities inventory 2006–2022 (data from MINEM, 2022a).

According to the inventory of 2022, in which 6,903 mining environmental liabilities have been identified, they are classified into 4 main types: infrastructure, mining works, waste (mining and other types) and chemical substances. Mining works represent 58.95% (4069), waste 33.38% (2304), infrastructure about 7.63% (527) and chemical substances 0.04% (3).

The progress of the mining environmental liabilities inventory in Peru from 2006 to 2016 shows a steady increase in the number of identified mining environmental liabilities. However, since 2016 to the last inventory in 2022, 1,951 of them have been excluded, which shows that the identification is a dynamic activity over time. The numbers revealed in this chapter are variable as the inventory is updated, according to the technical criteria established by the General Directorate of Mining. In this regard, the General Directorate of Mining has a Management Plan for mining environmental liabilities, in which four phases were identified, as seen in Figure 10.2.

It should be noted that, in the process of inventorying mining environmental liabilities, not only is the identification made, but also the characterization and prioritization according to risk estimation. Here the General Directorate of Mining collects data on human health, environmental quality, wildlife and conservation through six technical files (MINEM, 2009).

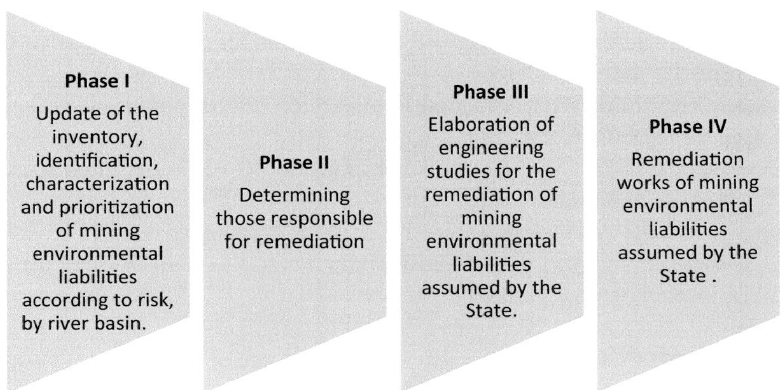

FIGURE 10.2 Phases of the management plan for mining environmental liabilities (MINEM, 2012).

From the data obtained in the cards, numerical values are assigned to each of the GPAs, and then, according to the results obtained, the GPAs are categorized into five categories: insignificant, low, medium, high and very high. Depending on the categorization of mining environmental liabilities, the prioritization of related activities will be accomplished, from the identification of those responsible, the engineering studies associated with the closure of the mining environmental liabilities, the remediation works as well as the subsequent control and inspection, prior to completion of the closure and post-closure activities.

10.4.2 Identification of Responsible Parties

As described in Article 3 of the Rules of the Law regulating mining environmental liabilities, the person or entity generating a mining environmental liability is responsible for carrying out all remediation activities, for which it empowers the General Directorate of Mining to execute the actions leading to the identification of the generators of these liabilities. Likewise, the regulation of mining environmental liabilities establishes that any person or entity can devise a voluntary remediation, without necessarily holding mining concessions, as well as in certain cases as referred to in Article 20 of the aforementioned regulation. The state can assume responsibility for the closure of mining environmental liabilities that have not identified or voluntary responsible parties (MINEM, 2009).

For the identification of who or what generates the mining environmental liability, the General Directorate of Mining in the Management Plan for Mining Environmental Liabilities, as presented in Figure 10.2, established in Phase II that, those responsible will be identified. This will help to determine those mining environmental liabilities to be assumed by individuals or private companies, and those that will be taken up by the state.

Based on this, after conducting a historical review of documents such as mining concession titles, the current status of the mining concession holder, environmental studies conducted in concessions with the presence of mining environmental liabilities, benefit concessions, among others, the General Directorate of Mining determines the person or company responsible for the remediation. In this regard, according to the 2022 inventory, 2,527 mining environmental liabilities have identified responsible, while 4,376 mining environmental liabilities do not have identified responsible for their remediation efforts.

10.4.3 Remediation of Mining Environmental Liabilities by the State

As stated in the previous paragraph, the state can assume responsibility for remediating mining environmental liabilities, for which the regulation establishes the following assumptions:

- The state has the authority to proceed with remediation of areas with mining environmental liabilities where a state enterprise is responsible for not less than two-thirds of the amount of remediation.
- When areas with mining environmental liabilities do not have identified responsible parties or voluntary remediation parties.
- When areas with mining environmental liabilities do not have identified responsible parties or voluntary remediation parties.
- In the opinion of the competent authority, if there are high-risk mining environmental liabilities and the person responsible has not remediated the area, or when the closure plan has not been implemented repeatedly.

For mining environmental liabilities where the state is in charge of remediation, closure will be paid for by trust funds managed by MINEM. If the General Direction of Mining becomes responsible for the remediation of mining environmental liabilities, it can directly hire a specialized company to carry out the engineering studies and the closure or commission the state company Activos Mineros

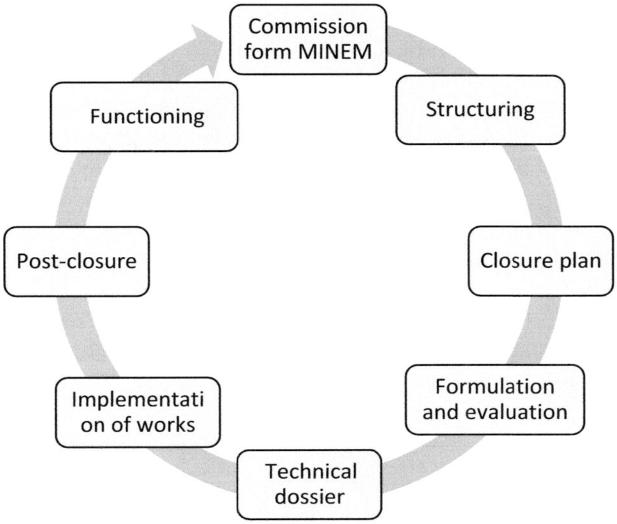

FIGURE 10.3 Remediation phases of mining environmental liabilities (Activos Mineros S.A.C., 2019).

S.A.C. to carry out the management of these liabilities. The period covered is from the presentation of the Closure Plan to obtaining the final closure certificate (MINEM, 2006).

The assignment to Activos Mineros S.A.C. is being carried out through Ministerial Resolutions of MINEM, whereby economic resources are transferred so that the company can manage certain mining environmental liabilities. Once the mining environmental liability has been transferred to Activos Mineros S.A.C., it carries out various procedures. Figure 10.3 illustrates the environmental remediation phases involved in mining environmental liabilities.

In that sense, according to the 2022 inventory, Activos Mineros S.A.C. is responsible for the environmental remediation of 1,103 mining environmental liabilities, having developed the engineering study to carry out the closure plan of 780, leaving 321 still to be dealt with.

10.4.4 Elaboration and Certification for the Closure Plan of Mining Environmental Liabilities

In order to carry out the remediation of a mining environmental liability, the regulation in its Article 12 states the remediation modalities, which are:

- Elaborate a closure plan of mining environmental liabilities.
- Inclusion of mining environmental liabilities in a mine closure plan.
- Reuse of mining environmental liabilities.
- Reclamation of mining environmental liabilities.

On the other hand, articles 4° and 64° of the regulation broach the concept of alternative use that can be implemented in an area with mining environmental liabilities, depending on a request to do so by the local government. It should be suitable for production activities, tourism, culture, recreation, sport or others. Currently, the most common ways to carry out the remediation of a mining environmental liability are inclusion in a mine closure plan or in a closure plan of mining environmental liabilities.

The mine closure plan and the closure plan of mining environmental liabilities are technical documents at the feasibility level, detailing the remediation activities for the mining environmental liability, the schedule for execution of closure activities and post-closure management measures that will ensure the effectiveness of the remediation. Both the mine closure plan and the closure plan for mining environmental liabilities are subject to assessment by the certifying authority, namely MINEM. It is applicable to medium- and large-scale mining holders, while for small-scale and artisanal mining holders the certifying authority is the regional government.

MINEM reviews mine closure plans or environmental liability closure plans through the General Directorate of Mining Environmental Affairs, which may request opinions from various related entities in the evaluation of the environmental management instrument, prior to its approval. In this regard, according to the 2022 inventory, 351 mining environmental liabilities are included in mine closure plans of existing mining units, while 1,646 are included in closure plans of mining environmental liabilities.

On the other hand, regarding the reclamation method, which deals with the extraction of minerals from mining environmental liabilities that have an economic value, article 61 of the regulations indicates that, as the first step, a reclamation request must be issued to the General Directorate of Mining, which must go through an approval process. Here, according to MINEM, 32 reclamation rights have been approved, involving a total of 110 mining environmental liabilities (MINEM, 2023b).

After obtaining the reclamation rights, the interested party has one year to submit a detailed or semi-detailed environmental impact assessment to the General Directorate of Mining Environmental Affairs (for medium- and large-scale mining holders) or the regional government (small-scale and artisanal mining holders). The environmental study must include a closure plan at the feasibility level or modify an existing mine closure plan if the mining environmental liabilities to be reused are in the area concerned.

From the review of the 2022 inventory, for large- and medium-scale mining, 3 reclamation environmental impact assessments have been identified, which includes 26 mining environmental liabilities. Likewise, with respect to small-scale and artisanal mining, with information from regional governments up to 2015, 5 approved environmental management instruments were identified (Ombudsman's Office, 2015).

Finally, with respect to the type of reuse, according to Article 4 of the regulations, this refers to the use that a holder of a mining concession can do with mining environmental liabilities that are within its concession, which can be incorporated as part of current or future mining activities. In this regard, according to the Ombudsman's Office, as of 2015 there were no mining environmental liabilities approved for reuse.

10.4.5 Environmental Inspection of Closure Plans of Mining Environmental Liabilities

In accordance with the provisions of the SINEFA Law and as seen in Table 10.3, environmental inspection of the closure plans of mining environmental liabilities is subject to the stratum in which the mining concession holder responsible for the remediation belongs. Considering that the vast majority of remediation plans for such liabilities correspond to large- and medium-scale mining strata, it is necessary to focus on the competencies and functions of the Environmental Evaluation and Enforcement Agency (OEFA, using its Spanish acronym).

OEFA is a specialized technical public agency, attached to MINAM, and in charge of environmental inspection and to ensure the appropriate balance between private sector investment in the economy and environmental protection. OEFA is the designated authority for monitoring and verifying compliance with the auditable obligations of mining activities, under the regime of large and medium mining, from July 22, 2010 (OEFA, 2010). In this sense, OEFA, within the framework of its supervisory function, must have the competence to control the compliance of environmental and social obligations of the mining holders of large- and medium-scale mining strata. These obligations are contained in the approved environmental management instrument, provisions issued by OEFA and the environmental regulations in force.

In this respect, in the case of the closure plans of mining environmental liabilities developed by mining owners of medium- and large-scale mining strata (the vast majority of cases), the OEFA is responsible for verifying compliance with the environmental obligations contained in the closure plan approved by the competent authority, and for sanctioning non-compliance with such auditable obligations. It should be noted that, if a mining environmental liability does not have a closure plan yet or is not included in a mine closure plan, OEFA would not be able to undertake environmental monitoring activities.

On the other hand, in the Annual Environmental Audit Plan of OEFA, it is observed that mining environmental liabilities that have closure plans are being audited; however, it does not specify whether mining environmental liabilities included in the mine closure plans are being audited (OEFA, 2019). By 2019, OEFA had 50 auditable units with closure plans of mining environmental liabilities, corresponding to 22 mine holders, of which 7 auditable units are in the closure stage, while 43 auditable units are in the post-closure stage.

10.4.6 Certification of Final Closure

As described in Article 46 of the regulations, after complying with the execution of closure, maintenance and post-closure measures, according to the commitments assumed in the closure plan of mining environmental liabilities, the mining holder in charge of the remediation may request the General Directorate of Mining Environmental Affairs to issue a "Certificate of Final Closure". This confirms compliance with all the measures stated in the approved closure plan.

It should be noted that the regulation provides for the issuance of this certificate only for the closure plans of mining environmental liabilities, and not for other types of remediation instruments, which would be overseen by the applicable legal rule. According to the regulation, the issuance of the Certification of Final Closure is subject to an audit in order to verify full compliance with the remediation of mining environmental liabilities, according to the commitments and schedule of activities documented in the closure plan.

It should be noted that, by 2019, the General Directorate of Mining Environmental Affairs had not issued the Certificate of Final Closure for mining environmental liabilities that have an approved closure plan and whose schedule of activities had already been completed. It is done with the understanding that OEFA has not yet agreed to the remediation works of the aforementioned liabilities contained in the closure plans.

10.5 SUCCESSFUL OR EMBLEMATIC CLOSURES (REHABILITATIONS) – CASE STUDY

As discussed in the previous section, environmental remediation is being carried out in Peru at various mining sites, which, although they have not yet obtained a certification of final closure, the environmental remediation of these old mining components has led to great environmental and visual effects. This is evident in the closure of Tablachaca, Antuquito, Bellavista and Casapalca tailings deposits. Furthermore, the state continues to close mining environmental liabilities, some of which are representative because of their size and volume, such as the Excelsior waste rock dump, where in 2017 remediation works began, valued at US$ 44.15 million, benefiting nearly 9,000 inhabitants of the communities of Champamarca and Quiulacocha.

10.5.1 Closure (Rehabilitation) of Tablachaca, Antuquito, Bellavista and Casapalca Ttailings Deposits

These old tailings deposits were a constant source of contamination to the Rimac River basin, which is the main supply source of drinking water for the city of Lima with approximately 10 million

FIGURE 10.4 Tailings deposit Tablachaca - 2001 (from Activos Mineros S.A.C., 2018).

inhabitants. The environmental remediation was carried out by Centromin Peru (formerly a Peruvian state-owned company), through the Program of Environmental Adjustment and Management of the former Casapalca mining unit. Subsequently, in 2006, the state-owned company Activos Mineros S.A.C. took charge of maintenance and monitoring of the environmental remediation.

In general terms, the remediation of the old tailings deposits consisted of profiling, covering with geo-membrane, placing a layer of limestone and organic soil, as well as hydraulic structures for the derivation of surface water and works for physical stabilization, mainly gabions. Post-closure maintenance and monitoring activities are aimed at ensuring that closure measures are maintained over time. Therefore, Activos Mineros S.A.C. has inspected and maintained the hydraulic structures, including verification of the coverage and revegetation, as well as monitoring the physical stability of the remediated tailings deposits.

Graphically, Figures 10.4, 10.5, 10.6 and 10.7 show the remediation activities carried out at the Tablachaca and Casapalca tailings deposits. The first photograph from 2001 depicts the accumulation of tailings on the north bank of the Rimac River and these tailings are contaminating the river. The second photograph from 2006 shows the area of tailings accumulation after remediation works, with closure measures implemented on the mining environmental liability. Meanwhile the third photograph from 2000 shows the tailings accumulation on the south bank of the Rimac River. The fourth photograph from 2004 shows the area of tailings accumulation after environmental remediation measures were implemented.

In the figures shown above, at first glance, what is noticeable is the great visual changes in the remediation works, and the environmental control that currently exists. This should guarantee that the closure carried out in the tailings accumulations does not lead to any environmental danger in the future. For that reason, it is important to put in place a proper post-closure maintenance and monitoring program.

FIGURE 10.5 Area of former tailings deposit Tablachaca – 2006 (from Activos Mineros S.A.C., 2018).

FIGURE 10.6 Tailings deposit Casapalca – 2000 (from Activos Mineros S.A.C., 2018).

10.5.2 Closure (Rehabilitation) of Excelsior Waste Rock Dump

According to the 2018 Annual Report of Activos Mineros S.A.C., the Excelsior waste dump consists of waste rock from the Raul Rojas open pit, located in Cerro de Pasco. Here, approximately 50 million tons of waste rock from the Cerro de Pasco Copper Corporation and Centromin Peru operations accumulated over an area of approximately 69 hectares from 1956 to 2000. The accumulation of waste rock can be seen in Figure 10.8.

FIGURE 10.7 Area of former tailings deposit Casapalca – 2004 (from Activos Mineros S.A.C., 2018).

FIGURE 10.8 Excelsior waste rock dump (from Activos Mineros S.A.C., 2018).

The remediation work done on this mining environmental liability, which was in response to the numerous requests of the population, aims to restore its physical, chemical and hydrological condition and stability. It also constitutes an important intervention by the state to help with ecosystem recovery and prevent people's ill-health. Taken into account here is the waste rock of the dump registers, among other pollutants, where there is a large amount of pyrite that turns it into a potential source of acidic water.

FIGURE 10.9 Remediation progress of Excelsior waste rock dump (from Activos Mineros S.A.C., 2018).

In the social aspect, the remediation works demanded goods, services and local labor from the Champamarca Family Group, Pasco region. During 2018, the remediation of the mining environmental liabilities employed an average of 175 inhabitants per month. Activos Mineros S.A.C. carried out campaigns in various medical facilities for more than 450 members of the Champamarca Family Group. By the end of 2018, the project had made more than 36% progress. Figure 10.9 shows the remediation progress by the end of November 2019.

According to the 2022 Annual Report of Activos Mineros S.A.C., in late 2022, the Excélsior project reached the final stages of its construction phase, attaining an 84.47% physical completion. Key activities involved the application of geosynthetics, crushed stone and topsoil as the final layer on excavated areas. Furthermore, the project included the construction of hydraulic structures to effectively manage non-contact water, thereby preserving the integrity of the Ragra River watershed.

The project is structured into three distinct zones. The section adjacent to the Champamarca community achieved full revegetation of its slopes, while substantial progress was made in waterproofing and ground coverage in the remaining two zones. Figure 10.10 shows the remediation progress by the end of 2022.

10.6 CONCLUDING REMARKS

Mining has existed in Peru for much of its history, and plays an important role in the country's economy. As a result of this legacy, Peru has currently identified 6,903 mining environmental liabilities that need to be rehabilitated and remediated. Environmental regulation of mining in Peru started with the General Mining Law in the 1980s, and environmental impact studies became mandatory in the early 1990s. Subsequently, in 2003, the obligation to rehabilitate the areas affected by mining was established through the Mine Closure Law, which led to creating a regulatory framework for the remediation of environmental liabilities left by mining operations.

The regulatory frameworks for mine closure and closure of mining environmental liabilities establish that both definitions – "rehabilitation" in the case of mine closure and "remediation" in the case of closure of mining environmental liabilities – are aimed at physical, chemical and hydrological stabilization, ecosystem recovery, reducing health risks and alternative future use of the infrastructure and areas involved. The management of these liabilities in mining is divided into public

FIGURE 10.10 Remediation progress of Excelsior waste rock dump (from Activos Mineros S.A.C., 2022).

and private sector institutions, from the identification of responsibility by the Ministry of Energy and Mines, remediation by mining holders and the state, as well as inspections carried out by the Environmental Evaluation and Enforcement Agency and the Regional Directorates of Energy and Mines.

With regard to the remediation of environmental liabilities of mining as directed by the state, most of them are being remedied by the state-owned company Activos Mineros S.A.C. By 2022, it has been carrying out remediation of 782 environmental liabilities through their respective closure plans. The most commonly used closure alternative is remediation, either through mine closure plans or through closure plans of mining environmental liabilities. This opens up the opportunity to improve other closure options, such as reuse and reclamation, with the objective that the closure would be self-financed and make viable the economic use of mining environmental liabilities.

It is necessary to replicate successful models of closure of environmental liabilities from other mining countries, in which added value is given to closure. To replace mining, tourism, site museums or facilities for new activities are recommended. It is necessary to improve the participation of local communities in planning the remediation of mining environmental liabilities, considering that the only way to make remediation sustainable is to involve the population. Remediation aims at achieving the desired and feasible future use that people want for the recovered area or infrastructure.

REFERENCES

Activos Mineros S.A.C. (2018). Memoria Anual 2017. pp. 33–35. Seen: https://www.amsac.pe/publicaciones/memorias-anuales/

Activos Mineros S.A.C. (2019). Memoria Anual 2019. pp. 14, 15. Seen: https://www.amsac.pe/publicaciones/memorias-anuales/

Agency for Environmental Evaluation and Enforcement - OEFA. (2010). *Resolución de Consejo Directivo N° 003-2010-OEFA/CD*. Seen: https://www.oefa.gob.pe/?wpfb_dl=26118

Agency for Environmental Evaluation and Enforcement - OEFA. (2019). *Plan Anual de Fiscalización Ambiental del OEFA*. Seen: https://www.oefa.gob.pe/planefa/planefa-2020/

Alfredo Mendiola et al. (2017). *Perspectivas de reestructuración del Complejo Metalúrgico de La Oroya mediante un análisis ambiental y económico. Universidad ESAN.* pp. 16. Seen: https://www.esan.edu.pe/publicaciones/2017/12/14/la_oroya_analisis_ambiental_economico_seria_gerancia_desarrollo.pdf
Energy and Mining Investment Supervisory Agency - OSINERGMIN. (2017). *La industria de la minería en el Perú.* pp. 124, 125. Seen: https://www.osinergmin.gob.pe/seccion/centro_documental/mineria/Documentos/Publicaciones/Osinergmin-Industria-Mineria-Peru-20anios.pdf
José Deustua. (1986). *Producción minera y circulación monetaria en una economía andina: El Perú del siglo XIX.* pp. 325–329 Revista Andina. Seen: http://www.revistaandinacbc.com/wp-content/uploads/2016/ra08/ra-08-1986-01.pdf
Ministry of Energy and Mines - MINEM (2005a). *Decreto Supremo N° 033-2005-EM. Seen: https://www.minem.gob.pe/_legislacionM.php?idSector=4&idLegislacion=5282*
Ministry of Energy and Mines - MINEM . (2005b). *Decreto Supremo N° 059-2005-EM.* Seen: *https://www.minem.gob.pe/_legislacionM.php?idSector=4&idLegislacion=5284*
Ministry of Energy and Mines - MINEM. (2006). *Decreto Supremo N° 058-2006-EM.* Seen: http://www.minem.gob.pe/_legislacion.php?idSector=10&idLegislacion=5682
Ministry of Energy and Mines - MINEM. (2009a). Decreto Supremo N° 003-2009-EM, *que modificó al Reglamento de Pasivos Ambientales de la Minería. Artículo 12°.- Modalidades de Remediación voluntaria.* Seen: https://www.minem.gob.pe/_legislacionM.php?idSector=4&idLegislacion=5299
Ministry of Energy and Mines - MINEM. (2009b). *Resolución Directoral N° 173-2009-MEM-DGM.* Seen: http://www.minem.gob.pe/_legislacionM.php?idSector=1&idLegislacion=4123
Ministry of Energy and Mines - MINEM. (2012). *Resolución Directoral N° 088-2012-MEM/DGM.* Seen: http://spijlibre.minjus.gob.pe/normativa_libre/login.asp
Ministry of Energy and Mines - MINEM. (2014). *Reglamento de Protección y Gestión Ambiental para las Actividades de Explotación, Beneficio, Labor General, Transporte y Almacenamiento Minero.* Seen: http://spijlibre.minjus.gob.pe/normativa_libre/login.asp
Ministry of Energy and Mines - MINEM. (2015). *Resolución Ministerial N° 116-2015-MEM/DM.* Seen: https://sinia.minam.gob.pe/normas/aprueban-terminos-referencia-comunes-elaboracion-eia-detallados
Ministry of Energy and Mines - MINEM. (2017). *Reglamento de Protección y Gestión Ambiental para las Actividades de Exploración Minera.* Seen: http://www.minem.gob.pe/_detallenoticia.php?idSector=1&idTitular=8233
Ministry of Energy and Mines - MINEM. (2018). *Resolución Ministerial N° 108-2018-MEM/DM.* Seen: https://sinia.minam.gob.pe/normas/aprueban-formato-ficha-tecnica-ambiental-guia-contenido-asi-terminos
Ministry of Energy and Mines - MINEM. (2022a). *Actualización del Inventario de Pasivos Ambientales Mineros.* Seen: http://www.minem.gob.pe/_detalle.php?idSector=1&idTitular=5769&idMenu=sub5768&idCateg=961
Ministry of Energy and Mines - MINEM. (2022b). *Texto Único Ordenado de la Ley General de Minería, actualizada y anotada.* Seen: https://www.gob.pe/institucion/minem/informes-publicaciones/3304361-tuo-de-ley-general-de-mineria-ed-2022
Ministry of Energy and Mines - MINEM (2023a). *Anuario Minero 2022.* pp. 48. Seen: https://cdn.www.gob.pe/uploads/document/file/4700376/2022.pdf?v=1689975935
Ministry of Energy and Mines - MINEM. (2023b). *Derechos de Reaprovechamiento de Pasivos Ambientales Mineros* [Portal web]. Seen: http://www.minem.gob.pe/_detalle.php?idSector=1&idTitular=9489&idMenu=sub5768&idCateg=1793
Ministry of Environment - MINAM. (2009). *Reglamento de la Ley del Sistema Nacional de Evaluación y Fiscalización Ambiental.* Seen: https://www.gob.pe/institucion/minam/normas-legales/317424-022-2009-minam
Ministry of Environment - MINAM. (2011). *Reglamento de la Ley del Sistema Nacional de Evaluación de Impacto Ambiental y su Reglamento.* Seen: http://www.minam.gob.pe/seia/wp-content/uploads/sites/39/2013/10/Ley-y-reglamento-del-SEIA.pdf
Ministry of Environment - MINAM. (2016a). *La lucha por la legalidad en la actividad minera.* pp. 38. Seen: http://www.minam.gob.pe/informessectoriales/wp-content/uploads/sites/112/2016/02/12-La-lucha-por-la-legalidad-en-la-actividad-minera.pdf
Ministry of Environment - MINAM. (2016b). *Historia ambiental del Perú. Siglos XVIII y XIX.* pp. 33, 35. Seen: http://www.minam.gob.pe/wp-content/uploads/2016/07/Historia-ambiental-del-Per%C3%BA.-Siglos-XVIII-y-XIX.pdf
Ministry of Environment - MINAM. (2017). *Aprueban Criterios para la Gestión de Sitios Contaminados.* Seen: http://www.minam.gob.pe/wp-content/uploads/2017/12/ds_012-2017-minam.pdf

National Service of Natural Areas Protected by the State - SERNANP. (2016). *Perú: País Megadiverso.* [Website]. Seen: https://www.sernanp.gob.pe/documents/10181/88081/Peru-+Pais+Megadiverso.pdf/4f361370-434d-405f-986e-2b4052219abf

Ombudsman Office. (2015). *¡Un llamado a la remediación! Avances y Pendientes en la gestión estatal frente a los pasivos ambientales mineros e hidrocarburíferos.* Seen: https://www.defensoria.gob.pe/modules/Downloads/informes/defensoriales/informe-Defensorial-171.pdf

GLOSSARY

MINAM:	Ministry of Environment
MINEM:	Ministry of Energy and Mines
OEFA:	Agency for Environmental Evaluation and Enforcement
OSINERGMIN:	Energy and Mining Investment Supervisory Agency
SEIA:	National System of Environmental Impact Assessment
SERNANP:	National Service of Natural Areas Protected by the State
SINEFA:	National System of Environmental Evaluation and Enforcement

11 Abandoned and Inactive Mines in the Philippines

Current Environmental Conditions, Ongoing Rehabilitation Efforts and Future Opportunities

Jessie O. Samaniego, Cris Reven L. Gibaga, Alexandria M. Tanciongco and Rico Neil M. Quierrez

11.1 ABANDONED AND INACTIVE MINES ISSUES

Abandoned mines are defined as those which originally operated with a mining lease contract or mineral agreement in place, but are now otherwise closed without proper decommissioning. Furthermore, the mining owner and/or contractor cannot be found or is not willing to carry out the rehabilitation work. Some abandoned mines in the country are now being operated by informal subcontract small-scale miners (Ocampo and Schmitz, 2023). Almost all abandoned and inactive mines pose high risks to human health and the environment, due to the disturbed and exposed heavy metal-laden soils and sediments and water-filled open-pit mines. Mine tailings that were produced and accumulated during the mine operations in these abandoned mines contain elevated amount of potentially toxic elements (PTE) (Gibaga et al., 2022b). This may result in the pollution of agricultural soil and various health problems experienced by nearby residents. People living in abandoned mine areas may directly ingest heavy metal-contaminated soils through plants and vegetables or consume contaminated water. Contaminated soils can be leached out and by precipitation runoff and can be eroded to the rivers and lakes, eventually being deposited into nearby coastal areas and other aquatic ecosystems that serve as the sink for such pollutants (Corales-Ultra et al., 2019; Gigantone et al., 2020). In coastal environments, heavy metals may become available and bioaccumulate to biota, causing biomagnification of the contaminants as they ultimately enter the food chain (Akcali and Kucuksezgin, 2011; Ali et al., 2019). Unvegetated soils and tailings in a mining area contribute to the pollution in this area, brought about by the particle-bound heavy metals suspended in the air, which can be inhaled by people living near the site (Fortoul et al., 2015). Contaminated soils and water in abandoned mines also make a big impact on local wildlife, as these metal ions are bioavailable to animal body organs and tissues and can cause chronic illnesses (Lim et al., 2008). Abandoned and inactive mine sites have stockpiled mine wastes and tailings that are detrimental to natural plant growth due to their physico-chemical characteristics of high pH, high salinity, low water retention capacity, high heavy metal concentrations and deficiency in soil organic matter and fertility (Abanto et al., 2011; Domingo and David, 2014; De Campos Favas et al., 2018; Wang et al., 2017; Samaniego et al., 2020a).

In the Philippines, there are around twenty-seven (27) abandoned and inactive mines as presented in Table 11.1 (Aggangan et al., 2019; Tanciongco et al., 2022). Among the minerals extracted from these mines were gold, copper, chromite, mercury, silica, iron, pyrite and marble. Most of the copper mines, such as the Kennon Mine of Black Mountain, the Tapian and Ino-Capayang Mines,

DOI: 10.1201/9781315141855-11

TABLE 11.1
List of Abandoned Mines in the Philippines (Adapted from Aggangan et al. (2019) and Tanciongco et al. (2022))

Abandoned Mines	Location	Mineral Commodity
Benguet Corporation (formerly Benguet Consolidated, Incorporated)	Balabac Island, Palawan	Iron, etc.
Philex Mining Corporation (Sto. Nino Mines formerly Baguio Gold Mining Company)	Tublay, Benguet	Copper, gold
Unidos Mining Corporation	Nabas, Aklan	Silica
Western Minolco Mining Company	Kapangan, Baguio	Copper
Zambales Base Metals, Incorporated	Upper Baluno, Zamboanga City	Copper
Barlo Mines	Mabini, Pangasinan	Copper
Black Mountain, Incorporated	Tuba, Benguet	Gold
Benguet Exploration, Incorporated	Tuba, Benguet	Gold
Palawan Quicksilver Mines, Incorporated	Puerto Princesa City, Palawan	Mercury
Romblon Marble Mines	Romblon, Romblon	Marble
Silica Sand Mines	Roxas, Palawan	Silica
Acoje Mining Company, Incorporated (AMCI)	Sta. Cruz, Zambales	Metallurgical chromite
Batong Buhay Gold Mines, Incorporated	Pasil, Kalinga	Gold
Benguet Corporation – Antamok Mines (BC-BAGO)	Itogon, Benguet	Gold
Benguet Consolidated Mining Corporation (BCMC)	Itogon, Benguet	Gold
Benguet Corporation –Dizon Copper/Gold Mines (BCD)	San Marcelino, Zambales	Copper
Masinloc Chromite Mines –Benguet Corporation (MCM-BC)	Masinloc, Zambales	Refractory chromite
Filminera Resources Corporation (FRC)	Rosario and Bunawan, Agusan del Sur	Gold
CDCP-Basay Mine	Basay, Negros	Copper
Ino-Capayang Mines (ICM)	Mogpog, Marinduque	Copper
Heritage Mining Corporation Alamag Processing Group	Llorente, Eastern Samar	Chemical chromite
Hixbar Gold Mining	Rapu-Rapu Island, Albay	Gold
Philippine Iron Mines	Jose Panganiban, Camarines Sur	Iron
Philippine Pyrite Corporation	Bagacay, Samar	Pyrite
Surigao Consolidated Mining Corporation, Incorporated	Siana and Mapawa, Surigao del Norte	Gold
United Paragon Mining Corporation	Paracale, Camarines Norte	Gold
Vulcan Mining Corporation	Cordon, Isabela	Gold

Sto. Nino Mine and Boneng-Lobo Mine of Western Minolco, started operation in the 1960s due to the worldwide demand for copper but declined in the latter part of the 1970s due to high inflationary pressures. The closure of several copper mines continued due to falling copper prices globally up until the 1990s. Similar trends occurred for other metallic mineral commodities like gold, mercury and chromite. Since it was only in 1995 when a law that governs all mining activities in the Philippines was legislated (Philippine Mining Act of 1995), all mines that stopped operating before 1995 were mostly abandoned without rehabilitation. Some of these mines, up to the present day, are still not remediated. The regulations and guidelines on mine rehabilitation and decommissioning were incorporated into the implementing rules and regulations of the Philippine Mining Act only in 1996. Meanwhile, the active pursuit to examine the state of the areas with abandoned mine sites only came in later. In early 2001, the Mines and Geosciences Bureau (MGB) commissioned Tetra Tech EM to conduct a preliminary site assessment of twenty-one (21) selected abandoned mines throughout the country (Aggangan et al., 2019; Samaniego et al., 2019). The outcome of the semi-detailed assessment showed that all abandoned mine sites pose environmental risks of varying degrees – from unstable tailings ponds to toxic contaminants in the soil and water. This may harm human health and the aquatic, terrestrial and wildlife.

Among the above-mentioned abandoned mine sites, this chapter will only discuss the status of selected and recently revisited mines by the authors, namely Palawan Quicksilver Mines, Inc. (PQMI) in Palawan, Philippine Iron Mines (PIM) in Camarines Norte, Barlo Mining Corp. (BMC) in Pangasinan and Ino-Capayang Mine (ICM) in Marinduque (Figure 11.1). The current environmental conditions of these mines, especially on PQMI, and the assessment of potentially toxic elements (PTE) in mine wastes and tailings in the other mines will also be presented in this chapter.

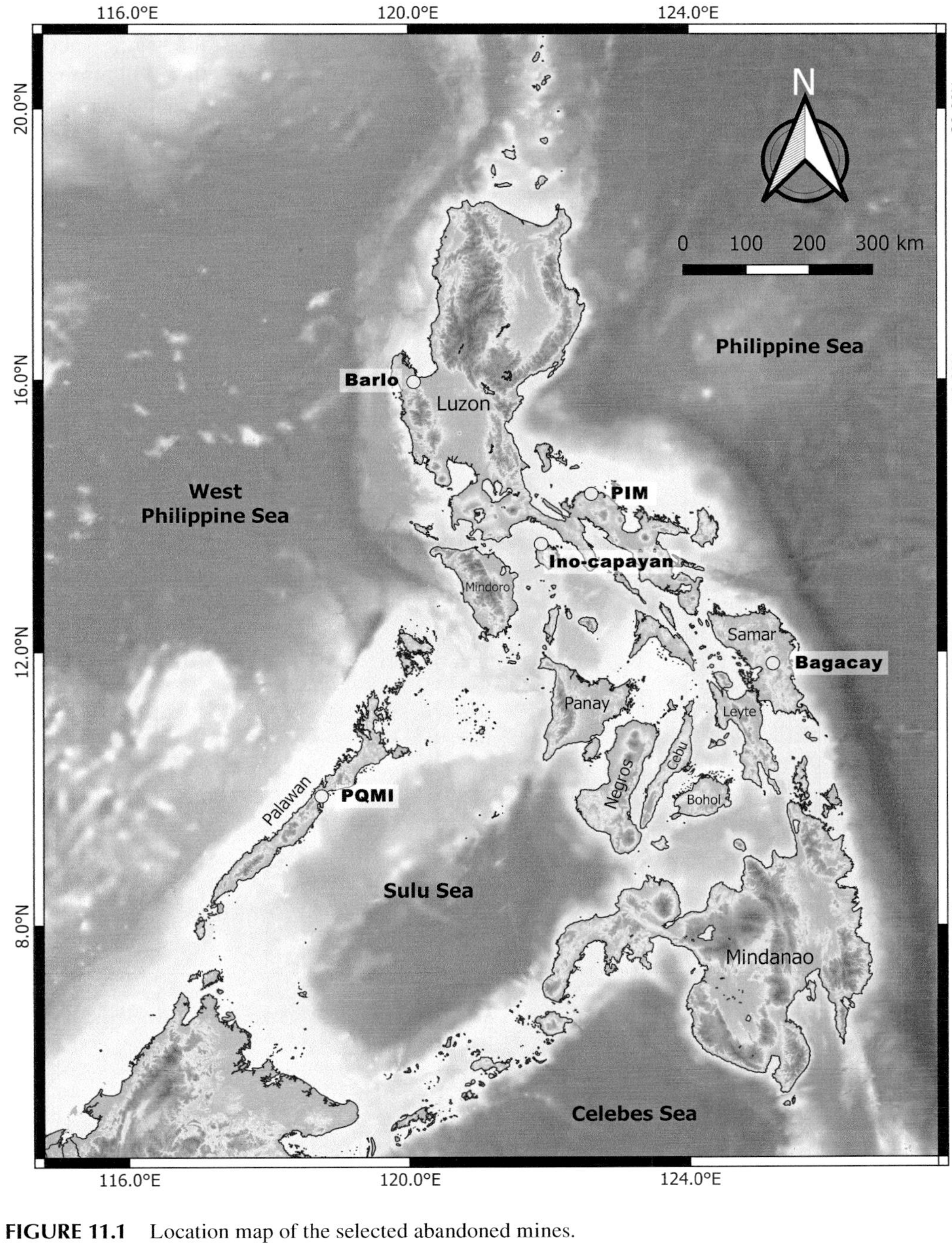

FIGURE 11.1 Location map of the selected abandoned mines.

Current government efforts to rehabilitate the abandoned mine sites will also be discussed as well as the future works to be done in these mines.

11.2 CURRENT ENVIRONMENTAL CONDITIONS OF SELECTED ABANDONED MINES

11.2.1 Palawan Quicksilver Mines

One of the most known abandoned mines in the Philippines is the Palawan Quicksilver Mines, Inc. (PQMI) in Puerto Princesa City. It is the host of the only naturally occurring deposit of Hg in the country. It mostly occurs in cinnabar (HgS) ores hosted by opalite bodies deposited along fractures and weak zones in the volcanic rocks of Central Palawan Ophiolite (Irving, 1954; Williams et al., 1999; Gibaga et al., 2020). From 1953 to 1976, PQMI mined the Hg deposit in the village areas of Sta. Lourdes and produced around 2,900 metric tons of mercury (Gray et al., 2003). Today, the abandoned open-pit mine is filled with water, and transformed into a pit lake located about 3 km west from the Honda Bay coast (Figure 11.2a). During the mining operations of PQMI, around 2,000,000 metric tons of cinnabar mine waste calcines were produced (Gray et al., 2003). Part of those mine waste calcines served to construct a 600 m x 50 m x 3 m thick wharf at the nearby Honda Bay to serve as a port for mining operations from 1953 to 1976. The wharf is now used as a port for tourists going island-hopping and for fishing boats. The rest of the mine waste calcines were stocked northeast near the pit lake shore, which was bare and had little vegetation until now. For the

FIGURE 11.2 Photos of selected abandoned mines in the Philippines. (a) Palawan Quicksilver Mines, Inc. pit lake and stocked mine waste calcines; (b) Philippine Iron Mine pit lake and slope; (c) Barlo Mine; and (d) Ino-Capayang Mine Lake.

past four decades, mine rehabilitation and remediation works have not been carried out at this site, resulting in weathering and erosion of bare mine tailings. Surface runoff and infiltration contributed to the discharge of PTE especially Hg deposited in the sediment of Honda Bay (Samaniego et al., 2020a).

An assessment of the PQMI mined-out area and vicinities in 2018–2019 was conducted to trace the pathways of Hg concentration in different environmental media in the area, including surface waters, groundwater, soil and sediment and air (Diaz, 2019; Samaniego et al., 2019; Samaniego et al., 2020a; Samaniego et al., 2020b; Samaniego et al., 2020c; Langcay et al., 2021; Posadas et al., 2022; Samaniego et al., 2022). Water samples collected from pit lake, river, coast, other nearby streams, leachate from landfill, and hot spring were analyzed for physico-chemical parameters and concentrations of Hg and other PTEs. Mercury and other PTE concentrations in soils and sediments were measured to assess their possible release from the site into the surrounding natural environment. Mercury in the ambient air in PQMI mined-out area was measured to assess human risk of Hg exposure. Groundwater from existing water wells, drilled boreholes and hand water pump were analyzed for Hg and other heavy metals concentrations. Environmental isotopes (^{3}H, ^{2}H, ^{18}O) in the groundwater and other surface waters were measured to determine the quality of water and the presence of Hg contamination.

Results showed that the physico-chemical characteristics of freshwater (pit lake and river) and coastal water were generally within the water quality guidelines (Samaniego et al., 2019). Heavy metals in the pit lake and river – except for Mn and Ni, Fe and Mn – were measured within the guidelines, respectively. Mercury concentrations in pit lake and river were not detected while small Hg concentrations were measured in the coastal water near the jetty (0.001 mg/L) and in the hot spring ($\leq$0.0004 mg/L). A landfill near the pit lake was releasing partially treated leachate with high total suspended solids and heavy metal concentrations that contribute to the area's pollution (Samaniego et al., 2019). While in the study of Garagara et al. (2015), water of the pit lake was measured positive for total and fecal coliform.

Total Hg concentrations found in soils (0.04 – 67.5 mg/kg), mine waste calcines (52.7 – 924.2 mg/kg), river sediments (1.8 – 119 mg/kg), and marine sediments (0.04 – 12.7 mg/kg) were elevated compared to the global background of 0.045 – 0.16 mg/kg (Samaniego et al., 2020a). The large concentrations of Hg in soils and river sediments were influenced by the different pathways for the release of Hg from its mine operations; while Hg in marine sediments was caused by the erosion of mine waste calcines near the pit lake and calcines used to construct a wharf at the nearby Honda Bay. Based on statistical analyses, it has been found that mercury has a negative bias with higher REE concentration but is positively correlated with other heavy metals, in particular As, Sb, Fe and Tl (Samaniego et al., 2020b).

The site with the highest Hg concentration (2,035 – 2,138 ng/m^{3}) in the air was located at the calcine deposit, northeast of the PQMI Lake (Samaniego et al., 2020c). Air collected from the perimeter of PQMI Lake has Hg concentrations of 8.44 - 61.63 ng/m^{3} which is the same (64.9 ng/m^{3}) reported in 2002. Concentrations of Hg in the air collected from the rest of the mined-out area, including Honda Bay, have relatively smaller concentrations (4.07 – 12.72 ng/m^{3}), yet are still high compared to the global background of total gaseous mercury, which is set at 1.0 – 3.0 ng/m^{3}. These elevated Hg concentrations in the area were attributed to the exposed Hg mine waste calcines that are possibly releasing gaseous Hg from the site to the surrounding natural environment (Samaniego et al., 2020c).

Mercury concentrations of the groundwater samples collected from the drilled boreholes and private wells have measured total mercury (THg) concentrations that are below the international and Philippine standards for drinking water of 1 µg/L (Samaniego et al., 2020d).

Mercury is the main contributor of non-carcinogenic risk in the area with a hazard index (HI) of 3.1×10^{1} and 1.3×10^{1} for children and adults, respectively (Samaniego et al., 2022). Aside from the ingestion pathway, inhalation of mercury vapor from the soils and sediments is found to be significant in addition to the health risks evident in the area. Other heavy metals such as Co, Fe, Ni and Tl were calculated with HI > 1, which pose a serious health risk to children due to the ingestion pathway. All carcinogenic heavy metals assessed were below the recommended limit of Total RISK = $1.0x10^{-6}$ indicating no significant carcinogenic health risks for both adults and children in the area. Of all the metals assessed, As can potentially cause cancer in both children and adults through ingestion since its Total RISK value falls within the $1.0x10^{-4}$ to $1.0x10^{-6}$ guidance value (Samaniego et al., 2022).

Results from the physical and environmental hazard assessment of the mine using the Historic Mine Site Scoring System (HMS-SS) identified the pit lake as extremely hazardous physically, while the liquid material in the lake, which is accumulated rain, yielded very low environmental hazard scores. The mine tailings stockpiled near the lake were found to be physically moderately hazardous but are categorized as extremely hazardous from the environmental perspective (Tanciongco et al., 2022).

11.2.2 Philippine Iron Mines

The abandoned iron mine in Jose Panganiban, Camarines Norte, was operated by Philippine Iron Mines (PIM) from 1925 to 1975 and it used to be the biggest iron mine in Asia during its operating days. The operation left two pit lakes in the area, the Main Pit Lake and the Bessemer Pit Lake (Figure 11.2b). The Bessemer deposit, which was mined by PIM, is a Fe-Cu-Mo skarn deposit associated mainly with pyrite (FeS_2) and magnetite (Fe_3O_4) mineralization, with small amounts of molybdenite (MoS_2) and chalcopyrite ($CuFeS_2$) disseminated in the host rock localized in calcareous sediments (Aurelio and Peña, 2004). The estimated size of the deposit is around 12.59 million metric tons with 10% Fe and 0.291% Cu (Aurelio and Peña, 2004). Low total rare earth element concentrations (42 mg/kg) are reported in the ore samples of PIM (Gibaga et al., 2022a).

Mine tailing samples from the abandoned iron mine of the Philippine Iron Mine (PIM) in Jose Panganiban, Camarines Norte were collected in 2021 to determine the concentrations of several potentially toxic elements (PTE). In the recent study of Gibaga et al. (2022b), the average As (110 mg/kg) and Cu content (553.3 mg/kg) concentrations of the PIM tailing samples are 9 times higher than the Canadian Soil Quality Guidelines (12 mg/kg and 63 mg/kg) (CCME, 2007), respectively. Based on the average enrichment factor (EF) (Buat-Menard and Chesselet, 1979), the PIM tailings have extremely high enrichment of As (62.53) and Cu (53.82). The average geoaccumulation index (Igeo) (Muller, 1969) of the PIM tailings shows that the samples are strongly contaminated by As (3.94) and Cu (3.72). It is also identified that the PIM tailings have elevated uncontaminated to moderately contaminated by these PTEs. While the high Cu content of the PIM tailings is related to the chalcopyrite ($CuFeS_2$) mineralization in the Bessemer deposit (Aurelio & Peña 2004), the elevated As concentration could be related to the by-products of the smelting process of Fe ores (Artiola et al., 2019) during the operation of PIM.

11.2.3 Barlo Mines

The abandoned copper mine in Mabini, Pangasinan, was operated by Barlo Mining Corporation until 1981 and it left tailings rich in massive sulfide minerals (Figure 11.2c). The copper-zinc geological deposit mined in this abandoned mine is a Cyprus-type volcanogenic massive sulfide (VMS) deposit, in which base metal sulfide ores are associated with rocks of mafic-ultramafic association of ophiolitic affinity (Aurelio and Peña 2004). The mineralization in the deposit is related to chalcopyrite ($CuFeS_2$), sphalerite (ZnS), pyrite (FeS_2), chalcocite (Cu2S) and traces of galena (PbS) with very small amounts of gold and silver (Aurelio and Peña 2004).

Tailing samples from the abandoned Cu mine of Barlo Mining Corporation (BMC) in Mabini, Pangasinan from Gibaga et al. (2022b) show that the average Pb (264.8 mg/kg), Cd (10.32 mg/kg), Zn (2,572 mg/kg), Tl (27.27 mg/kg), As (582 mg/kg) and Cu (9,109 mg/kg) are 4 times, 7 times, 13 times, 27 times, 48 times and 145 times higher than the Canadian Soil Quality Guidelines (70 mg/kg, 1.4 mg/kg, 200 mg/kg, 1 mg/kg, 12 mg/kg and 63 mg/kg) (CCME 2007), respectively. The BMC tailings are extremely highly enriched by Pb (82.81), Tl (161.1), Zn (204.1), Cd (609.6), As (645) and Cu (1,730) based on its average EF. Meanwhile, its I_{geo} index indicates that the samples are strongly contaminated by Pb (3.38), strongly to extremely contaminated by Tl (4.34) and Zn (4.68) and extremely contaminated by Cd (6.26), As (6.34) and Cu (7.76). The high Zn content in BMC tailings is related to the occurrence of sphalerite (ZnS), the high Pb concentration is associated with trace of galena (PbS), and the high Cu content is related to the presence of chalcopyrite ($CuFeS_2$) and chalcocite (Cu_2S) minerals in the deposit (Aurelio and Peña 2004). The elevated contents of the other PTEs (Tl, Cd and As) could possibly be linked to the by-products of the smelting process of Cu ores (Artiola et al., 2019) during the operation of BMC.

11.2.4 Ino-Capayang Mines

The abandoned copper mine in Ino and Capayang Villages in Mogpog, Marinduque, was operated by Consolidated Mines Inc. from 1968 and abandoned in 1980 due to low copper prices. After abandonment, Ino-Capayang, one of the open pits created by CMI was filled with water and became a mine pit lake, as shown in Figure 11.2d (Lanot et al., 2020).

In 2006, a group of researchers from the University of the Philippines Los Baños Bioremediation Group identified the uppermost part of the mined-out area in the village of Capayang for their rehabilitation project and applied soil amelioration interventions with the use of mycorrhizal fungi, lime and compost during field planting in addition with NPK fertilizer (Aggangan et al., 2017). Despite amelioration added in the soil, only small changes were observed in soil physical characteristics between the rehabilitated area and unrehabilitated area. The soil bulk density for the rehabilitated area and unrehabilitated area was 1.25 g/cm^3 and 1.41 g/cm^3, respectively. Both results fall under the classification described as suitable for plant growth according to USDA (2008). The water-holding capacity of the rehabilitated area and unrehabilitated area was 29.99% and 27.07%, respectively. The organic matter (OM) analysis in the rehabilitated area was higher (0.50%) than in the unrehabilitated area (0.39%), and soil pH in both areas was acidic. However, in terms of plant growth, the mean species richness was higher in the rehabilitated area than in the unrehabilitated area. Vegetation cover of the rehabilitated area showed a higher basal area for seedlings and saplings than the unrehabilitated area. Vines and trees were observed in the rehabilitated area but were absent in the unrehabilitated area. Soil amendments used during the rehabilitation efforts have helped increase the growth of plant species (Mante et al., 2019).

In Delos Angeles et al. (2018), examination of the pit lake water, soils, microalgae and floral composition of the area revealed higher Cu concentrations and pH in soil than in water. Their assessment of microalgal and plant diversity in the Ino-Capayang open-pit lake also presented potential phytoremediator species.

For microalgal species in the water, diatoms were found dominant, including *Fragilaria sp.*, *Rhopalodia sp.*, *Navicula sp.*, *Hantzschia sp.* and *Surirella sp.*, which can be potential phytoremediators of copper in water. Some of the species are reported to be indicators of organic and nutrient pollution, but the low diversity of microalgae in the pond may still be due to high copper concentrations and high total suspended solids (TSS) in the water.

The riparian vegetation was reported to be composed of 19 genera from Division Ancerophyta and 3 genera from Division Pteridophyta. Among the angiosperm families observed, the most represented families were Poaceae (5 spp.), Compositae (4 spp.), Leguminosae (4 spp.), Convulvulaceae (4 spp.) and Cyperaceae (3 spp.). Identified grass species that are capable of dominating the site are Bamboo, *Synedrella nodiflora*, *Axonopus compressus*, *Mimosa pudica* and *Poaceae sp.* These identified plant species have the potential to be used for phytoremediation due to their high Importance Values and their capability to tolerate soils with high copper concentrations (delos Angeles et al., 2018).

The recent physico-chemical characteristics of the pit lake waters were investigated by Lanot et al. (2020) and described to be slightly alkaline (pH 7.83) with a mean temperature of 31.74°C.

The average heavy metal concentrations of Cd (<0.006 mg/L), Cu (0.018 mg/L), Fe (0.151 mg/L), Pb (0.0089 mg/L), Mn (0.1404 mg/L) and Zn (0.1507 mg/L) in the lake waters are within the regulated levels for freshwater and can be used for the propagation of fish and other aquatic resources, recreational activities such as boating, fishing and for agricultural use. While ORP measurements indicate that the lake water has a high reducing agent.

The mined-out area, excluding the lake, at present is barely covered with vegetation and it has been a site for outdoor activities such as biking, walking and shooting. It is also reported that the municipal government of Mogpog, Marinduque, is planning to convert the lake and surrounding area into an eco-tourism park, where people can enjoy activities, such as walking around the lake, boating or fishing, and other outdoor activities (Lanot et al., 2020).

11.3 DEVELOPMENT AND CHALLENGES IN TTHE MANAGEMENT OF ABANDONED MINES

11.3.1 Development in the Management of Abandoned Mines

Recently, the MGB started to conduct further assessment of heavy metal contaminations and environmental conditions of several mines throughout the country and subsequently started the rehabilitation by planting vegetation and trees. While other researchers independently conducted studies on mine soil characterization (Posadas et al., 2022), stream sediments (Zuluaga et al., 2017), identification of native metallophytes thriving in the mine area (Claveria et al., 2019), and the phytoremediation potential of some known heavy metal tolerant plants (Fernando et al., 2013; De la Torre et al., 2016; Delos Angeles et al., 2020).

Among the mines that were assessed and started their rehabilitation are Philippine Pyrite Mines in Bagacay, Samar (Bueno, 2018), Ino-Capayang Mines in Mogpog, Marinduque (Aggangan et al., 2019; Lanot et al., 2020) and Palawan Quicksilver Mines, Inc. (Samaniego et al., 2019), while there were no recent reported rehabilitation works in Philippine Iron Mines and Barlo Mines.

At Ino-Capayang Mines, which is the well-studied abandoned mine site, several researchers from academia and research institutions conducted experiments to determine the potential treatments in revegetating the mined-out area. Hyperaccumulator plants such as ferns and grass species were tested, thriving and found to be abundant in the area (Claveria et al., 2020; Delos Angeles et al., 2018). Meanwhile, other studies assessed the application of microbial biofertilizers and soil amendments to the growth and survival of tree species in barren mined-out areas (Aggangan and Aggangan, 2012; Aggangan and Cortes, 2018; Aggangan and Anarna, 2019; Belarmino et al., 2021; Trocio and Paguntalan, 2023).

Based on the latest survey, the abovementioned rehabilitation efforts being conducted in Ino-Capayang mine are very much supported by the community with a very high awareness of its danger to human health and the environment (Mante and Cadiz, 2021). Additionally, surveys conducted in the households near the mine showed that bioremediation is acceptable as a solution to remove the toxic waste in lake water for domestic use (Nelson et al., 2020).

In PQMI area, rehabilitation works are being undertaken by various stakeholders in the area including MGB, the local government unit of Puerto Princesa City, private companies, and non-governmental organizations. Tree-planting activities were undertaken in the mined-out area which aimed to provide aesthetic pleasure, improve land stability, and create an ecosystem within the high-risk Hg mine area. In 2018, MGB started fencing off the area to prevent human intrusion in the lake. While there were already grown trees and other vegetations intended for phytoremediation, until now there are still no published results presenting its effect in reducing mercury content of the mine waste and surrounding soils.

At PIM, natural regeneration was studied, and the results indicate that regreening the mined soil is possible if the natural process of ecological succession is allowed (Abanto et al., 2011).

11.3.2 Challenges in Managing Abandoned Mines

While several mining companies in the Philippines that once operated the abandoned mines now cannot be found or are not willing to carry out rehabilitation, the burden now lies with the local and national government to rehabilitate these mine sites and protect the environment and nearby communities. Mine rehabilitation is a very complex process in which several issues need to be considered to make it safe and limit the adverse effects for local people. It ranges from preventing them from entering the pit lake to the need to assess the PTE concentration in the surrounding soils and to conduct health and environmental risk assessment. In the Philippines, opportunities to commercialize abandoned mine waste and tailings or repurpose the contaminated lands for agriculture or other land-use are still in the assessment and investigation phase.

11.4 ABANDONED MINES REHABILITATION POLICIES

Mining laws in the country, before the enactment of the Philippine Mining Act of 1995 (Republic Act No. 7942), did not provided adequate requirements for the rehabilitation of mine sites after their operation. With the enactment of the Philippine Mining Act in 1995, several mines were considered abandoned after their permit holders and responsible parties could not be identified or were not financially or technically capable of undertaking complete rehabilitation of mine sites (Tanciongco et al., 2022). While other permit holders of abandoned and inactive mines that are identified were compelled to continue proper care and maintenance for the mines.

The Philippine Mining Act of 1995 is the current main policy that governs all mining operations in the country, and it provides various measures to protect the environment. Under this law, rehabilitation and other mine-closure obligations of the mining permit holders were established in accordance with international best practices. Specifically, mine permittees are required to present a work program on mining operations and plans on rehabilitation, regeneration, revegetation and reforestation of mineralized areas; slope stabilization of mined-out and tailings covered areas; aquaculture; watershed development and water conservation; and socioeconomic development. Among its important features to address the environmental issues in mines is the mandatory establishment of Mine Rehabilitation Fund that is composed of monitoring trust fund and rehabilitation cash fund to meet the complete rehabilitation and other mine closure obligations.

Another important mining law in the country is the Executive Order No. 270 enforced in 2004 to revitalize the mining industry in the Philippines and address the environmental issues by strengthening the implementation of remediation and rehabilitation measures. The Executive Order was accorded the prioritization of remediation and rehabilitation of abandoned mines to address the negative impacts of past mining projects (Tanciongco et al., 2022).

As part of the action plans of the government through MGB for the abandoned mines, mining permit holders were compelled to undertake remediation and rehabilitation of the mine sites. Assessments of the abandoned mines were conducted for rehabilitation and redevelopment of the land for other uses, such as Philippine Pyrite Mine in Bagacay, Samar, and Palawan Quicksilver Mine in Puerto Princesa City, Palawan.

Transforming the mine waste and tailings material into a more viable material for plant growth should be a priority (Domingo and David, 2014). Abandoned and inactive mine sites have stockpiled mine wastes and tailings that are detrimental to natural plant growth due to their physico-chemical characteristics of high pH, high salinity, low water retention capacity, high heavy metal concentrations, and deficient in soil organic matter and fertility (Abanto et al., 2011; Domingo and David, 2014; De Campos Favas et al., 2018; Wang et al., 2017; Samaniego et al., 2020a; Posadas et al., 2022). The latter is the biggest challenge in the revegetation and rehabilitation of the mine waste stockpiles and exposed slopes.

Another consideration in evaluating the usability of abandoned mines is radioactivity in the area. Soils and rock in mining areas, especially Cu mines, may contain naturally occurring radioactive materials (NORM) such as potassium (K), uranium (U), thorium (Th) and radium (Ra) (US EPA, n.d.). Mining and extraction of Cu and other valuable metals by common surface methods can concentrate and expose radionuclides in the waste rocks or tailings. This is the same with the in-situ leaching process, which can transport U and Th into the groundwater and surface water at the site. Both methods expose and/or concentrate NORM and can further alter the properties of radioactive materials by processes, or beneficiates, or disturbing the natural properties that can increase the potential for human and/or environmental exposures.

11.4.1 Abandoned Mine Rehabilitation Prioritization and Integrated Mine Scoring

One of the tools developed in other countries for management and rehabilitation of abandoned mines is using mine hazard scoring. This system can be used as part of a decision-making tool to determine which abandoned mine in a particular area of interest requires the most immediate restoration efforts through inventory and ranking system. In Ireland, a comprehensive survey and step-by-step implementation of mine scoring of historic mines using Historic Mine Site Scoring System (HMS-SS) was reported by Stanley et al. (2019) under the United States Environmental Protection Agency (US-EPA). Several works by Mhlongo et al. (2019a, 2019b, 2021) also studied the methods for hazard ranking systems for abandoned mines in South Africa by assigning numeric standards both based on environmental and physical conditions of the former mine site. An attempt for mine scoring in an abandoned mercury mine in the Philippines was conducted by Tanciongco et al. (2022) by integrating the comprehensive assessment methods by Samaniego et al. (2020a) and modified HMS-SS. In their study, they utilized visual, physical, environmental and geochemical data and incorporated the information to the modified HMS-SS in the source-pathway-receptor diagram to map the area and give ranking which part of the site needs immediate monitoring activities and rehabilitation efforts. Results from the physical and environmental hazard assessment of the abandoned mercury mine site identified the pit lake as extremely hazardous physically, while the liquid material in the lake, which is accumulated rain, yielded very low environmental hazard scores. The mine tailings stockpiled near the lake were found to be physically moderately hazardous but are categorized as extremely hazardous from the environmental perspective (Tanciongco et al., 2022).

While the study of Tanciongco et al. (2022) focused only on one abandoned mine, the same process should be conducted in ideally all abandoned mines in the Philippines to rank which should be prioritized for rehabilitation and/or remediation. The results of this abandoned mine scoring can be communicated with the public and policymakers to aid in the decision-making and to assist in the sustainable management of treatment of abandoned mines (Mhlongo et al., 2019b).

11.5 ABANDONED MINES REHABILITATION EFFORTS

11.5.1 Abandoned Pyrite Mine in Bagacay, Samar

According to its recorded chronological timeline (Table 11.2), the Bagacay pyrite mine was mined during 1956 and abandoned in 1992, after which several important legislations relating to mining were enacted in the Philippines (Bueno, 2018). The rehabilitation of Bagacay Mine is funded by the World Bank and MGB was the lead agency in implementing the project in coordination with the MGB- Regional Office VIII, DENR-Regional Office VIII with the assistance of the Ecosystems Research Development Bureau (ERDB) as well. The Bagacay Rehabilitation Project concentrated on efforts that addressed the Bagacay Mine's environmental challenges, including Acid Mine Drainage, Visual/Aesthetic Impact and the mine's location being situated within Samar Island National Park. Within the mine site, interim environmental remediation initiatives included

TABLE 11.2
Chronology on Mining in Bagacay and Jurisdictional Legislations in the Philippines (Bueno, 2018)

Year	Activity
1956	Copper mining in Bagacay Mine under the Marinduque Mine Industrial Corporation (MMIC)
1986	Pyrite mining under the Philippine Pyrite Corporation (PPC)
1992	Termination of PPC operations due to high recovery cost and labor disputes
1995	Enactment of the Philippine Mining Act of 1995
1996	Declaration of Samar Island Natural Park (SINP) as a forest reserve
2003	Establishment of SINP as a natural park
2004	Issuance of National Policy Agenda on Revitalizing Mining in the Philippines (E.O. 270)
2005	Privatization and Management Office (PMO) under the Department of Finance (DOF) took the management of the former Bagacay Mine
2007	Limestone rock dam was installed by the MGB in the Bagacay Mine as an initial pollution control measure
2009	Phytoremediation research of the ERDB commenced

phytoremediation studies, reforestation on the peripheral areas, upkeep of previously reforested areas and the installation of a diversion channel or pipeline. A panel of experts was also contracted to undertake a geotechnical assessment and select the best approach for Bagacay Mine rehabilitation and mitigation (MGB, 2014).

Whilst DENR, through its two bureaus: MGB and ERDB, is the lead implementing agency of the rehabilitation activities (Table 11.3) with its organizational chart shown in Figure 11.3a. While the Department of Finance (DOF) through its Privatization and Management Office (PMO) became the overall management and decision-making body (Bueno 2018) as shown in Figure 11.3b.

Moreover, Bueno (2018) evaluated the Bagacay Rehabilitation Project in terms of project success and local community perception and concluded that abandoned mine rehabilitation in the Philippines is a relatively new concept compared to our international counterparts. She also argued that the Philippines is still in a long way of strong environmental governance that is willing to take abandoned mine rehabilitation responsibilities. Bagacay Mine Rehabilitation Project was terminated in 2018.

11.5.2 Abandoned Mercury Mine in Puerto Princesa City, Philippines

The abandoned mercury mine in Puerto Princesa City is one of the recently most studied abandoned mines in the Philippines. This site was once operated by the Palawan Quicksilver Mines Inc. (PQMI) from 1953 to 1976 and was only recently gaining attention. Several reports of mercury poisoning in the nearby communities arose in 2017 (Anda, 2017; Fabro, 2019) and the claimed cause of these events were pointed to the exposed mine waste and unprocessed ores and the 3-hectare pit lake that was once the open-pit mine of PQMI. In 2003, Gray et al. analyzed the pathway of elevated methylmercury levels in surface and groundwater from the former PQMI site to Honda Bay, a nearby tourist and fishing spot about 3 kilometers away. These results were complemented with the case studies on effects of mercury and its derivatives on humans, which were conducted by Maramba et al. (2006). Their findings showed a statistically significant increase in methylmercury levels in the hair of infants and pregnant women, and in fish species found in Honda Bay. The Puerto Princesa City government is currently putting up phytoremediation measures along with the MGB-Region IV-B rehabilitation efforts.

TABLE 11.3
Highlights of the Rehabilitation Efforts in Bagacay Mine (Bueno, 2018)

Year	Activity
1999	Municipal Council of Taft, Eastern Samar conducted a study that recognized the degradation of Taft River; sought budget for the rehabilitation of the river
2000	MGB RO No. VIII submitted a "Report on the Assessment of Mine Affected/Abandoned Areas of Philippines Pyrite Corporation"
2001	MGB-DENR commissioned Tetra Tech EM, Inc. to conduct a "Semi-detailed Assessment of Abandoned Mines in the Philippines"
2003	MGB conducted a "River Assessment and Water Quality Monitoring of Taft River"; suggestions include silt dredging and planting of appropriate tree species"
2004	MGB-VIII conducted a rapid appraisal of the Bagacay Mine, as per DENR Special Order No. 2004-74, "Rehabilitation of Abandoned Mined-out Areas in Samar Island
2005	MGSD, MGB Central Office (CO) submitted a "Preliminary Report on Taft River and the Coastal and Nearshore/Offshore Environment of Taft, Eastern Samar" from their geo-environmental study along the downstream portion of Taft River and the adjacent Coastal and Nearshore/Offshore areas which are the main receiving bodies of water from the Bagacay Mine
2006	World Bank consultants visited Bagacay Mines and suggested MGB-CO to assess the abandoned Bagacay Mine; Geotechnical Assessment of the Tailings Dam of Bagacay Mine was undertaken
2007	"Detailed Engineering, Planning and Design for the Rehabilitation of Bagacay Mine" was completed by SR Metals Inc. (hired by MGB) but was evaluated as unfeasible
2009	ERDB started its research on phytoremediation
2010	ERDB submitted the "Hydrologic Behavior of Bagacay Watershed" report; MGB completed its geophysical survey: "Resistivity Profiling and VLF-EM Surveys at Bagacay Mine, Hinabangan Samar"
2011	Geohazard assessment for the possible route of the diversion channel/pipeline for drinking water was conducted
2012	Actual geotechnical study of the mine pits, waste dumps, tailings pond, and other structures; monitoring and evaluation of the accomplishment of the Bagacay Mine Rehabilitation Project through fieldwork conducted by World Bank representatives

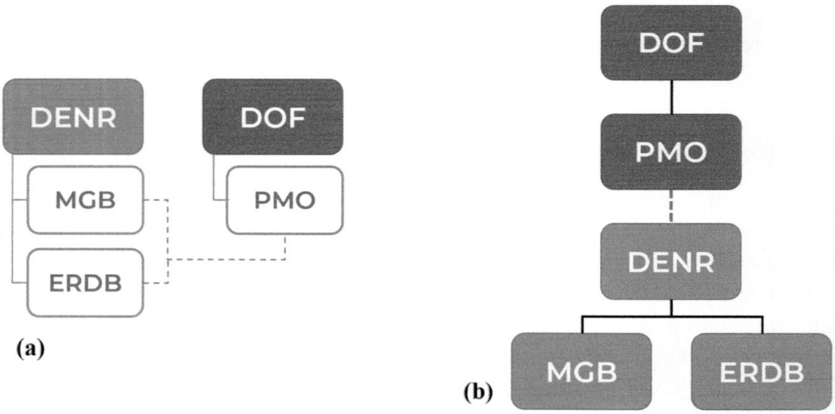

FIGURE 11.3 Figure 11.3 (A) Organizational chart of Bagacay Rehabilitation while (B) shows the power relationships with the government agencies involved in the rehabilitation adapted from Bueno (2018).

Samaniego et al. (2020b) proposed a detailed, graded procedure for the evaluation of environmental conditions of abandoned and inactive mines in the Philippines. Their approach focused on environmental in-situ and laboratory analysis, paired with health risk assessment based on inorganic pollutants. The authors conducted case studies with the abandoned PQMI mine and the abandoned copper mine in Mogpog, Marinduque. Moreover, to further aid the prioritization of the rehabilitation of various abandoned mines throughout the country, Tanciongco et al. (2022) utilized the Historic Mine Site Scoring System (HMS-SS) to be adapted as a ranking tool and conducted a case study with the PQMI site.

In 2013, PQMI was identified as a priority site for rehabilitation during the assessment commissioned by MGB. MGB Regional Office IV-B and the City Government of Puerto Princesa, consequently, entered a MOA in December 2014 with the objective to prevent the further deterioration of the PQMI abandoned mine site by addressing the physical and environmental threats in the area, while promoting the sustainable utilization of the resource for socioeconomic and environmental significance (Palawan News, 2019).

The first phase of the rehabilitation with initial PhP15-million funding from MGB centered on the isolation of the area with the construction of outer and inner fences within the identified 20–40 meters buffer zone/easement of the pit lake. Vegetation was also followed by introducing carabao grass, bamboo and other ornamental plants in preparation of the area as an ecotourism site. In 2019, MGB Regional Office IV-B initiated the phase II of the interim rehabilitation program of PQMI with plans of turning the mine site into an ecotourism destination. Scheduled activities include topsoil capping, benching for the slope stabilization, additional landscaping, perimeter lighting with underground wiring, additional construction of a three-meter-wide concrete pathway and fence to contain the 60m-wide abandoned national road; installation of water systems; repainting of existing inner and outer fence, and construction of driveway parking, guard house and mini-information center building. DENR-Ecosystem Research and Development Bureau plans to introduce phytoremediation measures within the buffer zone (Palawan News, 2019).

PQMI Rehabilitation is under MGB's Mine Rehabilitation Program as espoused by Subsector Outcome 2 (Environmental Quality Improved) of Chapter 20 (Ensuring Ecological Integrity, Clean and Healthy Environment) of the Philippine Development Plan (PDP) 2017–2022, one of the targets of the PDP is to transform abandoned mines and mined-out areas into land use that is beneficial to the communities (NEDA, 2021).

11.5.3 Ino-Capayang Mines in Mogpog, Marinduque

In 2006, University of the Philippines Los Baños bioremediation researchers identified the uppermost part of the mined-out area in the village of Capayang for their rehabilitation project and research study. The researchers applied soil amelioration interventions with the use of mycorrhizal fungi, lime and compost during field planting of *Jathropa curcas* in addition with NPK fertilizer (Aggangan et al. 2017), compost and *Trichoderma pseudokoningii* Rifai (Fontanilla and Cuevas, 2010) and narra tree (*Pterocarpus indicus*) with arbuscular mycorrhizal fungi (AMFs) and nitrogen-fixing bacteria (NFBs) as microbial biofertilizers (Aggangan et al., 2019). In their 2016–2017 revisit, they found out that despite amelioration added in the soil, only small changes were observed in soil physical characteristics between the rehabilitated area and unrehabilitated area. However, in terms of plant growth, they observed success where mean species richness is higher in the rehabilitated area than in the unrehabilitated area. The vegetation cover of the rehabilitated area also showed a higher basal area for seedlings and saplings than the unrehabilitated area and reported that the soil amendments used during the rehabilitation efforts helped increase the survival of plant species (Mante et al., 2019).

Hyperaccumulators such as ferns (*Pityrogramma calomelanos*, *Pteris vittata* and *Pteris melanocaulon*) that thrive very well in different terrains and found to be abundant in abandoned Cu-Au mining areas (Claveria et al., 2020). The identified ferns can be used to partly reduce the Cu and As content of the soil and can be used to clean up the Cu and As in tailings of PIM and BMC. Other researchers identified grass species as having the potential to be used for phytoremediation due to its high importance values and its capability to tolerate soils with high copper concentrations (Delos Angeles et al., 2018). Meanwhile, other studies assessed the application of microbial biofertilizers and soil amendments to the growth and survival of tree species in the barren mined-out areas (Aggangan and Aggangan, 2012; Aggangan and Cortes, 2018; Aggangan and Anarna, 2019; Belarmino et al., 2021). *Jatropha curcas* plant was tested in Cu-rich soil because of its potential as a source of biodiesel (Fontanilla and Cuevas, 2010; Aggangan et al., 2017). Based on the latest survey, the abovementioned rehabilitation efforts being conducted in Ino-Capayang mine are very much supported by the community with a very high awareness of its danger to human health and the environment (Mante and Cadiz, 2021). Additionally, surveys conducted in the households near the mine showed that bioremediation is acceptable as a solution to remove the toxic waste in lake water for domestic use (Nelson et al., 2020).

11.6 RECOMMENDATION AND FUTURE WORKS

Rehabilitation of abandoned mines can be effectively done when there is a comprehensive assessment of the environmental conditions and contamination in the area. Here we propose a simple and rapid assessment of the environmental conditions of all the environmental compartments in the

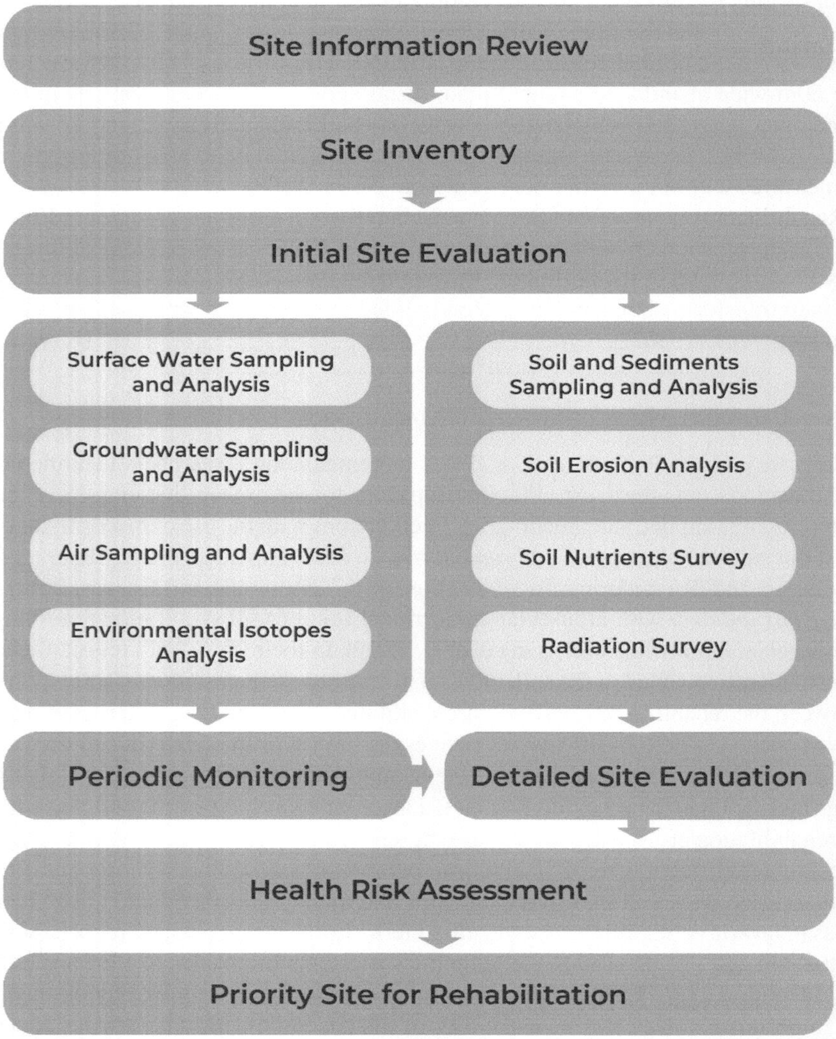

FIGURE 11.4 Process flow chart of rapid environmental assessment of abandoned and inactive mines. Adapted from Samaniego et al. (2020b).

mine area that is possible to be done in abandoned mines. Proper assessment of PTE in air, soil, sediments, surface water and groundwater can be used as a guide in designing a rehabilitation plan as well as its rehabilitation prioritization. Presented in Figure 11.4 is a flow chart of the methods for rapid assessment of any abandoned mine sites in the country. The methods include analyses of surface water and groundwater quality assessment, environmental isotopes, analysis of PTE and anomalous heavy metals, erosion rate determination, nutrient availability, air quality and radiometric surveys. Potentially toxic elements in soils, sediments and water can be used in the calculation to determine health risks in the area, both carcinogenic and non-carcinogenic in character. A detailed assessment of the different environmental media related to a mine site and subsequent mine site scoring based on the compiled data from the assessment will narrow down which particular portion needs more attention and focus during rehabilitation. So far, the model has been employed effectively in a case study of the PMQI site and is currently employed at the PIM mine site.

While there was substantial data available on the environmental conditions of PQMI, there are still a lot of things to do to further expand the tracing of Hg contamination in the area. The following future works are recommended:

1. Conduct aan Hg concentration analysis in sediment profile in Honda Bay to determine the deposition of cinnabar particles from non-processed ores and mine waste calcines.
2. Determine the erosion rates in the whole abandoned mine area, preferably using nuclear techniques (^{137}Cs, ^{210}Pb) to calculate the total volume deposited in Honda Bay since the start of mining operations.
3. Identifying the species of Hg present in soil and sediment, and
4. Determine the mercury stable isotopes present in marine sediments since it would determine if the fish in Honda Bay are contaminated by Hg that comes from the PQMI area by comparing its isotopic signatures.

11.7 CONCLUSIONS

This chapter provides an extensive assessment of the environmental challenges faced by some abandoned mines in the Philippines, namely Barlo Mines, Philippine Iron Mines, Palawan Quicksilver Mines Inc., Ino-Capayang Mines, and Philippine Pyrite Corporation. These mines have been left untreated contamination because of past policy inadequacies, which has had significant adverse effects on the surrounding communities. This chapter addresses critical issues such as soil, water, and air contamination, as well as the presence of potentially toxic elements in the mine wastes. Recent government-led management efforts, existing policies, ongoing rehabilitation endeavors were underscored. To effectively address these challenges, the chapter proposes the implementation of an integrated mine scoring system, which aims to prioritize the rehabilitation efforts. Subsequently, this chapter recommends conducting comprehensive site-specific assessments to identify the most appropriate remediation strategies. By adopting these proposed methodologies for dealing with abandoned mines, stakeholders and communities can collaborate to mitigate the negative impacts and work towards restoring the affected regions' environmental health in the Philippines.

ACKNOWLEDGMENTS

The authors acknowledge the Department of Science and Technology; the Philippine Council for Industry, Energy, and Emerging Technology Research and Development (DOST-PCIEERD) for the research grant.

AUTHOR CONTRIBUTIONS

Conceptualization was done by JS and CRG, while data curation was done by JS, CRG, AT and RNQ. Writing of the original draft was done by JS and CRG; review and editing were done by JS,

CRG, AT and RNQ and visualization was done by CRG, AT and RNQ. Funding acquisition was initiated by JS. All authors have read and agreed to the published version of the manuscript.

CONFLICT OF INTERESTS

The authors declare no conflict of interest.

REFERENCES

Abanto, R.G., Florece, L.M., Rebancos, C.M., Briones, N.D., & Castillo, A.S.A., 2011. Natural regeneration and carbon budget of rock dumpsite of an ex-mine site in Jose Panganiban, Camarines Norte, Philippines. *Journal of Environmental Science and Management*, 14(2), 60–70.

Aggangan, N.S., & Aggangan, B.J.S., 2012. Selection of ectomycorrhizal fungi and tree species for rehabilitation of Cu mine tailings in the Philippines. *Journal of Environmental Science and Management*, 15(1), 59–71.

Aggangan, N.S., & Anarna, J.A., 2019. Microbial biofertilizers and soil amendments enhanced tree growth and survival in a barren mined-out area in Marinduque, Philippines. *Journal of Environmental Science and Management*, 22(2), 77–88.

Aggangan, N.S., & Cortes, A.D., 2018. Screening mined-out indigenous mycorrhizal fungi for the rehabilitation of mine tailing areas in the Philippines. *Reforesta*, 6, 71–85.

Aggangan, N.S., Cadiz, N., Llamado, A., & Raymundo, A., 2017. Jatropha Curcas for bioenergy and bioremediation in mine tailing area in Mogpog, Marinduque, Philippines. *Energy Procedia*, 110, 471–478.

Aggangan, N.S., Anarna, J.A., & Cadiz, N.M., 2019. Tree legume – Microbial symbiosis and other soil amendments as rehabilitation strategies in mine tailings in the Philippines. *Philippine Journal of Science*, 148(3), 481–491.

Akcali, I., & Kucuksezgin, F., 2011. A biomonitoring study: Heavy metals in macroalgae from eastern Aegean coastal areas. *Marine Pollution Bulletin*, 62, 637–645.

Ali, H., Khan, E., & Ilahi, I., 2019. Environmental chemistry and ecotoxicology of hazardous heavy metals: Environmental persistence, toxicity, and bioaccumulation. *Journal of Chemistry*, 6730305, 1–14.

Anda, R.D., 2017. Widespread mercury poisoning detected in 2 Palawan villages. *Philippine Daily Inquirer.* http://newsinfo.inquirer.net/903525/widespread-mercury-poisoning-detected-in-2-palawan-villages. Accessed 16 October 2017.

Artiola, J.F., Walworth, J.L., Musil, S.A., & Crimmins, M.A., 2019. Soil and land pollution. In: Brusseau, M.L., Pepper, I.L., & Gerba, C.P. (Eds.), *Environmental and Pollution Science*. Academic Press, New York, 219–235.

Aurelio, M.A., & Peña, R.E., 2004. *Geology and Mineral Resources of the Philippines, Vol 1, Geology.* Department of Environment and Natural Resources, Mines and Geoscience Bureau, Manila, 18 p.

Belarmino, A.J.D., Pampolina, N.M., & Coracero, E.E., 2021. Growth response of Tabernaemontana pandacaqui Poir in soil from a former mining site in Marinduque, Philippines. *Ukrainian Journal of Ecology*, 11(2), 111–121.

Bueno, M., 2018. Evaluation of abandoned mine lands rehabilitation: The case of Bagacay mine, Western Samar, Philippines [Master's thesis]. Tokyo: University of Tokyo.

Buat-Menard, P., & Chesselet, R., 1979. Variable influence of the atmospheric flux on the trace metal chemistry of oceanic suspended matter. *Earth and Planetary Science Letters*, 42(3), 399–411.

CCME, 2007. Canadian soil quality guidelines for the protection of environmental and human health. Winnipeg: Canada Council of Ministers of the Environment.

Claveria, R.J.R., Perez, T.R., Perez, R.E.C., Algo, J.L.C., & Robles, P.Q., 2019. The identification of indigenous Cu and As metallophytes in the Lepanto Cu-Au Mine, Luzon, Philippines. *Environmental Monitoring and Assessment*, 191, 185.

Claveria, R.J.R., Perez, T.R., Navarrete, I.A., Perez, R.E.C., & Lim, B.C.C., 2020. The identification of heavy metal accumulator ferns in abandoned mines in the Philippines with applications to mine rehabilitation and metal recovery. *Journal of Sustainable Mining*, 19(1), 45–57.

Corales-Ultra, O.G., Peja, R.P., & Casas, E.V., 2019. Baseline study on the levels of heavy metals in seawater and macroalgae near an abandoned mine in Manicani, Guiuan, Eastern Samar, Philippines. *Marine Pollution Bulletin*, 149, 110549.

De Campos Favas, P.J., Martino, L.E., & Prasad, M.N.V., 2018. Abandoned mine land reclamation-challenges and opportunities (holistic approach). In: Prasad, M.N.V, De Campos Favas, P.J., & Maiti, S.K. (Eds.), *Bio-geotechnologies for Mine Site Rehabilitation*. Elsevier B.V., Amsterdam, 3–31.

De la Torre, J.B.B., Claveria, R.J.R., Perez, R.E.C., Perez, T.R., & Doronila, A.I., 2016. Copper uptake by Pteris melanocaulon Fée from a Copper-Gold mine in Surigao del Norte, Philippines. *International Journal of Phytoremediation*, 18(5), 435–441.

Delos Angeles, M.D., Buot Jr, I.E., Cuevas, V.C., & Sanchez, P.B., 2020. Phytoremediating capacity of copper tolerant plants in mine tailing soil materials with compost amendment in Mankayan, Benguet, Philippines. *Environment Asia*, 13(1), 86–98.

Delos Angeles, M.D., Merilles, M.L.D., Silva, B.B., Masungsong, L.A., & Cadiz, N.M., 2018. Floral and algal species composition in an abandoned mine tailings pond at Bgy. Mogpog, Marinduque, Philippines. *Environment Asia*, 11(2), 91–108.

Diaz, 2019. Mercury contamination and community awareness on the abandoned Palawan Quicksilver Mine in Puerto Princesa City,Philippines. *Asian Journal of Resilience*, 1(1), 9–14.

Domingo, J.P.T., & David, C.P.C., 2014. Soil amelioration potential of legumes for mine tailings. Philippine Journal of Sc*ience*, 143(1), 1–8.

Fabro, K.A., 2019. Widespread mercury poisoning recorded in 2 Palawan villages. *Rappler.* https://www.rappler.com/nation/172410-widespread-mercury-poisoning-puerto-princesa-villages-palawan/. Accessed 16 October 2017.

Fernando, E.S., Quimado, M.O., Trinidad, L.C., & Doronila, A.I., 2013. The potential use of indigenous nickel hyperaccumulators for small scale mining in the Philippines. *Journal of Degraded and Mining Lands Management*, 1(1), 21–26. .

Fontanilla, C.S., & Cuevas, V.C., 2010. Growth of Jatropha curcas L. seedlings in copper-contaminated soils amended with compost and Trichoderma pseudokoningii Rifai. *The Philippine Agricultural Scientist*, 93(4), 369–376.

Fortoul, T.I., Rodriguez-Lara, V., Gonzalez-Villalva, A., Rojas-Lemus, M., Colin-Barenque, L., Bizarro-Nevares, P., García-Peláez, I., Ustarroz-Cano, M., López-Zepeda, S., Cervantes-Yépez, S., López-Valdez, N., Meléndez-García, N., Espinosa-Zurutuza, M., Cano-Gutierrez, G., & Cano-Rodríguez, M.C., 2015. Health effects of metals in particulate matter. In: Nejadkoorki, F. (Ed.), *Current Air Quality Issues*. InTechOpen, Vienna.

Garagara, L.P., Dolorosa, R.G., & Baaco, A.T., 2015. Water parameters of Pulang Lupa Lake, an abandoned open pit mine in Puerto Princesa City, Palawan, Philippines and its potential as bird watching destination. *The Palawan Scientist*, 7, 12–26.

Gibaga, C.R.L., Arcilla, C.A., & Hoang, N., 2020. Volcanic rocks from the Central and Southern Palawan Ophiolites, Philippines: Tectonic and mantle heterogeneity constraints. *Journal of Asian Earth Sciences: X*, 4, 100038.

Gibaga, C.R.L., Samaniego, J.O., Tanciongco, A.M., Quierrez, R.N.M., Montano, M.O., Gervasio, J.H.C., Reyes, R.C.G., & Peralta, M.J.V., 2022a. The rare earth element (REE) potential of the Philippines. *Journal of Geochemical Exploration*, 107082.

Gibaga, C.R.L., Samaniego, J.O., Tanciongco, A.M., Quierrez, R.N.M., Montano, M.O., Gervasio, J.H.C., Reyes, R.C.G., & Peralta, M.J.V., 2022b. Pollution and radiological risk assessments of mine wastes from selected legacy and active mines in the Philippines. *Journal of Degraded and Mining Lands Management*, 9(4), 3621–3633.

Gigantone, C.B., Sobremisana, M.J., Trinidad, L.C., & Migo, V.P., 2020. Impact of abandoned mining facility wastes on the aquatic ecosystem of the Mogpog River, Marinduque, Philippines. *Journal of Health and Pollution*, 26, 200611.

Gray, J.E., Greaves, I.A., Bustos, D.M., & Krabbenhoft, D.P., 2003. Mercury and methylmercury contents in mine-waste calcine, water, and sediment collected from the Palawan Quicksilver Mine, Philippines. *Environmental Geology*, 43, 298–307.

Irving, E.M., 1954. Note on the first discovery of commercial mercury in the Philippines. *Philippine Geologist*, 9, 1–8.

Langcay, M.L., Clemente, E.D., & Arranz, C.G., 2021. Risk assessment of mercury in soil and surface water in Brgy. Santa Lourdes, Puerto Princesa City, Palawan. *IOP Conference Series: Earth and Environmental Science*, 690, 012043.

Lanot, J.L., Lawig, J.A.L., Lecaros, J.A., Malagotnot, P.J.L., Labay, P.M., & Samaniego, J.O., 2020. Physico-chemical properties and heavy metal contents of Ino-capayang mine-made lake in Marinduque, Philippines. *International Journal of Engineering Research and Technology*, 13(6), 1493–1496.

Lim, H.S., Lee, J.S., Chon, H.T., & Sager, M., 2008. Heavy metal contamination and health risk assessment in the vicinity of the abandoned Songcheon Au-Ag mine in Korea. *Journal of Geochemical Exploration*, 96(2–3), 223–230.

Mante, K.M.B., Cadiz, N.M., Cuevas, V.C., & Rebancos, C.M., 2019. Soil and vegetation analysis of rehabilitated and unrehabilitated area in an inactive copper mined-out site in Mogpog, Marinduque, Philippines. *Journal of the International Society for Southeast Asian Agricultural Sciences*, 25(2), 118–129.

Mante, K.M.B., & Cadiz, N.M., 2021. Social acceptability of the rehabilitation efforts conducted in an inactive mined-out area in Marinduque, Philippines. *Journal of Degraded and Mining Lands Management*, 8(4), 2931–2936.

Maramba, N.P.C., Reyes, J.P., Francisco-Rivera, A.T., Panganiban, L.C.R., Dioquino, C., Dando, N., Timbang, R., Akagi, H., Ma Castillo, M.T., Quitoriano, C., Afuang, M., Matsuyama, A., Eguchi, T., & Fuchigami, Y., 2006. Environmental and human exposure assessment monitoring of communities near an abandoned mercury mine in the Philippines: A toxic legacy. *Journal of Environmental Management*, 81(2), 135–145.

Mines and Geosciences Bureau (MGB). National program support for environment and natural resources management project. 2014. Project Completion Report Annex C: Remediation Risk. https://faspselib.denr.gov.ph/sites/default/files//Publication%20Files/PCR%20Annex%20C%20Bagacay%20Remediation.pdf. Accessed 1 March 2023.

Mhlongo, S.E., Amponsah-Dacosta, F., & Kadyamatimba, A., 2019a. Development and application of a methodological tool for prioritization of rehabilitation of abandoned tailings dumps in the Giyani and Musina areas of South Africa. *Cogent Engineering*, 6, 1.

Mhlongo, S.E., Amponsah-Dacosta, F., & Kadyamatimba, A., 2019b. Incorporation of the method of ranking the hazards of abandoned mine entries into a rule-based expert system. *Minerals*, 9(10), 600.

Mhlongo, S.E., & Sigxashe, S., 2021. The criteria for ranking and prioritization of rehabilitation of abandoned mines. *Journal of Degraded and Mining Lands Management*, 8(4), 2947–2956.

Muller, G., 1969. Index of geoaccumulation in sediments of the Rhine River. *GeoJournal*, 2(3), 108–118.

National Economic and Development Authority (NEDA), 2021. Updated Philippine Development Plan 2017-2022, Chapter 20: Ensuring ecological integrity, clean and healthy environment. https://pdp.neda.gov.ph/wp-content/uploads/2021/12/Chapter-20.pdf. Accessed: 15 January 2022.

Nelson, G.L.M., Abrigo, G.N.A., & Raymundo, A.K., 2020. The acceptability of using bioremediation in Mogpog, Marinduque, a mined-out community in the Philippines. *Journal of Nature Studies*, 19(1), 81–104.

Ocampo, L.A., & Schmitz, S., 2023. Accumulation by dispossession and hazardscape production in post-corporate gold mining in Itogon, Philippines. *Geographical Research*, 61(1), 44–57.

Palawan News, 2019. Abandoned PQMI mine rehab to focus on treatment of Contaminated Lake. *Palawan News*. https://palawan-news.com/abandoned-pqmi-mine-rehab-to-focus-on-treatment-of-contaminated-lake/. Accessed 5 April 2019.

Posadas, R.C., Salting, A.A.A., Yamas, J.G.T., & Patayon, E.S., 2022. Mercury concentration of the soil profiles of the abandoned mercury mine in Sta. Lourdes, Puerto Princesa City, Palawan, Philippines. *Philippine Journal of Science*, 151(3), 895–899..

Samaniego, J.O., Gibaga, C.R.L., Tanciongco, A.M., Rastrullo, R.M., & Costa, M.A.V., 2019. Surface water characteristics in the vicinity of abandoned mercury mine site in Puerto Princesa City, Philippines. *Philippine Journal of Science*, 148(3), 493–498.

Samaniego, J., Gibaga, C.R., Tanciongco, A., & Rastrullo R., 2020a. Total mercury in soils and sediments in the vicinity of abandoned mercury mine area in Puerto Princesa City, Philippines. *Applied Sciences*, 10(13), 4599.

Samaniego, J., Gibaga, C.R., Tanciongco, A., Rastrullo, R., Mendoza, N., & Racadio, C.D., 2020b. Comprehensive assessment on the environmental conditions of abandoned and inactive mines in the Philippines. *ASEAN Journal on Science & Technology for Development*, 37(2), 81–86.

Samaniego, J.O., Gibaga, C.R.L., Tanciongco, A.M., & Rastrullo, R.M., 2020c. Monitoring of mercury in air from the abandoned mercury mine area using direct mercury analyzer. *International Journal of Engineering Research and Technology*, 13(6), 1373–1378.

Samaniego, J.O., Gibaga, C.R.L., Mendoza, N.D., Racadio, C.D.T., Tanciongco, A.M., & Rastrullo, R.M., 2020d. Mercury and other heavy metals in groundwater in the abandoned mercury mine in Puerto Princesa city, Philippines. *Philippine Journal of Science*, 149(3a), 897–801.

Samaniego, J.O., Gibaga, C.R.L., Tanciongco, A.M., Devanadera, M.C.E., Paro, F.R.C., Adil, J.H., Montano, M.O., Quierrez, R.N.M., & Gutierrez, A.C.S., 2022. Health risk assessment of trace metals in the vicinity of an abandoned mercury mine in Puerto Princesa city, Philippines. *Philippine Journal of Science*, 151(2), 671–682.

Stanley, G., Gallagher, V., Ni Mhairtín, F., Brogan, J., Lally, P., Doyle, E., & Farrell, L., 2019. *Historic Mine Sites – Inventory and Risk Classification, Vol 1.* U.S. Environmental Protection Agency and The Geological Survey of Ireland, 75 p.

Tanciongco, A., Quierrez, R.N., Samaniego, J., Gibaga, C.R., & Montano, M., 2022. Integrating historic mine hazard scoring for comprehensive assessment of abandoned mine rehabilitation in the Philippines. *International Journal of Environmental Science and Development*, 13(4), 110–117.

Trocio, D.Y.C., & Paguntalan, D.P., 2023. Review on the use of arbuscular mycorrhizal fungi in bioremediation of heavy metal contaminated soils in the Philippines. *Philippine Journal of Science*, 152(3), 1139–1159.

United States Department of Agriculture (USDA) - Natural Resources Conservation Service, 2008. Soil Quality Indicators. https://www.nrcs.usda.gov/sites/default/files/2023-01/Soil%20Quality-Indicators-Bulk20%Density.pdf. Accessed 1 February 2023.

United States Environmental Protection Agency (US EPA), n.d. *TENORM: Copper Mining and Production Wastes.* https://www.epa.gov/radiation/tenorm-copper-mining-and-production-wastes. Accessed 16 October 2017.

Wang, L., Ji, B., Hu, Y., Liu, R., & Sun, W., 2017. A review on in situ phytoremediation of mine tailings. *Chemosphere*, 184, 594–600.

Williams, T.M., Weeks, J.M., Apostol, A.N., & Miranda, C.R., 1999. Assessment of mercury contamination and human exposure associated with coastal disposal of waste from a cinnabar mining operation, Palawan, Philippines. *Environmental Geology*, 39, 51–60.

Zuluaga, M.C., Norini, G., Lima, A., Albanese, S., David, C.P., & De Vivo, B., 2017. Stream sediment geochemical mapping of the Mount Pinatubo-Dizon Mine area, the Philippines: Implications for mineral exploration and environmental risk. *Journal of Geochemical Exploration*, 175, 18–35.

12 Abandoned Open Cast Coal Mine Waste in Russia – Emphasizing the Approaches for Ural Region Mine Waste Rehabilitation via Biological Recultivation

Tamara S. Chibrik, Natalia V. Lukina, Margarita A. Glazyrina, Elena I. Filimonova, Evgeniy A. Rakov and Majeti N.V. Prasad

12.1 INTRODUCTION

The complex fuel and energy concerns in Russia play an important role in the country's economy. In fact, coal is Russia's fifth base export product. The Russian coal industry has proven to have more than 193 billion tons of coal reserves, including anthracite and brown coal deposits. Currently, in the Russian Federation coal is mined in 16 coal basins and the territories involved cover 85 municipalities (Anonymous, Ministry of Energy of the Russian Federation, 2010; Kuznetsov and Ilyushechkin, 2013; State Report, 2017). In 2014, 165.1 million tons of coal were mined in the country (BP Statistical Review, 2014), accounting for 4.3% of global production and Russia stands in 6th place after China, the United States, Australia, Indonesia and India. More than half of mined coal (56.6%) is consumed in Russia, while all the rest (43.4%) is exported (Golitsyn et al., 2015).

Success in the coal mining industry in Russia was preceded by the restructuring of the coal industry, which began in 1994 and is unparalleled in scale and timing in terms of domestic and international practice. Restructuring is characterized by the fact that in the shortest possible time a massive, one-time closure of unprofitable mines in mining regions was done, and also taking into account the critical state of the environment. Liquidation of especially unprofitable enterprises was carried out in 24 constituent entities of the Russian Federation and 78 mining municipalities (Gorbacheva and Sovacool, 2015). The elimination program included 188 mines and 15 open pit mines (Stoyanova and Kharchenko, 2011). However, despite significant costs of eliminating the evident problems in areas where coal mining enterprises were closed, many issues about preserving and restoring the environment remain unresolved (Andreyashkina, 2012; Bojko and Kabala, 2016; Eremin, 2014; Eremin and Shakhova, 2010). One of these problems stems from the fact that the work of such enterprises in the coal industry is inextricably linked to the creation of waste, most of which is stored on the surface in the form of waste heaps. They occupy potentially useful territories, thereby excluding them from economic circulation, and this, in turn, also has a negative effect. Impact on environments around coal mining areas is evident, taking the form of the destruction of soil cover, natural vegetation, disrupted groundwater levels, pollution of water bodies and air basins

 DOI: 10.1201/9781315141855-12

and intensification of erosion processes both on dumping grounds and in the adjacent areas (Carras et al., 2009; Finkelman, 2004; Misz-Kennan and Fabianska, 2010).

Elimination of harmful effects of industrial dumps and their expedient use have become a significant issue, and a solution is required to recultivate dumps (Howrald and Karim, 2015; Tichanek and Tichanek, 2014). It is, however, expensive, laborious and not always a possible process to carry out. Development of methods of biological recultivation is preceded by research regarding dump formations, their agrochemical composition, as well as the process of vegetation for the restoring on dumps of land in a more natural way. These processes were studied by scientists at the Ural Federal University (Bolshakov and Chibrik, 2007; Chibrik, 2008, 2012; Chibrik et al., 2016, 2018; Rakov and Chibrik, 2009) (Table 12.1).

TABLE 12.1
Abandoned Open Cast Mines and Mine Waste in Russia – Major Environmental Challenges and Approaches for Rehabilitation

Salient Research Findings	Author(s)	Year
Geochemical partitioning of heavy metals and metalloids in the ecosystems of abandoned mine sites within the Moscow brown coal basin.	Semenkov et al.	2022
Mining wastes of the Russian North-West: microbiological prerequisites of reclamation.	Abakumov et al.	2021
Potential environmental pollution from copper metallurgy and methods of management.	Izydorczyk et al.	2021
The ecological state and the problems of recultivation of man-made disturbed irrigated soils.	Jobborov et al.	2021
The influence of mining wastes from the Russia's Arctic zone on the plants development.	Krasavtseva,. and Maksimova.	2021
Involving small and medium-sized mining industry businesses in mining waste processing in the Russian Federation.	Nevskaya et al.	2021
Predictive assessment of toxicants' migration from technogenic gold-mining wastes (case study of the tailings management facility of Tokur mill, Amur region, Russia).	Radomskaya et al.	2021
Organic matter accumulation by alkaline-constructed soils in heavily metal-polluted area of Subarctic zone.	Slukovskaya et al.	2021
Small mining and exploration enterprises in the Urals.	Boltyrov	2020
Recultivation of post-mining disturbed land.	Ignatyeva et al.	2020
Manganese biomining from manganese-bearing industrial wastes of Georgia.	Jandieri et al.	2020
Nephrite-bearing mining waste as a promising mineral additive in the production of new cement types.	Khudyakova et al.	2020
Geochemical transformation of soil cover and vegetation in a drained floodplain lake affected by long-term discharge of effluents from rayon industry plants, lower Don River Basin, Southern Russia.	Konstantinova et al.	2020
Toxic metal (loid) s contamination and potential human health risk assessment in the vicinity of century-old copper smelter, Karabash, Russia.	Kumar et al.	2020
Experiences on establishment of Scots pine (*Pinus sylvestris* L.) plantation in ash dump sites of Reftinskaya HPES, Russia.	Zalesov et al.	2020
Impact of mine waters on chemical composition of soil in the Partizansk Coal Basin, Russia.	Arefieva et al.	2019
Basic methodology and modern technological solutions for reclamation of disturbed lands and soils.	Banov et al.	2019
Using the industrial wastes for remediation of sites affected by acid mine water dumping..	Blinov and Krasilnikova	2019
Rare-earth elements in groundwater of the abandoned Levikha copper mine (Middle Urals, Russia..	Rybnikov and Rybnikova	2019.
Regularities in the evolution of groundwater quality at abandoned copper sulfide mines at the Levikha ore field, Central Urals, Russia.	Rybnikov and Rybnikova	2019

(*Continued*)

TABLE 12.1 (CONTINUED)
Abandoned Open Cast Mines and Mine Waste in Russia – Major Environmental Challenges and Approaches for Rehabilitation

Salient Research Findings	Author(s)	Year
Geological structure and mining development of the Urals.	Rybnikova et al.	2019
Elements distribution in soil and plants of an old copper slag dump in the Middle Urals, Russia.	Zolotova and Ryabinin	2019
Environmental impact of disposal of coal mining wastes on soils and plants in Rostov Oblast, Russia.	Alekseenko et al..	2018
Environmental problems of mining waste disposal sites in Russia.	Azarova et al..	2018
Establishment of Phytocenosis on Brown Coal Mine Waste in Urals Sverdlovsk and Chelyabinsk Regions, Russia – Drivers, Constraints, and Trade-Offs.	Chibrik et al.	2018
Remediation of fly ash dumpsites through bioenergy crop plantation and generation.	Madhumita et al.	2018
Hydrometallurgical processing of low-grade sulfide ore and mine waste in the Arctic regions: Perspectives and challenges.	Masloboev et al.	2018
Analysis of metal content in soils near abandoned mines of Bashkir Trans-Urals and in the hair of children living in this territory.	Semenova et al.	2018
Copper recycle from sulfide tailings using combined leaching of ammonia solution and alkaline bacteria.	Yin et al.	2018
Analysis of mine waste by geocheimical and geophysical methods (a case study of the mine tailing dump of the Salair ore-processing plant).	Epov et al.	2017
Water quality and Hydrogeochemistry of the abandoned sulfide mines of the Middle Urals (Russia).	Rybnikov and Rybnikova	2017
Geoecological challenges of mined open pit area use in the Ural.	Rybnikova et al.	2017
Production activity analysis Methodology for open pit coal mines	Zhaglovskaya et al.	2017
Biological recultivation of mine industry deserts: facilitating the formation of phytocoenosis in the Middle Ural region, Russia,	Chibrik et al.	2016
Energy-efficient technologies of integrated coal and mining waste development in the Moscow Coal Basin.	Griazev et al.	2016
The Kizel Coal Basin (the Western Urals, Russia): Environmental problems and Solutions.	Khayrulina et al.	2016
Restoration of smelter industrial barrens following pollution reduction drives economic recovery.	Koptsik et al.	2016
Using of soda production waste.	Krasilnikova, and Blinov	2016
Academic science and new industrialization (example of the Komi republic).	Lazhentsev	2016
Prospects of technological use of coals in Russia,	Golitsyn et al.	2015
Reviewing the risks and rewards of investing in Russian coal-fired electricity.	Gorbacheva and Sovacool	2015
An acidophilic desulfosporosinus isolated from the oxidized mining wastes in the Transbaikal area.	Karnachuk et al.	2015
Integrated geophysical and geochemical methods for the environmental assessment of active smelter operations and abandoned mines in Karabash, Ural Mountains, Russia.	Korneeva et al.	2015
Environmental problems of coal mining in the Southern Far East of Russia and soil remediation on mining dumps.	Krupskaya	2015
Rational development of noble metal placer mining waste in the east of Russia.	Litvintsev	2015
Technogenically polluted lands of the city of Karabash and methods for their reclamation.	Lonzinger et al.	2015
Regulatory aspects of mining waste management in the Russian Federation.	Nevskaya and Marinina	2015

(*Continued*)

TABLE 12.1 (CONTINUED)
Abandoned Open Cast Mines and Mine Waste in Russia – Major Environmental Challenges and Approaches for Rehabilitation

Salient Research Findings	Author(s)	Year
Deposit as a means of restoring the content and reserves of humus of long-arable chernozems of Northern Zauralye forest-steppe zone,	Eremin	2014
Problems and prospects concerning the phytoremediation of heavy metal polluted soils: a review.	Koptsik	2014
The application of coal combustion by-products in mine site rehabilitation.	Park et al.	2014
Contribution to the solution of thermally active reclamation of coal waste heaps,	Tichanek and Tichanek	2014
The current status of nature, and ecological problems of the Middle Onon region.	Vyrkin et al.	2014
An environmental history of Russia.	Josephson et al.	2013
Coal resources, production and use in the Russian Federation.	Kuznetsov and Ilyushechkin	2013
Resource potential of placer mining waste.	Litvintsev	2013
The composition of plant communities of natural and anthropogenically disturbed ecotypes on watersheds of Yamal Peninsula.	Andreyashkina	2012
Some aspects of assessing the experience of biological recultivation in Ural coal fields.	Chibrik	2012
Ecorestoration of the coalmine degraded lands.	Maiti	2012
Ecological and economic problems in the regions of mass closure of coal mines.	Stoyanova and Kharchenko	2011
The dynamics of the leached Chernozem moisture content of under different tillage systems for spring wheat in the conditions of Northern Zauralye,	Eremin DI, Shakhova OA	2010
Power strategy of Russia for the period up to 2030.	Anonymous	2009
On the problem of flora formation in industrially disturbed land areas,	Rakov and Chibrik	2009
Phytocenosis formation in Urals anthropogenic landscapes.	Chibrik	2008
Biological recultivation: Ural approach.	Bolshakov and Chibrik	2007
Industrial barrens: extreme habitats created by non-ferrous metallurgy.	Kozlov and Zvereva	2007
Assessment of environmental impacts of active smelter operations and abandoned mines in Karabash, Ural Mountains of Russia.	Udachin et al.	2003
Effects of mining waste waters on fish in lakes of NW Russia.	Tkatcheva et al.	2002
Trace metal accumulation and fish pathologies in areas affected by mining in the Kola Region, Russia.	Moiseenko, and Kudryavtseva	2001

The Urals constitute an old industrial region in the Russian Federation, and it has a variety of mineral resources, including coal and lignite. These coal reserves in the Urals are insignificant – less than 1% of the country's total reserves. Main deposits are found in the Kizel coal basin, Chelyabinsk, and South Ural coal basins, Bulanash coal deposit, etc. (Fig. 12.1). The Urals do not have good prospects for future coal mining enterprises. Furthermore, any progress made in the coal industry may be associated with use of resources to a greater extent rather than being fuel-specific, but more significantly as a raw material for producing mineral fertilizers, carbon-alkaline reagents and other chemical products (Fig. 12.2). Coal mining in the Urals today has virtually ceased and subsequently most coal is imported. The Ural Federal district includes the Republic of Bashkortostan (142,947 square kiliometers), the Sverdlovskregion (194,307 square kiliometers), the Chelyabinsk region (88,529 square kiliometers), the Kurgan region (71,488 square kiliometers), the

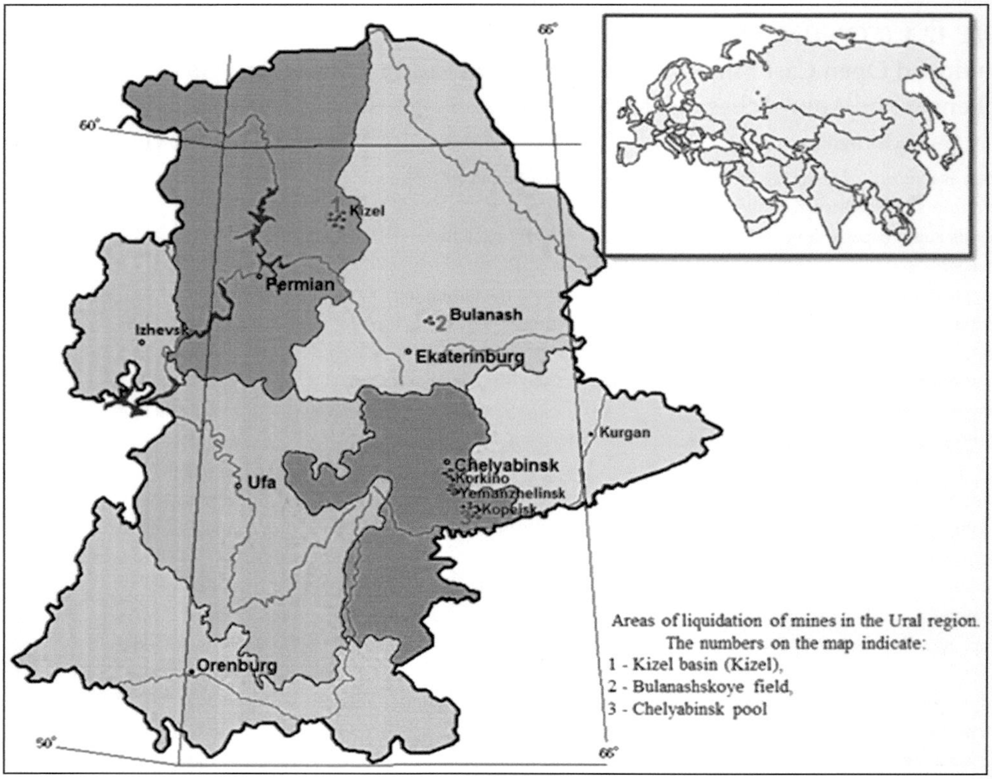

FIGURE 12.1 Areas of abandoned coal mines and overgurdens in the Urals region.

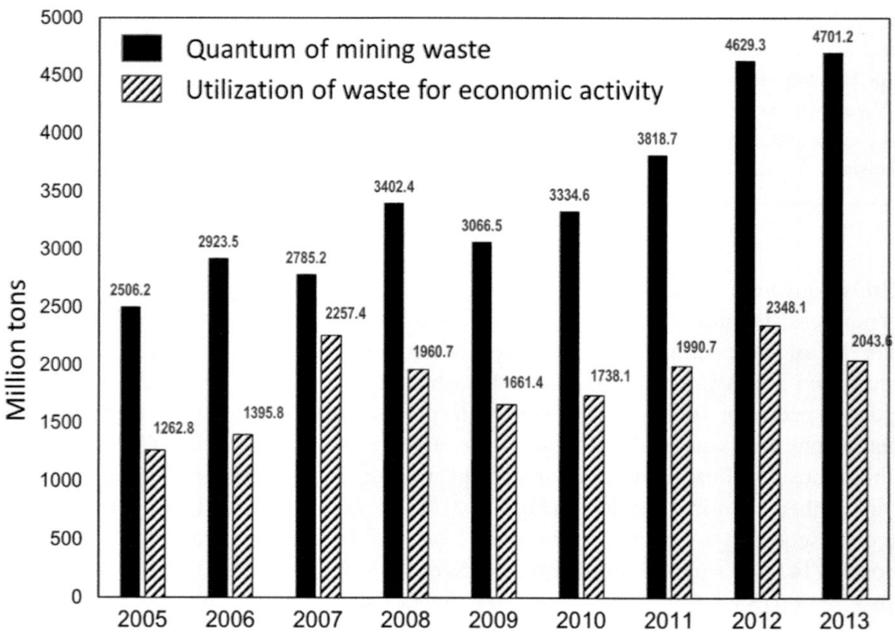

FIGURE 12.2 Quantum of waste formation in mining and waste utilization for economic activity in the Russian Federation. Figure drawn based on the data of Nevskaya and Marinina (2015)

Tuymenregion (without KhMAD and YaNAD) (160,122 square kiliometers), the Khanty-Mansiysk Autonomous District – Yugra (KhMAD -Yugra) (534,801 square kiliometers) and the Yamal-Nenets Autonomous District (YaNAD) (769,250 square kiliometers) (Kolchina et al., 2019)

12.2 GENERAL CHARACTERISTICS OF THE STUDIED OBJECTS

Kizelovsky Coal Basin (KCB) is located on the western slope of the Middle Ural, in the Perm region of the Russian Federation (Taiga zone). Coal-bearing sediments are located on a narrow strip (5–20 km) along the western slope of the Ural Mountains at a distance of about 150 km, which is why it is also called West Ural basin. The basin area covers about 200 km^2; coal reserves amount to 464 million tons

Bulanashsky Coal Deposit (BCD) is located in the center of the industrial Trans-Urals, at the junction of Trans-Ural forest, agricultural areas and mining zone of the Sverdlovsk region, on the eastern slope of the Ural Mountains. It is located 120 km northeast of Ekaterinburg (Taiga zone). Mines of BCD are found in Bulanash village, Artyomovsky District and Sverdlovsk Region.

Chelyabinsk Brown Coal Basin (ChCB) is situated on the eastern slope of the Ural Mountains in Chelyabinsk Region (forest-steppe zone). The area of the lignite basin is 1300 km^2. Coal mining in Chelyabinsk coal basin has been carried out since 1907 to the present day. Coal mining amounted to 1 billion tons. From 1957 to 1975 total coal production was 20 million tons per year.

12.3 KIZELOVSKY COAL BASIN (KCB)

In the KCB, coal mining began in 1797 in "Zaprudnaya" mine. At the beginning of the 20th century, 36 mines operated in this basin; coal mining in 1915 amounted to 976 thousand tons. In 1956, there were 37 mines in the basin and production amounted to 11.4 million tons. In 1959, coal mining in the KCB reached a maximum of 12 million tons (Fig. 12.3). After this it continued to decline. In connection with development of reserves in the upper horizons of deposits, coal production decreased, and in 1984 this amounted to 3.7 million tons. In 1995, a program was adopted for ending the "Kizelugol" company and restructuring of the KCB industry was done due to its unprofitability. By 2002, the KCB completely ceased coal mining and nearly all the mines were flooded.

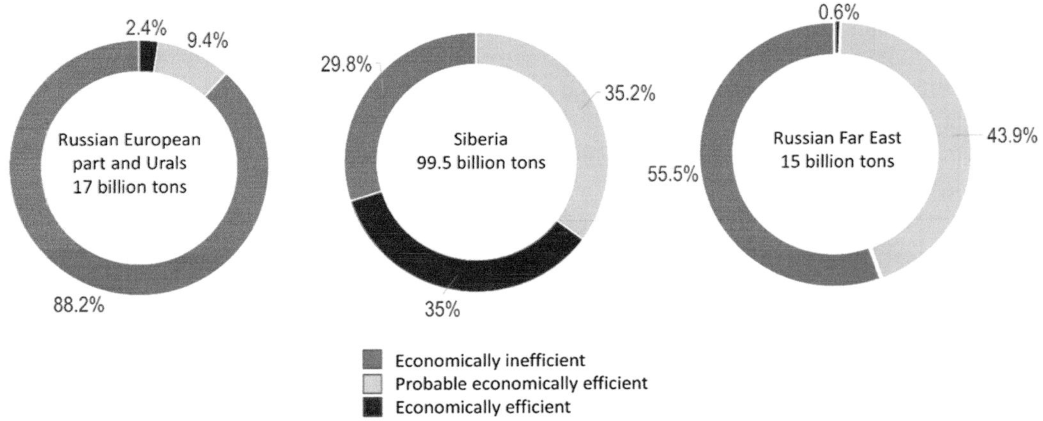

FIGURE 12.3 Coal reserves of Russia. (Redrawn based on the data Gorbacheva and Sovacool 2015)

Among the activities implemented in the framework of joint projects of the "Kizelugol" company liquidation, environmental programs became priorities. The most important of these is research related to studies of the damage caused by industrial development of coal deposits, on the basis of which recommendations are being developed for their elimination. Studies have shown that the greatest danger in KCB formation of earth surface dips above the mining locations. As well, flooding of mines caused the formation of technogenic horizons of mine waters, and this represents the danger of much larger-scale pollution of the groundwater, including that used for drinking water in the cities and villages of the Kizelovsky basin. Mine water pouring onto the surface pollutes soil and surface water, and this has serious implications for people, flora and fauna.

Another big danger is represented by environmental pollution by dumps, which are formed during coal mining. Rock heaps consist of fragments of argillite, sandstone and limestone with inclusions of coal. Weathering, oxidation, hydrolysis, etc. occur here. All lithological varieties of coal strata include pyrite and organic sulfur, which plays a crucial role in the formation of acid mine water. The amount of sulfide and organic sulfur reaches 12–15%, with an average level of 5–8%. The main form of sulfur is pyrite, which is unstable in oxidizing conditions. Oxidation of pyrite contained in dumps, with formation of sulfuric acid, oxides and hydroxides of iron occur and this is accompanied by heat and spontaneous combustion dumps. Atmospheric precipitation flowing down from dumps is a major source of pollution of groundwater and surface water.

KCB coals are very rich in Be, Sc, Ti, V, Co, Ni, Cu, Ga, Ge, Y and Zr. In the KCB before 1968–1970 waste rocks were stored in heaps that burned intensely. In order to prevent future spontaneous combustion of rocks, flat heaps were formed at most mines. We have conducted a survey of waste heaps and flat dumps of 24 mines in the KCB. Geobotanical and agrochemical characteristics of some mine heaps are given below.

The "Shirokovskaya" mine operated from 1945 to 1996. At the "Shirokovskaya" mine, a heap of waste (dumped from 1946–1958) and a flat blade (dumped from 1968) were examined. Rocks of slag are very stony; strongly acidic; poor in phosphates, potassium and sodium; slightly saline, with the type of salinity being chloride-sulfate; and having a large amount of carbon of coal origin. According to agrochemical indices, flat dump rocks are similar to slagging rocks, but they are not saline (Table 12.2). In 1963–1964 after spontaneous combustion of a slagheap, it was quenched with a clay suspension. A few years after that, in places where clay was deposited, settlement of single individual species of *Betula pendula* Roth, *Populus tremula* L., *Salix caprea* L., *Alnus incana* (L.) Moench and *Epilobium angustifolium* L. was evident.

The mine named after Lenin (Lenina mine) operated from 1881 to 1997, and the depth of operations reached 925 m. The mine dumps are flat and occupy a considerable area. Their height is approximately up to 7 m. Many of these are also subject to greater erosion. Analysis of fresh rocks entering the dumps, as well as rocks of 20- and 40-year-old dumps, revealed (Table 12.2) that they are strongly amenable, strongly acidic (40-year-old-slightly acidic), poor in mobile forms of potassium and sodium, low in phosphorus and medium in salinities. The saline components are mainly chloride-sulfate. Rocks contain mostly carbon originating from coal. *Populus balsamifera* L. was planted on 20- and 40-year-old dumps with the addition of soils into landing pits. By the time of the survey, 15-year-old plantings remained and attained a height of up to 4–5 m. Some individuals showed signs of drying out.

The Cospashskaya mine operated from 1949 to 1998. Rocks piled into waste mine heaps are strong like stones, acidic, non-saline, poor in mobile forms of phosphorus, and they contain potassium and sodium. A high carbon content is evident in them. There is no vegetation growing on "young" dumps. At the foot of the old waste heap, settlement of sparse groups of woody species, i.e. *Betula pendula* Roth, *Populus tremula* L., *Salix caprea* L. was observed.

The "Lunevsky" mine is one of the oldest in the KCB. Heaps are 3–5 m in height with gentle slopes (slope steepness up to 30°). Rocks of "Lunyevsky" heaps are very stony, not saline, strongly acidic and poor in mobile forms of phosphorus as well as sodium. Dumps are located on the outskirts of a forest dominated by *Picea obovata* Ledeb. and *Abies sibirica* Ledeb., which is a source of

TABLE 12.2
Agrochemical Characteristics of Rocks in Some Waste Fields of Kizelovsky Deposit

Dump	Particles Portion >10 mm, %	pH_{H_2O}	P_2O_5	K_2O	Na_2O	C, %	Dry Residue, %	Volume of Salinity
			mg per 100 g of Rocks					
"Shirikoskaya" mine heap	65	2,2–2,4	0,18	2,42	3,75	14,74	0,72	Weakly saline (salinity of chloride-sulfate)
"Shirikoskaya" mine flat dump	53	2,3–2,5	0,23	3,75	2,33	15,79	0,24	Not salty
Lenina mine flat dump ("fresh" rocks)	42	2,4	0,02	4,50	1,25	7,41	1,78	Medium saline (salinity of chloride-sulfate)
Lenina mine flat dump (20 years)	50	2,4–3,0	2,61	6,66	1,41	8,13	0,51–1,18	Salinity from the weak to the middle (salinity of chloride-sulfate)
Lenina mine flat dump (40 years)	37	6,1–6,8	0,28	24,00	7,41	4,45	0,09–1,29	The upper layers are not saline, the lower ones are medium salinized (salinity of chloride-sulfate)
Kospashskaya mime heap ("fresh" rocks)	69	3,7	1,32	23,25	10,50	18,91	0,20	Not salty
Kospashskaya mime heap (foots)	54	3,55	0,70	22,5	2,51	4,16	0,11	Not salty
"Lunevskaya" mine heap	31	3,1–3,6	0,37	8,68	0,92	4,46	0,03	Not salty
"Severnaya" mine flat dump	61	6,6–7,2	0,04	17,33	5,91	3,89	0,66	Salinity from the weak to the middle (salinity of chloride-sulfate)
"Volodarsky" mine heap	55	2,3–2,9	0,11	3,12	0,75	9,46	0,10	Not salty
"Vladimirovskaya" mine heap	52	2,4–2,5	0,02	13,6	1,8	7,9	0,10	Not salty
"Klyuchevskaya" mine Flat dump	63	3,1–5,9	0,37	0,65	2,5	3,33	0,84	Heavily salty (salty sulfate)
Klyuchevskaya mine heap	45	3,6–3,8	0,5	1,5	2,0	2,3	0,06	Medium saline (saline sulfate)

seed drift. On gentle slopes of "young" heaps, shoots of woody species (*Betula pendula* Roth, *Salix caprea* L.) and small groups of *Poa pratensis* L . are found. Meanwhile, at the foot, mass shoots of *Betula pendula* Roth and *Picea obovata* Ledeb. did in fact eventually die. On "old" 50–70-year-old dumps, peaks are not overgrown. The lower section of the slopes is overgrown with *Epilobium angustifolium* L. (sp), *Agrostis gigantea* Roth, *Antennaria dioica* (L.) Gaertn., *Fragaria vesca* L. (sol – sp gr), shoots of *Betula pendula* Roth (sp).

The "Severnaya" mine was worked from 1942 to 1998. This mine has a heap and a flat dump and the latter is composed of slightly saline (a type of salinity chloride-sulfate) neutral rocks. Rocks can sufficiently provide mobile forms of potassium but are indeed poor in phosphorus. On the old part of the heap, settlement of the following single individuals of grassy species was observed: *Tussilago farfara* L., *Epilobium angustifolium* L., *Poa trivialis* L. and *Salix caprea* L. shrubs.

The Volodarsky mine is one of the oldest mines, and it operated for around 150 years. Rocks are strongly stone argillites, siltstones, clay and sandy shales stored in two heaps and a flat heap. According to the agrochemical characteristics, they are strongly acidic, poor in mobile forms of phosphorus, potassium and sodium, and they are non-saline. Carbon content in rocks varies greatly. Overgrowth of pure rocks was not observed.

The "Vladimirovskaya" mine existed from 1954 to 1998. According to agrochemical indicators, rocks at this mine are similar to those in the Volodarsky mine. Mine dump processes of vegetation overgrowth were not observed. In the lower part of the southern slope, seedlings of *Populus balsamifera* L. (height 1.0–1.5 m) were planted with soil introduced into planting pits. Trees were depressed with leaf shredding and yellowing being observed.

The Klyuchevskaya mine operated from 1941 to 1997. Not far from the mine were two large flat heaps and two waste heaps. Fresh rocks taken from a flat dump are slightly acidic, poor in mobile phosphates and rich in potassium and sodium, very strongly salted with sulfates containing a lot of carbon of coal origin. On the old flat heap of rock, more acidic, mobile forms of phosphorus, potassium and sodium are few, salinization is strong, and there are sulfates. Plant populations on flat dumps were not observed.

The height of heap № 1 is about 40 m. According to agrochemical parameters,rocks from heaps are close to rocks of old flat blade, and they differ from them only by alternation of non-saline layers with saline ones. During spontaneous combustion, the heap accumulated clay silting. After some time, in places where clay was deposited, a settlement of grassy plants was observed: first, *Polygonum aviculare* L., *Agrostis gigantea* Roth, *Rumex acetosella* L., *Oxybasis rubra* (L.) S. Fuentes, Uotila & Borsch and, second, tree species such as *Betula pendula* Roth, *Populus tremula* L., *Salix caprea* L., etc.

Thus, the agrochemical and geobotanical survey of dumps and waste heaps of KCB revealed that virtually all rocks are very acidic, mostly saline with sulfates, poor in nutrients and strongly amenable. There is practically no natural overgrowing of dumps even 50–70 years after dumping finished. The presence of vegetation on them is associated with the introduction of clay, soil or household garbage. Rocks at all the surveyed dumps are not suitable for biological recultivation unless their properties are greatly improved.

12.4 BULANASH COAL DEPOSIT

Egorshinsky Mine Administration of Vakhrushevugol (Bulanash settlement, Artyomovsky District, Sverdlovsk Region, the Russian Federation) manages the Bulanash Coal Deposit (BCD). Rocks of productive strata are represented by: gravelites, conglomerates, sandstones (20–45%), aleurolites (45–65%). Coal seams are separated by layers of aleurolite, sandstone and, less often, siderite.. Then the mines known as "Bulanash-3" were opened in 1947, "Bulanash-4" in 1953 and "Bulanash-5" in 1951. Then in 1957, mines 1/2 and 5 were merged into one, known as Bulanash-2/5. In 1970, all the mines of Bulanash were incorporated into one enterprise; Egorshinsky mine department with two production units – the Bulanash-2/5 and Bulanash-3/4 mines. Annual coal production was about 600,000 tons.

In 1999, mines "Bulanash-2/5" and "Bulanash-3/4" were merged into one, which was renamed the "Egorshinskaya" mine. In 2003, the mine was closed and flooded. Currently, this mine operates in drainage mode only. Rocks that ended up on the surface from mines (siltstones, argillites, sandstones, etc.) were subsequently used for dumping of water-protective dams, for dumping and repair of roads, and they were stored in paving stones and flat dumps. In 1992, 35.68 thousand tons of waste were used for dumping roads; in 1998, 15 thousand tons of rocks were transferred to construction organizations. Total volume of waste from the Egorshinsky mine administration in dumps as of 01/01/1999 amounted to 1,036.2 thousand tons.

In order to develop viable methods of recultivation, we surveyed waste heaps from the Egorshinsky mine (Table 12.3). Rocks' composition of all surveyed heaps proved to be similar. These are siltstones, argillites and sandstones with inclusions of coal in various proportions. Most waste heaps are mainly stone. According to agrochemical indices, heaps of rocks of BCD are neutral or weakly alkaline, mostly unsalted, with low content of mobile phosphorus, high - potassium, and medium and high - sodium, i.e., more suitable for plant growth than rocks of CCB (Table 12.4).

In 1978 we examined mines heaps at the Bulanash coal deposit. The heap of the Bulanash mine is the highest (52 m), and erosion was observed on its slopes. Vegetation was distributed very unevenly. Overgrowth of the lower (looped) part of slopes was the best – the total projective cover by vegetation (TPC) was 60–80%. Species composition was dominated by: *Festuca rubra* L., *Tussilago farfara* L., *Artemisia absinthium* L. (sol1), *Calamagrostis epigejos* (L.) Roth, *Poa pratensis* L. and *Rumex confertus* Willd. (sp gr). In the upper part of slope vegetation cover was more sparse, and plants were distributed in small groups: *Tussilago farfara* L., *Lappula squarrosa* (Retz.) Dumort., *Lepidium ruderale* L., and *Rumex confertus* Willd.

Slagheap of Bulanash-2 mine is characterized by very steep slopes, subject to severe erosion. Vegetation overgrowth was observed only on the southern, more gentle slope. On the lower part of the slope some species were more abundant: *Festuca rubra* L., *Elytrigia repens* (L.) Desv. ex Nevski, *Artemisia vulgaris* L., *Artemisia marschalliana* Spreng., Gypsophila paniculata L., Tussilago

TABLE 12.3
General Characteristics of Some Waste Heaps at Bulanash Coal Deposit

Dump	Height, m	Base Area, m²	Volume, Thousand m³	Years of Work	Thermal Facts	Predominant Rocks Composition
mine № 2 вump	29,0	20,000	204,000	1916–1956	absent	Sandstone, limestone
Bulanash Mine 3:					absent	Mudstones (45%), siltstone (30%), sandstones (20%), coal (5%)
north dump	29,0	17,600	141,000	1947–1977		
Bulanash Mine 3: southern dump	39,5	39,750	429,300	1947	Partially burned out	Mudstones (25%), siltstone (40%), sandstones (25%), coal (10%)
Bulanash Mine 4						Mudstones (30%), siltstone (40%), sandstones (20%), coal (10%)
heap	37,5	20,000	217,800	1953	absent	
flat dump	13,0	–	27,000	1957	There are pockets of fire	
Bulanash 2/5 mine dump	52,0	32,500	540,000	1951–1992	absent	

TABLE 12.4
Agrochemical Characteristics of Rocks on Some Waste Heaps at Bulanash Coal Deposit

Dump	Portion of Particles >10 mm, %	pH_{H2O}	P_2O_5	K_2O	Na_2O	C, %	Dry Residue, %	Volume of Salinity
			mg per 100 g of Rocks					
Heap of mine "Bulanash" -2	43,6	6,6–7,3	6,5	21,6	5,8	27,2	0,06	Not salty
Heap of mine "Bulanash" -3 (top)	19	6,8–7,8	2,5	24,1	10,0	5,6	0,19	Not salty
Heap of mine "Bulanash" -3 (foot)	9	7,8–7,9	3,3	32,5	16,3	2,5	0,036	Not salty
Heap of mine "Bulanash" -4	65	7,2	7,5	80,0	135,0	8,7	0,84	Highly salty (mixed salinity type)
Heap of mine "Bulanash" -2/5 (northern slope)	52	6,1–8,3	3,0–8,2	35,1	12,8	7,2	0,06–2,56	Not salty (there are saline areas)

farfara L., *Pastinaca sylvestris* Mill. (cop1). *Gypsophila paniculata* L., *Convolvulus arvensis* L., *Centaurea scabiosa* L. (sp) and others grew on the middle and upper part of the slope.

Slagheap No. 1 of Bulanash-3 mine at the time of the survey is active, and its height is 39 m. Stubs of its southern slope were overgrown with *Tussilago farfara* L., *Poa pratensis* L., *Trifolium repens.* L. and *Crepis tectorum* L. (cop1– sp gr). *Pinus sylvestris* L. and Salix caprea L. were singularly encountered. Only 11 species were found faurther up of the slope of which only two, *Atriplex tatarica* L. and *Tussilago far*fara L., were abundant (sp). On the northern slope plants are isolated, very small and confined to depressions. These are mainly weed-ruderal species, specifically *Tussilago farfara* L., *Sonchus arvensis* L., *Atriplex tatarica* L., *Lepidium ruderale* L., *Polygonum aviculare* L. and *Taraxacum officinale* F.H. Wigg.

Overgrowing of heap No. 2 of Bulanash-3 mine was more intensive with *Elytrigia repens* (L.) Desv. ex Nevski, Festuca rubra L., Artemisia absinthium L., Artemisia vulgaris L. (sol2), *Sonchus arvensis* L., *Pastinaca sylvestris* Mill. and *Taraxacum officinale* F.H. Wigg. (sp). The slagheap of Bulanash-4 mine was burnt out, operating, had a height 37.5 m, area of 2 ha, and a part of it was disassembled for construction purposes. Vegetation on the slopes of the slag heap was not found. At its foot were found the following: *Poa pratensis* L., and *Puccinellia distans* (Jacq.) Parl. (cop1 – sol2), and small groups from *Taraxacum officinale* Wigg., *Artemisia absinthium* L., *Calamagrostis epigejos* (L.) Roth. (sp gr).

Monitoring studies on the formation of vegetation was undertaken after 25 years. Revealed here was that in waste heaps of Bulanash 2/5 and Bulanash 3 mines there was an active settlement of woody species. So on these heaps of "Bulanash 2/5" mine a tree layer (up to 3 m in height) consisting of *Pinus sylvestris* L. (cop1), *Betula pendula* Roth, *Betula pubescens* Ehrh. (sol), *Populus tremula* L., *Salix caprea* L. and *Salix myrsinifolia* Salisb. (sol – sp) and others had formed. Species of herbs such as *Epilobium angustifolium* L. (cop2), *Melilotus albus* Medik. (cop1), *Trifolium repens* L., *Medicago lupulina* L., *Potentilla norvegica* L. (cop1 gr) and *Fragaria vesca* L., (sp gr) averaged 80% of the vegetation.

On the slag heap of Bulanash 3 mine the vegetation cover was more sparse, GPC – 50–60%. Tree species were represented by rare *Hippophaë rhamnoides* L. bushes, from grassy species both on the northern and on the southern slopes *Trifolium hybridum* L. (copl gr), *Artemisia absinthium* L. had a high abundance (sp – sol). The restraining factors are rocky nature of the substratum, spontaneous combustion and the shape of waste heaps. The significant height and steepness of their slopes create peculiar ecotopic conditions. On waste heaps of many mines, strong erosion is observed, which leads to washing away of the surface layers of rocks and disruption of formed vegetation cover.

12.5 CHELYABINSK BROWN COAL (LIGNITE) BASIN

ChCB is located to the east of Chelyabinsk, covering an area of 1,300 km^2 and is the main mine on the eastern side of the Urals. The field was discovered in 1832 and industrial development began in 1907; its reserves in 1935 amounted to 1800 million tons. The peak of coal mining was in the 1960s. Coal mining was carried out in the Kopeysk, Kamyshinsky, Korkinsky and Emanzhelinsky districts. Remaining coal reserves in the basin at the end of the 20th century amounted to 523 million tons. The thickest layers (up to 200 m) were created by the Korkinsky open pit (Korkino). In the Kopeisk open-pit mining was carried out.

Along with extensive open cast mines (Korkino, Kopeisk), underground mining of coal (Korkino, Yemanzhelinsk, Kopeisk) was conducted at ChCB. Slag heaps of mines are located in the cities of Korkino, Emanzhelinsk and Kopeisk. We conducted a survey of heaps of 17 mines. We describe some of these and a summary is presented in Table 12.5. In Korkino, two mine heaps – Korkinsky and Kalachevskaya – were examined.

Korkinsky mine (Korkino) is an open-pit mine, and it operated from 1930 to 2013. In the territory of the mine there is an old heap. Its rocks are mainly stony, neutral or slightly alkaline, poor in mobile forms of phosphorus, with a very high content of mobile potassium and sodium. Rocks are medium and strongly saline. The type of salinization is mixed, mainly sulphatic. Carbon content decreases sharply with depth (Table 12.6). A survey of vegetation displayed that the northeast slope of the heap was the most intensively overgrown. In the upper part of slope, high abundance (sol2 – sol3) was evident in *Berteroa incana* (L.) DC., *Bassia scoparia* (L.) A. J. Scott, *Lappula squarrosa* (Retz.) Dumort., *Linaria vulgaris* Mill. *Erysimum cheiranthoides* L. In the middle parts of the slope other species were growing: *Poa pratensis* L., *Elytrigia repens* (L.) Desv. ex Nevski, and *Artemisia absinthium* L. TPG of vegetation in the lower and middle parts of slope reached 70–80%. Groups of

TABLE 12.5
General Characteristics of Some Mine Heaps in the Chelyabinsk Coal Basin

Dump	Height, m	Base Areas	Volume, Thousand m^3	Years of Work	Thermal Facts
Terrikoniki mine	59	60,000	1000	1948–1969	weak
"Korkinskaya" (Korkino) mine heaps	52	7,660	80	1969–н.д.	weak
"Prigorodnaya" (Korkino) mine heap	36	9,000	108	1949–1976	moderate
(Kopeisk) mine heaps	49,6	68,831	828	1960–1982	weak
	48	28,693	458	1963–1980	weak
	56	49,279	929	1963–1980	weak
"№23" (Kopeisk) mine heaps	37,5	15,260	214	1953–1964	weak
"Miass" (Kopeisk) mine heaps	37	36,000	814	1950–1977	weak

TABLE 12.6
Agrochemical Characteristics of Rocks at Some Mine Heaps in Chelyabinsk Coal Basin

Dump	Portion of Particles >10 mm, %	pH_{H2O}	P_2O_5	K_2O	Na_2O	C, %	Dry Residue %	Volume of Salinity
			mg per 100 g of Rocks					
Korkino								
“Korkinskaya” mine top of heap	12,0	6,8–7,4	4,1	104,2	118,3	4,5	0,19	Medium saline (mixed salinization)
“Korkinskaya”: mine heap foot (south-western slope)	34,6	6,9–7,1	7,6	73,3	104,6	10,9	0,61	Highly saline (mixed salinization)
“Kalachevskaya”mine heap	13,0	7,2–7,3	5,9	57,5	61,0	5,1	0,30	Not salted
Kopeysk								
”Podozernaya” mine heap	37,7	6,3–7,5	1,1	16,0	39,4	5,4	1,14	Highly saline (mixed salinization)
“№23” mine heap	22,1	4,1–4,8	0,9	17,3	25,6	8,3	1,3	Highly saline (mixed salinization)
“Tsentralnaya” mine heap	61,1	5,2–7,5	2,9	6,2	43,8	15,7	0,53	Highly saline (mixed salinization)
“Glubokaya” mine heap	34,0	6,1–6,5	0,8	18,6	40,0	5,3	0,56	Highly saline (sulfate type of salinization)
“Komsomolskaya” mine heap	50,5	4,5–4,8	0,7	11,7	57,5	6,8	6,35	Very saline (sulfate type of salinization)
Yemanzhelinsk								
“Vostochnaya” Mine heap								
top (northern slope)	43,5	7,4–8,1	2,1	39,1	149,2	3,4	0,43	Medium saline (mixed salinization)
foot (northern slope)	19,3	7,3–8,1	2,3	80,0	173,3	4,6	0,22	Not salted
“Kullarskaya” mine heap	23,0	6,9–7,3	3,6	50,0	102,3	7,3	0,39	Medium saline (mixed salinization)

Linaria vulgaris Mill. *Bassia scoparia* (L.) A. J. Scott., *Lappula squarrosa* (Retz.) Dumort., *Erysimum cheiranthoides* L. (sp gr) are noted on the southern slope.

Kalachevskaya mine (Chumlyak, Korkino) was commissioned in 1953 and closed in 1995. During 1975–1979 the average daily coal production was 1,300 tons, and this annually reached 570,000 tons. Kalachevskaya mine is composed of strong stone-rich, slightly alkaline, not saline rocks, with a small amount of mobile phosphates and a larger content of potassium and sodium. Carbon content decreases with increasing depth (Table 12.5). An examination of the heap demonstrated that vegetation grew only in the lower sections of the slopes. On the southern slope, only two species were found: *Salsola collina* Pall. (sol2) and *Chenopodium album* L. (sp). Vegetation of the northern slope was more diverse. Some 29 species growing here include: *Poa pratensis* L., *Lactuca tatarica* (L.) C.A. Mey., *Convolvulus arvensis* L. (soll2); *Elytrigia repens* (L.) Desv. ex Nevski; *Artemisia vulgaris* L. and *Polygonum aviculare* L. (copl gr).

In Kopeysk, mine heaps were surveyed, specifically "Glubokoye", No. 23, "Central", "Komsomolskaya" and "Podozernoy". The heap belong to the "Glubokaya" mine is composed of strongly rocky, slightly acidic rocks. According to the agrochemical properties of rocks of heap, they are poor in mobile forms of phosphates and rich in potassium and sodium. Sulfate salinization is noted to varying degrees (Table 12.5). On the southern slope of the heap, only isolated individuals of *Salsola collina* Pall. were noted. On remaining slopes, *Bassia scoparia* (L.) A. J. Scott., *Salsola collina* Pall. (sp gr), *Elytrigia repens* (L.) Desv. ex Nevski and *Puccinellia distans* (Jacq.) Parl. (sol gr) were found. Meanwhile the TPC of vegetation averaged 60–80%.

The heap of "Glubokoye", No. 23, is not suitable for recultivation. Rocks are very stony, acidic, highly saline (mixed salinity type, mainly sulfate), with small amounts of mobile forms of phosphorus, potassium (except for the upper 0–2 cm layer) and high in sodium. *Calamagrostis epigejos* (L.) Roth. (sp gr – copl) grows on the eastern and western slopes of the heap. TPC of the lower part of slopes comprises 40% vegetation. Single or individual groups of 15 species were encountered

Tsentralnaya mine operated from 1932 to 2006. Its mine field is confined to the southern part of the Kopeysk coal-bearing region (the northern half of Chelyabinsk brown coal basin). Mine worked out in deep horizons (reservoirs with a 2.2–3.5 m thick). The Tsentralnaya mine heap rocks' agrochemical properties are very strong, acidic, generally poor in mobile phosphate and potassium but rich in sodium. Salinization of rocks is strongly evident and the type of salinization is mixed, mainly sulfate.

"Komsomolskaya" mine operated from 1952 to 2008. In 1970, 15 million tons of coal were mined and this mine's heap is replete with very boulder-like rocks. At the top of heap to the depth of 40 cm, rocks are strongly acidic, while deeper it is alkaline, and in the middle part of slope they are neutral, poor in mobile forms of phosphorus. They are sufficiently provided with potassium and excess sodium. Rocks are very saline and the type of salinization is sulfate. Carbon content diminishes dramatically as depth increases. Self-growing vegetation on the waste heaps is very slow. *Salsola collina* Pall. is marked on slopes while at the foot of these slopes, plant groups are predominantly *Artemisia absinthium* L., *Salsola collina* Pall., *Calamagrostis epigejos* (L.) Roth. and *Elytrigia repens* (L.) Desv. ex Nevski.

The heap belonging **"Podozernaya" mine** consists of neutral rocks, they contain little phosphorus and a lot of potassium and sodium. Rocks are highly saline (the type of salinization is mixed, mainly sulfate). Self-growing of vegetation is weak. On the southeastern slope there are small groups of *Salsola collina* Pall. Foot dominated by *Salsola collina* Pall. Also evident are *Artemisia absinthium* L. *Linaria vulgaris* Mill., *Carduus crispus* L., *Taraxacum officinale* F.H. Wigg. and *Onopordum acanthium* L.

In the city of Emanzhelinsk, heaps belonging to the Vostochnaya and Kullarskaya mines were examined. **Vostochnaya mine** (Emanzhelinsk) began operating in 1962. In 1989, it was merged with Baturinskaya mine. During its 28 years of operation, the Vostochnaya mine extracted 14.5 million tons of coal. The peak production was achieved in 1971 (955.9 thousand tons). The heap of **Vostochnaya mine** is old, strongly stone-like and stretches from west to east. Rocks are slightly alkaline, poor in mobile forms of phosphorus, rich in potassium and sodium and saline to varying degrees. The type of salinization is mixed and mainly sulfate. There are separate layers of non-saline rocks.

A survey of the local vegetation showed that at the foot of the heap it was overgrown: from the southern side by *Atriplex tatarica* L. (sol2), while *Sisymbrium loeselii* L. (cop1) was dominated by *Bassia scoparia* (L.) A. J. Scott (sp gr). The foot of the northern and eastern slopes was dominated by *Artemisia absinthium* L., (sol2), *Salsola collina* Pall. (sol1-3), *Berteroa incana* (L.) DC. (cop1), *Convolvulus arvensis* L. (sp gr – sol1), *Calamagrostis epigejos* (L.) Roth. (sp gr). The remaining species were sporadic. Above the slopes thinning of vegetation was observed. On the southern slope only 7 species were recorded: *Salsola collina* Pall. (sol1), *Sisymbrium loeselii* L. (sp gr) and others. Other species encountered were also sporadic. On the northern slope there were 21 species and the dominant ones were *Erysimum cheiranthoides* L., *Lactuca tatarica* (L.) C. A. Mey., *Atriplex sagittata* Borkh., *Artemisia absinthium* L., *Artemisia vulgaris* L., *Convolvulus arvensis* L. (cop1). Other species were rare and only single samples were found.

The **"Kullarskaya Mine"** (Emanzhelinsk) has the longest service life among the mines near this city. It produced its first coal in 1956 and the last batch was dug in 1998. During the period of operation, 22,887,200 tons of coal were produced. The peak year of production was 1977 (801.7 thousand tons). The Kullarskaya mine heap is composed of very stony, neutral rocks with large amounts of mobile forms of potassium and sodium. Rocks are medium and strong in saline, the type of salinization is mixed and mainly sulfate. Carbon content is high and in some places declines as the depth deepens. Initial stages of vegetation formation were observed on the northeastern slope of the heap. Despite the large species diversity (32 in total), only *Convolvulus arvensis* L. and *Erysimum cheiranthoides* L. had a high level of abundance (cop1), while other species were rare.

Thus, studies have shown that rocks of mines at the ChCB are neutral or slightly alkaline, poor in mobile forms of phosphorus and rich in potassium and sodium. Most of these rocks are medium and strongly saline. The type of salinization is mixed, with a predominance of sulfates. Among saline rocks at different depths, there are non-saline layers with a small thickness. Carbon content decreases with depth (Table 12.5). In almost all heaps, the initial stages of **vegetation overgrowth** was observed, depending largely on rocks' composition, exposure and steepness of the slope. The initial stages of vegetation formation on most heaps revealed high abundance, and the following species are found: *Bassia scoparia* (L.) A. J. Scott., *Salsola collina* Pall., *Linaria vulgaris* Mill. Lappula squarrosa (Retz.) Dumort., *Artemisia absinthium* L., *Taraxacum officinale* F.H. Wigg., *Convolvulus arvensis* L. and *Elytrigia repens* (L.) Desv. ex Nevski.

Repeated monitoring studies of vegetation formation on heaps in the ChCB, covering a period of 25 years, showed that on areas not affected by erosion, as well as anthropogenic influence, a further transformation of vegetation occurred. GPC of plants depends on the slope position. In the lower sections of heaps, GPC varies from 40 to 70%, while in the upper part it ranges from 30 to 50% .

In plant communities growing on heaps, the following species are evident but mostly in small numbers: *Artemisia absinthium* L., *Artemisia vulgaris* L., *Calamagrostis epigejos* (L.) Roth. and *Poa trivialis* L. High abundance is found for *Linaria vulgaris* Mill., *Achillea millefolium* L., *Artemisia dracunculus* L., *Festuca pseudovina* Hack. ex Wiesb., *Hordeum jubatum* L., *Poa angustifolia* L. and *Poa trivialis* L. So, studies have shown that vegetation formation on mine heaps in the Chelyabinsk region takes the form of simple plant groups ranging from young weed-ruderal species to plant communities with a high proportion of zonal meadow-steppe, polyzonal weed-ruderal and adventive species.

12.6 DEVELOPMENT OF RECULTIVATION SCHEMES

Studies have shown that rocks found at the heaps of different coal deposits in the Urals have little use for biological recultivation due to their many characteristics. All of them are strongly amenable but have little in the way of nutrients. In addition, rocks of heaps at the Kizelovsky coal basin are extremely acidic and saline (chloride-sulfate salinization). Added to this, the salted rocks of mine heaps at the Chelyabinsk lignite basin have mixed levels of salinity with a predominance of sulfates. At the Bulanash deposit, mine heaps are more intensely overgrown, but the process of self-overgrowing does not ensure the creation of a coherent (dense) vegetation cover. For successful propagation, they require specially devised and economical methods that make biological recultivation possible. This is due to the specific nature of environmental conditions of locales. Spontaneous combustion of rocks is observed on all the existing dumps.

There is no reserve of fertile soil at all the mine enterprises sites, and there are very few potentially fertile rocks. Most of them are poor in nutrients. The urgent need for recultivation is determined by requirements for improving sanitary conditions and protecting the environment. Taking into account the realistic remedial work that enterprises can do, we have developed a flowchart of biological reclamation of dumps using the Kizelovsky coal basin (Fig. 12.4) as an example. In the developed flowchart, instructions for recultivating disturbed or damaged land caused by the Russian coal industry are taken into account.

To neutralize phytotoxic rocks, it is necessary to overlap the newly formed surface with a layer of neutral heavy clay and its thorough compaction. This creates a waterproof layer and also helps to protect the underlying rocks from spontaneous combustion. Creation of an insulating layer can be replaced (or combined with) by liming the formed surface to a depth of 0.5–0.8 m with subsequent thorough compaction (with the aim of deoxidation of rocks). The required dosage of lime is calculated depending on the hydrolytic acidity of rocks.

The next stage is application of potentially fertile, and in their absence, non-toxic, unsuitable and low-fertile rocks in order to create an impermeable layer. Thickness of the applied layer will depend on the rationale of biological remediation: for forest plantations – 1.0–1.5 m; and for perennial grasses — 0.4 m. In the absence of a sufficient number of rocks suitable for application, it is possible to reduce the recultivation layer's thickness to 0.7 m by planting trees and shrubs of sanitary-greening type. However, it will require preliminary deep liming of this surface layer.

After reshaping the dumps or preserving the existing form during the technical stage of recultivation, it is necessary to create around the dumps a directional flow of acid mine waters. This will help to greatly reduce any damage being done to the surrounding areas. In the case of absence or deficiency of suitable rocks, it is possible to reduce the phytotoxicity of the surfaces of old dumps (flat and conical-shaped) by silting them with clay pulp. This technique is widely used in eliminating self-igniting dumps.

Technical surface preparation makes biological recultivation possible, but it does not eliminate all the adverse properties of waste dumps as specific ecotopes. Overcoming or minimizing adverse environmental conditions is possible in two ways: by improving the properties of substrate (water-physical, agrochemical, etc.) and by selecting the appropriate range of species for these conditions. Measures to improve the properties of substrate are usually reduced to inoculation of mineral and organic fertilizers complex. For perennial grasses, it is recommended to apply mineral fertilizers at the rate of nitrogen of 90 kg, phosphorus 60 kg and potassium 60 kg of active substance per hectare during seeding and annual fertilizing of crops. For forest plantations that have an economic use, it is advisable to apply nitrogen at a rate of 3 c / ha, phosphorus and potassium at 2 c / ha (active ingredient). *Melilotus albus* Medik. is recommended on margins and *Melilotus officinalis* (L.) Lam. between strips.

Rocks found on mine heaps are mostly poor in nutrients, so treating plants with sludge may actually be beneficial. An accelerated method of dump recultivation using a microbiological reagent was developed and tested on dumps at Kizelovsky basin and indicated good results (Krasavin and Safin 2005 in Russian "Ekologicheskaya reabilitatsiya uglepromyshlennykh territoriy Kizelovskogo basseyna v svyazi s zakrytiyem shakht" - "Ecological rehabilitation of coal-mining territories of the Kizelovsky basin in connection with the closure of mines"). The biological stage of recultivation involves perennial grasses seeding and planting trees and shrubs. It is important to select a range of plants that can resist the high acidity caused by substrates and gases. For tree plantings in

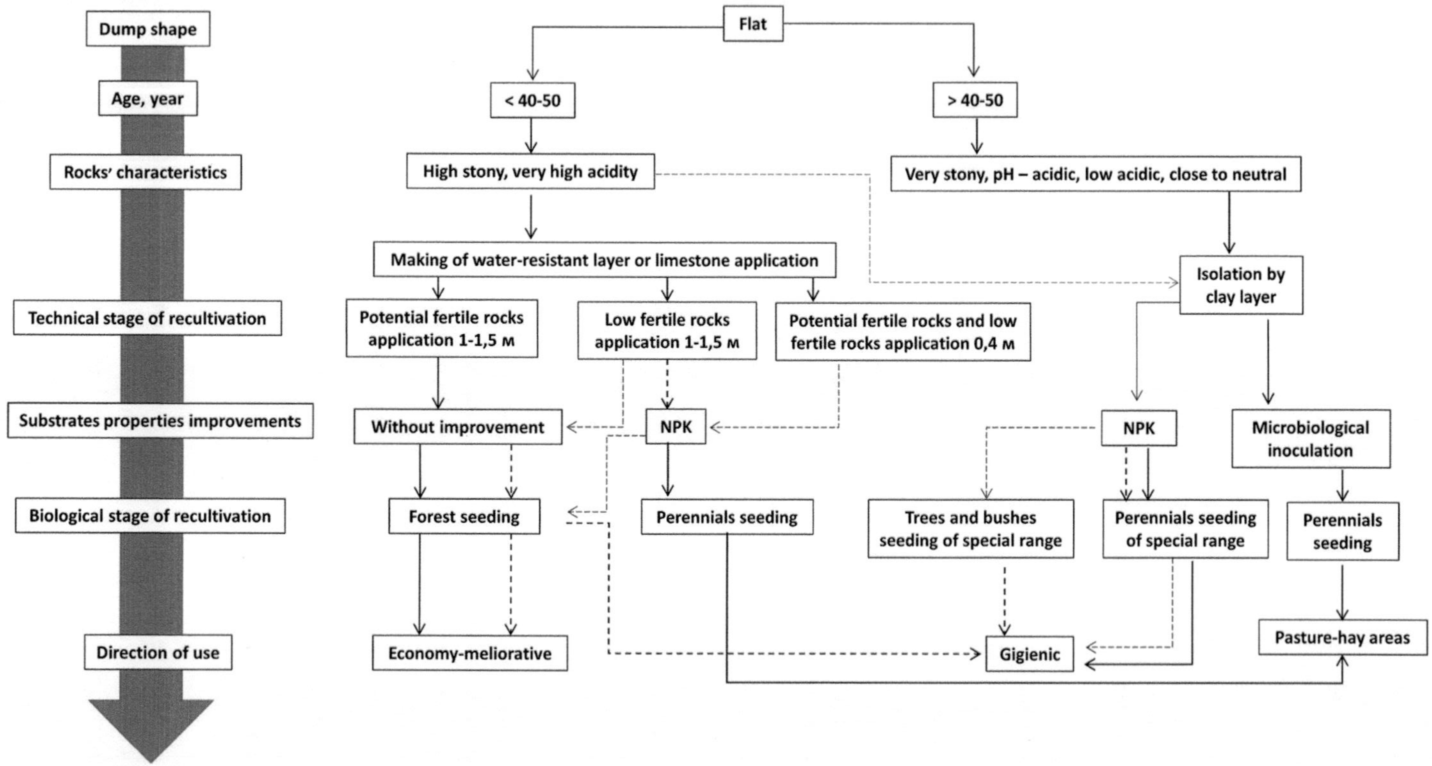

FIGURE 12.4 Flow chart of Kizelovsky coal dump's biological recultivation →Main variant→ Extra variant.

the taiga zone, the following species are recommended: *Pinus sylvestris* L. (in the absence of gas content), *Larix sibirica* Ledeb., *Betula pendula* Roth and *Betula pubescens* Ehrh., *Alnus incana* (L.) Moench, *Salix caprea* L., *Populus tremula* L. et al. *Bromus inermis* Leyss., *Festuca rubra* L., *Melilotus albus* Medik., while *Melilotus officinalis* (L.) Lam. The range of perennial grasses can be expanded by carrying out additional testing of zonal varieties. Double seeding rate is recommended in seeding.

Directions for recultivating territories are determined by the required technical and biological measures (Fig. 2). In any case, biological reclamation neutralizes the hazards posed by dumps for the environment and contributes to its improvement or allows some form of viable economic and ecological value to be returned to waste areas. The heaps at the Bulanash deposit and Chelyabinsk coal basin can be recultivated (planted) by overlapping their surface with non-toxic and potentially fertile rocks and planting species able to cope with these heaps' conditions. Landscaping steep slopes using a terracing approach can produce good and attractive results.

12.7 CONCLUSION

Among the activities implemented as projects to remediate closed mines in Russia's mining regions, environmental programs are now very important. Priority is given to research that explains the dangers and risks of extracting coal deposits, and subsequently recommendations are proposed for their elimination and to protect the environment as much as possible.

Assessments to analyze and eliminate the technogenic outcomes are currently focused on three areas: (i) geodynamic monitoring; (ii) monitoring of earth's surface movement and (iii) monitoring of waste dumps status. The greatest danger is failure of the earth's surface above the mine workings. This process which began with the closure of mines continues even more intensively. It is particularly relevant in the Kizelovsky and Chelyabinsk coal basins. In addition to the plateau-like and cone-shaped heaps in mining works, there are relief features such as cracks, landslides, craters, dips, etc., which require a specific strategy that is consistent with the conditions of the local territory.

The value of social and environmental monitoring is not only in the implementation of specific projects, but also in collecting important data through applied science. This makes it possible, along with recultivation of mine waste heaps, to suggest other useful appliance of heap material: (i) burnt waste heaps for road construction; (ii) to fill the voids within mines and (iii) as building materials. In addition, establishing geochemical barriers for purification of acid waste from waste dumps (heaps) is justified and viable.

Over 35 million m^3 of rocks, half of which are not burnt, were stored in more than 70 dumps in the Russian Ural region. They can be used as secondary raw materials to create or consolidate a building material (rubble, sand, etc.). What has been very promising is the creation of building ceramics and putting the conservation of heaps to good use. Ultimately, despite the lost productivity of disturbed lands and other negative consequences, their recultivation is one of the best ways to save the environment and design a new, aesthetically filled landscape.

ACKNOWLEDGMENTS

The study was funded by the Ministry of Science and Higher Education of the Russian Federation as part of state task of the Ural Federal University, FEUZ-2023-0019.

REFERENCES

Abakumov, E., Kimeklis, A., Gladkov, G., Andronov, E. and Zverev, A., 2021, December. Mining wastes of the Russian North-West: Microbiological prerequisites of reclamation. In *IOP Conference Series: Earth and Environmental Science* (Vol. 938, No. 1, p. 012015). IOP Publishing.

Alekseenko, V.A., Bech, J., Alekseenko, A.V., Shvydkaya, N.V. and Roca, N., 2018. Environmental impact of disposal of coal mining wastes on soils and plants in Rostov Oblast, Russia. *Journal of Geochemical Exploration*, 184, pp.261–270.

Andreyashkina, N.I., 2012. The composition of plant communities of natural and anthropogenically disturbed ecotopes on watersheds of Yamal Peninsula: Floristic diversity. *Ecology,* 1, pp.22–26.

Anonymous: Ministry of Energy of the Russian Federation 2010 Power Strategy of Russia for the Period up to 2030. Approved by Decree n 1715-r of the Government of the Russian Federation, dated 13 November 2009. Moscow.

Arefieva, O., Nazarkina, A.V., Gruschakova, N.V., Skurikhina, J.E. and Kolycheva, V.B., 2019. Impact of mine waters on chemical composition of soil in the Partizansk Coal Basin, Russia. *International Soil and Water Conservation Research*, 7(1), pp.57–63.

Azarova, S.V., Usmanova, T.V. and Mezhibor, A.M., 2018. Environmental problems of mining waste disposal sites in Russia. *Reviews of Environmental Contamination and Toxicology,* 247, pp.59–84.

Banov, M., Tsolova, V. and Zhelezov, G., 2019. Basic methodology and modern technological solutions for reclamation of disturbed lands and soils. *Problems of Geography*, 1, pp.3–17.

Blinov, S. and Krasilnikova, S., 2019. Using the industrial wastes for remediation of sites affected by acid mine water dumping. In *Proceedings of the International Mine Water Association Conference*, Perm, Russia (pp. 15–19).

Bojko, O. and Kabala, C., 2016. Transformation of physicochemical soil properties along a mountain slope due to land management and climate changes – A case study from the Karkonosze Mountains. *SW Poland,* 140, pp.43–54.

Bolshakov, V. and Chibrik, T., 2007. Biological recultivation: Ural approach. *Science Russia,* 3, pp.106–112.

Boltyrov, V., Storozhenko, L., Degtyarev, S. and Bobina, T., 2020. Small mining and exploration enterprises in the Urals. In *E3S Web of Conferences* (Vol. 177, p. 05015). EDP Sciences.

BP Statistical Review of World Energy. June, 2014. http://www.bp.com/content/dam/bp/pdf/Energy-economics/statistical-review-2014/BP-statistical-review-of-world-energy-2014-full-report.pdf.

Carras, J., Day, S., Saghafi, A. and Williams, D., 2009. Greenhouse gas emissions from low temperature oxidation and spontaneous combustion at open-cut coal mines in Australia. *International Journal of Coal Geology,* 78, 161–168. https://doi.org/10.1007/s10661-009-1021-7.

Chibrik, T.S., 2008. Phytocenosis formation in Urals anthropogenic landscapes. *Odesa National University Herald Biology*, 13(16), pp.130–136.

Chibrik, T.S., 2012. Some aspects of assessing the experience of biological recultivation in Ural coal fields. *Orenburg State Agrarian University,* 5(37), pp.216–218.

Chibrik, T.S., Lukina, N.V., Filimonova, E., Glazyrina, M.A., Rakov, E.A., Maleva, M.G. and Prasad, M.N.V., 2016. Biological recultivation of mine industry deserts: Facilitating the formation of phytocoenosis in the middle Middle Ural region, Russia. *Bioremediation and Bioeconomy,* pp.389–418.

Chibrik, T.S., Lukina, N.V., Filimonova, E., Glazyrina, M.A., Rakov, E.A. and Prasad, M.N.V., 2018. Establishment of phytocenosis on brown coal mine waste in Urals Sverdlovsk and Chelyabinsk regions, Russia – Drivers, constraints, and trade-offs. In *Bio-Geotechnologies for Mine Site Rehabilitation.* Amsterdam: Elsevier Inc. (pp. 529–546).

Epov, M.I., Yurkevich, N.V., Bortnikova, S.B., Karin, Y.G. and Saeva, O.P., 2017. Analysis of mine waste by geocheimical and geophysical methods (a case study of the mine tailing dump of the Salair ore-processing plant). *Russian Geology and Geophysics*, 58(12), pp.1543–1552.

Eremin, D.I., 2014. Deposit as a means of restoring the content and reserves of humus of long-arable chernozems of Northern Zauralye forest-steppe zone. *Fertility,* 1(76), pp.24–26.

Eremin, D.I. and Shakhova, O.A., 2010. The dynamics of the leached Chernozem moisture content of under different tillage systems for spring wheat in the conditions of Northern Zauralye. *Agrarian Bulletin of the Urals*, 1, pp.38–40.

Finkelman, R., 2004. Potential health impacts of burning coal beds and waste banks. *International Journal of Coal Geology,* 59, pp.19–24. https://doi.org/ 10.1016/j.coal.2003.11.002.

Golitsyn, M.V., Vyalov, V.I., Bogomolov, A.K., Pronina, N.V., Makarova, E.Y., Mitronov, D.V., Kuzevanova, E.V. and Makarov, D.V., 2015. Prospects of technological use of coals in Russia. *Georesursy,* 2(61), pp.41–53. https://doi.org/doi.org/10.18599/grs.61.2.4.

Gorbacheva, N.V. and Sovacool, B.K., 2015. Pain without gain? Reviewing the risks and rewards of investing in Russian coal-fired electricity. *Applied Energy,* 154, pp.970–986.

Griazev, M.V., Kachurin, N.M. and Spirin, V.I., 2016. Energy-efficient technologies of integrated coal and mining waste development in the Moscow Coal Basin in the context of secure and sustainable supply of row materials in Central Russia. *Eurasian Mining*, 2, pp.15–19.

Howrald, F. and Karim, M., 2015. The selection of backfill materials for Barapukuria underground coal mine, Dinajpur, Bangladesh: Insight from the assessments of engineering properties of some selective materials. *Environmental Earth Sciences,* 73(10), 6153–6156. https://doi.org/10.1007/s12665-014-3841-1.

Ignatyeva, M., Yurak, V. and Pustokhina, N., 2020. Recultivation of post-mining disturbed land: Review of content and comparative law and feasibility study. *Resources*, 9(6), p.73.

Izydorczyk, G., Mikula, K., Skrzypczak, D., Moustakas, K., Witek-Krowiak, A. and Chojnacka, K., 2021. Potential environmental pollution from copper metallurgy and methods of management. *Environmental Research*, 197, p.111050.

Jandieri, G., Sakhvadze, D. and Raphava, A., 2020. Manganese biomining from manganese-bearing industrial wastes of Georgia. *Journal of The Institution of Engineers (India): Series D*, 101(2), pp.303–316.

Jobborov, B.T., Alikarieva, D.M., Kamalova, M.D. and Adilova, N.A., 2021. The ecological state and the problems of recultivation of man-made disturbed irrigated soils. *Annals of the Romanian Society for Cell Biology*, pp.4477–4492.

Josephson, P., Dronin, N., Mnatsakanian, R., Cherp, A., Efremenko, D. and Larin, V., 2013. *An Environmental History of Russia*. Cambridge University Press.

Karnachuk, O.V., Kurganskaya, I.A., Avakyan, M.R., Frank, Y.A., Ikkert, O.P., Filenko, R.A., Danilova, E.V. and Pimenov, N.V., 2015. An acidophilic Desulfosporosinus isolated from the oxidized mining wastes in the Transbaikal area. *Microbiology*, 84(5), pp.677–686.

Khayrulina, E.A., Khmurchik, V.T. and Maksimovich, N.G., 2016, July. The Kizel Coal Basin (the Western Urals, Russia): Environmental problems and solutions. In *Mining Meets Water-Conflicts and Solutions*. In *Proceedings of the IMWA2016 Annual Conference*, Leipzig (pp. 11–15).

Khudyakova, L.I., Kislov, E.V., Paleev, P.L. and Kotova, I.Y., 2020. Nephrite-bearing mining waste as a promising mineral additive in the production of new cement types. *Minerals*, 10(5), p.394.

Kolchina, M., Konovalov, V. and Kolchina, N., 2019. Analyzing the state of mining towns in the Ural region. In *E3S Web of Conferences* (Vol. 135, p. 04015). EDP Sciences.

Konstantinova, E., Burachevskaya, M., Mandzhieva, S., Bauer, T., Minkina, T., Chaplygin, V., Zamulina, I., Konstantinov, A. and Sushkova, S., 2020. Geochemical transformation of soil cover and vegetation in a drained floodplain lake affected by long-term discharge of effluents from rayon industry plants, lower Don River Basin, Southern Russia. *Environmental Geochemistry and Health*, pp.1–20.

Koptsik, G., Koptsik, S., Spiers, G. and Beckett, P., 2016. Restoration of smelter industrial barrens following pollution reduction drives economic recovery. In *Environmental Materials and Waste* (pp. 463–483). Academic Press.

Koptsik, G.N., 2014. Problems and prospects concerning the phytoremediation of heavy metal polluted soils: A review. *Eurasian Soil Science*, 47(9), pp.923–939.

Korneeva, T., Aminov, P. and Kucher, D., 2015. Integrated geophysical and geochemical methods for the environmental assessment of active smelter operations and abandoned mines in Karabash, Ural Mountains, Russia. *WIT Transactions on Ecology and the Environment*, 196, pp.459–469.

Kozlov, M.V. and Zvereva, E.L., 2007. Industrial barrens: Extreme habitats created by non-ferrous metallurgy. *Reviews in Environmental Science and Bio/Technology*, 6(1), pp.231–259.

Krasavin, A.P. and Safin R.T., 2005. Environmental rehabilitation of the coal mining areas of the Kizelovsky coal basin in connection with the closure of mines. Perm: Zvezda, p. 287.

Krasavtseva, E.A. and Maksimova, V.V., 2021, September. The influence of mining wastes from the Russia's Arctic zone on the plants development. In *IOP Conference Series: Earth and Environmental Science* (Vol. 839, No. 4, p. 042051). IOP Publishing.

Krasilnikova, S. and Blinov, S., 2016. Using of soda production waste. *International Multidisciplinary Scientific GeoConference: SGEM*, 3, pp.67–71.

Krupskaya, L.T., Gul, L.P., Zvereva, V.P. and Onishchenko, M.S., 2015, April. Environmental problems of coal mining in the Southern Far East of Russia and soil remediation on mining dumps. In *Future Communication Technology and Engineering: Proceedings of the 2014 International Conference on Future Communication Technology and Engineering (FCTE 2014)*, Shenzhen, China, 16–17 November 2014 (p. 295). CRC Press.

Kumar, A., Maleva, M., Kiseleva, I., Maiti, S.K. and Morozova, M., 2020. Toxic metal (loid) s contamination and potential human health risk assessment in the vicinity of century-old copper smelter, Karabash, Russia. *Environmental Geochemistry and Health*, 42(12), pp.4113–4124.

Kuznetsov, P.N. and Ilyushechkin, A.Y., 2013. *Coal Resources, Production and Use in the Russian Federation*. Woodhead Publishing Limited.

Lazhentsev, V.N., 2016. Academic science and new industrialization (example of the Komi republic). *R-Economy*, 2(4), pp.402–411.

Litvintsev, V.S., 2013. Resource potential of placer mining waste. *Journal of Mining Science*, 49(1), pp.99–105.

Litvintsev, V.S., 2015. Rational development of noble metal placer mining waste in the east of Russia. *Journal of Mining Science*, 51(1), pp.118–123.

Lonzinger, A.V., Mantorova, G.F. and Costin, A.M., 2015. Technogenically polluted lands of the city of Karabash and methods for their reclamation. *Russian Agricultural Sciences*, 41(5), pp.365–368.

Madhumita, R.O.Y., Roychowdhury, R. and Mukherjee, P., 2018. Remediation of fly ash dumpsites through bioenergy crop plantation and generation: A review. *Pedosphere*, 28(4), pp.561–580.

Maiti, S.K., 2012. *Ecorestoration of the Coalmine Degraded Lands*. Springer Science & Business Media.

Masloboev, V.A., Seleznev, S.G., Svetlov, A.V. and Makarov, D.V., 2018. Hydrometallurgical processing of low-grade sulfide ore and mine waste in the Arctic regions: Perspectives and challenges. *Minerals*, 8(10), p.436.

Misz-Kennan, M. and Fabianska, M., 2010. Thermal transformation of organic matter in coal waste from Rymer Cones (Upper Silesian Coal Basin, Poland). *International Journal of Coal Geology,* 81, pp.343–358. https://doi.org/10.1016/j.coal.2009.08.009.

Moiseenko, T.I. and Kudryavtseva, L.P., 2001. Trace metal accumulation and fish pathologies in areas affected by mining and metallurgical enterprises in the Kola Region, Russia. *Environmental Pollution*, 114(2), pp.285–297.

Nevskaya, M., Seleznev, S., Masloboev, V., Klyuchnikova, E. and Makarov, D., 2021. Involving small and medium-sized mining industry businesses in mining waste processing in the Russian Federation. *Mineral Economics*, 34(1), pp.81–86.

Nevskaya, M.A. and Marinina, O.A., 2015. Regulatory aspects of mining waste management in the Russian Federation. *Biosciences Biotechnology Research Asia*, 12(3), pp.2619–2628.

Park, J.H., Edraki, M., Mulligan, D. and Jang, H.S., 2014. The application of coal combustion by-products in mine site rehabilitation. *Journal of Cleaner Production*, 84, pp.761–772.

Radomskaya, V.I., Pavlova, L.M., Shumilova, L.P., Voropaeva, E.N. and Osipova, N.A., 2021. Predictive assessment of toxicants' migration from technogenic gold-mining wastes (case study of the tailings management facility of Tokur mill, Amur region, Russia). *Environmental Earth Sciences*, 80(23), pp.1–13.

Rakov, E.A. and Chibrik, T.S., 2009. On the problem of flora formation in industrially disturbed land areas. *Russian Journal of Ecology,* 40(6), pp. 448–451. DOI: 10.1134/S1067413609060125

Rybnikov, P.A. and Rybnikova, L.S., 2019. Rare-earth elements in groundwater of the abandoned Levikha copper mine (Middle Urals, Russia). In *E3S Web of Conferences* (Vol. 98, p. 01043). EDP Sciences.

Rybnikova, L. and Rybnikov, P., 2017, Water quality of the abandoned sulfide mines of the Middle Urals (Russia). In *Proceedings of the 13th International Mine Water Association Congress—Mine Water & Circular Economy*, Lappeenranta, Finland (pp. 25–30).

Rybnikova, L.S. and Rybnikov, P.A., 2019. Regularities in the evolution of groundwater quality at abandoned copper sulfide mines at the Levikha ore field, Central Urals, Russia. *Geochemistry International*, 57(3), pp.298–313.

Rybnikova, L.S., Naumova, O.B., Rybnikov, P.A., Naumov, V.A., Navolokina, V.Y., Sosnin, V.E. and Kopylov, I.S., 2019, July. Geological structure and mining development of the Urals. In *Post-conference Tour.* Perm-Yekaterinburg-Nizhny Tagil-Perm. 19–22 July 2019. Deposited manuscript. Editors: Rybnikova L.S., Naumova O.B. (No. 462283).

Rybnikova, L.S., Rybnikov, P.A. and Tarasova, I.V., 2017. Geoecological challenges of mined-put open pit area use in the Ural. *Journal of Mining Science*, 53(1), pp.181–190.

Semenkov, I., Sharapova, A., Lednev, S., Yudina, N., Karpachevskiy, A., Klink, G. and Koroleva, T., 2022. Geochemical partitioning of heavy metals and metalloids in the ecosystems of abandoned mine sites: A case study within the Moscow Brown Coal Basin. *Water,* 14(1), p.113.

Semenova, I.N., Rafikova, Y.S., Khasanova, R.F. and Suyundukov, Y.T., 2018. Analysis of metal content in soils near abandoned mines of Bashkir Trans-Urals and in the hair of children living in this territory. *Journal of Trace Elements in Medicine and Biology*, 50, pp.664–670.

Slukovskaya, M.V., Vasenev, V.I., Ivashchenko, K.V., Dolgikh, A.V., Novikov, A.I., Kremenetskaya, I.P., Ivanova, L.A. and Gubin, S.V., 2021. Organic matter accumulation by alkaline-constructed soils in heavily metal-polluted area of Subarctic zone. *Journal of Soils and Sediments*, 21(5), pp. 2071–2088.

State report 2017 On the state and protection of the environment of the Russian Federation in 2016. M.: Ministry of Natural Resources and Environment of Russia; NIA-Priroda. p. 760.

Stoyanova, I.A. and Kharchenko, A.V., 2011. Ecological and economic problems in the regions of mass closure of coal mines. *22nd World Mining Congress & Expo,* 2, p.3.

Tichanek, F. and Tichanek, R., 2014. Contribution to the solution of thermally active reclamation of coal waste heaps. *14th International Multidisciplinary Scientific GeoConference SGEM Proceedings*. Albena (pp. 777–791). https://doi.org/10.5593/SGEM2014/B13/S3.100.

Tkatcheva, V., Holopainen, I.J. and Hyvärinen, H., 2002. Effects of mining waste waters on fish in lakes of NW Russia. *Internationale Vereinigung für theoretische und angewandte Limnologie: Verhandlungen*, 28(1), pp.484–487.

Udachin, V., Williamson, B.J., Purvis, O.W., Spiro, B., Dubbin, W., Brooks, S., Coste, B., Herrington, R.J. and Mikhailova, I., 2003. Assessment of environmental impacts of active smelter operations and abandoned mines in Karabash, Ural Mountains of Russia. *Sustainable Development*, 11(3), pp.133–142.

Vyrkin, V.B., Plyusnin, V.M., Belozertseva, I.A., Shekhovtsov, A.I., Enushchenko, I.V. and Zakharov, V.V., 2014. The current status of nature, and ecological problems of the Middle Onon region. *Geography and Natural Resources*, 35(1), pp.18–26.

Yin, S., Wang, L., Wu, A., Kabwe, E., Chen, X. and Yan, R., 2018. Copper recycle from sulfide tailings using combined leaching of ammonia solution and alkaline bacteria. *Journal of Cleaner Production*, 189, pp.746–753.

Zalesov, S., Sezgin, A.Y.A.N., Zalesova, E. and Opletaev, A., 2020. Experiences on establishment of Scots pine (*Pinus sylvestris* L.) plantation in ash dump sites of Reftinskaya HPES, Russia. *Alinteri Journal of Agriculture Science*, 35(1), pp.7–14.

Zhaglovskaya, A.V., Savon, D.Y., Safronov, A. and Sidorova, E., 2017. Production activity analysis Methodology for open pit coal mines (in terms of Shestaki open pit mine). *Eurasian Mining*, 1, p.27.

Zolotova, E. and Ryabinin, V., 2019. Elements distribution in soil and plants of an old copper slag dump in the Middle Urals, Russia. *Ecological Questions*, 30(4), pp.41–47.

13 Abandoned and Derelict Mines of the Witwatersrand

Reclamation, Restoration and Reparation

Harold Annegarn

13.1 INTRODUCTION

Gold on the Witwatersrand was discovered in 1886 by George Harrison on an eroded quartz outcropping in a low set of hills close to what is today central Johannesburg. The range of hills came to be called the Witwatersrand (*Wit* – referring to the whitish quartz conglomerate rock containing the gold; and *watersrand* – an archaic German word for continental divide). Exploitation of these deposits over the ensuing century created vast wealth and laid the foundation for the modern nation-state of South Africa, which, in time, became the world's largest gold producer. However, the mining activity was accompanied by actions that, by modern standards, constitute dereliction towards the environment, the workers who extracted the gold from deep underground workings, human rights and norms of ethical and professional behaviour. The tale of this epoch is recounted through the lens of dust in its many manifestations when gold is being extracted and mining wastes disposed of. The history of dealing with derelict mines and dust through to the modern era is addressed through interlinking themes of Reclamation, Restoration and Reparation.

After presenting a brief history of gold mining on the Witwatersrand and the creation of hundreds of tailings storage facilities (TSFs) (sand dumps and slime dams), the section on *Reclamation* will deal with dust-related environmental impacts arising from the recovery and secondary processing of these TSFs to recover residual gold. It is argued that reclamation is the only effective permanent solution for the manifest environmental risks posed by the TSFs. The decline of the gold industry in South Africa to less than a tenth of its peak production has left no direct source of finance to pay for land *restoration*. Innovative partnerships between the private and public sectors and the gold reclamation industry have crafted viable solutions to the economic and technical challenges of restoring mining footprints to ratable land-use categories. The final section considers the enduring human tragedy visited on labourers exposed to large dust concentrations in South African mines and how recent legal processes and shifting societal values have led to *Reparations*, i.e. financial compensation for work-related silicosis and tuberculosis. Despite such advances, dealing with dereliction still presents many challenges involving a complex interplay between economics, the environment, human and political rights, science, medicine, technology, labour and corporate power.

13.2 RECLAMATION

13.2.1 A Brief History of the Witwatersrand Gold Field

Initial mining activities were surface operations along the shallow outcrops of the Witwatersrand ridge, extending in an east-west direction. The gold occurs in thin layers, <1 m thick, of consolidated

DOI: 10.1201/9781315141855-13

rock rich in quarzitic pebbles, termed conglomerate, believed to have been formed by hydraulic deposition. These ancient layers have been tilted and deformed by geological processes so that the gold-bearing horizons dip southwards at an angle of ~37°. Continued access to the ore soon required mining methods to resort to underground operations. Small-scale artisanal mining gave way to corporate operations that could source the capital and management skills for deep-level underground mining.

The mining industry expanded rapidly after World War II, reaching a peak annual gold production of ~1 000 t in 1970 (Hartnady 2009). Subsequently, as the accessible gold reserves were depleted and deep-level mining became correspondingly more expensive, production declined. Numerous underground mines reached their end-of-life operations and closed during 1970–2000. In 2018, South Africa produced only 120 tons of gold. After a period of rapid expansion of the economy and gold mining in the post–World War II years, the number and size of tailing deposits proliferated. Initially, these tailings deposits had been some kilometres from residential or commercial developments. However, as the urban margins of Johannesburg expanded, developments encroached on the tailing deposits. The residues of a century of mining have left several hundred million tons of tailings in 400 separate deposits covering 11,000 ha, including five mega-dumps, each of more than 1,000 ha. There is an additional 5,000 ha of sterilised land and 300 voids or open mine shafts, all embedded in the urban fabric of the Witwatersrand/Gauteng Mega-city region (Figure 13.1).

As the gold volumes and profits dwindled, the former large mining houses, such as Anglo American and Gencor, moved out of the gold sector and relocated their headquarters offshore, leaving the new owners with the liabilities of managing the defunct infrastructure and TSFs, often without the reserve funds supposedly set aside for post-closure rehabilitation or restoration. For ownerless mines, legislation required that the state take over the liability for derelict facilities. The South African government has been reluctant to issue mine closure certificates to avoid such eventualities. By 2017, the Department of Mineral Resources had issued only one full mine closure certificate. This certificate was for a small open-cast diamond mine in the Limpopo Province (Botham 2011). Such reserve funds set aside for rehabilitation of TSFs may legally only be accessed *after* formal closure. This conundrum freezes the rehabilitation funds and gives no credit or encouragement for progressive rehabilitation measures during the active life of a mine.

FIGURE 13.1 Image of Johannesburg from space, showing the legacy mine tailings deposits of 100 years of gold mining embedded in the urban fabric. (Image credit: NASA Earth Observatory Image of the Day for October 6, 2003. Instrument: ISS Digital Camera. https://earthobservatory.nasa.gov/images/3850/johannesburg-south-africa)

13.2.2 Tailings Storage Facilities – A 100-Year Legacy

The earliest tailings storage facilities were sand dumps, in which the spent ore, coarsely ground in stamping mills, was conveyed to the dump in trolleys and mechanically tipped. Such dumps are recognised by their conical shape, governed by the angle of repose of the mineral grains (~37°). Advances in milling resulted in finer grinding, and the spent ore could be pumped as a slurry and allowed to settle in large dams. These dams have a low, flat profile and typically a rectangular plan, the largest of which extend over 1 km^2 with elevations up to 200 m (Figure 13.1).

Wind-blown dust from unconsolidated tailings storage facilities is the most immediate manifestation of derelict mines and, on the Witwatersrand, constitutes the phenomenon that has elicited the most complaints and public health concerns. Many larger TSFs were established post–World War II in sparsely populated areas south of the Witwatersrand ridge. Dust from these TSFs made the southern areas less amenable for residential developments, and Johannesburg expanded north of the ridge. However, with the advent of the apartheid laws, new settlements were established, for instance, the township of Soweto (SOuth WEst TOwnship) 15 km SW of Johannesburg. As populations in these townships grew, urban expansion filled the 2–3 km spaces between settlements and TSFs. Eventually, settlements encroached on the boundaries of the mining reserves, often due to improper municipal land use authorisation, within the safety zones around the TSFs (Figure 13.2).

The mineral composition of the tailing material is 80–90% silica quartz (SiO_2), 5–10% chlorites and 2–5% iron pyrite (FeS_2). Of the minor and trace elements, a few are enriched relative to the global average crustal rock. Even after extraction, a residue of recalcitrant Au remains but typically ~0.5 ppm. It can reach 2.1 ppm, enriched by a factor x280. Maseki et al. (2017) reported the enrichment of heavy metals of potential environmental concern in tailings in the central and east Witwatersrand TSFs: As ×68 (116 ppm); Cd ×24 (2.4 ppm); Pb ×1.4 (13 ppm); and U ×8 (21 ppm). Uranium (and uranium decay products) occur at higher concentrations towards the west of the Witwatersrand ore deposits, to the extent that several West Rand mines were commissioned as uranium mines, with gold as a by-product. After the 28 March 1979 Three Mile Island nuclear accident and the subsequent collapse of the global uranium market, these mines discontinued uranium extraction and continued as gold-only producers. Uranium embedded in the milled ore was sent to the tailings. Silica quartz (SiO_2) dust is a regulated substance for mining and industrial exposure; however, nowhere has a standard been promulgated for ambient exposure.

Before the 2000s, the comminution of gold-bearing ore was to a median grain size of 75 μm in diameter. In the early 2000s, extraction processes were changed to increase yields – ore was ground to a finer median particle size (80% <25 μm). This finer grain size had implications for dust

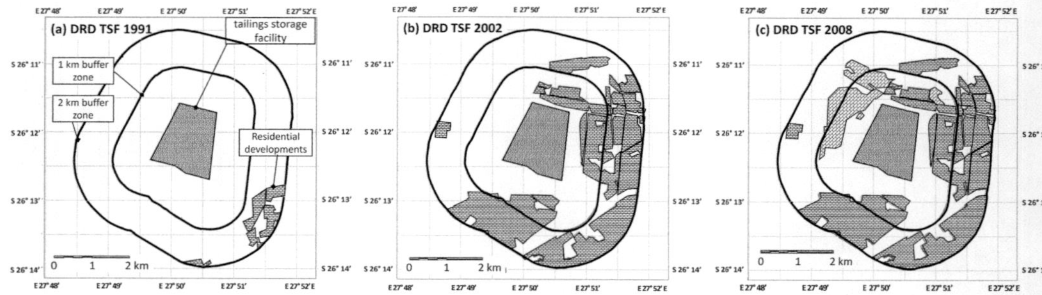

FIGURE 13.2 The extent of the aeolian dust problem: Urban encroachment on unconsolidated tailings storage facility over time, showing the expanding residential areas around the Durban Roodepoort Deep tailings storage facilities (inner polygon) with 1 km and 2 km buffer zone contours. (a) 1991; (b) 2002 and (c) 2008. (Source: Adapted from Kneen, M. A., M. E. Ojelede, and H. J. Annegarn 2015. Housing & population sprawl near tailings storage facilities: Witwatersrand, 1952 to current. *South African Journal of Science* 111:142–50. http://dx.doi.org/10.17159/sajs.2015/20140186.)

emissions, which now contained a higher fraction of respirable particles in the respirable, health-hazardous fraction less than 10 μm.

Besides dust emissions, there were other less obvious but possibly more severe environmental issues with the legacy TSFs. Earlier technical specifications for constructing the foundation of TSFs to limit water seepage were not as stringent as current requirements, nor were such specifications consistently implemented. Water seepage into the underground aquifers and leakages from the toe dams occurred. Failures of the slime dam impoundments resulted in the partial collapse of the TSFs, with large quantities of slime leaking into the downstream areas. Aerial photographs taken at various times reveal the extent of these tailing failures, which went largely uncommented by the public. Even without such extensive visible breakouts, airborne gamma radiation surveys by the South African Geological Survey revealed radioactive contamination in many waterways downstream from TSFs (Coetzee 2004).

In the 1970s and 1980s, the Chamber of Mines Vegetation Unit investigated methods to establish vegetation on the sterile silica rock dust of the TSFs (Reichardt 2013). An inimical factor was the high pyrite concentration (2-5% FeS_2), which reacts to form sulphuric acid and acid solutions with pH <1 in the presence of moisture and oxygen. They recommended a procedure starting with a prolonged period of up to two years of leaching using drip or spray irrigation. This was followed by liming, compost addition and planting of short-lived pioneer species and longer-lived grasses. The widespread and successful application of these methods largely suppressed dust emissions, enabling residential and business developments up to the toe dams of the TSFs. The long-term sustainability of the vegetation was not assured. Yet, the Chamber of Mines, always sensitive to containing costs, used the initial operational techniques but failed to support further research, and the Vegetation Unit was eventually closed.

In the 1970s, other economic factors had a bearing on the willingness of the mines to continue extensive grassing of the TSFs. In 1971, the USA moved away from the gold standard. The subsequent substantial rise in the gold price rendered the recovery of residual gold in the tailing deposits economically viable. Enhanced metallurgical processes were devised that enabled a recalcitrant component of the gold remaining in the tailings material to be extracted. The first operating plant of this type was opened in 1978 (Anglo American's East Rand Gold Operations – ERGO) in Springs, 40 km east of Johannesburg. The recovery process involved flotation in enriching the pyrite fraction of the reclaimed tailings; up to 40% of the gold was encapsulated in the pyrite grains, rendering it immune to cyanide leaching. The enriched fraction was subjected to fine grinding with 80% material <25 μm diameter, facilitating the subsequent leaching of the Au, improving recovery from 40% to 75%. The fine grinding is significant also for environmental dust reasons, as will be discussed in more detail below. The sulphide extracted was used to manufacture sulphuric acid, which, in turn, served to extract uranium. ERGO was soon followed, in 1980, by Crown Gold Recoveries, which commenced the reclamation of sand dumps in central Johannesburg.

Of necessity, the reclamation of TSFs to recover the tailings resulted in stripping the vegetation cover and exposing large areas of the tailings deposits to wind and water erosion. The prospect of near-term reclamation of TSFs made the mine owners/operators reluctant to further invest in research or implementation of grassing programmes. The location of some TSFs amid dense metropolitan infrastructure posed significant challenges for environmental control, considering the short distances (sometimes less than 20 m) from the base of the TSFs to prime residential and business properties. The largest of the sand dumps towered 250 m above surrounding buildings. Nor was this to be a short-term operation – reclamation of the largest of the central Johannesburg dumps took twenty years.

Fortunately, the Witwatersrand is located on the inland plateau and has a mild continental climate with low to moderate wind regimes. The driest months of the year (austral winter, May through mid-August) have mostly stagnant wind conditions. Summer months, November to April, have frequent rainfalls that mostly suppress aeolian dust generation despite higher mean and maximum wind speeds. The vulnerable periods for dust generation are mid-August to mid-October when high wind speeds may precede rainfall, and periods of maximum dustiness occur regionally, specifically from unconsolidated soil surfaces.

13.2.3 Dust from the Dumps

The worst episodes of dust from tailings occur during severe convective thunderstorms, which are frequent in summer. These storms are 15 to 20 km in diameter and generate strong up-and-down draughts. When the downdrafts reach the surface, they are deflected horizontally with severe turbulence and speeds of up to 50 km/h. If the centre of such a storm passes directly over a TSF reclamation site or an unconsolidated dump before rainfall wets the surface, extreme dust generation occurs. Such an event was recorded while monitoring with a particle counter-sizer, showing peak concentrations of 30,000 μg/m³ during an episode lasting less than 30 minutes. The sampling site was 1 km SW of the Durban Roodepoort Deep (DRD) TSF, as depicted in Figure 13.2. The DRD TSF is derelict, with ownership of the mining rights and obligations held by a company in receivership (in 2018). In the absence of ongoing maintenance during the 2000s, the entire 2 km² vegetated top surface became engulfed by a fresh layer of sand blown off the side slopes. The DRD TSF again became a prolific source of dust. Despite the storm being brief, of ~30 minutes duration (Figure 13.3), the 24-h average concentration of respirable particulate matter (PM10) was 2 160 μg/m³, which exceeded the South African limit value of 120 μg/m³ at that time by a factor ×18 (Ojelede 2012).

13.2.4 Monitoring and Mitigation Measures

In recognition of the hazard posed by wind-blown dust, Crown Gold Recoveries initiated a dust monitoring programme into its operations from the outset in 1980. The task was contracted to local research institutions, who, for the first five years, applied sophisticated airborne sampling and meteorological monitoring techniques. While these measurements fascinated academic researchers, it was realised over time that the results were not of great value to the company in managing

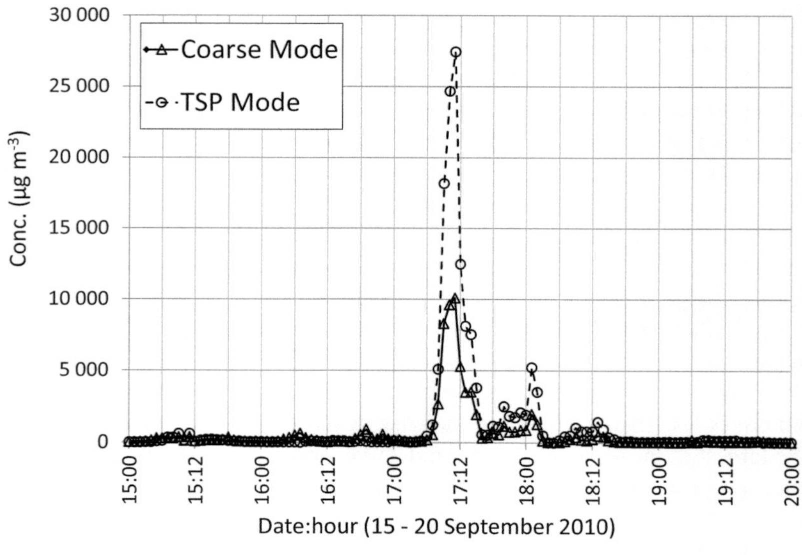

FIGURE 13.3 The capture of an extreme dust event close to an unconsolidated tailings storage facility during a thunderstorm. Winds above 7 m/s lifted respirable dust (coarse mode 2.5-10 μm diameter, and TSP - total suspended particles 10–30 μm). (Source: Ojelede, M. E. 2011. Risk Assessment of Atmospheric Emissions from Gold Mine Tailings on the Witwatersrand. PhD thesis, University of the Witwatersrand, Fig. 82, p. 166.).

the risk. In 1985, dust monitoring methods were altered by implementing the American Standard Test Method for Collection and Measurement of Dustfall (Settleable Particulate Matter) (ASTM D1739-1970). The sampling entails the deployment of a 5-L container supported on a 2-m high pole and basket and exposed for 30 days. The settleable matter is collected and weighed. Results are expressed in units [mg/m^2/day] on a 30-day average.

The method is inexpensive, robust and does not require electrical power. The disadvantage is that the method lacks precision and has variable capture efficiency dependent on wind speed. Nevertheless, it is useful in establishing long-term trends and, more importantly, for the mining industry, providing unambiguous signals if dustfall has deviated from the trends due to, for example, failure of mitigation measures. Chemical analysis of the deposited material makes it possible to establish the provenance of the dust. However, in the case of the gold mining residues, the whitish colour of silica sand provides a unique determinant by a simple visual examination of the sample. It differs significantly from the region's predominantly red and brown surface soils.

The fallout samplers' low capital and operating costs allow for arrays of multiple samplers around TSF reclamation operations, up and downwind and at varying distances from the source. What the technique lacks in precision and short time resolution is compensated for by a dense spatial network that enables robust determination of whether dust episodes are mine-related or related to local disturbances or regional meteorological events, such as the passage of frontal systems. Furthermore, the deployment of networks of dust fallout samplers has evolved and adapted as reclamation on various TSFs commenced and ceased operations. Individual reference sites have operated continuously from 1985 to the present day (2016). The reclamation of dumps on the Witwatersrand initiated in 1981, and the associated network of dust samplers, ~80 in number, is still being done. Although the ASTM 1739 method is not employed widely in the USA, the continuous operation of dustfall networks in gold and coal surface mining in South Africa has resulted in the method being adopted by the Department of Environmental Affairs in the National Air Quality Management Dust Control Regulations (NEMA: Air Quality Act 2004). Dustfall limit values have been established from residential and non-residential areas, and operators of dust-generating operations are required to run dustfall monitoring in areas surrounding their operations and to report results to the DEA routinely.

Typical dustfall results are illustrated in Figure 13.4, showing dustfall downwind of a previously vegetated dump which had been partially reclaimed and then abandoned without remedial vegetation, resulting in severe dust emissions during the July to November windy season. After community protests and intervention by the Legal Resource Centre, a public benefits law firm, the mining

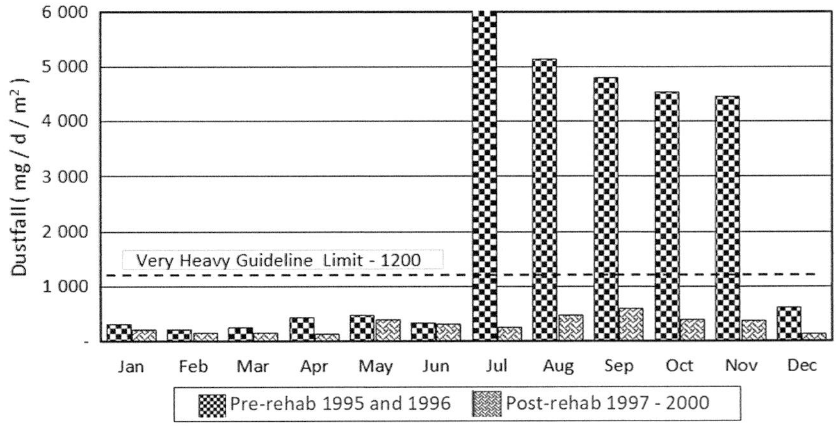

FIGURE 13.4 Monthly mean dustfall at Moroeroe Primary School, 500 m downwind of the Rand Leases dump, Soweto, in the years before and after the restoration of the vegetation cover. (Author's figure)

company commissioned an urgent vegetation intervention, which led to a subsequent reduction in aeolian emissions.

Criticism of the dustfall monitoring method is that it does not address concerns about inhalation health hazards, which are regulated by ambient concentrations of airborne particulate matter less than 10 µm diameter (PM10) and less than 2.5 µm diameter (PM2.5). For the gold tailings, there have also been public concerns about the heavy element enrichments, including arsenic and uranium (see Section 13.2.3 above). The concern is substantively valid – dustfall buckets and suspended samplers for PM10 and PM2.5 particulate matter measure distinctly different portions of the size distribution continuum of airborne particles. Dustfall buckets capture the bulk mass in size above ~30 µm. Aeolian or mechanical disturbances may generate dust across the full-size range, as shown in Figure 13.3 above. However, there is no simple, constant correlation possible between PM and dustfall since the differential settling rates of particles result in a rapid partitioning once airborne. In practice, dustfall from mine tailings becomes indistinguishable from background dustfall beyond ~3 km from sources, while fine particulate matter may remain airborne for hours and be transported many kilometres from the source.

Concern about the heavy metal concentrations was addressed by conducting a health risk assessment using current US EPA methods. As input data, samples of the source material were taken from the TSFs. For atmospheric concentrations, values were taken for the worst-day case (Figure 13.3) and applied conservatively for the full ninety days of the windy season. Risks for exposure by inhalation for all the measured elements (As, Cd, Pb and U) for carcinogenic and non-carcinogenic were within acceptable health limits.

The conclusions emerging from these diverse monitoring and academic studies of aeolian dust emissions from the gold mine tailings may be summarised as follows:

- For unconsolidated TSFs, there is a palpable risk of high dust emissions during the windy months of August to October; aeolian dust will exceed the 24-h average air quality limits for respirable dust. There is also a direct and severe loss of amenities due to soiling brought about by the deposition of *Total Suspended Particles* (10–80 µm diam.). These large particles settle out within the first 2 to 3 km from the source, causing soiling, structural damage and physical irritation to eyes and mucous membranes, particularly of infants who cannot clear their eyes or noses.
- Although dustfall sampling is a crude, imprecise monitoring technique, it provides robust and inexpensive management information on whether dust emissions from a TSF are under control. A quarterly Dust Forum was established by Crown Gold Recoveries (CRG) in 1981 to receive reports from consultants on dust and meteorological monitoring of the TSF reclamation projects. Meetings of the dust forum continue to the present day, open to all interested and affected parties, and they have proved an effective channel for sharing dust emissions with businesses, community groups and regulatory authorities.

13.3 RESTORATION

Dereliction has left no contemporary funds to pay for land *restoration*. Yet, innovative solutions have been implemented on the Witwatersrand to solve economic and technical challenges, enabling mining footprints to be remediated and reclassified to commercial and industrial land-use categories.

13.3.1 Mitigation in Practice

The holder of a mining permit is required by the Mineral and Petroleum and Resources Development Act of 2002 to conduct operations using integrated environmental management practices and, on closure, has a responsibility to remedy the affected environment.

(d) must, as far as it is reasonably practicable, rehabilitate the environment affected by the prospecting or mining operations to its natural or predetermined state or to a land use which conforms to the generally accepted principle of sustainable development. (MPRDA 2002)

This clause raises two issues: first, the sustainability of vegetation once regular watering and fertilisation cease, and, second, the hope that the sterile surfaces of acidic silica mineral sand can somehow be restored to a natural ecosystem type. Rossouw et al. (2010) reported that left untended and undisturbed by external mechanical forces, a vegetative cover of diverse – mainly non-indigenous species – could sustain itself over thirty years, sufficient to suppress wind and water erosion. In other instances, the short-term partial reclamation, vandalism, off-road motorcycle recreation and human foot traffic have damaged vegetation, resulting in either minor or severe degradation of TSFs and recurrence of high dust emissions and erosion.

While the worst effects of aeolian emissions could be mitigated by vegetating the tops and side slopes of TSFs, these measures did not solve all the water contamination risks. The total removal of the deposits and relocation of the material into fewer large, well-engineered deposits appears to be the only feasible long-term solution to mitigate both the dust and the water pollution risks. Containment of dust and water releases during reclamation required ongoing monitoring and management.

13.3.2 Reclamation at Risk

From 1980 to 2000, the two large operations for reclaiming derelict TSFs for further gold recovery, ERGO and Crown Gold Recoveries, had cleared large tonnages and hectares to ground level. All regulatory requirements had been met, including stringent specifications from the Nuclear Regulatory Agency for removing residual radioactive materials. This allowed the Department of Mineral Resources to issue closure certificates for individual TSF footprints, which were then reclassified for industrial or commercial use. However, falling prices for gold and specifically uranium, as well as a perceived deterioration of political risk, led Anglo-American to shut down its ERGO processing plant and all reclamation activities. Anglo mothballed the plant and disposed of the assets, including mining rights to the derelict dumps. The economic engine that had driven reclamation and land restoration had stalled and, with it, a risk of the remaining TSFs once again becoming derelict or a state liability. As part of closure operations, companies are legally obliged to remove surface mining infrastructure. Finally, in 2007, Anglo gave notice of its intention to dismantle the ERGO plant and the extensive associated network of pipelines used to convey slurry to the plant (Figure 13.5).

In addition to the mining rights held by the two large operations, multiple low-grade deposits were not included in the planning for reclamation. Several smaller reclamation operations had started and folded over the decades since 1980 – without economies of scale, such operations were not economically viable. Consequently, individual mining companies would likely choose to rehabilitate their tailings deposits *in situ*, leaving many dumps scattered throughout the area. In this scenario, the reclamation operations then in progress would have only partially mitigated the long-term environmental risks.

13.3.3 Project Hloekisa: Urban Development Implications

At this point, Andrew Barker, a town planning consultant employed by Crown Gold Recoveries, put forward a bold and far-reaching concept to, first, extend the life of the CGR operations and, second, address the challenges of the derelict dumps. *Project Hloekisa* (Sotho for “clean up”) was conceptualised to achieve the following:

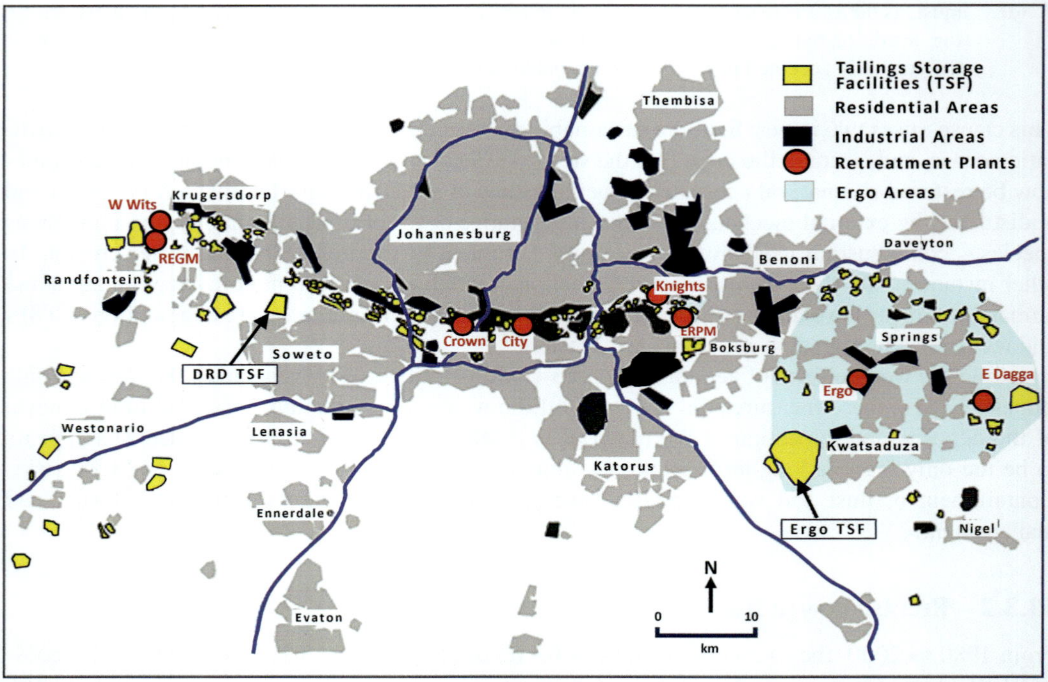

FIGURE 13.5 Location of mine tailings facilities and processing plants owned by Crown Gold Recoveries (Crown, City, Knights and ERPM) and ERGO (ERGO and E Dagga), respectively, circa 2007. (Source with permission: Barker, A. 2011. Project Hloekisa: Summary Conceptual Report, MWS001 Hloekisa Summary. 8 December 2011, Andrew Barker Development Consultant, Johannesburg)

> The safe, effective and eco-sustainable relocation of polluting mine dumps to release land for socio-economic development and growth and to address the health and safety of the greater Gauteng community. (Barker 2011)

The project envisaged a partnership of mining companies, government, local authorities, private sector and communities in a project that could lead to a collective solution resulting in a permanent reduction in air, water and land pollution from the derelict tailings. This would liberate large tracts of valuable land and result in a sustainable solution. The nature of the Hloekisa project was envisaged to consolidate the rights and obligations for *all* the mine residue deposits within the region and operation of a centralised processing plant. The processed material would then be deposited on one or, at most, a few super dumps, engineered to meet modern environmental and engineering standards. The essence of the consolidation was an operation that would commit to removing all tailings deposits, including smaller marginal or sub-marginal portions that would otherwise not be reclaimed. Financial viability was premised on acquiring existing processing and pipeline infrastructure. However, should the dismantling of the ERGO plant and pipeline infrastructure continue, a critical component of the financial and operational viability of the Hloekisa project would have been permanently lost.

Other innovative aspects of Project Hloekisa were a broad economic model that factored in externality benefits and cost savings from fewer environmental, health and safety problems associated with mine waste. The model envisaged increased local rates and services revenue, and other taxable income from businesses established on the restored land. It foresaw creating job opportunities within the boundaries of existing municipal infrastructure, especially near low-income, high-density residential areas beset by high unemployment. In a yet-to-be-tested instrument, the project speculatively proposed ring-fencing a portion of additional municipal rates and services revenues for financing the continuing management of the ultimate deposition areas.

Project Hloekisa was accepted by Crown Gold Recoveries and the parent company DRD Gold. The ERGO plant was purchased and renovated, but without the uranium and sulphuric acid sections. All processing was consolidated at the ERGO plant, and other processing plants were shut down. The ERGO Brakpan tailings dam (Figure 13.5, bottom right) was selected as the final deposition site. Further acquisitions resulted in the consolidation of mining rights for almost all tailings deposits on central and east Witwatersrand, an essential prerequisite to meet the obligation for clearing all the TSFs.

Once the footprints had been cleared, the released land was handed over to a partner company, iProp, that had acquired the surface rights (in South Africa, mineral rights and land ownership are not necessarily coincident). iProp has released the land at 30 to 40 ha per year, mainly for industrial and commercial usage. Although there is also intense political pressure for land to be set aside for low-income and subsidised housing, the sale price for residential land is insufficient to cover the costs of removing remaining mining infrastructure, which often requires extensive remedial work to secure land over shallow underground workings. The iconic mine tailings, which in the 1950s and 1960s had been the symbol of Johannesburg's wealth and prosperity, were gradually replaced by bulk supermarkets, warehouses, shipping depots and factories. For these land parcels, restoration and transformation are complete.

13.4 REPARATIONS

From the onset, labour for the underground mines was based on a migrant system, in which workers were contracted for fixed periods, interspersed by intervals during which they returned to their homes. In the early decades of the 1900s, labour was sourced from South Africa. After World War II, rising industrial wages proved attractive to labourers, and the mines set up systems to attract labour from adjacent territories, primarily Lesotho, Mozambique, Botswana and Malawi. After the USA left the gold standard in 1971, a shortage of labour and increased profitability enabled the mine owners to increase salaries, and an increasing number of labourers were again recruited from within South Africa, mainly migrant labourers from the Eastern Cape. Political tensions with neighbouring territories, particularly Malawi and Mozambique, led to a curtailment of labour recruitment from those territories.

13.4.1 Tuberculosis and Silicosis in the Gold Industry

Labourers in the mining industry constitute a component of dereliction seldom considered as part of a holistic consideration of mine closure. While working conditions in the Witwatersrand gold mines improved over the decades, the health of the workforce was always tempered by the need for productivity and profitability. Until 1971, when the USA went off the gold standard, the price of gold had been fixed relative to the US dollar for several decades. Constrained by a fixed price for their product against rising costs due to inflation and salary increases, the industry was myopically focused on cost containment. This affected the research and compensation for work-related diseases – tuberculosis and silicosis – and mining companies sought to limit their liability for treating employees and former employees who were no longer fit to work underground.

The underground workforce of the gold industry was divided into two groups – *miners*, exclusively *Whites* – persons of European extraction, and *labourers* – *Blacks,* generally of indigenous African heritage but including imported Chinese and Indian workers. This convenient fiction of racially based discrimination existed from the earliest days, which was formalised in law when the Nationalist Party came to power in 1948 and instituted the policies of *apartheid* (separateness).

Already in the early 1900s, it was known that exposure to dust resulted in silicosis, and there were links between dust exposure and pulmonary tuberculosis. Under pressure from the UK House of Commons, the mines instituted a system of health checks, at the minimum, at the start and termination of each contract. A compensation scheme was established whereby a worker found to have certified lung impairment resulting from risk work would be compensated, either during or post-employment or the estate would be compensated if a post-mortem examination of the miner's lungs showed symptoms of silicosis. As part of the contract, workers agreed to undergo regular medical checkups and post-mortem excision of their lungs for pathological evaluation by the National Institute for Occupational Health and to waive any legal claim for work-related industrial disease.

Medical checks and compensation for work-related diseases for the White miners were more comprehensive and regular. Medical checkups were conducted in government-funded institutions with adequate diagnostic and medical facilities. Compensation was in the form of a pension, which was also transferred to a surviving spouse.

Medical checks on entry and exit from each contract for Black labourers were conducted by medical staff on the mines or by the mine recruiting agency. Checking for signs of lung impairment relied on scanning "mini-X-rays" for lumps or nodules, which the medical fraternity themselves knew was an unreliable and incomplete diagnostic technique – such nodules become apparent only at advanced stages. The X-rays were supposed to be accompanied by a medical examination. Facilities for these medical checks were grossly inadequate to cope with the several hundred consultations per day – staff had no time for proper diagnoses or recording of work history. Management further pressured medical staff to contain costs so that, compared to White miners, relatively few labourers were certified for compensation. This is even though they did the bulk of underground tasks such as drilling and tramming (loading and moving blasted ore), which entailed a lot of dustiness.

Any compensation for labourers was a fixed sum related to the final salary. Labourers who exhibited symptoms of TB (generally manifest at an advanced stage) were *repatriated* to their homelands, where it was believed that rest and clean air would allow them to recuperate. Many men died on the trains on their way home or within weeks after their arrival home. The exit medical checks were often not done, removing any basis for a subsequent compensation claim. Over a prolonged period, the number of compensation claims awarded to labourers was 1/10th of those awarded to White miners, despite the ratio of labourers to miners being twenty to one. In any case, there was poor record-keeping, so those men returning for subsequent tours of duty did not have their medical records and record of work (and dust exposure) when they signed on for new contracts.

This was a cause of concern to the mines, as labourers who became unfit for underground work within weeks of signing on represented lost productivity and compensation liability. Despite seven commissions of enquiry over the decades, the industry took scant account of the few recommendations made to improve the health and living conditions of the labourers. One particularly damning estimate that did not see the light of day was an actuarial estimate of the cost of compensation to the industry. If all the "labourers" were to be compensated at the same rate as "miners", the liability was estimated to exceed the entire profit of the gold industry and render it economically unviable.

The role of silicosis as an adjunct factor was underplayed. For decades, the industry debated whether tuberculosis was imported to the mines from the rural areas or contracted in the mines (where workers lived in unhygienic conditions in crowded unisex hostels, with up to twenty men sharing communal sleeping quarters and were fed on a diet comprising mainly a coarse maize porridge). The labourers' predisposition to TB was sometimes attributed to their sexual or other behaviours, genetic susceptibility or infection from their home areas. At the end of the 20th century, the added confounding factor of HIV/AIDS was similarly and initially attributed to the moral depravity of the labouring class.

In practice, the compensation amount was inadequate to pay for the treatment and care of workers, who, after employment, would return to their remote rural homes, where clinics and medical treatment were not accessible. Workers were mainly unaware of their rights, nor had the means to appeal these arrangements' inadequacy or criminal neglect. Since diagnosis from the X-rays was often ambiguous, workers were frequently dismissed for malingering or not reemployed for subsequent contracts due to poor physical fitness related to lung function impairment. The formal medical modality for treating tuberculosis specified treatment over two years. After the initial intense six-week treatments, labourers were to be allowed to return home and continue treatment. However, few facilities existed in the remote rural areas of their home countries for this treatment to be continued. At a later stage, the mines changed the regimen to allow the workers to resume underground work as soon as treatment rendered them non-infectious, based on the argument that continued treatment would be more reliable if the workers were kept at the mine. This also ensured – so went the argument – continued productivity, rather than paying for the workers to recover in sanatoria.

During the last decades of the 20th century, the Chamber of Mines regularly reported that certified lung impairment (silicosis) had dropped to below 5% for workers employed in underground operations for 25 years. South Africa was lauded at international forums, such as conferences of the International Labour Organisation, for its advanced and progressive policies for industrial hygiene, worker protection and compensation for silicosis. This optimistic view was contradicted in McCulloch's comprehensive account titled *South Africa's Gold Mines & the Politics of Silicosis*:

> The failures of the South African system of medical surveillance, data collection and compensation underpinned the commercial success of the nation's most important industry. The costs of production were shifted to rural communities within and outside South Africa's borders, and it is only now, more than a century after the mines began production, that those costs may at last be called in. (McCulloch 2012)

13.4.2 Reparations

13.4.2.1 Asbestos-Related Disease (ARD)

Various types of asbestos were mined in South Africa between 1910 and 2002, of which crocidolite formed the bulk. At its peak in 1997, there were 40 operating mines, most of them in the Northern Cape Province, employing between 1,200 and 21,000 workers according to various estimates. South Africa was the world's largest producer and extracted 200,000 tons of crocidolite in 1997, most of which was exported. From the mid-1940s, clinical evidence had shown that exposure to asbestos fibres caused asbestosis. In 1960, a South African study established epidemiological proof of both industrial and environmental links between asbestos and mesothelioma – an aggressive form of cancer of the lining of the lungs. Although the disease has a latency period of decades after exposure, once the disease manifests and is diagnosed, the five-year life expectancy is estimated to be 9%.

Asbestos mining in South Africa was dominated by a UK-registered company, Cape plc. The social consequences of asbestos mining have been described by Mothemela:

> The asbestos mining areas in South Africa were poor and isolated, making it easy for the British companies and their subsidiaries to enforce poor working conditions on desperate communities. Since there were no forms of alternative employment, or knowledge about the risks they would be exposed to, the labourers accepted these conditions. Cobbing was a form of social work since women operated in groups, sharing food preparation and childminding duties. This was a perfect arrangement as families stayed together, avoiding the unpleasant conditions of migrant labour. However, this arrangement also meant that entire families were exposed to asbestos and therefore the risk of mesothelioma. The workers remained in poverty even though they were in some form of employment. Until 1950 Gefco and Cape Asbestos Pty. used the "tributer" system which meant that labourers were not formally or permanently employed. The companies bought fibre from tributers when the market was conducive. The informalities of the mining operations ensured that families fell outside the provisions of the Mine Acts. Mining operations also avoided the cost of providing accommodation or medical care. When labourers were eventually formally employed, and they fell ill, they would simply be fired. It was only in 1980, when trade unions were mobilised, that workers learned about the hazards of the asbestos fibres. (Mothemela 2012)

Despite the proven relationship, the asbestos industry in South Africa and Europe continued to deny or minimise the health risks or to take active steps to reduce exposure or ban the use of asbestos. In this, the industry was aided and abetted by the government (Flynn 1982). Nevertheless, as the links between asbestosis became more widely known, various countries imposed stricter industrial hygiene conditions, recognising asbestosis as a work-related disease. The use of sprayed asbestos cement was banned. In 1997, in a UK court, five former employees of a South African asbestos mining operation sued Cape plc – a British registered company and the owner of the most extensive asbestos mining operations in South Africa. They complained that their asbestosis resulted from exposure to fibre in an unsafe working environment. After protracted legal battles regarding jurisdiction in the United Kingdom or South Africa, the House of Lords ruled on appeal the suit would be allowed to proceed in UK courts. During this period, the number of claimants rose to 7,500. Eventually, in 2001, the case was settled out of court, but the agreed compensation was not finalised until further court action was enforced on Cape plc in 2003. The final amount of settlement, GBP 45.5 million from Cape plc and a subsequent owner, Gencor Ltd, was a compromise, as the original amount claimed would have resulted in the demise of Cape plc, in which there would not be any funds at all for compensation payments. Two trust funds were set up to administer the compensation claims and payouts to both workers and those environmentally exposed – the Asbestos Relief Trust (ART) and the Kgalagadi Relief Trust (KRT) (Business & Human Rights Resource Centre 2014).

The South African Government has taken over responsibility for the derelict asbestos mining sites and waste deposits, having already spent USD 7 million on rehabilitation, with a further USD 14 million required to complete the consolidation and maintenance of the remaining deposits. Even if the asbestos dumps are restored to an apparent ecologically stable state, the dumps will continue to pose a risk due to diverse forces that can disturb the soil and vegetative cover, including human and animal traffic, wildfires and insect (ant) tunnelling.

Although the total number of asbestos miners and surrounding community members were small compared to gold, coal and iron mining, this story about asbestos is seminal in several respects. It illustrates unequal knowledge, power and financial relationships between (mainly colonial) mining companies and indigenous labour pools. There was long-term collusion between industry and government to suppress what, in modern terms, are considered fundamental worker rights, and, most importantly, it set a precedent for successful retrospective class action suits by disabled workers against former employers for work-related diseases.

13.4.2.2 Silicosis

In 2003, fresh from the success of the asbestos suit, the two lawyers responsible for initiating and steering the asbestos claims, Richard Meeran from a UK law firm and Richard Spoor from South Africa, and others, turned their attention to silicosis sufferers who worked on the South African gold mines. Regarding the number of workers involved, this case was orders of magnitude larger in scope. At its peak, the gold mining industry employed more than 500,000 workers. The precedent circumstances concerning tuberculosis and silicosis in the gold mining industry have been discussed above. With the assistance of various research institutions and public interest groups, the lawyers compiled lists of current and former workers suffering from lung impairment cases. Through various legal skirmishes, a consolidated class action suit was eventually brought before a court. Not surprisingly, the case was vigorously opposed by the mining companies.

In May 2016, the High Court of South Africa gave a judgement that allowed the case to proceed as a class action. The judgement is voluminous and detailed in setting out the case for the long-term dereliction of duty of the mining industry to maintain safe working conditions and protect workers' health from known hazards. The court ordered costs against the mining companies in opposing the class action application. Preparation for the main application proceeded. In early 2018, faced with the prospect of lengthy and expensive litigation, continuing reputational damage (refer to the preceding remarks on Anglo American's *FutureSmart Mining Strategy*) and the possibility of eventually losing the case, most of the mining companies named in the suit agreed to an out of court settlement. This settlement was ratified and made a Judgement and Order of the High Court of South Africa on 13 December 2018 (High Court of Gauteng, Case number 2018/44060). According to the terms of the order, a trust fund was to be established (the Tshiamiso Trust) to administer compensation for present and past employees, living or deceased, who were medically certified to have silicosis or tuberculosis. The amount of trust collectively was R5 billion (USD 400 million). Several but not all the mining groups accepted joint responsibility, with the most substantial contribution from Anglo American plc. It was anticipated that claimants who were employed from 1965 to the date of settlement and who could satisfy the conditions of employment and proven medical impairment would benefit from R10,000 to R500,000 for the severest cases of advanced silicosis.

In reparation for past dereliction, Penelope Andrews, a Professor of Law at the University of Cape Town, noted five elements of significance in the settlement: "compensation, accountability, deterrence, education and reinforcing rights" (Andrews 2018). The judgement allows all eligible employees to apply for compensation and be included in the settlement unless an individual actively opts out. Provision is made for widespread publicity to notify prospective claimants, including those residing in remote locations. These provisions finally counteract the past practice of repatriating the ill – out of sight and out of mind. The compensation amounts are well above the compensation provided under previous arrangements and will provide a measure of relief for medical care and necessities of life.

The extended retrospective scope of the award is precedent-setting in that employers will be held accountable irrespective of the length of time elapsed since exposure occurred. The closing of mines, mergers and acquisitions and transfer of country of registration will no longer provide a protective shield for failures to provide safe working environments consistent with current knowledge and technology. Andrews commented that the retrospective nature and considerable awards would serve as a deterrent against future transgressions of duty of care to protect the health and safety of workers. In parallel, the detailed content of the two leading judgements will provide educational material and examples to organised labour, employers, the legal profession and the public, specifically on the importance of industrial hygiene and the right of workers to establish and use channels of communication to air deviations or grievances, without fear of victimisation by employers. Finally, the judgement is a milestone in reinforcing the rights embodied in the post-1994 Constitution, demonstrating that "the law can be used as an instrument of justice" by workers and others (Andrews 2018).

13.4.3 Failure of the Johannesburg Plan of Implementation of the WSSD 2002

In 2002, South Africa hosted the World Summit on Sustainable Development (WSSD) in Johannesburg. This premier global event was billed as the successor to the 1992 United Nations "Conference on Environment and Development" (UNCED) held in Rio de Janeiro, otherwise known as the Earth Summit. The significant contribution of the WSSD was to incorporate a social dimension to complement the environmental and economic sustainability framework established during the Earth Summit. South Africa's president Thabo Mbeki was the leading proponent of this innovation to the global sustainability agenda.

Realising that conference declarations were necessary but insufficient to result in action, the WSSD released a mandate to guide implementation, the Johannesburg Plan of Implementation (JPI). Clause 46 of the JPI deals with mining:

> 46. Mining, minerals and metals are important to the economic and social development of many countries. Minerals are essential for modern living. Enhancing the contribution of mining, minerals and metals to sustainable development includes actions at all levels to:
>
> a. Support efforts to address the environmental, economic, health and social impacts and benefits of mining, minerals and metals throughout their life cycle, including workers health and safety, and use a range of partnerships, furthering existing activities at the national and international levels

> among interested Governments, intergovernmental organizations, mining companies and workers and other stakeholders to promote transparency and accountability for sustainable mining and minerals development;
> b. Enhance the participation of stakeholders, including local and indigenous communities and women, to play an active role in minerals, metals and mining development throughout the life cycles of mining operations, including after closure for rehabilitation purposes. (United Nations 2002)

In the sustainability euphoria following the WSSD, global and South African local mining companies, large firms such as BHP Billiton embraced the notions of environmental stewardship, including creating related posts in the senior executive teams of companies. In the future, environmental sustainability reports were to be included in the annual company reports or commissioned separate annual environmental reports.

The sustainability impetus was itself not sustainable. From a peak impact circa 2004, environmental and sustainability members of executive teams dropped to second, then third levels of management. By 2010, environmental officers were back at the mine operational level, with only the public relations aspect functioning at full impetus. This is reflected in documents such as Anglo American's 2018 sustainability strategy, "*Our Sustainability Strategy FutureSmart Mining™: Our blueprint for the future of our business*". The strategy sets ambitious environmental and social targets to be met at interim stages by 2025 and fully by 2030. The glossy pamphlet is an attempt to change the public perception of the company's failures over the past 100 years; ironically, the concluding remarks of the strategy state:

> BUILDING ON 100 YEARS OF LEADERSHIP Our goals for the future are firmly rooted in the values of our past. As our founder Sir Ernest Oppenheimer said: The aim of this Group is, and will remain, to earn profits for our shareholders, but to do so in such a way as to make a real and lasting contribution to the communities in which we operate. It's still true today. Our Sustainability Strategy is one way we are re-imagining mining to improve people's lives. (Anglo American 2018)

Although South Africa hosted the WSSD in 2002 and was one of the key role-players in introducing extended commitments to poverty alleviation and social responsibility, the South African government has not been successful in getting mining companies in the host country to implement the principles and practices implicit in the Johannesburg Plan of Implementation. The reality is that the South African mining sector did not match the ambitious goals of the WSSD. A study of the conditions in the platinum mining region of Rustenburg, South Africa, is presented as evidence. The author set out to evaluate the extent to which the goals of the JPI were being achieved. Her sobering conclusion was:

> According to the evidence presented in this thesis, Sustainable Development is **not** being successfully applied in the Rustenburg region, and specifically not within the confines of the Rustenburg platinum-mining districts. The changes in the environmental stewardship and social responsibility programmes, as envisaged in the Agenda 21 Declaration, the Earth Summit in Rio de Janeiro (United Nations, 2002), and the Johannesburg Declaration on the World Summit on Sustainable Development (2002), have **not** been reflected in the mining practices in the Rustenburg region. The Johannesburg Declaration is nothing more than a fine declaration. (Ololade 2011).

A tragic manifestation of this failure followed in 2012. A confluence of labour unrest over wages, poor working and living conditions in the Rustenburg platinum mines, neglect of the social compact explicit in the Johannesburg Plan of Implementation, and a police force acting outside the bounds of their mandate resulted in what became known as the Marikana massacre. On 12 August 2012, 34 striking miners were killed, and 78 were seriously injured when police opened fire on them. A lengthy commission of enquiry followed, and it chronicled the events leading up to the massacre and identified contributory causes and culpability for failures (Farlam et al. 2015). The police were severely criticised for lying to the commission and attempting to cover up their real intent and actions before, during and after the massacre. More to the point, for this chapter, were the commission's findings of the mining company's dereliction:

> **Findings about Lonmin**
>
> **Lonmin's protection of its workers**
>
> Lonmin called on its employees to go to work during the strike, even though it knew that it was not safe. Lonmin knew that it was unable to protect its workers and did not warn its workers about the danger. …
>
> Lonmin's reckless actions in urging employees to come to work in circumstances where they were aware of the potential dangers to them and in the full knowledge that they could not protect them, falls to be condemned in the strongest terms.
>
> **Lonmin's housing obligations**
>
> The Commission found that Lonmin's failure to fulfil its housing obligations did create "an environment conducive to the creation of tension, labour unrest, disunity among its employees or other harmful conduct." Lonmin did not provide the houses which it had promised. They say that they could not afford to build the houses, but the commission found this to be a lie. (Marikana Justice 2016)

It was stated that this event "will go down in the annals of South African policing history as one of the incomprehensible calamities of our times" (PRU 2015). Despite numerous commissions of enquiry into mining safety and health, political and societal changes in perceptions of the humane and proper treatment of labourers, the mining industry of South Africa has not yet freed itself from the dereliction of its human capital.

13.4.4 Contemporary Issues

13.4.4.1 Informal Settlements on Mining Land

State-enforced restrictions on the movement and residence of Black citizens between rural and urban areas, which had become increasingly untenable during the 1980s, were formally terminated with the inauguration of the new majority-based government in South Africa in 1994. The dammed-up poverty of enforced separation in politically defined Black homeland areas resulted in a surge of internal migration to the cities. City and provincial authorities were overwhelmed and unable to provide land and infrastructure for housing development to meet this influx of internal migrants. Informal settlements burgeoned, often through illegal land occupation. Within the urban conglomeration of the Witwatersrand, extensive areas of unoccupied land lay adjacent to derelict mines and tailings storage facilities, including exclusion zones restricted in terms of safety, ground with shallow underground workings or land susceptible to dolomitic sinkholes. Many informal settlements were established in such inappropriate areas. In extreme cases, informal dwellings were erected in spaces carved out of the sides of derelict TSFs.

The contradictions between social needs, rights under the Constitution, government responsibilities to provide land and housing and public health are palpable. Politically, forced removals of people and shelters are no longer permitted – unless responded to within 48 hours, the remedy against illegal land occupation requires formal and lengthy legal processes, even in cases of high-risk collapsing ground or exposure to dust and radioactive residues.

A paradoxical case occurred in the West Rand Town of Krugersdorp, where an informal settlement was built at a location referred to as Tudor Shaft. The settlement touches on the side of a small tailings deposit, and a pathway across the deposit provides the shortest route to industrial and commercial suburbs. The tailings deposit at Tudor Shaft was found to contain elevated concentrations of naturally occurring radioactive elements (NORMs) from samples taken by environmental NGOs on behalf of the communities. Legal ownership of the Tudor Shaft dump was indeterminate, so as per prescription, the state assumed responsibility and commenced a clearing operation. However, environmental activists objected that the removal process had not been approved through consultations with the affected communities, and the removal was exposing residents to exposure hazards from radioactive dust. Court interdicts were obtained that halted further clearing of the material (estimated to have taken only a few weeks). Two years later, after additional specialist (and disputed) studies and legal proceedings, the dump remained in place, and the community remained exposed to the adjacent hazard.

13.4.4.2 Residential Exposure to Derelict TSFs

Under the doctrine of apartheid, enforced segregation also occurred within urban spaces. People of Indian ethnicity, or mixed race – classified as *Coloured* – were forced in terms of the Group Areas Act No. 41 of 1950 (A key piece of apartheid legislation demarcating areas for either Whites only or Black, Coloured or Indian residential and business purposes) (O'Malley n.d.) to relocate from suburbs designated for Whites-only to newly created suburbs demarcated for Coloureds-only. One such instance was the small village Davidsonville on the West Rand. The settlement, comprising self-standing formal housing and three-storey apartment blocks, is located with the closest residences 10 m north of a dormant, unremediated slimes dam and 20 m west of a sand dump. Complaints by Davidsonville residents concerning dust emissions resulted in the commissioning of independent studies and the incorporation of the complaints into a class action suit brought by the Legal Resource Centre against the mine owners, the Department of Minerals and the Department of Environmental Affairs. The DRD mining company acknowledged its ownership of the *sand* deposit, the sand was trucked out for reprocessing and the footprint was levelled and grassed. However, the ownership of the *slimes tailing* portion was unidentified, so no remediation was done. Within two years, runoff from the tailings portion had covered the previous sand dump footprint, providing a new area of unconsolidated slime at the mercy of wind erosion. Without conceding liability, DRD cleared and consolidated the site for a second time. Eventually, ownership of the land and liability for the slime tailings was established as the City Council of Johannesburg, which then, reluctantly, took steps to consolidate the surface and attend to acid drainage that was seeping into an adjacent housing estate.

13.4.4.3 Informal Miners – The Zama-Zamas. Whose Gold Is It, Anyhow?

High levels of unemployment, exacerbated by the closure of underground mines, have led to the decline of whole towns whose creation and prosperity derived almost exclusively from underground gold mining. Stabilisation of the workforce, through providing long-term employment and housing benefits rather than fixed-term migrant contracts, left communities of migrant miners in remote rural areas and from neighbouring countries with no jobs and no prospects. Groups of former miners began entering closed or abandoned workings and conducting informal artisanal mining. These informal miners, termed *zama-zama* (a colloquial Isi-Zulu term meaning "take a chance"), spend days or weeks underground, supported by a network of suppliers and dealers, who somehow filter the illicit gold into the formal markets (Minerals Council n.d.) The Minister of Mineral Resources has estimated that illegal gold mining costs companies over USD 5 billion annually.

The *zama-zama* face the hazards of operating in uncontrolled underground workings without any of the safety provisions enforced in formal mines. Artisanal gold extraction by manual crushing and extraction using mercury amalgamation and distillation over open fires is an additional hazard. Underground accidents, conflicts between different *zama-zama* groups and clashes with mine security and police have led to several deaths. The *zama-zama* readily bypass standard fencing, steelwork or shallow concrete plugs to reach the closed-off underground workings. Such illegal reopening of sealed shafts sabotages company efforts to remediate residual hazards associated with closed underground operations. The Department of Minerals created a fund to seal the six hundred identified derelict openings on the Witwatersrand. Inexplicably, after ~250 shafts were sealed with 3-m thick concrete plugs, no further funds were forthcoming to complete the task, and the remaining shafts have remained unattended for almost a decade.

Security enforcement over the widely dispersed mine areas is challenging. Communities for whom this is a significant source of alternative income are reluctant to report these informal mining operations. In a bizarre twist of logic, it is asked: "Whose gold is it anyhow?" A disinformation campaign created by the ruling African National Congress party fostered the notion of "White Monopoly Capital" as a significant reason for the failure of the ANC-led government to deliver substantive economic improvements. In this political climate, the right of large mining houses to have the exclusive right to mine gold has been challenged, and in the minds of the unemployed former miners, legitimated their right to informal access to the resources. In the diamond mining sector, government discussions with artisanal miners have resulted in an arrangement for formalising and issuing small mining permits to several hundred informal miners working the residues of the Kimberly (Northern Cape Province) diamond field. As of 2019, no such arrangement exists to bring the gold mine *zama-zamas* into the formal economy. This unfortunate consequence of the decline of the gold mining industry and the coincident political climate is one of the intractable consequences of mine dereliction.

13.5 CONCLUSIONS

The above account of over a century of dust and dereliction in South African mining has shown that dereliction is not only a matter of dealing with the aftermath of mineral extraction, mine residues and residual environmental damage. Rather, dereliction may be a constant companion to mining, present at all stages from exploration to post-closure. In South Africa, the human dimensions of dereliction have been severe, with the collusion of economic and political interests resulting in pervasive neglect of workplace health and safety. The recent and belated compensation agreements for asbestosis and silicosis constitute some form of reparation for this past neglect. Reclamation and restoration of the land covered by gold mine tailings of the Witwatersrand have already been in progress for forty years, with at least another twenty years planned for continued operation. The means and costs of restoration of and further reparation to communities adversely affected by the respiratory diseases of their breadwinners are intangible and will persist for decades longer.

Dealing with dereliction is more than a technical-scientific matter – it has substantial ethical and moral dimensions shaped by societal values towards fellow citizens of different classes and racial origins and by dominant political forces. Large mining corporations are attempting to redefine their responsibilities, practices and image in line with contemporary human rights and environmental values. Even the efforts of high-profile international conventions, such as the 2002 World Summit on Sustainable Development, have proved elusive and unsustainable. *Homo sapiens*, taking a millennial scale perspective, has wielded a profound impact on the planet and all living species (Harari 2011). Coming to terms with mining dereliction is one small step in our resurgent awareness of the changes needed to make our sojourn on Earth sustainable.

REFERENCES

Andrews, P. 2018. South Africa's historic silicosis class action: Why the settlement matters. *The Conversation*, May 9, 2018. https://theconversation.com/south-africas-historic-silicosis-class-action-why-the-settlement-matters-96234.

Anglo American. 2018. Anglo American Platinum/our blueprint for the future of sustainable mining. https://www.angloamericanplatinum.com/~/media/Files/A/Anglo-American-Platinum/our-blueprint-for-the-future-of-sustainable-mining.pdf.

ASTM D1739-1970, Standard Test Method for Collection and Measurement of Dustfall (Settleable Particulate Matter), ASTM International, West Conshohocken, PA, 2017. Superseded by ASTM-98 (2017) www.astm.org.

Barker, A. 2011. Project Hloekisa: Summary Conceptual Report, MWS001 Hloekisa Summary. December 8, 2011. Andrew Barker Development Consultant, Johannesburg. www.andrewbarker.co.za. Section 3.3 is based on this unpublished report by Andrew Barker, which is used with his permission.

Botham, N. D., C. J. Kelso, and H. J. Annegarn. 2011. Best practice to acquire a mine closure certificate: A critical analysis of the De Beers Oaks Diamond Mine, Limpopo Province, South Africa. *Mine Closure: Proceedings of 6th International Conference on Mine Closure*, Canada, 18–21 September 2011. Eds. A. Fourie, M. Tibbett, and A. Beersing. Crawley: Australian Centre for Geomechanics, University of Western Australia, Vol. 2:401–10.

Business & Human Rights Resource Centre. 2014. Cape/Gencor lawsuits (re So. Africa). February 18, 2014. https://www.business-humanrights.org/en/capegencor-lawsuits-re-so-africa-0.

Coetzee, H. 2004. An assessment of sources, pathways, mechanisms, and risks of current and potential future pollution of water and sediments in gold-mining areas of the Wonderfonteinspruit catchment. Water Research Commission, WRC Report No. 1214/1/06, Pretoria, 266 pp.

Farlam, I. G., P. D. Hemraj, and B. R. Tokoto. 2015. Marikana commission of inquiry: Report on matters of public, national and international concern arising out of the tragic incidents at the Lonmin Mine in Marikana, in the North West Province. Farlam Commission Report. https://www.sahistory.org.za/archive/marikana-commission-inquiry-report-matters-public-national-and-international-concern-arising.

Flynn, L. 1982. South Africa blacks out blue asbestos risk. *New Scientist* 1982:237–9.

Harari, Y. N. 2011. *Sapiens – A Brief History of Humankind*. London: Penguin Random House.

Hartnady, C. J. H. 2009. South Africa's gold production and reserves. *South African Journal of Science* 105:328–9.

Marikana Justice. 2016. Summary of the Farlam Commission Report, 2016. http://www.marikanajustice.co.za/wp-content/uploads/2016/10/Summary-of-the-Farlam-Commission-Report.pdf.

Maseki, J. B., H. J. Annegarn, and G. Spiers. 2017. Health risk of enriched heavy metals (As, Cd and Cr) in airborne particles from Witwatersrand gold tailings. *Journal of the South African Institute of Mining and Metallurgy* 117:663–9.

McCulloch, J. 2012. *South Africa's Gold Mines & the Politics of Silicosis*. Johannesburg: Jacana Media.

Minerals Council. n.d. Illegal mining. Minerals Council of South Africa. https://www.mineralscouncil.org.za/work/illegal-mining.

Mothemela, M. 2012. Mesothelioma in South Africa 3 decades post peak of asbestos production: An analysis of a claims database of asbestos ex-miners. MMed dissertation, University of Cape Town.

MPRDA. 2002. Mineral and Petroleum and Resources Development Act 2002 (Act No. 28 of 2002), clause 38(d). Pretoria: Government of South Africa.

NEMA: Air Quality Act. 2004. National Environmental Management: Air Quality Act 2004 (Act 39 of 2004). National Dust Control Regulations. Government Gazette, 1 November 2013. Pretoria: Government of South Africa.

Ojelede, M. E. 2011. Risk assessment of atmospheric emissions from gold mine tailings on the Witwatersrand. PhD thesis, University of the Witwatersrand.

Ojelede, M. E., H. J. Annegarn, and M. A. Kneen. 2012. Evaluation of aeolian emissions from gold mine tailings on the Witwatersrand. *Aeolian Research* 3:477–86.

Ololade, O. O. 2011. Evaluation of the sustainability and environmental impacts of mining in the Rustenburg Region, PhD thesis, University of Johannesburg.

O'Malley, P. n.d. The heart of hope, Subsection *1950*. Group Areas Act No 41. https://omalley.nelsonmandela.org/index.php/site/q/03lv01538/04lv01828.htm (Accessed September 29, 2023).

PRU. 2015. An overview of the Farlam Commission's recommendations, Parliament Research Unit report, August 19, 2015. http://pmg-assets.s3-website-eu-west-1.amazonaws.com/15081919FarlamCommission.pdf.

Reichardt, M. 2013. The Wasted Years: A history of mine waste rehabilitation methodology in the South African mining industry from its origins to 1991. PhD thesis, University of the Witwatersrand, Johannesburg.

Rossouw, A. S., H. J. Annegarn, I. M. Weiersbye, and D. G. Furniss. 2010. Evaluating the functional status of a rehabilitated gold tailings storage facility—A case study in the Witwatersrand. *South African Journal of Botany* 76:402. doi: 10.1016/j.sajb.2010.02.041.

United Nations. 2002. *Plan of Implementation of the World Summit on Sustainable Development - A/CONF.199/20 Chapter 1, Resolution 2 - UN Document*. Geneva: United Nations. www.un-documents.net/jburgpln.htm.

United Nations. 2002. Johannesburg Declaration on Sustainable Development. World Summit on Sustainable Development, 4 September 2002. United Nations A/CONF.199/20. http://www.un-documents.net/jburgdec.htm.

14 Cleaning Up Centuries of Pollution

The UK Approach to Managing Abandoned Mines

Hugh Alastair Brymer Potter

14.1 INTRODUCTION

The UK has a long history of mining with evidence of Bronze Age, Roman and medieval activity. Monks are reported to have collected coal from the beaches of North East England in the twelfth century (Younger *et al.*, 2002). The most productive period was during the 18th and 19th centuries; however, almost none of the more than 10,000 mines remain active today. In 2023, there are a few surface mines for coal and fluorspar, and no large scale underground mines for coal or metals – although de-watering of the South Crofty tin mine in Cornwall, which closed in 1998, has started with the goal to restart production in 2026 (Cornish Metals, 2023), and there are several companies prospecting for lithium. Potash and rock salt continue to be extracted from beneath the North Sea. Figure 14.1 is a map of England and Wales showing the rivers that are polluted by abandoned metal and coal mines.

The peak of coal extraction was in 1913 when a million people worked in 1,600 mines and produced nearly 300 million tonnes of coal per year (Environment Agency, 2021a). Major closures of coal mines took place in the 1980s and rebounding mine water caused significant pollution of rivers, turning them orange and smothering the river bed with iron hydroxides. Almost all the metal mines closed in the early 20th century, but they continue to damage aquatic life in more than 2,000 km of rivers in England, Wales and, to a much smaller extent, Scotland. About half the cadmium, lead and zinc found in UK rivers come from abandoned mines; as much as is discharged by all industrial activities (Mayes *et al.*, 2010, 2013). The high metal concentrations left in spoil heaps and downstream river sediments have encouraged the development of distinctive metallophyte floras, many of which are protected as "Sites of Special Scientific Interest (SSSI)" (Batty, 2005). The ecological importance of abandoned mine sites extends beyond plants and lichens, supporting metal tolerant invertebrates and providing widespread habitats for bats.

A prime example of a long-abandoned metal mine is Mynydd Parys (Parys Mountain) on the island of Anglesey, Wales (Figure 14.2). Mining started over 3,000 years ago and at its peak in the 18th century, this single mine controlled the world copper price. It was originally dug as an underground mine but progressed into a large open pit, excavated almost entirely by hand (Younger & Potter, 2012).

Mining ceased about 100 years ago and yet this site is still the largest single source of metals to the Irish Sea, discharging about 10 tonnes of copper, 20 tonnes of zinc and 50 kg of cadmium each year with an average flow of 10 L/s. The discharge is only a couple of meters across but this bright orange, extremely acidic (~pH2) stream then passes very close to a children's playground in the village of Amlwch before entering the sea.

DOI: 10.1201/9781315141855-14

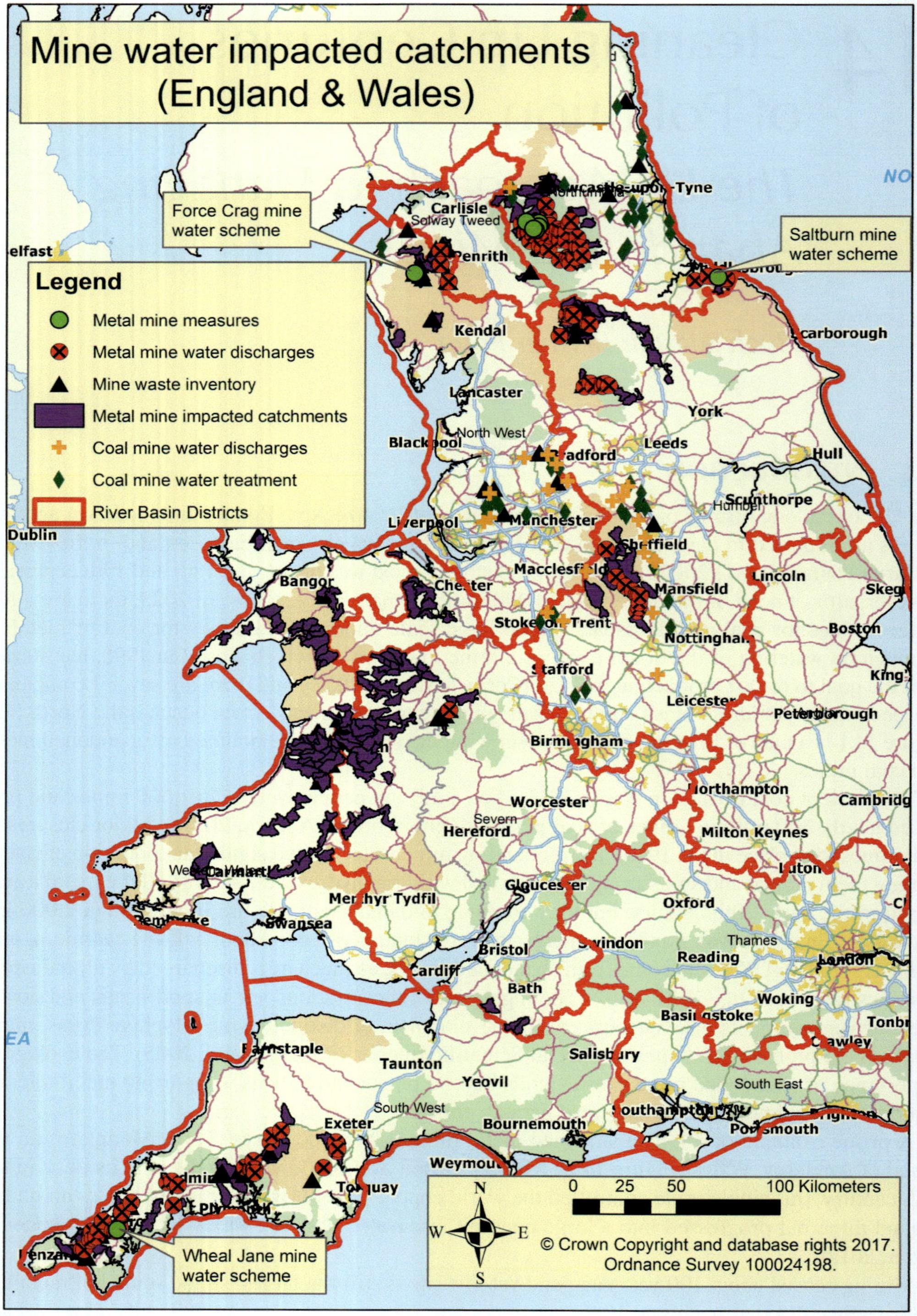

FIGURE 14.1 Map of England and Wales showing the location of abandoned coal mine water discharges (orange crosses) and treatment schemes (green diamonds), the river catchments which are polluted by abandoned metal mines (red) and remedial measures at metal mines (green circles), dated 2017.

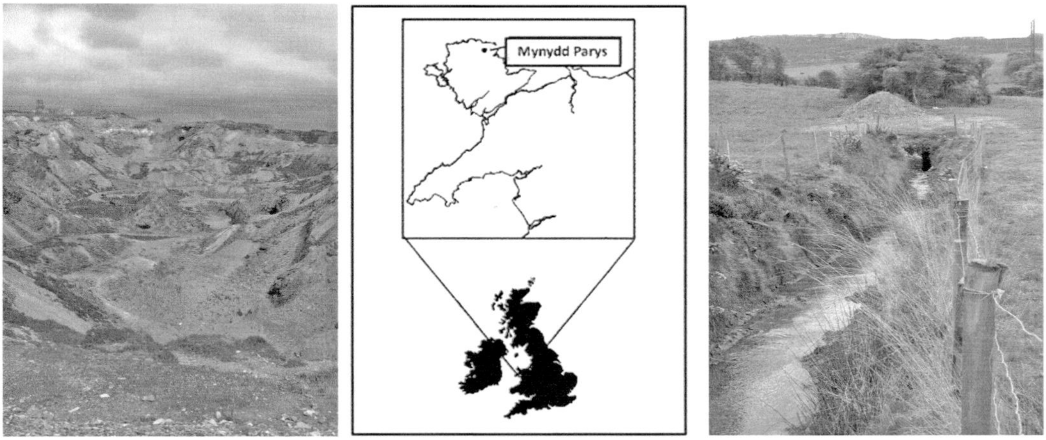

FIGURE 14.2 Mynydd Parys: Great Open Cast (left), location (centre) and Dyffryn Adda portal (right).

In Yorkshire, the Old Gang Smelt Mill stands testament to the ingenuity and resilience of Victorian miners (Figure 14.3). Abandoned in the 19th century, the extensive spoil heaps and buildings are now protected "Scheduled Monuments" and SSSI for their industrial archaeological interest and unusual metal-tolerant vegetation. These remote upland mine sites have no electrical supply and are popular with hikers walking through the moors of the Yorkshire Dales National Park.

The scale of landscape disturbance can be appreciated properly only from the air, for example from aerial photographs in popular Internet-based mapping services, as they cover many square kilometers. Lead and other metals were originally dug from the surface before underground workings, and the wastes have often resisted natural revegetation due to the extremely high concentrations of lead, zinc and cadmium. Until recently in the Yorkshire Dales National Park, these wastes were sometimes used by local landowners to create tracks since very few plants will grow on them.

However, these widespread wastes are a key source of diffuse pollution to the rivers (Mayes *et al.*, 2015; Mighanetara *et al.*, 2009), and along with drainage tunnels, such as the Hard Level, cause significant exceedances of environmental quality standards and severely impact aquatic life, such as fish and river flies (Jones *et al.*, 2018). The Hard Level was driven to drain many interconnected

FIGURE 14.3 Old Gang Smelt Mill and associated spoil heaps (left) and Hard Level discharge (right).

mines by gravity, and, today, it provides very significant baseflow to the river, often greater than 50% of the total flows.

Construction of the longest drainage tunnel in the UK, the Great County Adit, started in 1748 and eventually reached 65 km length, connecting more than 100 mines. At its peak in 1839, it discharged over 66 million liters of mine water each day into the Carnon River and had more steam engines pumping into it than were used by the whole of continental Europe and America combined (Cornwall Heritage, 2023). The County Adit mine water remains the single largest source of metals to rivers in the UK. In 2006, selected mining landscapes, including the Great County Adit, across Cornwall and west Devon were inscribed as the Cornish Mining World Heritage Site by UNESCO (Cornish Mining World Heritage, 2023).

In some parts of the country, these levels are regularly used by mine explorers to gain access to the underground workings. Sometimes these enthusiasts build dams to alter underground water flows and remove rock falls to give them better access without worrying about causing pollution, changing mine drainage patterns or their personal safety. A number of former mines, both for coal and for metal, are operated as tourist mines (Nenthead Mines, 2023); to comply with health and safety obligations, these sites must have a qualified Mine Manager and are regulated by the Mines Inspectorate. Museums like the National Coal Mining Museum for England (NCM, 2023) have an important role in educating children and adults about the lives of those people who extracted the raw materials that underpinned Britain's industrial revolution. In this chapter, the approach that the UK is taking to cleaning up the environmental legacy of centuries of mineral extraction will be described.

14.2 LEGAL AND ADMINISTRATIVE FRAMEWORK

Writing in 1874, the Rivers Pollution Commissioners recognized that serious pollution of rivers and agricultural land was being caused, especially from lead mines, describing affected streams as "turbid, whitened by the waste of the lead mines in their courses, bringing down 'slimes' at flood time which were spread over the fields, fouling and destroying the grass", further that fish were all killed along with cattle, chickens, ducks and geese (Erichsen Jones, 1964). This led to a change in mining practices, and miners began to build tailings ponds to retain wastes rather than dumping them straight into the nearest river. Vast quantities of metals were discharged in the preceding centuries, which has led to a large reservoir of contaminated sediments and floodplain soils many tens of kilometers downstream of the mines (Macklin *et al.*, 2006, 2023). These metals will be retained in these floodplains for hundreds to thousands of years depending on the rates of erosion and remobilization (Macklin, 1992) whilst in shorter river systems (e.g. in Devon and Cornwall) storage is primarily in estuarine sediments (Pirrie *et al.*, 2002).

Despite specific exemptions from key water pollution regulations, mine operators were occasionally prosecuted for causing fish kills, for example in 1890 at the Threlkeld lead mine in Cumbria. This lead and zinc mine operated from 1661 until finally closing in 1928; however, it continues to be one of the most significant sources of metals to UK rivers today, polluting about 25 km of rivers with cadmium, nickel, lead and zinc (Figure 14.4).

The legal loophole that allowed mine operators to close their mines without taking any responsibility for post-closure water pollution was finally removed for mines that closed after 31 December 1999. Coincidentally, the last of the deep fluorspar and tin mines ceased operating shortly before this date. As a result, none of the vast wealth that was generated from centuries of mineral extraction in the UK is available or has been used to clean up this polluting legacy.

The Environment Agency is the primary government regulator for the water environment in England (equivalent agencies are Natural Resources Wales and the Scottish Environmental Protection Agency). The Environment Agency is sponsored by the UK Government's Department for Environment, Food and Rural Affairs (Defra), which provides the overall strategic policy direction, most recently set out in the Environmental Improvement Plan (Defra, 2023a), and the Plan for Water (Defra, 2023b). These plans explain the Government's goals, such as achieving clean and

FIGURE 14.4 The Woodend Low Level, Threlkeld Mine, Cumbria.

plentiful water by improving at least 75% of England's waters to as close to their natural state as soon as is practicable.

14.3 METAL MINES

Although pollution from abandoned metal mines was known to be a widespread issue, and there had been some spectacular pollution incidents, such as at the Wheal Jane tin mine in 1991 (Younger *et al.*, 2005; Coal Authority, 2020a), it is only in the last decade that the scale of the problem was systematically assessed. Unlike coal extraction, the metal mines were all privately operated and so Government was reluctant to take action (Environment Agency, 2008a, 2021a). However, it was recognized that chronic pollution would undermine efforts to improve rivers to the "good" status by 2015 that was required by the European Union Water Framework Directive. In 2007, the Environment Agency set up a project to bring together existing data on water quality impacts and known or suspected mine locations so as to prioritize catchments for interventions based on the severity of water pollution and estimate the potential costs of remediation. The "Prioritisation of abandoned non-coal mine impacts on the environment" study was led by Newcastle University with the Coal Authority and Atkins and is published in a series of reports and journal articles (Environment Agency, 2012a, b; Mayes *et al.*, 2009a, 2010). Key findings from these and subsequent investigations include:

- At least 1,500 km of rivers in England are polluted along with a further 1,000 km in Wales (Figure 14.1).
- About 50% of the cadmium, lead and zinc found in rivers in England and Wales is discharged from abandoned metal mines; as much as discharged from all industrial effluents permitted (and hence measured) by regulators.

FIGURE 14.5 Wastes at the long abandoned West Chiverton Mine, Cornwall

- Around 250 mine water discharges have been identified, but at only ~10% have the flows and metal concentrations been monitored, and rarely in sufficient detail to design treatment schemes.

A parallel study by the Environment Agency, carried out to comply with the European Union Mining Waste Directive, identified more than 100 abandoned mine waste facilities that are causing serious harm to the environment or human health (Environment Agency, 2014). For example, between 1859 and 1886, the West Chiverton lead mine in Cornwall (Figure 14.5) produced over 45,000 tonnes of high grade lead ore, over 1 million ounces of silver as well as large amounts of zinc and copper (Explore Cornwall, 2023). Downstream of the mine, the river exceeds the EQS for Cd (710 times EQS), Zn (2,200 times EQS), Pb (215 times EQS) and Cu (44 times EQS), and it is still up to 40 times the Zn EQS when it reaches the sea 6 km away (Environment Agency, 2023a). Merrington and Alloway (1994) estimated there is about 75,000 tonnes of wastes, covering over 6,000 square meters and still containing 500 tonnes of lead and 100 tonnes of zinc. Each year, 0.5 tonne of lead and 1 tonne of zinc is released to the environment and despite activity ceasing at the mine in the 1920s, the wastes are so contaminated that no vegetation has grown.

14.3.1 Environmental Harm: Point Sources versus Diffuse Inputs

Numerous monitoring investigations by the Environment Agency and academic studies (e.g. Mighanetara *et al.*, 2009; Jarvis *et al.*, 2019) demonstrate that the severity of environmental harm varies significantly with rainfall and river flows. "Point sources", where drainage tunnels or mine entries discharge metal rich mine waters all year round, cause the most severe river pollution at lower river flows when comparing in-river metal concentrations with toxicologically derived Environmental Quality Standards. "Diffuse sources", such as mine wastes, metal-contaminated sediments and groundwater, contribute increasing amounts of metals as river flows increase, for example after heavy rainfall, and lead to the highest flux of metals (Gozzard *et al.*, 2011).

Climate change will worsen these environmental impacts because drier summers mean there will be less dilution of polluting mine water but conversely, more intense rainfall will wash even more metals into rivers from wastes.

14.3.2 The Water and Abandoned Metal Mines Programme

The outcomes of these national scale research projects convinced Defra to make a policy commitment to begin to address this source of river pollution through the 2011 Water for Life White Paper (Defra, 2011). This led to the creation of the Water and Abandoned Metal Mines (WAMM) Programme, which is a partnership between the Environment Agency, the Coal Authority and Defra (Coal Authority, 2016a). The objective is to:

> "clean up pollution from abandoned mines to make our rivers cleaner and healthier to benefit current and future generations, wildlife and the economy".

This is a risk-based program which focuses on minimizing environmental impacts on defined receptors (water quality and river wildlife) rather than creating an inventory of the many thousands of abandoned mines.

The Environment Agency's role is to gather data to allow key sources of metals to be identified and to prioritize which sources should be remediated. There is a long-established extensive network of water quality monitoring locations that confirms which rivers are polluted, but these data do not allow the contributions from different sources of metals to be quantified. It is essential to gather data on metal flux from multiple locations in a catchment in as short a time as possible with ideally stable flow conditions. Detailed monitoring of metal flux from mine water discharges and in-river locations enables the significance of various point and diffuse sources to be estimated (Environment Agency, 2012b). A simple spreadsheet tool, developed for WAMM by the Universities of Newcastle and Hull, is then used to predict the likely improvement in water quality from managing these sources.

The Coal Authority leads the development of measures to decrease metal inputs, through both mine water treatment schemes and control of diffuse sources, builds and operates interventions. The decision to take a particular remedial measure forward depends on technical feasibility, the benefits of the intervention and the whole life cycle costs. Schemes will be built only if the benefits outweigh the costs. In 2023, four metal mine water treatment schemes are in operation, each year capturing about 750 tonnes of pollutants, including iron, zinc, cadmium, copper, nickel and arsenic (Coal Authority, 2016a). A fifth treatment scheme will start operating in 2024 (Environment Agency, 2023c).

14.3.3 A Legally Binding Target to Clean Up Rivers Polluted by Abandoned Metal Mines

In January 2023, the UK Parliament approved thirteen new, legally binding long-term targets to tackle some of the biggest pressures facing the environment in England and improve water quality and availability, biodiversity on land and in seas, woodland cover, air quality and decrease residual wastes (UK Parliament, 2023). One target is to halve the length of rivers polluted by harmful metals from abandoned metal mines by 2038.

The Government's Environmental Improvement Plan (Defra, 2023a) and Plan for Water (Defra, 2023b) explain the basis for these targets, along with interim targets, for example to build eight new mine water treatment schemes and 20 interventions to control metal pollution from diffuse wastes by 2028. This ground-breaking and ambitious new long-term target requires a step-change in the pace of mine water remediation in England, increasing the number of mine water treatment schemes from four to about 40 over the next 15 years. Passive technologies utilizing nature-based solutions are preferred due to lower whole-life cycle costs and a smaller carbon impact than active technologies. This requires innovation in treatment technologies to minimize whole life-cycle cost and several small and large field scale pilot trials are underway and planned.

The Environment Act metal mines target will deliver a clean water environment, for people and nature, by decreasing metal pollution of rivers whilst increasing biodiversity and supporting natural flood management, e.g., by planting thousands of trees and sequestering carbon in soils. The Government estimates that the long-term costs (to 2100) to build and operate remedial interventions at metal mines is approximately £276M (discounted present value: Defra, 2023c).

14.3.4 Force Crag Mine Water Scheme: Passive Bacterial Sulfate Reduction Treatment Ponds

The Force Crag mine, worked for zinc, lead and barytes from 1835 until 1991, was famously the last working mine in the Lake District, North West England. It is a major source of heavy metal pollution, with two mine water discharges from the mine as well as diffuse pollution from waste heaps and an old tailings lagoon. The cadmium, zinc and lead in these discharges pollute at least 10 km of rivers and Bassenthwaite Lake, which is designated as a Special Area of Conservation. There is a significant impact on river insects and fish since zinc and cadmium exceed the EQS by up to 60 times and 40 times, respectively.

The site is owned by the National Trust and run as a visitor attraction. It is within the Lake District World Heritage Site and High Fells Special Area of Conservation, is a Site of Special Scientific Interest and is also a Scheduled Monument due to the mining heritage. The Environment Agency, Coal Authority, National Trust and Newcastle University worked in partnership to develop a remediation scheme for this site with funding from Defra. In March 2014, the '*compost-based treatment pond*' (also known as a vertical flow pond or biochemical reactor) began operating (Jarvis *et al.,* 2015; Mayes *et al.,* 2021; Moorhouse *et al.,* 2015). This passive system harnesses natural chemical and biological processes to remove metals from the water (Figure 14.6). The design by Newcastle University followed four years of research, funded by Defra, during which the system was scaled up from laboratory columns (0.003 m^3) through a 2 m^3 field pilot plant (Gandy *et al.,* 2016) to the full-scale 800 m^3 system treating 6 L/s with a relatively short hydraulic residence time of about 15 hours. This was the first full-scale bacterial sulfate-reduction based treatment pond for a European metal mine. Most of the mine water base flow from the main drainage tunnel (Level 1: ~1 tonne zinc/yr) is diverted into two treatment ponds containing compost and woodchips. Without any need to add chemicals or energy, reactions in the compost mixture capture more than 90% of the cadmium and zinc, primarily as sulfides, and so decrease the metal concentrations in the water that flows back into the Coledale Beck.

An important feature of the scheme is that the treatment system was built on the former tailings lagoons that were used to store wastes as an integral part of the metal mining process (Figure 14.7).

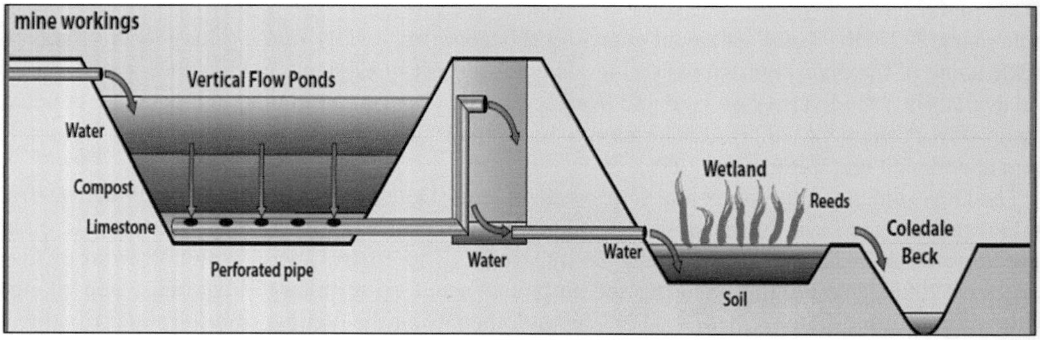

FIGURE 14.6 Schematic of the Force Crag compost-based treatment ponds.

FIGURE 14.7 The Force Crag compost-based treatment ponds and mine buildings (left); Rory Stewart MP (& Defra Water Minister) formally opening the scheme in 2015 (right).

Historic England agreed to this since they considered it to be the logical next development in the industrial heritage of the mine.

The system treats about 190,000 cubic meters of contaminated groundwater each year and has decreased the severity of pollution in the downstream rivers despite only treating a portion of the mine water flow. Further work is investigating how to deal with diffuse sources of metals in the catchment, which cause most of the pollution when it rains and river flows are higher (DIFFUSE, 2023).

In the nine years that the scheme has been operating, there have been some operational difficulties, primarily related to a decrease in permeability of the treatment material, which is quite fine-grained. There is some odor caused by the release of excess hydrogen sulfide and, whilst this does not cause a nuisance due to the remote location, it will need to be managed for treatment schemes closer to residential properties. The scheme cost around £1m to develop and build.

14.3.5 The Tyne Catchment

Intensive mining took place in the upper part of the Tyne catchment during the 18th and 19th centuries (and coal mining was widespread in the lower catchment). Today, around 150 km of rivers fail to meet the EQS for zinc, cadmium and/or lead, meaning aquatic life is being harmed. The centuries of metal pollution has led to the development of *calaminarian* (metal-tolerant) grasslands in river gravels and floodplain soils many tens of kilometers downstream of the mines (Hudson-Edwards *et al.*, 1996). Many of these sites have legal conservation protection since the soil metal concentrations are so high that common plant growth is restricted, allowing metal-tolerant plants such as Alpine penny-cress and Spring sandwort to flourish (Batty, 2005). Using synchronous monitoring water quality and flow at several permanent gauging stations, it has been calculated that about 127 tonnes of zinc and 40 tonnes of lead continues to enter the estuary each year (Figure 14.8, Environment Agency, 2021b). The majority of this comes from two small sub-catchments, the Nent and West Allen, which are now the focus for a suite of interventions.

14.3.5.1 West Allen: Carrshield Tailings Dam

This waste tailings facility was an important diffuse source of metals. In the 1950s, approximately 48,000 cubic meters of highly contaminated mineral processing wastes were tipped at the site. In common with many abandoned mines, the site is designated as a Scheduled Monument, particularly for the 300 meter long dry stone wall built in the 1800s to keep the river out of the mineral processing area. This wall was beginning to fail and with the increasing intensity of rainfall and river flows, there was a heightened risk of the wastes, which contain about 1,000 tonnes of lead, zinc

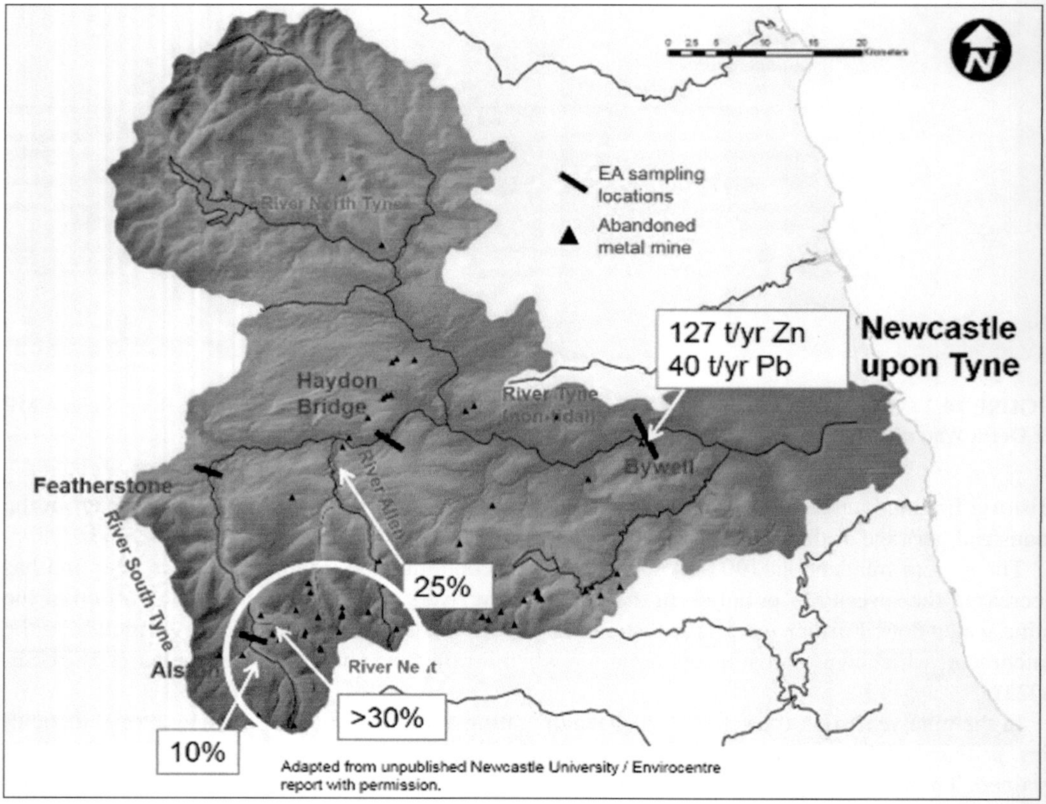

FIGURE 14.8 Metal flux and sources in the Tyne catchment.

and cadmium, collapsing into the river. Under dry conditions, it causes little pollution but following rainfall, there is a lot of erosion and flushing out of metals.

A £2M remedial intervention was completed in 2019 which re-profiled the wastes to a more stable landform, repaired the drystone wall by installing pre-cast concrete blocks and then facing with local stone and capped the wastes to prevent rainfall infiltrating and mobilizing the metals (see Figure 14.9). This design followed extensive consultation with Historic England to ensure the nationally important industrial archaeological interest was protected, and the works were carried

FIGURE 14.9 Carrshield Tailings Dam by the River West Allen, Northumberland, before (left) and after (right) remediation.

out with close supervision by an archaeologist. Prior to work starting, several areas of calaminarian grassland had to be relocated temporarily, and then brought back to the site. The final surface is a thin layer of metal-contaminated wastes placed on top of the cap, to encourage calaminarian plants and ensure that the final landform continues to look like a traditional tailings facility. On completion, Historic England was able to remove the site from their list of Scheduled Monuments at Risk (Coal Authority, 2020b; Culledge *et al.*, 2019).

14.3.5.2 River Nent: Mine Water Discharges and Diffuse Metal Inputs

The River Nent is the second most metal polluted river in England with a particularly intense and long history of metal extraction (Figure 14.10). Lead and silver were mined from at least the 12th century (Wilson, 1963), and possibly by the Romans, until underground mining ceased in 1958 whilst reprocessing of the existing waste heaps for zinc and fluorspar continued until 1963 (Historic England, 1997). The mine entrances, watercourses, dressing floors and buildings on main mine site at Nenthead have been designated as a "Scheduled Monument" since it is the most intact mining landscape within the North Pennines. The mine site is also protected as a "Site of Special Scientific Interest" for geology and metal tolerant lichens and plants.

On average, water quality at the catchment outlet (5km downstream of the mines) exceeds the EQS by nearly 60 times for zinc, 30 times for cadmium and 3 times for lead. However, at lower river flows when there is less dilution, the EQS can be exceeded by 200 times for zinc. The impacts from the many sources of metals extend all the way to the sea in Newcastle, more than 100km away. The Environment Agency's 30-year monitoring records show no change in average metal concentrations over this period (Environment Agency, 2023b). Aquatic life is severely impacted with about half the fish (Brown trout) and river flies compared with similar unpolluted rivers (Environment Agency, 2021b). No young trout are found in the main river, presumably because the metal concentrations restrict them to less polluted tributaries. Intensive monitoring of metal flux has identified where remediation will have the most impact and that this depends on the driver for interventions. The greatest improvement in water quality will come from treating the three main mine water drainage discharges since these contribute much of the base flow at low river flows and cause the highest metal concentrations and hence the most severe environmental harm. However, to decrease the metal flux down the river, the focus needs to be on diffuse sources, such as spoil heaps and in-river sediments. Whilst the point sources contribute a relatively constant mass of metal to the river, the input from diffuse sources increases hugely at higher river flows. Unpublished data show that more

FIGURE 14.10 Sources of metals in the River Nent, Cumbria: Nenthead spoil heaps (left); Haggs mine water discharge (right).

than half the zinc and 80% of the lead which enter the Tyne estuary do so in the highest 5% of river flows, and when these metals accumulate in deep-water shipping berths, they significantly increase the costs and difficulty when disposing of dredged sediments.

14.3.5.3 The Nent Haggs Mine Water Treatment Scheme

The Haggs mine drainage adit is one of the key sources of metal pollution, releasing about 3 tonnes of zinc and cadmium each year (Coal Authority, 2016b). A proposal to build a treatment scheme utilizing bacterial sulfate reduction to capture 90% of the zinc and cadmium was presented to the public in 2015 and met with considerable resistance and concern, primarily about the potential for odor nuisance. An extensive public engagement process was carried out over the following three years involving numerous public question and answer sessions, further explanation of the severity of river pollution and completely revisiting the process of treatment site selection to enable substantial input from the community on the criteria used to assess various locations. This led to the scheme being fundamentally redesigned in a new location considered acceptable to the public which involves a 2.5km pipeline to transport the mine water from the adit to treatment ponds. Planning permission was granted in 2019 and construction started in July 2020 for an expected 18-month period. Unfortunately, the COVID-19 pandemic significantly delayed progress and increased costs, but the scheme will start operating in 2024 (Environment Agency, 2023c).

The scheme is designed to treat the total Haggs mine water flow, which averages 10 l/s, in three downflow ponds containing a mixture of limestone, wood bark and straw (Figure 14.11). Most of the zinc and cadmium will be removed by encouraging natural biogeochemical processes to convert some of the sulfate in the mine water into dissolved sulfide. This will bind to the metals to form solids that are retained in the treatment layer. To prevent any residual sulfide being released as hydrogen sulfide and potentially causing an odor nuisance, hydrogen peroxide is added to the treated mine water to convert sulfide back into sulfate and sulfur. The dosing rate is automatically adjusted to ensure full oxidation of the treated mine water, and several sensors monitor hydrogen sulfide concentrations in air at potential emission points.

Once fully operating, this 'nature-based' scheme is expected to improve river water quality by between 20 and 45% in up to 60km of rivers. As part of the scheme, a new wildlife habitat and community open space was created around the mine water pumping station, in partnership with the Tyne Rivers Trust and Northumbrian Water, where hundreds of trees have been planted along with a wildflower meadow and wetland (Environment Agency, 2023d).

14.3.6 Saltburn Ironstone Mine

Saltburn Gill is a narrow stream in a wooded valley which leads to one of the most popular surfing beaches on England's north east coast. This area of East Cleveland saw large-scale deep mining

FIGURE 14.11 The Nent Haggs mine water treatment scheme. Artists impression (left) and aerial photo in May 2023 (right).

FIGURE 14.12 Ochreous mine water discharges (left) and discoloration of Saltburn beach (right) before remediation.

for well over 100 years. When local ironstone mining operations stopped in the 1960s, the pumps that removed water from the mines were turned off and removed. Water made its way into the old mine workings and eventually reached the surface. In 1999, a large, iron-rich outbreak occurred that caused significant environmental damage to Saltburn Gill and Skelton Beck (Figure 14.12), with severe effects on fish and invertebrates, and discolored the beach (Lord & Mason, 2011). Iron concentrations in the stream rose to over 1,000 times the Environmental Quality Standard with more than 100 tonnes of iron entering the stream each year.

14.3.6.1 Community Group, Funding and Benefits

Local residents started the efforts to clean up the pollution by forming the Saltburn Gill Action Group in 2005. They secured initial funding from the Landfill Communities Fund and the Environment Agency to carry out a feasibility study. This demonstrated that in addition to the existing pollution, there was a serious risk of a major mine water outbreak that would further harm the local environment and economy. Following an intervention in Parliament by Tom Blenkinsop, then the local Member of Parliament (Hansard, 2011), that helped extend the powers of the Coal Authority, Defra secured funding to allow the Coal Authority and Environment Agency to begin to develop a long-term treatment solution. A key driver for the treatment scheme was a benefits assessment, which calculated that cleaning up the pollution has the potential to add £10.5 million into the local economy over the next 25 years. This significantly outweighs the predicted £5.4 million lifetime cost of building and running the treatment plant (Defra, 2023c).

14.3.6.2 The Treatment Scheme – A Staged Approach

To stop the existing mine water discharges, groundwater levels in the mine workings have to be lowered by pumping. Following extensive pumping tests, groundwater levels have to be controlled at about 50m below ground in the abstraction borehole with a long-term pumping rate of about 25 l/s. The Coal Authority built the first stage of the treatment scheme in 2013: a settlement lagoon where iron precipitates out of the pumped mine water to form a sludge at the bottom. A temporary chemical dosing system helped to remove more iron while the second stage was being built and from January 2014, the stream was no longer orange (Figure 14.13).

The success of this first stage confirmed that a larger passive treatment scheme would work, thus avoiding the need for an active chemical treatment with much higher operating costs. Construction of a new abstraction borehole, a further three settlement ponds, a sludge drying bed and a reed bed wetland was completed in May 2015 (Figure 14.14). In 2019, a 100kW solar photovoltaic array was installed to off-set some of the financial and carbon costs of pumping; this is helping to ease the impact of energy inflation on operational costs (Coal Authority, 2019).

FIGURE 14.13 Saltburn Gill in 2009 (left), and in 2015 after remediation (right).

FIGURE 14.14 The settlement ponds (left), the cascade (centre) and reedbed (right).

Today, more than 1 million cubic meters of mine water is pumped out of the mine each year, which first flows over an aeration cascade that adds oxygen to the water and makes the dissolved iron precipitate to form tiny particles. In the settlement ponds, these particles settle out over a couple of days and leave a sludge, iron ochre, which accumulates in the ponds. After the mine water has passed through the settlement ponds, it then flows by gravity into the reed bed wetland (Figure 14.14). The reeds act like a natural filter, removing the majority of the remaining iron so that clean water is then returned to the stream. As well as filtering and cleaning the mine water, the reed beds provide a valuable new habitat for wildlife; ducks, moorhens (including chicks) and geese took up residence in 2016.

14.3.6.3 Environmental Improvements

The Environment Agency carried out ecological surveys in 2015 which showed a "significant ecological recovery" for invertebrates and fish following the first stage of treatment. Further surveys in 2016 and 2017 confirmed that the stream continues to move towards good status from the "bad" condition it was in prior to the treatment scheme being built with increasing numbers and diversity of invertebrates and fish. In 2016, a local resident photographed sea trout migrating up the river.

14.4 CHALLENGES WITH METAL MINES

Most English rivers that are polluted by abandoned metal mines remain to be cleaned up and adoption of the new Environment Act target in January 2023 will mean (and require) a step-change in progress. Minimizing costs to taxpayers is essential and so investigations continue to find more cost-effective ways of quantifying the metal inputs from the multiple sources and then implementing remedial measures.

14.4.1 Ecological Impacts and Recovery

Environmental quality standards are used as targets for clean up since compliance means the risk of harming aquatic life is minimal. However, in many rivers, it will be technically infeasible to return the water environment to a pre-mining condition after centuries of exposure to elevated metals and fundamental alteration to groundwater hydrology. In some rivers, the ecological community has adapted to centuries of exposure to elevated metals (Environment Agency, 2012c). The standard UK metrics used to assess the health of aquatic biology were devised for sanitary and organic pollution, and are less good at identifying metal impacts. Furthermore, standards are focused only on the water column and do not take account of harm from exposure to contaminated sediments. Sediment metal (Zn, Cd, Pb, Cu) concentrations are regularly in the tens of thousands downstream of mining areas (Environment Agency, 2008b; Gozzard *et al.,* 2011). Recent research has demonstrated that uptake from such sediments is the main exposure pathway for some macro-invertebrates (Jones *et al.,* 2018), which are at the base of the aquatic food chain. It seems likely that further consideration of "how clean is clean enough?" will be needed in the UK and across the world to inform practical decisions about how to remediate rivers polluted by abandoned metal mines.

14.4.2 Benefits and Stakeholder Engagement

A key test for any remedial measure is whether the benefits outweigh the costs. The UK Government considers that the water environment is a public good that provides services to society and ecosystems, and therefore it needs to be protected from deterioration and enhanced. A standardized approach to monetizing the benefits of a healthier more resilient water environment has been developed (Environment Agency, 2015).

Since almost all UK metal mines were abandoned by the early 20th century, communities often do not recognize the pollution in their local rivers since it is rarely visible to the naked eye. This lack of understanding, coupled with a desire to avoid major developments in rural upland locations, has led to an unexpected degree of resistance to proposed mine water treatment schemes in some places. This has been compounded by concerns that the currently preferred bacterial sulfate reduction treatment ponds may generate odor nuisance. The methodology for selecting potential treatment sites and how engagement takes place with local communities have been changed to encourage substantially more consultation and follow the Environment Agency's "Working with Others" approach. Short videos and simple websites have been created to explain the pollution issue and how remediation schemes will operate, for example https://bit.ly/WAMM_youtube and https://consult.environment-agency.gov.uk/north-east/nenthead-mwts/. It is to be hoped that the successful operation of the Nent Haggs scheme (Environment Agency, 2023c) will demonstrate that these nature-based solutions can effectively remove metals and clean up river pollution without causing odor nuisance.

14.4.3 Treatment Technology Development

The WAMM Programme carries out R&D with an overarching objective to minimize the whole life cycle financial and environmental costs of remedial interventions, particularly mine water treatment schemes. Trials have been carried out in the UK with several potential passive or semi-passive technologies for removing metals like zinc and cadmium from mine water, including ochre pellets (Mayes *et al.,* 2009b), Biochar (Hodgson *et al.,* 2016), hydrous ferric oxide pellets and vertical flow reactors (that co-precipitate iron, manganese and other metals), compost-based reactive media (Gandy *et al.,* 2016), ion exchange fabrics, pH adjustment with waste materials (steel slags, fish shells) and sorption by organic wastes from the whisky industry.

A large pilot scale passive vertical flow reactor, based on a concept by Cardiff University (Kennedy *et al.,* 2023), started operating in 2023 in Cornwall (Coal Authority, 2022) treating up to 8 l/s (about 50% of the average mine water flow). The mine water passes down through a bed of gravel that encourages the iron and manganese in the water to precipitate out and coat the gravel with iron-manganese hydroxides. This coating is very effective at capturing the cadmium, copper and zinc which pollute about 14km of rivers. Further trials of various technologies are planned and in progress, particularly how to manage odors and minimize the land required for treatment schemes (Bamforth *et al.,* 2023).

14.4.3.1 Diffuse Inputs

Monitoring has demonstrated that diffuse inputs from waste heaps and groundwater are a significant source of metals, particularly as river flows increase (Gozzard *et al.,* 2011; Jarvis *et al.,* 2019; Jones *et al.,* 2013; Mayes *et al.,* 2021; Tame *et al.,* 2017). Climate change will lead to increasing intensity of rainfall and river flows. Investigations continue into sustainable, long-term interventions to decrease the mobilization of metals into rivers. One promising approach is to improve soil conditions and encourage plants to grow on highly contaminated mine wastes as this will minimize infiltration and erosion by rainfall, which mobilizes dissolved and particulate metals into rivers. Several large-scale vegetation trials are in progress in which different soil amendments and seed mixtures have been applied by hydroseeding. In the most successful applications, the mine wastes have been transformed into organic rich soils with good plant growth, increasing biodiversity, sequestering carbon dioxide whilst also decreasing the release of metals to rivers. Examples include the Garrigill mine waste project in Cumbria (Figure 14.15, Coal Authority, 2020c), where 100,000 cubic meters of wastes were re-profiled before vegetation trials; this project was highly commended in the 2022 Chartered Institute of Ecology and Environmental Management Awards (CIEEM, 2022).

14.5 COAL MINES

The UK coalfields cover about 25,000 square kilometers and contain seven million properties (about 25% of all properties). There are 170,000 recorded coal mine entries, and 130,000 properties are within 20 meters of a mine entry. Therefore, the coal mining legacy affects the public, planners, developers, infrastructure and landowners. Between 1948 and 1994, deep coal mines were run by the UK Government. When the last few mines were privatized in 1994, the Coal Authority was created to manage liabilities associated with the former nationalized industry (Environment Agency, 2021a). Between 1985 and 2003, more than 150 collieries closed with the loss of more than 250,000 jobs (Hansard, 2003). Water pollution was not initially recognized as being a significant issue (the focus was primarily on the effects of subsidence on buildings and public safety). However, as dewatering pumps had been switched off in many deep mines in the previous decade, rising mine waters began to pollute rivers with bright orange discharges (Figure 14.16).

FIGURE 14.15 The Garrigill mine site: after re-profiling (left) and 6 months after vegetation trials started (right).

FIGURE 14.16 Jackson Bridge (West Yorkshire) discharge (left) in February 2014, and the polluted river 12 km downstream (right).

Pressure from coalfield communities to deal with the wider legacy of closure of the mines, supported through technical investigations led by Professor Paul Younger (then at Newcastle University), identified the potential scale of water pollution to rivers and groundwater drinking water supplies (Younger *et al.*, 2002). The UK Government accepted their obligation to manage this issue on behalf of the UK and therefore provides funding to the Coal Authority via the Department for Energy Security and Net Zero (DESNZ).

14.5.1 Prioritization of Environmental Impacts

The first step in dealing with existing visible river pollution was to assess mine water discharges in a methodical way so that remedial measures could be prioritized and planned. In the 1990s, the predecessor to the Environment Agency (the National Rivers Authority) assessed hundreds of known coal mine discharges using a standardized protocol based on environmental impacts rather than create a comprehensive inventory of the thousands of known mine sites. If more than 500 meters of river was polluted, these discharges were included on the priority list for action. This list forms the basis of the Coal Authority's remediation program.

There is a parallel preventative program to manage water in those coalfields where mine waters are still rising but have not yet caused pollution of rivers, or the groundwater that overlies some coalfields. In some areas, drinking water supplies come from aquifers that lie stratigraphically above the abandoned coal mines. Since closure, water in the mines has been rebounding and if not controlled through renewed pumping, the aquifers will be polluted, usually with hypersaline water that would be very expensive to treat. Currently, the Coal Authority is protecting the drinking water supplies for about 500,000 people.

14.5.2 Treatment of Coal Mine Discharges

The Coal Authority operates more than 70 treatment systems in the UK, protecting over 350 km of rivers and cleaning up about 120 billion liters of mine water each year. Almost all of these are "passive" systems relying on aeration and settlement of iron hydroxides followed by aerobic reedbeds, although many require the mine water to be pumped to the treatment system (Figure 14.17).

Design is largely based on long-standing guidance developed by the United States Bureau of Mines (Hedin *et al.*, 1994). These schemes are extremely effective at removing iron from the mine water and also provide significant amenity and biodiversity benefits. A number of schemes have won awards for creating new habitats for insects, birds and wildflowers such as the Mouse Water scheme in Scotland (Environment Agency, 2008a). Where the mine water is particularly aggressive or has higher flows, active treatment plants have been built in which chemicals are added to raise the

FIGURE 14.17 Coal Authority passive treatment scheme: Taff Merthyr settlement ponds (left) and Lamesley aerobic reedbeds (right).

pH and precipitate the iron and other metals as hydroxides, usually as a high-density sludge. Such systems are considerably more expensive to build and operate than passive schemes, although they generally require a much smaller land area.

14.5.3 Mine Water Heat

A relatively new area of work for the Coal Authority is to harness heat stored in abandoned coal mine waters to generate sustainable low carbon heat for homes and workplaces and help tackle climate change and support the transition to net zero. Several projects are already in operation, with more in development (Coal Authority, 2021).

14.6 CONCLUSIONS

The UK is managing the pollution legacy created by centuries of mining for coal, metals and industrial minerals by focusing on the most significant environmental impacts and where the environmental and economic benefits are greatest. More than 2,000 km of rivers and several major public water supply aquifers are polluted or threatened by abandoned mines. More than 75 mine water treatment schemes are operating at abandoned metal and coal mines, primarily applying passive technologies, such as settlement lagoons, compost-based ponds and reedbeds. Measures to control inputs of diffuse metals are being installed.

For the formerly publicly owned and operated abandoned coal mines, the UK Government has always recognized its liabilities to deal with pollution and other impacts. The Coal Authority is working with partners to unlock heat within abandoned coal mines, to help transform how homes and workplaces are heated and to deliver net zero targets.

For abandoned metal mines, the recent adoption by Parliament of a legally binding target to halve the length of rivers polluted by harmful metals from these mines in England by 2038 has strengthened the work being done to clean up this legacy of Britain's industrial past whilst requiring a step-change in the pace of mine water remediation. The number of metal mine water treatment schemes will need to increase from four to about 40 by 2038, with a similar number of diffuse interventions. Passive technologies utilizing nature-based solutions are preferred due to lower whole-life cycle costs and a smaller carbon impact than active technologies.

The main challenges to cleaning up rivers polluted by abandoned metal mines are developing and implementing cost-effective treatment technologies, clarifying "how clean is clean enough?" in rivers that have been polluted for centuries, securing funding (primarily from Government) to operate these schemes in perpetuity and simplifying the process of finding suitable locations to build treatment schemes that are acceptable to local communities, particularly in rural upland environments.

The groundbreaking Environment Act metal mines target requires a step-change in the pace of mine water remediation in England, driving innovation in treatment and demonstrating transferable technologies to support remediation of the global problem of rivers polluted by abandoned metal mines.

ACKNOWLEDGEMENTS

The work described here stems from a partnership between the Environment Agency, the Coal Authority and others with funding from Defra, the North East Local Enterprise Partnership and DESNZ. Particular thanks are due to Professor Adam Jarvis (Newcastle University), Nick Cox (Coal Authority) and many Environment Agency staff. The views expressed are those of the author and not necessarily the Environment Agency or other partners.

REFERENCES

Bamforth, S.M., Cox., N., & Potter, H.A.B. 2023. *Innovative passive treatment technologies for UK metal mine waters: results from field scale trials*. Proceedings of the International Mine Water Association Conference. 17–21 July 2023, Newport, Wales.

Batty, L.C. 2005. The potential importance of mine sites for biodiversity. *Mine Water and the Environment,* **24(2),** 101–103.

CIEEM, 2022. Water and abandoned metal mines: Calaminarian grassland mitigation in the North Pennine Moor mines. https://cieem.net/wp-content/uploads/2023/06/2022-Awards-Winners-and-Finalists.pdf (accessed 26 November 2023).

Coal Authority. 2016a. Metal mine water treatment. https://www.gov.uk/government/collections/metal-mine-water-treatment (accessed 25 November 2023).

Coal Authority. 2016b. Haggs mine water treatment scheme. https://www.gov.uk/government/collections/nent-haggs-mine-water-treatment-scheme-proposed (accessed 26 November 2023).

Coal Authority. 2019. Solar energy landmark for the Coal Authority. https://www.gov.uk/government/news/solar-energy-landmark-for-the-coal-authority (accessed 26 November 2023).

Coal Authority. 2020a. Wheal Jane mine water treatment scheme. https://www.gov.uk/government/case-studies/wheal-jane-mine-water-treatment-scheme (accessed 25 November 2023).

Coal Authority. 2020b. Project cuts pollution in Northumberland. https://www.gov.uk/government/news/project-cuts-pollution-in-northumberland (accessed 25 November 2023).

Coal Authority. 2020c. Garrigill Culvert project in North Pennines tackles pollution caused by historical metal mining. https://youtu.be/kCqSGyektgg?si=Gq21Zm6Cf79zH_VK (accessed 26 November 2023).

Coal Authority. 2021. Mine water heat. https://www.gov.uk/government/collections/mine-water-heat (accessed 26 November 2023).

Coal Authority. 2022. Coombe mine water treatment scheme. https://www.gov.uk/government/collections/coombe-mine-water-treatment-scheme (accessed 26 November 2023).

Cornish Metals. 2023. South Crofty project. https://cornishmetals.com/projects/uk/south-crofty/ (accessed 25 November 2023).

Cornwall Heritage. 2023. The Great County Adit: Cornwall's marvel beneath our feet. https://www.cornwall-heritage.com/ertach-kernow-blogs/ertach-kernow-the-great-county-adit-cornwalls-marvel-beneath-our-feet/ (accessed 25 November 2023).

Cornish Mining World Heritage Site. 2023. The Cornwall and West Devon mining landscape world heritage site. https://www.cornishmining.org.uk/ (accessed 23 November 2023).

Culledge, R., Potter, H., & Cox, N. 2019. Improving water quality in the metal polluted River Nent, West Allen and South Tyne. https://waterprojectsonline.com/custom_case_study/wamm-water-abandoned-metal-mines-programme/ (accessed 26 November 2023).

Defra. 2011. Water for life. https://www.gov.uk/government/publications/water-for-life (accessed 26 November 2023).

Defra. 2023a. Environmental improvement plan 2023. https://www.gov.uk/government/publications/environmental-improvement-plan (accessed 25 November 2023).

Defra. 2023b. Plan for water: Our integrated plan for delivering clean and plentiful water. https://www.gov.uk/government/publications/plan-for-water-our-integrated-plan-for-delivering-clean-and-plentiful-water (accessed 25 November 2023).

Defra. 2023c. The Environmental Targets (Water) (England) Regulations 2023 – Impact assessment. https://www.legislation.gov.uk/uksi/2023/93/impacts (accessed 25 November 2023).

DIFFUSE. 2023. DIFFUSE pollution from historic metal mining in the UK. https://www2.bgs.ac.uk/diffuse/index.html (accessed 25 November 2023).

Environment Agency. 2008a. Abandoned mines and the water environment. https://www.gov.uk/government/publications/abandoned-mines-and-the-water-environment (accessed 25 November 2023).

Environment Agency. 2008b. Assessment of metal mining contaminated river sediments in England and Wales. https://www.gov.uk/government/publications/assessment-of-metal-mining-contaminated-river-sediments-in-england-and-wales (accessed 26 November 2023).

Environment Agency. 2012a. Prioritisation of abandoned non-coal mine impacts on the environment. II. The national picture. https://www.gov.uk/government/uploads/system/uploads/attachment_data/file/290866/scho1111bubx-e-e.pdf (accessed 25 November 2023).

Environment Agency. 2012b. Prioritisation of abandoned non-coal mine impacts on the environment. XII. Future management of abandoned non-coal mine water discharges. https://www.gov.uk/government/uploads/system/uploads/attachment_data/file/290871/scho1111buvh-e-e.pdf (accessed 25 November 2023).

Environment Agency. 2012c. Ecological indicators for abandoned mines. https://www.gov.uk/government/publications/ecological-indicators-for-abandoned-mines (accessed 26 November 2023).

Environment Agency. 2014. Inventory of closed and abandoned mining waste facilities causing serious environmental impacts. https://www.gov.uk/government/publications/inventory-of-closed-mining-waste-facilities (accessed 25 November 2023).

Environment Agency. 2015. Update to the river basin management plans: impact assessment. https://www.gov.uk/government/uploads/system/uploads/attachment_data/file/500583/Impact_assessment_update_to_the_RBMPs_for_England_s_water_environment__2015_.pdf (accessed 26 November 2023).

Environment Agency. 2021a. Mine waters: Challenges for the water environment. https://www.gov.uk/government/publications/mine-waters-challenges-for-the-water-environment (accessed 25 November 2023).

Environment Agency. 2021b. Nenthead mine water treatment scheme – Supporting reports and surveys. https://consult.environment-agency.gov.uk/north-east/nenthead-mine-water-treatment-scheme-evidence/ (accessed 25 November 2023).

Environment Agency. 2023a. Bolingey Stream at Bolingey: Water quality. https://environment.data.gov.uk/water-quality/view/sampling-point/SW-82311912 (accessed 25 November 2023).

Environment Agency. 2023b. Nent at Alston: Water quality. https://environment.data.gov.uk/water-quality/view/sampling-point/NE-43200046 (accessed 26 November 2023).

Environment Agency. 2023c. Major £9m project to reduce abandoned metal mine pollution. https://www.gov.uk/government/news/major-9m-project-to-reduce-abandoned-metal-mine-pollution (accessed 26 November 2023).

Environment Agency. 2023d. Tackling the metal pollution legacy of Britain's mining history. https://environmentagency.blog.gov.uk/2023/06/06/tackling-the-metal-pollution-legacy-of-britains-mining-history/ (accessed 26 November 2023).

Erichsen Jones, J.R. 1964. *Fish and river pollution*. Butterworths, London. 200pp.

Explore Cornwall. 2023. https://explorecornwall.org/west-chiverton-mine/ (accessed 23 November 2023).

Gandy, C.J., Davis, J.E., Orme, P.H.A., Potter, H.A.B., & Jarvis, A.P. 2016. Metal removal mechanisms in a short hydraulic residence time subsurface flow compost wetland for mine drainage treatment. *Ecological Engineering,* **97,** 170–185.

Gozzard, E., Mayes, W.M., Potter, H.A.B., & Jarvis, A.P. 2011. Seasonal and spatial variation of diffuse (non-point) source zinc pollution in a historically metal mined river catchment, UK. *Environmental Pollution,* **159,** 3113–3122. http://dx.doi.org/10.1016/j.envpol.2011.02.010

Hansard. 2003. Memorandum by the Coalfields Communities Campaign. https://publications.parliament.uk/pa/cm200203/cmselect/cmodpm/1169/1169we04.htm (accessed 26 November 2023).

Hansard. 2011. Former metal mines. https://hansard.parliament.uk/Commons/2011-02-09/debates/11020962000001/FormerMetalMines (accessed 26 November 2023).

Hedin, R.S., Nairn, R.W., & Kleinmann, R.L.P. 1994. *Passive treatment of coal mine drainage.* US Bureau of Mines Information Circular No. 9389, United States Department of the Interior, 35 pp.

Historic England. 1997. Lead mines, ore works and smeltmill at Nenthead. https://historicengland.org.uk/listing/the-list/list-entry/1015858 (accessed 26 November 2023).

Hodgson, E., James, A., Rao Ravella, S., Thomas Jones, S., Perkins, W., & Gallagher, J. 2016. Optimisation of slow-pyrolysis process conditions to maximise char yield and heavy metal adsorption of biochar produced from different feedstocks. *Bioresource Technology,* **214,** 574–581.

Hudson-Edwards, K.A., Macklin, M.G., Curtis, C.D., & Vaughan, D.J. 1996. Processes of formation and distribution of Pb-, Zn-, Cd-, and Cu-bearing minerals in the Tyne Basin, Northeast England: Implications for metal-contaminated river systems. *Environmental Science and Technology,* **30,** 72–80.

Jarvis, A., Gandy, C., Bailey, M., Davis, J., Orme, P., Malley, J., Potter, H., & Moorhouse, A. 2015. *Metal removal and secondary contamination in a passive metal mine drainage treatment system.* Proceedings of the 10th International Conference on Acid Rock Drainage (ICARD). 21–24 April 2015, Santiago, Chile.

Jarvis, A.P., Davis, J., Orme, P., Potter, H.A.B., & Gandy, C. 2019. Predicting the benefits of mine water treatment under varying hydrological conditions using a synoptic mass balance approach. *Environmental Science & Technology,* **53(2),** 702–709.

Jones, A., Rogerson, M., Greenway, G., Potter, H.A.B., & Mayes, W.M. 2013. Mine water geochemistry and metal flux in a major historical Pb-Zn-F orefield, the Yorkshire Pennines, UK. *Environmental Science and Pollution Research,* **20(11),** 7570–7581.

Jones, J.I., Murphy, J.F., Collins, A.L., Spencer, K.L., Rainbow, P.S., Arnold, A., Pretty, J.L., Moorhouse, A.M.L., Aguilera, V., Edwards, P., Parsonage, F., Potter, H.A.B., & Whitehouse, P. 2018. The impact of metal-rich sediments derived from mining on freshwater stream life. *Reviews of Environmental Contamination and Toxicology,* **248,** 111–189.

Kennedy, J., Okeme, I., & Sapsford, D. 2023. Novel approaches to passive and semi-passive treatment of zinc-bearing circumneutral mine waters in England and Wales. In Fosso-Kankeu, E., & Mamba, B.B., Eds. *Hybridized technologies for the treatment of mining effluents.* Scrivener Publishing LLC, 147–175.

Lord, R., & Mason, D. 2011. Metal mining - who pays for the clean-up? *Geoscientist, June 2011.* The Geological Society of London.

Macklin, M.G. 1992. Metal contaminated soils and sediment: A geographical perspective. In Newson, M.D., Ed. *Managing the human impact on the natural environment: Patterns and processes.* Belhaven Press, London, 174–195.

Macklin, M.G., Brewer, P.A., Hudson-Edwards, K.A., Bird, G., Coulthard, T.J., Dennis, I.A., Lechler, P.J., Miller, J.R., & Turner, J.N. 2006. A geomorphological approach to the management of rivers contaminated by metal mining. *Geomorphology,* **79,** 423–447.

Macklin, M.G., Thomas, C.J., Mudbhatkal, A., Brewer, P.A., Hudson-Edwards, K.A., Lewin, J., Scussolini, P., Eilander, D., Lechner, A., Owen, J., Bird, G., Kemp, D., & Mangalaa, K.R. 2023. Impacts of metal mining on river systems: a global assessment. *Science,* **381(6664),** 1345–1350.

Mayes, W.M., Johnston, D., Potter, H.A.B., & Jarvis, A.P. 2009a. A national strategy for identification, prioritisation and management of pollution from abandoned non-coal mine sites in England and Wales. I. Methodology development and initial results. *Science of the Total Environment,* **407(21),** 5435–5447.

Mayes, W.M., Potter, H.A.B., & Jarvis, A.P. 2009b. Novel approach to zinc removal from circum-neutral mine waters using pelletised recovered hydrous ferric oxide. *Journal of Hazardous Materials,* **162,** 512–520.

Mayes, W.M., Potter, H.A.B., & Jarvis, A.P. 2010. Inventory of aquatic contaminant flux arising from historical metal mining in England and Wales. *Science of the Total Environment,* **408,** 3576–3583.

Mayes, W.M., Potter, H.A.B., & Jarvis, A.P. 2013. Riverine flux of metals from historically mined orefields in England and Wales. *Water, Air, & Soil Pollution,* **224,** 1–14.

Mayes, W., Anton, A., Thomas, C., Potter, H., Rudall, S., Amezaga, J., Gandy, C., & Jarvis, A. 2015. *National assessment of sediment-related diffuse mining pollution in England and Wales.* Proceedings of the 10th International Conference on Acid Rock Drainage (ICARD). 21–24 April 2015, Santiago, Chile.

Mayes, W.M., Perks, M.T., Large, A.R.G., Davis, J.E., Gandy, C.J., Orme, P.A.H., & Jarvis, A.P. 2021. Effect of an extreme flood event on solute transport and resilience of a mine water treatment system in a mineralised catchment. *Science of The Total Environment,* **750,** 141693.

Merrington, G., & Alloway, B.J. 1994. The transfer and fate of Cd, Cu, Pb and Zn from two historic metalliferous mine sites in the U.K. *Applied Geochemistry,* **9(6),** 677–687.

Mighanetara, K., Braungardt, C.B., Rieuwerts, J.S., & Azizi, F. 2009. Contaminant fluxes from point and diffuse sources from abandoned mines in the River Tamar catchment, UK. *Journal of Geochemical Exploration,* **100,** 116–124.

Moorhouse, A., Harris, R., Wyatt, L., Hill, S., & Potter, H. 2015. *Up-scaling from trial to operation: practicalities of construction and permitting, Force Crag mine.* Proceedings of the 10th International Conference on Acid Rock Drainage (ICARD). 21–24 April 2015, Santiago, Chile.

NCM. 2023. National coal mining museum. https://www.ncm.org.uk/ (accessed 25 November 2023).

Nenthead Mines. 2023. http://www.nentheadmines.com/ (accessed 25 November 2023).

Pirrie, D., Power, M.R., Wheeler, P.D., Cundy, A., Bridges, C., & Davey, G. 2002. Geochemical signature of historical mining: Fowey Estuary, Cornwall, UK. *Journal of Geochemical Exploration,* **76,** 31–43.

Tame, C., Hudson-Edwards, K.A., & Potter, H.A.B. 2017. Weathering of zinc (Zn)-bearing mine wastes in a neutral mine drainage setting, Gunnerside Gill, Yorkshire. *Procedia Earth and Planetary Science*, **17,** 284–287.

UK Parliament. 2023. https://questions-statements.parliament.uk/written-statements/detail/2022-12-16/hlws449 (accessed 25 November 2023).

Wilson, P.N. 1963. The Nent Force level. *Transactions of the Cumberland and Westmorland Antiquarian and Archaeological Society,* **63,** 253–280.

Younger, P.L., Banwart, S.A., & Hedin, R.S. 2002. *Mine water: hydrology, pollution, remediation.* Kluwer Academic Publishers, Dordrecht.

Younger, P.L., Coulton, R.H., & Froggatt, E.C. 2005. The contribution of science to risk-based decision-making: lessons from the development of full-scale treatment measures for acidic mine waters at Wheal Jane, UK. *Science of The Total Environment,* **338(1–2),** 137–154.

Younger, P.L., & Potter, H.A.B. 2012. *Parys in Springtime: Hazard management and steps towards remediation of the UK's most polluted acidic mine discharge.* Proceedings of the 9th International Conference on Acid Rock Drainage (ICARD). 20–26 May 2012, Ottawa, Canada.

15 Mining in the United States

Kenneth S. Sajwan

15.1 INTRODUCTION

Mining occurs in many places around the world including the United States. In South America, mining is particularly active in the Amazon region, Guyana, Suriname and other countries. Mining in the United States has been active since the beginning of colonial times, becoming a major industry in the 19th century with a number of new mineral discoveries causing a series of mining rushes. In 2015, the value of coal, metals and industrial minerals mined in the United States was US$109.6 billion (USGS, 2015).[1]

The US mining industry consists of the search for, extraction, beneficiation and processing of naturally occurring solid minerals from the earth. These mined minerals include coal and metals such as iron, copper or zinc, while industrial minerals like potash, limestone and other crushed rocks are particularly sought after. In the year 2018, the United States had approximately 6,299 sand and gravel mines that were active. Furthermore, the number of active mines in the United States in 2018, by commodity* amounted to 905 for non-metal and 905 and 295 for metal.[2]

There are six general types of surface mining techniques being used in the United States: pit/quarry, area (modification of open pit), strip, contour, mountain top removal and placer. The choice as to what type is employed depends on the terrain, nature and size of the deposit. Area and strip-mining techniques disturb the greatest tracks of land because they are generally used on relatively thin, shallow lands, and much depends on the cover of extensive areas.[3]

It has been reported that US mines produced approximately $90.4 billion in mineral commodities in 2021. This was a $9.7 billion increase over the 2020 revised total of $80.7 billion, which was announced by the US Geological Survey in 2022.[4] The US Mines produced approximately $82.3 billion in minerals in 2020, which was about $1.5 billion lower than the 2019 revised total of $83.7 billion.[5] The United States has 25,521 prospective mines. Meanwhile, 73,499 mines were producing at the time the data was entered into USGS records.[6]

California, Nevada, and Colorado are the states with the most mines. The most important mining area in the United States is Nevada as it has an industry value of $7.65 billion (10.26% of the total value for the United States). Principal minerals in order of value are as follows: gold, copper, sand and gravel (construction), stone (crushed) and silver. The largest mine in the United States is Bingham Canyon, also known as Kennecott copper mine, which is located near Salt Lake City, Utah. It is one of the world's biggest man-made excavation open-pit copper mines. The most used minerals are coal, iron and bauxite, which top the list of minerals in the United States. Annual production of these minerals is estimated at: Coal: 7.4 billion tons; Iron: 4.6 billion tons; Bauxite: 289 million tons; Phosphate rock: 276 million tons; and Gypsum: 267.1 million tons. Currently, most US-produced gold is extracted from Nevada's large open-pit heap-leach mines. The remainder is taken from several lode mines and placer mines in Alaska and the western states.[7]

DOI: 10.1201/9781315141855-15

15.2 MAJOR SURFACE-MINEABLE MINERALS

15.2.1 Fuel Minerals

Surface mining and reclamation activities associated with fuel minerals (coal, oil shale and uranium) involve large tracts of land. Further mining activity is not expected to diminish greatly, particularly for coal despite the recent emphasis on getting energy from renewables or environmentally friendly sources.[8]

Coal: Coal is an organic sedimentary rock formed from the compression and alteration of plant residues which accumulated in prehistoric fresh or brackish water swamp. Coal is classified by rank according to its percentage of field carbon and heat content, which is calculated on a mineral-free basis.[9] The differences between ranks are derived from the increased temperature and pressure associated with burial depth. An increased depth of burial causes a progressive decline in the amount of volatile matter and moisture content and subsequently causes an increase in the fixed carbon and heat content of coal. The four classes of heat and fixed carbon content are anthracite, bituminous, subbituminous and lignite.[10]

The world's identified coal resources have been estimated to total 8.6 million metric tons (Table 15.1).[11] It has been reported that the United States contains 25% of these resources[12] The locations of the coal deposits in the United States are shown in Figure 15.1. Of the coal resources in the United States, 396 billion metric tons are classified as reserve base. Montana, Wyoming, North Dakota and Illinois contain 68% of the 124 billion metric tons of reserve base coal which is surface mineable. The data in Tables 15.2 and 15.3 indicates that all of the surface-mineable coal is bituminous or less.

15.2.2 Oil Shale

Oil shale is a fine-grained sedimentary rock containing insoluble organic matter which can yield substantial amounts of oil. Currently, the United States[13] does not produce any oil from oil shale. However, other countries have burned oil shale as a fuel or produced oil or a combustible gas from it for many years. Currently, it has been noted that the world has resources of oil shale totaling approximately 3 trillion barrels of oil. The United States has 67% of the global identified oil shale resources, and the country's shale yields an average of 15 gal or more of oil per ton. Most of these resources are in the facustrine rocks of the Eocene Green River formation located in Colorado, Utah and Wyoming.[14]

Uranium: Uranium is silvery white metal that consists of three semi-stable isotopes ^{238}U, ^{235}U, and ^{234}U. Currently ^{235}U is an important energy source because it releases large amounts of energy upon fission, but it only constitutes 0.7% of natural uranium. The isotope ^{238}U makes up most of the remaining natural uranium (99.3%); however, it is not readily fissionable. ^{238}U can be used after

TABLE 15.1
World Identified Coal Resources[5]

Continent	Identified Resources (million metric tons)
Asia	6349
North America	1560
Europe	562
Africa	73
Oceania	54
South and Central America	18
Total	8616

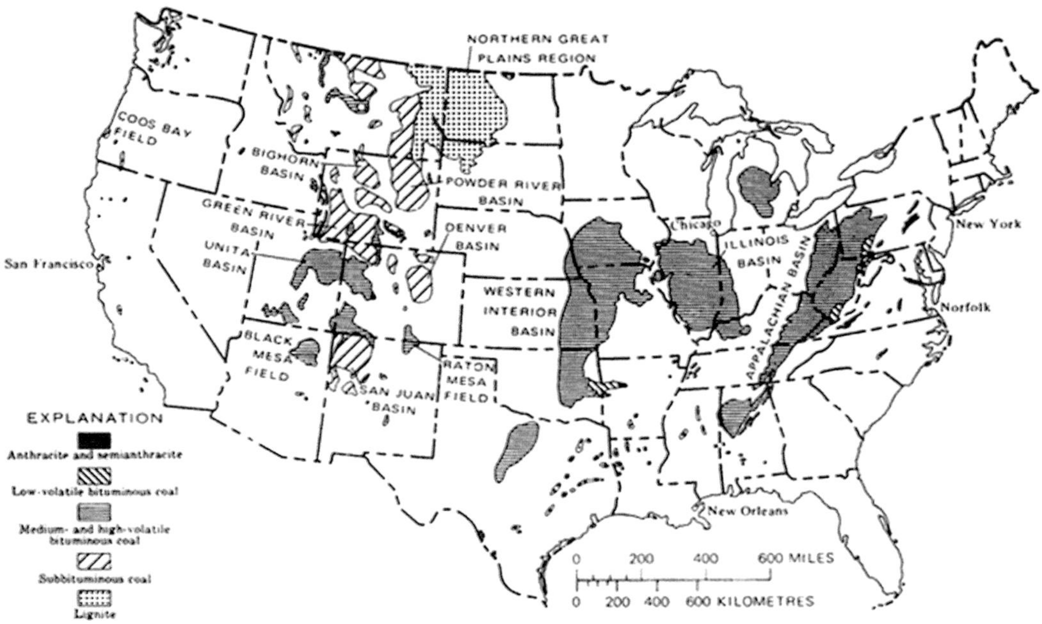

FIGURE 15.1 Major coal deposits in the United States, adopted from Averitt.[7]

TABLE 15.2
Coal Rank by Method of Mining[6,7]

Rank of Coal	Surface Mined (%)	Deep Mined (%)	Total (%)
Anthracite	0.0	1.7	1.7
Bituminous	14.6	38.8	53.4
Subbituminous	15.0	23.5	38.5
Lignite	6.4	0.0	6.4
Total	36.0	64.0	100.0

TABLE 15.3
World Oil Shale Resources[13]

Continent	25-100 (gal/ton)	10-25 (gal/ton)
United States	418	1600
Other	Small	Small
South America	Small	800
Europe	70	6
Africa	100	Small
Asia	90	14
Oceania	Small	1
Total	678	2421

neutron bombardment converts it to ^{239}PU, which is fissionable.[15] The United States has 17% of the world's 1.5 million metric tons of identified recoverable uranium. In 1969 the United States had classified 146,000 metric tons of this resource as the country's reserve.[16] Over 80% of these reserves are located in New Mexico and Wyoming. Over 50% of the reserves are mined by surface-type operations. The surface mining of these reserves will result in the disturbance of approximately 1,700 ha of land. This estimate is based on an average ore grade of 0.23% U3O8 and an average seam thickness of 3 ft.[17]

15.2.3 Non-Fuel Minerals

The amount of area which will be disturbed by the surface mining of non-fuel minerals is more difficult to predict than that of fuel minerals. Information about their lateral extent and thickness is not easily obtained. Reserve base tonnage, however, is available and inferences about surface mine activity can be made.

Aluminum: Aluminum is second only to iron in the world's consumption of metals. It has a high strength-to-weight ratio. This makes it ideal for aircraft, missiles and marine vessels. Aluminum also can also be applied in the major component of abrasives for precision grinding and polishing. Because of its importance to national security, aluminum is one of the four "controlled minerals" in the US Department of Commerce Defense Priorities and Allocation System, which maintain stockpiles.[18]

Clay: Clays are naturally occurring rocks. They can be both unconsolidated and consolidated, composed of very fine particles of minerals and other minerals that are, in fact, impurities. The major components of these clay deposits are kaolinite, halloysite, montmorillonite, palygorskite and illite. The US Bureau of Mines has classified the clay deposits into six categories based on what is produced rather than the raw material. Even if the data are not consistent, there are sufficient clay resources in the United States to meet demands in the foreseeable future.

Copper: Copper deposits can be classified into four types: porphyry, strata bound, massive sulfide and other. In 1983 the ore grade was about 0.65% copper, which was 30% lower than the world average.[19] Even if the new mines contain low grade ore, however, when copper prices are sufficient, mining companies have the ability to produce a profit from these lower-grade ores.[20] The land-based copper resources in the world are approximately 1.6 billion metric tons.[21] Of these resources, 500 million metric tons are considered to be the reserve base.

Gypsum: Gypsum, a dehydrated calcium sulfate with the designation $CaSO_4.2H_2O$. It is the most common naturally occurring sulfate mineral found in sedimentary deposits. Gypsum is also a generic term used for a series of calcium products used in construction, agriculture and other industries. These products include naturally occurring selenite; finely crystalized gypsum; alabaster; massively crystalline gypsum, which is easily carved; satin spar; fibrous silky gypsum; and gypsite. Gypsum crystals are found mixed with clay, and anhydride, $CaSO_4$. Also included is Plaster of Paris, which is manufactured by partially dehydrating gypsum, $CaSO_41/2H_2O$.[22] US domestic and foreign reserves of gypsum are large and considered to be adequate for the foreseeable future. Of the total world reserves, the United States has 726 million metric tons, which are concentrated in five main areas (Table 15.4).

Iron Ore: Iron is the most widely used metal throughout the world because of its abundance in the earth's crust, its great strength and the ability to economically produce it in large amounts. The world's estimated iron resources are 813 billion metric tons, containing approximately 181 billion metric tons of iron.[23] The world reserve base includes 48% of this resource. The grade of US iron ore has declined due to the effects of long-term mining. In 1963, the amount of raw ore needed to produce 1 ton of marketable ore, through beneficiation, had doubled from the amount needed in 1953.[24] This fact, coupled with the fact that 97% of the domestic iron ore is mined by open pit techniques[25] – will amplify future reclamation demands.

TABLE 15.4
World Gypsum Resources[22]

Continent	Reserves	Reserve Base
	(Million Metric Tons)	(Million Metric Tons)
North America		
United States	726	Large
Other	544	na
Total	1,270	
North America	36	na
Europe	816	Large
Africa	73	Large
Asia	91	Large
Oceania	73	na
World Total	2,359	na

Limestone and Dolomite: Limestone and dolomite are carbonate rocks which are mined and used for their physical or chemical properties. The physical properties make them useful as aggregates in construction. Their chemical properties are put to use in the production of lime, Portland cement, and glass, as a flux in metallurgy and in agriculture. Limestone is composed of the mineral calcite, $CaCO_3$, while dolomite rock is composed of the mineral dolomite, Ca $Mg(CO_2)_2$. Industry commonly refers to limestone as dolomite rock while its intermediates are collectively as limestone. Limestone and dolomite occur in more than 10% of the continental surface of the world. In the United States, nearly every state has deposits that are of adequate quality for construction or uses that do not require chemically pure stone. High-calcium limestone and high-purity dolomite make up only a small fraction of all limestone and dolomite formations, and they occur in restricted areas.[26]

15.2.4 Molybdenum

Molybdenum is a refractory metallic element used as an alloying agent in steel, cast iron and superalloys. It is used to improve the hardenability, strength, toughness, wear and corrosion resistance. Molybdenum is also employed in chemical applications, such as catalysts, lubricants and pigments.[26] The world's molybdenum reserves base contains 11 million metric tons and 85% of it is located in the United States, Chile, Canada and Russia. The United States has 50% of the world's reserves, so it is self-sufficient in molybdenum. A total of 90% of the US reserves is contained in low-grade porphyry deposits containing 0.2 to 0.5% molybdenite, MoS, which are mined exclusively for molybdenum.[25] These mines are located in Alaska, Colorado, Idaho, Nevada, New Mexico and Utah. Surface mining techniques are used in 45% of these mines.[26, 27]

15.2.5 Sand and Gravel

In 1982 the total production of sand and gravel in the United States amounted to 563 million metric tons. This was the second-highest annual production tonnage of all non-fuel minerals. Construction sand and gravel account for 90% of the total tonnage. This mineral is mainly used as an aggregate material in concrete, as road base material or as fill. The remaining 4% is industrial sand and gravel which is utilized in glassmaking, foundry work and abrasive materials. Construction sand and gravel have lower grade specifications than industrial sand and gravel.[28] Sand and gravel resources in the United States are large and should be adequate for future needs. The construction sand and gravel reserve base is estimated to be 45 billion metric tons of which 14 billion metric tons are classified as reserves. This is the only mineral commodity produced in all 50 states.[28] Industrial sand and gravel reserves are estimated to total 3 to 4 billion metric tons. This higher-grade material is mined in 40 states, but the majority is produced in the eastern United States, where deposits are the largest. Most of the sand and gravel is mined from open pits, but a few underground industrial sand mines do exist.[28]

15.2.6 Titanium

Titanium is an important strategic material because of its high strength-to-weight ratio and its resistance to hear and corrosion. However, only 5% of the world's titanium is used for metal. The majority of the titanium mined and processed is used as a white pigment in paints, paper, plastics and rubber.[29] The United States has 8% of the world's 272 million metric tons of titanium reserve base, of which 7.3 million metric tons, which are classified as a reserve are located in Florida, New York and Tennessee. Only 2% of these reserves are in the form of rutile. Other US resources of titanium include additional rock deposits of ilmenite found in New York, Minnesota, Wyoming and California and sand deposits of ilmenite and/or rutile in Georgia, Idaho, North Carolina, Oregon and South Carolina, as well as large quantities of accessory rutile in porphyry copper ores and tailings.[29]

15.3 ABANDONED MINES IN THE UNITED STATES

There are approximately 500,000 abandoned hard rock mines in the United States, with estimated cleanup costs as high as $54 billion. Abandoned mines are a problem of increasing magnitude. With coal production at its lowest level in 30 years, abandoned mines throughout the country are causing major environmental problems. They can catch fire, and debris from them can contaminate the water supply. In the United States, most abandoned mines are located in the western states, concentrated in Nevada, Colorado and Arizona. Abandoned uranium mines, on the other hand, are most often located on tribal lands, especially in the "Four Corners" of Colorado, New Mexico, Arizona and Utah. The abandoned mines could be very dangerous, such as drowning in water-filled pits and quarries; it is the most common cause of death in abandoned mines. Other dangers include exposure to radioactive gas and toxic chemicals, being toxic to wildlife dwelling in mines and injury from collapsing surface structures and equipment. There are about 48,529 abandoned coal mines in the United States which are known to pose a threat to the public and the environment.[30]

Oil Shale: Oil shale is a fine-grained sedimentary rock, containing insoluble organic matter which can yield substantial amounts of oil through destructive distillation. Although, there are major oil shale deposits in the United States (Figure 15.2), currently the United States does not produce any oil from oil shale. However, other countries have for many years used oil shale as a fuel or produced oil or a combustible gas from it.[31]

Uranium: Uranium is a silvery white metal that consists of three semi-stable radioisotopes, i.e. ^{238}U, ^{235}U and ^{234}U. Currently, ^{235}U is important because it releases large amounts of energy upon fission, but it only constitutes 0.7% of natural uranium. The ^{238}U makes up most of the remaining natural uranium (99.3%), but it is not readily fissionable. ^{238}U can be used after neutron bombardment converts it to ^{239}Pu, which is fissionable.[32] Major uranium processing facilities in the United States are included in Figure 15.3.

15.4 ABANDONED MINE LANDS: SITE INFORMATION

This section presents information about abandoned hard rock mines and mineral processing sites listed in the Superfund Enterprise Management System (SEMS). This includes sites on Superfund's National Priorities List (NPL).

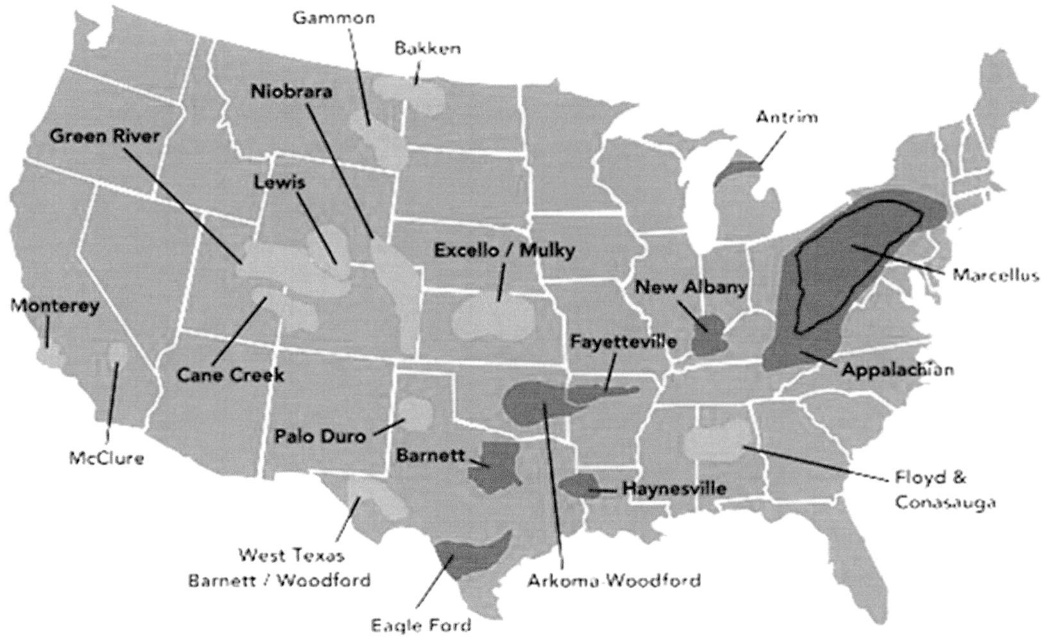

FIGURE 15.2 Major oil shale deposits in the United States.

FIGURE 15.3 Major uranium processing facilities in the United States.

15.5 NPL AND SAA SITES

Table 15.5 summarize the information about mining sites proposed for and listed on the National Priorities List (NPL), as well as mining sites being cleaned up using the Superfund Alternative Approach (SAA). Information provided for each site highlights the state and EPA region in which

TABLE 15.5
Mining Sites: Proposed, Find and Deleted NPL Sites and Non-NPL SAA Sites

Site Name	Region	State	EPA ID	NPL Status	SAA?
ACM Smelter and Refinery	8	Montana	MTD093291599	Final	No
Alcoa (Vancouver Smelter)	10	Washington	WAD009045279	Deleted	No
Alcoa Properties	5	Illinois	ILSFN0508010	non-NPL	Yes
Ambler Asbestos Piles	3	Pennsylvania	PAD000436436	Deleted	No
Anaconda Aluminum Co Columbia Falls Reduction Plant	8	Montana	MTD057561763	Proposed	No
Anaconda Co. Smelter	8	Montana	MTD093291656	Final	No
Annapolis Lead Mine	7	Missouri	MO0000958611	Final	No
Argonaut Mine	9	California	CAD983650011	Final	No
Arsenic Mine	2	New York	NYD982531469	Final	No
Asarco Hayden Plant	9	Arizona	AZD008397127	non-NPL	Yes

the site is located. It provides a current status update of the site and links to additional information about the site.

15.6 WHAT HAPPENS TO ABANDONED COAL MINES?

Throughout the United States, two out of every three acres of reclaimed coal mines have been turned into some form of grassland, which regulators say is often disused. Only 18% was converted into farming, housing, industry or recreation, while 16% was left in a wilder state as forest or undeveloped land. The OSMRE (Surface Mining Control and Reclamation Act) expects to recover $40,000 in permitting fees in 2019. Abandoned Mine Land Reclamation – Under Title IV of SMCRA, effective state, tribal and federal abandoned mine land programs help to ensure the expeditious return of pre-SMCRA mined lands to environmentally friendly post-mining uses.

15.7 WHY ARE ABANDONED MINES A PROBLEM?

With coal production at its lowest level in 30 years, abandoned mines throughout the United States are causing major environmental problems. They can catch fire and debris from them can contaminate the water supply. Many surface and underground abandoned mines, and their associated spoil and refuse piles, provide an ongoing source of acid mine drainage and toxic heavy metals that can have long-term devastating effects on groundwater, communities' water supplies, rivers, streams and aquatic life. Most are in the western states, concentrated in Nevada, Colorado and Arizona. Abandoned uranium mines, on the other hand, are most often on tribal lands, especially in the "Four Corners" of Colorado, New Mexico, Arizona and Utah.

Biological reclamation methods are used to enhance soil quality and regulate environmental pollution and soil pollution caused by mining; and ecological reclamation methods are used to improve land so that it recovers ecologically. The Surface Mining Control and Reclamation Act of 1977 (SMCRA) is the main federal law that regulates the environmental outcomes of coal mining in the United States. SMCRA created two programs: one for regulating active coal mines and a second for reclaiming abandoned mine lands.

15.8 RECLAMATION OF ABANDONED MINE LANDS

Mine reclamation is the process of restoring land that has been mined to a natural or economically usable state. In the United States, mine reclamation is a regular aspect of modern mining practices. Modern mine reclamation minimizes and mitigates the environmental effects of mining. Two centuries of coal mining occurred before the passage of the Surface Mining Control and Reclamation Act (SMCRA) of 1977. As a result, millions of Americans live less than a mile from an abandoned coal mine. OSMRE, through its Abandoned Mine Land (AML) Reclamation Program, addresses the hazards and environmental issues such as degradation posed by these legacy mine sites. Reclamation is the combined process by which adverse environmental damage of surface mining are minimized and mined lands are returned to a beneficial end use. Some components of reclamation include practices that control erosion and sedimentation, stabilize slopes and avoid threats to the habitats of wildlife. Abandoned mines affect the environment in many ways. Many surface and underground abandoned mines, and their associated spoil and refuse piles, provide an ongoing source of acid mine drainage and toxic heavy metals that can have long-term devastating impacts on groundwater, communities' water supplies, rivers, streams and aquatic life. There are clearly numerous environmental problems.[33]

The environmental challenges from coal mining include coal mine accidents, land subsidence, damage to the water environment and sources of water, mining waste disposal and air pollution. These take the form of either environmental pollution or landscape change. OSMRE partners with state and tribal governments, community groups and mining companies to protect the public from mining-related dangers. The return of various types of wildlife to reclaimed sites is one aspect that OSMRE technical experts use to evaluate the conditions of the environment. OSMRE uses this information to accurately assess how well the methods used in a particular project worked for the respective region. Developing and training referring knowledge and technology is an integral part of OSMRE's mission to provide the most up-to-date tools to federal and state government staff, to efficiently and effectively oversee current coal mining practices and to undertake reclamation of abandoned mine lands and economic development. OSMRE has funded projects that take growth throughout the regions and local communities into account. The partnerships and alliances that OSMRE builds upon for remaining and reclamation efforts have proved to be a natural fit and ensure that reclamation of abandoned coal mines and regulation of current coal mining is successful.

15.9 MINE CLOSURE PLANNING

Mine closures require effective and well-targeted planning for the after-mining landscape – all activities required before, during and after the operating life of a mine that are needed to produce an acceptable post-operation landscape economically. Closure performance refers to the activities near and after mine closure and how well activities listed in the closure plan are executed. A mine begins to close the day it opens. Decisions made during the mine planning and development phase – and even earlier, during the exploration phase – have profound effects on the ultimate closure plan, its cost and the resulting landscape remediation performance. These decisions need to be made within a framework of closure planning to realize successful land reclamation.[34]

Activities related to closure planning often include preparation of detailed drawings of disturbed landscape, compilation of baseline information, discussions with regulators and stakeholders on end land use issues, devising supporting research programs and preparation of budgets and schedules. For a mine nearing closure, the closure plan takes the form of a decommissioning process, and it includes details and selection of mitigative technologies (especially for acid rock drainage) and other specific reclamation and closure activities.

A critical element of successful reclamation and of good closure planning is stakeholder involvement. Generally, mines do a poor job of ongoing, meaningful stakeholder consultation, and many would benefit from professional assistance. A particularly useful consultation process involves forming a local committee to provide guidance to one or more mines in the region in their reclamation and closure activities. Here the focus should be on providing ongoing dialogue with stakeholders and identifying the goals important to stakeholders.

The most important benefit of closure planning is identification of critical activities to achieve successful reclamation. Closure planning usually identifies areas of needed research. It also identifies planning constraints (and sometimes opportunities), especially identifying safe methods and locations for tailings storage. These plans provide some assurance that the mine is not "painting itself into a corner" and provide a starting point to estimate financial assurance levels – important to both mines and regulators. It also forms a base case against which future planning changes can be compared. Much of this work falls under the concept of "design for closure" which was introduced 30 years ago.[35]

Typical closure plans are 100 to 200 page documents that require several months to prepare and cost approximately US$200,000. They are usually developed by the supervisor of the relevant environmental department or agency or by a mining, environmental or geotechnical consultant. They are submitted at the request of the regulatory authority and are used by the mine for life-of-mine costing.

Most closure planning documents share the same format:

i. Introduction, Predevelopment conditions.
ii. Mine operations explained.
iii. Landscape performance goals (and regulatory requirements).
iv. Reclamation technologies and methods.
v. Reclamation plans, balances, waste isolation methods.
vi. Long-term monitoring and maintenance plans.
vii. Landscape performance prediction.
viii. Comparison of the performance prediction relative to the goals.
ix. Schedules costs and financial assistance.

By the end of 2020, the number of operating coal mines in the United States fell to 551 mines, the lowest number since coal production peaked in 2008 (Figure 15.4).[36] In 2020, 40 coal mines

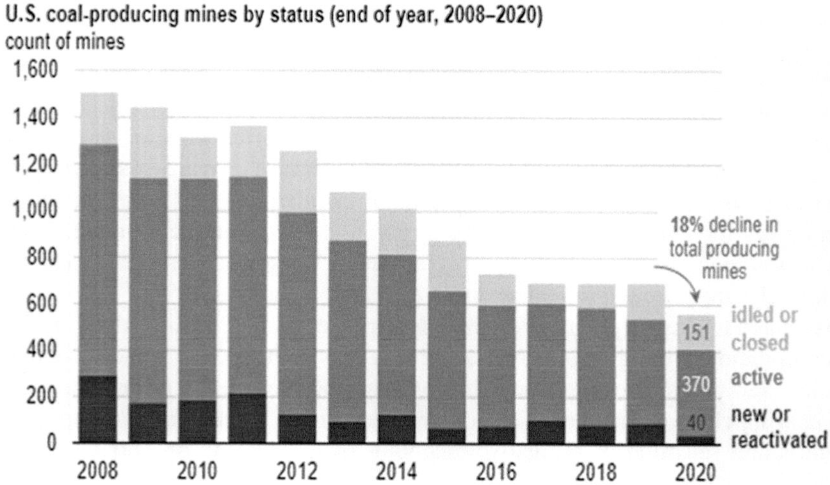

FIGURE 15.4 Major coal-producing mines by status (end of year, 2008–2020) in the United States. **Source:** Graph produced by the US Energy Information Administration (EIA), based on US Department of Labor, Mine Safety and Health Administration, Form 7000-2.[36]

were opened or reactivated, and 151 mines were idled or closed. This overall decline resulted in an 18% annual drop in the total number of producing coal mines from 2019 and a 62% decline since 2008. Shutting down less efficient mines while adding relatively few new mines and reactivating a few idled mines resulted in the reduction in 2020. A comparable drop in producing US coal mines (17%) occurred in 2016, when coal producers faced challenging economic conditions that resulted in numerous bankruptcies and industry consolidation. The declining number of new mines in recent years reflects reduced investment in the coal industry in the United States, less demand for coal internationally and less demand for coal in the American electric power sector. Lower natural gas prices in 2020 made coal less competitive than natural gas for power generation. Because natural gas was cheaper than coal and electricity consumption in general dropped in 2020, US coal-fired generation fell by 20% from 2019. The number of producing coal mines in the United States fell in 2020.

15.10 BIDEN CLOSING THE DOOR ON US MINING

Early in his presidency, Joe Biden promised many "good-paying, union jobs" to fuel his massive push for wind and solar-powered electric vehicles, EV charging stations and the tiny portion of his "infrastructure" package that actually involves road and bridge construction. However, it now appears many, if not most, of those "good-paying, union jobs" will be sacrificed on the altar of assumed "climate catastrophe" and Native American rights. Already Biden's EPA has "reopened" permits issued for two major copper mines, barred mining operations on federal lands and given radical preservationists oversight of mining operations. When he was still a candidate for the White House, Joe Biden had reportedly sent signals to US mine workers that his administration "would support boosting domestic production of metals used to make electric vehicles, solar panels, and other products crucial to his climate plan." According to Reuters' Ernest Scheyder, Biden at the time supported "bipartisan" efforts to foster a domestic supply chain for lithium, copper, rare-earths, nickel and other materials identified by the Interior Department as "critical" for meeting domestic raw materials needs.[37] However, Scheyder, noting that "the Obama Administration enacted rigorous environmental regulations that slowed the U.S. mining sector growth," added the caveat that Biden "has been expected to continue in that vein."

We have the 60-day (or 90-day) "permit review." On his first day in office, President Biden issued a 60-day moratorium on new oil and gas permits on public lands and waters; the order also blocked any new major mining actions as well as all transfers of public lands back to the states.[37] Biden further promised a permanent ban on minerals extraction on public lands. Then, in March 2021, Biden complied with a demand by Arizona Rep. Paul Grijalva (D, AZ) to block a federal land exchange for the Rio Tinto/BHP Group Resolution copper joint venture in Arizona. The decision casts grave doubt on the project, which was planned to produce as much as *40 billion pounds* of copper over 40 years. The Forest Service's revocation of its environmental impact statement, just hours before a deadline, was billed as "a victory for Indigenous rights and conservation activists" – but a loss for US consumers of copper.

15.11 CONCLUSION

The number of abandoned mines in the United States exceeds 500,000, and while every effort is directed toward the rehabilitation of such sites, the associated costs plus the closure of mining companies mean a significant challenge with the effective closure of such mines. Going forward with new mines, the policies linked to closure will ensure no such operations that lead to abandoned mines in the United States.

REFERENCES

1. https://en.wikipedia.org/wiki/United_States. 2015.

2. https://en.wikipedia.org/wiki/United_States. 2018.
3. Hossner, L. R. 1988. *Reclamation of Surface-Mined Lands*. CRC Press.
4. USGS. 2022. *Science of Changing World*. U.S. Department of Interior.
5. Foundation Coal Operations Overview. Foundation Coal Website. Foundation Coal. Retrieved 2009-04-20.
6. https://doi.org/10.1002/ldr.2412, 24 June 2015.
7. State of Nevada Commission on Mineral Resources Division of Minerals. https://ndep.nv.gov/land/mining/mining-info-faq.
8. Doolittle, J. J. and Hossner, L. R. 1988. Resources recoverable by surface mines, in *Reclamation of Surface Mine Lands*, Vol. 1. ERC Press, 2–16.
9. Averitt, P. 1975. *Coal Resources of the United States*, January 1, 1974. Geol, Survey Bull. 1412. U.S. Department of Interior, Washington, DC, I.
10. Averitt, P. 1975. *Coal Resources of United States*, January 1, 1975. Geol. Surv. Bull, 1412. U.S. Department of Interior, Washington, DC, I.
11. Averitt, P. 1973. Coal, in *United States Minerals Resources*. Geol. Surv. Pap. 820. Prof. Brobst, D. A. and Pratt, W. P., Eds. U.S. Department of Interior, Washington, DC, 133.
12. Culbertson, W. C. and Pitman, J. K. 1973. Oil shale, in *The United States Mineral Resources*. Geol. Surv. Prof. Pap. 820. Brobst, D. A. and Pratt, W. P., Eds. U.S. Department of Interior, Washington, DC, 497.
13. Finch, W. I., Butler, A. P., Jr., Armstrong, F. C., Weissenborn, A. F., Staatz, M. H. and Olsen, J. C. 1973. Nuclear fuels, in *United States Mineral Resources*. Geol. Surv. Prof. Pap. 820. Brobst, D. A. and Pratt, Eds. U.S. Department of Interior, Washington, DC, 455.
14. Bieniewski, C. I., Persse, F. H. and Brauch, E. F. 1971. *Availability of Uranium at Various Prices from Resources in the United States*. Bureau of Mines Inf. Circ. 8501. U.D. Department of Interior, Washington, DC, I.
15. Baumgardener, I. H. and McCawley, F. X. 1983. *Minerals Commodity Profiles: Aluminum*. Bureau of Mines. U.S. Department of Interior, Washington, DC, I.
16. Fisher, R. P. 1968. Uranium and vanadium deposits of the Colorado plateau region, in *Ore Deposits of the United States. 1933–1967. Vol. 1*, Ridge, J.D., Ed. American Institute of Mining, Metallurgical, and Petroleum Engineers, Inc., New York, 735.
17. Bienlewski, C. L., Persse, F. H. and Brauch, F. F. 1971. *Availability of Uranium at Various Prices from Resources in the United States*. Bureau of Mines Inf. Circ. 8501. U.S. Department of Interior, Washington, DC, I.
18. Baumgarderner, L. H. and McCawley, F. X. 1983. *Mineral Commodity Profiles: Aluminum*. Bureau of Mines. U.S. Department of Interior, Washington, DC, I.
19. Fisher, R. P. 1968. Uranium and Vanadium deposits of the Colorado Plateau regions, in *Ore Deposits of the Unites States, 1933–1967, Vol. 1*, Ridge, J. W., Ed. American Institute of Mining. Metallurgical, and Petroleum Engineers, Inc., New York, 735.
20. Butterman, W. C. 1983. *Mineral Commodity Profiles*. U.S. Department of Interior. Bureau of Mines, Washington, DC, I.
21. Jolley, J. L. W. 1985. Copper, in *Mineral Facts and Problems*. 1985 Ed. Bureau of Mines Bull. 675. Knoerr, A. W., Ed. U.S. Department of the Interior, Washington, DC, 197.
22. Pressler, J. W. 1979. *Mineral Commodity Profiles: Gypsum*. Bureau of Mines. U.S. Department of Interior, Washington, DC, I.
23. Pressler, J. 1985. *Gypsum in Mineral Facts and Problems*. 1985 Ed. Bureau of Mines Bull. 675. Knoerr, A. W., Ed. U.S. Department of Interior, Washington, DC, 349.
24. Jolley, J. L. W. 1985. Copper, in *Mineral Facts and Problems*, 1985 Ed. Bureau of Mines Bull. 675. Knoeer, A. W., Ed. U.S. Department of Interior, Washington, DC, 197.
25. Klinger, F. L. 1983. *Mineral Commodity Profiles: Iron Ore*. Bureau of Mines, U.S. Department of Interior, Washington, DC, I.
26. Klinger, F. J. 1985. Iron ore, in *Mineral Facts and Problems*. 1985 Ed. Bureau of Mine Bull. 675. Knoeer, A. W. U.S. Department of Interior, Washington, DC, 185.
27. Kummer, J. T. 1979. *Minerals Commodity Profiles: Molybdenum*. Bureau of Mines, U.S. Department of the Interior, Washington, DC, I.
28. Blossom, J. W. 1985. Molybdenum, in *Mineral Facts and Problem*. 1985 Ed. Bureau of Mines Bull. 675. Knoerr, A. W., Ed. U.S. Department of Interior, Washington, DC, 521.

29. Hubbard, H. A. and Erickson, G. E. 1973. Limestone and dolomite, in *United States Mineral Resources*. Geol. Surv. Prof. Pap. 820. Brobst, D. A. and Pratt, W. P., Ed. U.S. Department of Interior, Washington, DC, 357.
30. David, L. L. and Tepordei, V. V. 1985. Sand and gravel, in *Mineral Facts and Problems*. 1985 Ed. Bureau of Mines Bull. 675. Knoerr, A. W., Ed. U.S. Department of Interior, Washington, DC, 689.
31. Lynd, L. E. 1983. Mineral commodity profiles: Titanium. Bureau of Mines, U.S Department of Interior, Washington, DC, I.
32. Owens, L. H. 2017. Final evaluation report – Office of surface mining reclamation and enforcement's oversight of the abandoned mine lands program report. U.S. Department of Interior, March 30.
33. https://en.communications@revealnews.org
34. Reinsalu, E. and Aarna, I. 2015. About technical terms of oil shale and shale oil. *Oil Shale. A Scientific-Technical Journal*, 32(4): 291–292.
35. Abdelouas, A. 2006. Uranium mill tailings: Geochemistry, mineralogy, and environmental impact. *Elements*, 2: 335–341.
36. Rosalyn, B. 2020. The number of producing coal mining fell in 2020. Short Energy Outlook, U.S. Energy Information Administration, Washington, DC.
37. Flanakin, D. 2021. Biden slamming the door shut on much-needed U.S. mining. Climate Change Dispatch, Matthew Brown Associated Press.

16 Mine Wastelands and Abandoned Mines in Zambia

Pumulo Mukube and Stephen Syampungani

16.1 INTRODUCTION

The mining industry has continued to play an important role in the development of civilization throughout Africa. As a result of the long mining history in Africa, most copper and gold mines in countries like the Democratic Republic of Congo (DRC), Zimbabwe and Zambia are located on historical mine sites (Mwitwa et al., 2012). The global acceleration of industrialization during the past century has resulted in the exponential growth of various indicators of human activity and prosperity (Vidal et al., 2017). This has been followed by increased demand for mineral resources as population growth is directly linked to raw material consumption. The current projections on global population growth suggests high raw material consumption and, therefore, more mineral production by 2050 compared to the last 100 years (Busia and Akong, 2017; Humphreys, 2010). This implies an economic boom for mining continents such as Africa and their respective countries like Zambia as they will attract tens of billions of dollars in foreign investment (Wojcik, 2021). This is expected to translate into substantial economic growth and development as well as a significant amount of risk on human communities and the environment (Edwards et al., 2014).

Zambia's economy strongly depends on mining as it accounts for 12% of its gross domestic product (GDP), 70% of total export value and 62% of foreign direct investment (Kolala and Dokowe, 2021). While Zambia has a wide range of mineral commodities exploited in various parts of the country (Table 16.1), it has greatly depended on copper mining, and it remains as one of the essential drivers for Zambia's economic growth. Some of the mines highlighted in Table 16.1, such as the Luanshya copper mines, made announcements recently regarding mine closure plans. These mines have become dormant as a result of resource depletion and plans for the decommissioning process are underway. Most of the copper and cobalt deposits are located within the world-class metallogenic province called the Lufilian fold and thrust belt, which extends into the neighboring DRC (Figure 16.1). The deposits have a northwest-southeast (NW-SE) trend and mainly related to the domes region. In Zambia, large-scale copper mining takes place in the Copperbelt and Northwestern Provinces with high production output coming from three open-pit mines, namely, Kansanshi, Lumwana and Sentinel (Crundwell et al., 2020; Figure 16.1). These mines significantly contribute to the provision of socioeconomic activities including employment, education and health. The closure of any of these mines would therefore have a huge bearing on the well-being of the local communities and national economy. It is worth noting that mines usually close for two major reasons i.e., the reserves are depleted and are all mined out or they have become unprofitable due to the global market forces and technical aspects of the ore bodies.

Zambia has from time immemorial depended on the exploitation of its mineral resources. Mining activities have generated a vast amount of mining generated wastelands. Notably among these include tailings dams, waste rocks and overburden materials. Ultimately, this has resulted in large-scale ecosystem disturbance in form of landscape destruction or agro-ecosystem pollution arising from either mining generated wastelands or metallurgical processing facilities, such as smelters (Figure 16.2).

 DOI: 10.1201/9781315141855-16

TABLE 16.1
Prominent Mineral Commodities and Mining Firms in Zambia

Mineral Commodity	Mining Company	Location	Status
Copper, Cobalt	Mopani Copper Mines	Kitwe, Mufulira	Active
Copper, Cobalt	Konkola Copper Mines	Nampundwe, Chingola, Chililabombwe	Active
Copper, Cobalt	Lubambe Mine	Chililabombwe	Active
Copper, Gold	Kansanshi Mine	Solwezi	Active
Copper, Nickel	Kalumbila Mine	Kalumbila	Active
Copper, Uranium	Lumwana Mine	Lumwana	Active
Copper	NFCA Mining	Chambishi	Active
Copper	Chibuluma Mine	Chibuluma	Dormant
Copper	Luanshya Mine	Baluba Mine Shaft	Dormant
Manganese	Musamu Mine	Luapula	Active
	Kaboko Mine	Mansa	Active
Tin, Tantalum	Chisuki Mine	Choma	Active
Coal	Maamba Colleries	Maamba	Active
	Collum Coal	Sinazongwe	Active
	Kagem Mine	Lufwanyama	Active
Emerald	Grizzly Mine	Lufwanyama	Active
	Pridegems Mine	Musakashi	Active

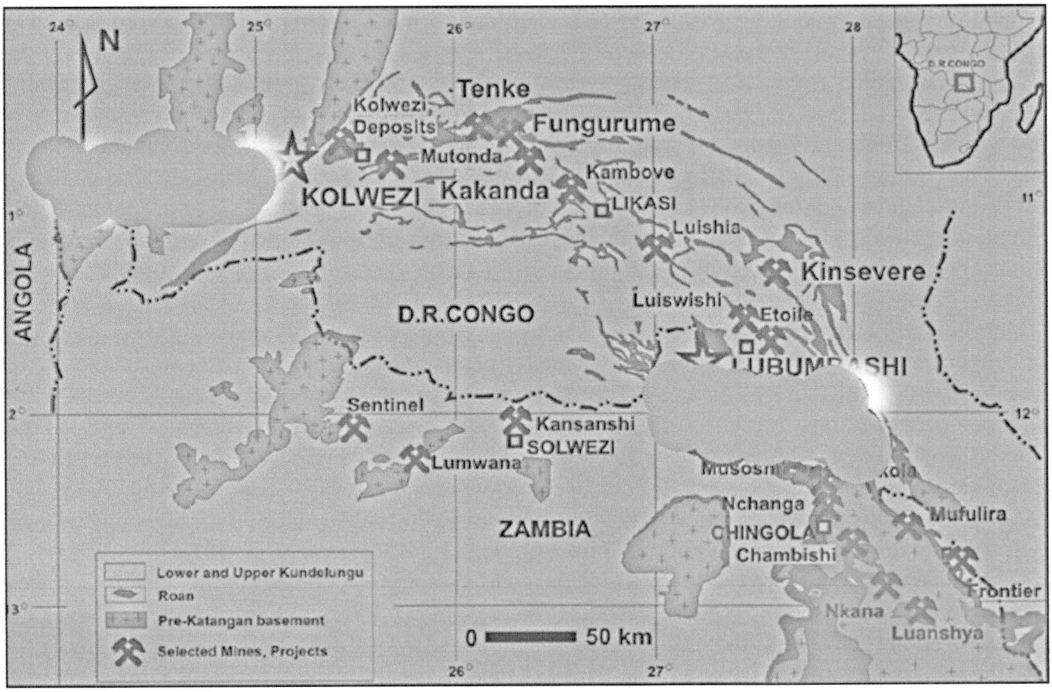

FIGURE 16.1 Locations of Kansanshi, Lumwana and Sentinel Copper Mines, Northwestern, Zambia.

FIGURE 16.2 Landscape disturbance arising from mining and metallurgical processing facilities.

The Copperbelt and Northwestern regions in Zambia are geologically located in a world-renowned copper district, the Lufilian fold and thrust belt, that has been mined since the 1900s (Kambani, 2003; Figure 16.3). As a result of long-term mining activities, large areas of land around the Copperbelt are covered by various kinds of mine waste (Ettler et al., 2011). The wastelands are frequently abandoned open cast mines, unconsolidated materials, waste dumps, tailings dams, areas stripped of vegetation and other types of degraded land through mining facilities (Venkateswarlu et al., 2016)

Zambia, a country rich in a diverse range of mineral commodities, has seen rapid changes during the past century on land use and landscape integrity. Large areas of land in mining towns are covered with different kinds of mine waste, including tailings dams, slag and waste dumps with no or limited restoration attempts (Festin, 2020; Mwitwa et al., 2012). Unrestored landscapes, particularly post-mining landscapes, negatively impacts health and environment because they are contaminated with elevated levels of heavy metals that cause various kinds of environmental pollution. The pollutants could accumulate in water and food chains, making agricultural crops unsuitable for consumption (Festin, 2020).

Metallurgical processing facilities have been observed to be sources of heavy metal pollution in the copper ore district. Fallout from the Cu smelter plants and mining generated wastelands have been observed to drastically degrade the Copperbelt region landscapes. A mosaic of eroded bare soil and short grasslands consisting of metal-tolerant plant species (metallophytes) have replaced the miombo woodland (the dominant open forest of southern Africa). The soil is acidic (mean pH 4.6) and contains elevated Cu, Zn, Co and Pb concentrations (Mpundu, 2010; Shutcha et al., 2015). For example, total Cu soil concentration may be as high as 50,000 mg kg-1 i.e., 250 times higher than non-contaminated soils from the miombo forest (220 mg kg_1 (Faucon et al., 2011).

Heavy metal contaminated soils represent a secondary source of contamination and are a major environmental issue due to low pH of Ferralsol (Ngongo et al., 2009; Pourret et al., 2015; Mwangi, 2016; Kourouma, 2018), thus increasing heavy metals mobility and provides for ease of rainfall and wind erosion in the rainy and dry seasons. This has resulted in chronically exposing human populations to metal-rich dust that represents a serious health hazard. Additionally, it also tends to expose other populations in general through either food chain or transport mechanism that results in pollutants being transferred away from the point of source (Kourouma, 2018). Unacceptably high metal concentrations (S, Cr, Cd) have been detected in Kitwe District on wood sawdust and tree barks

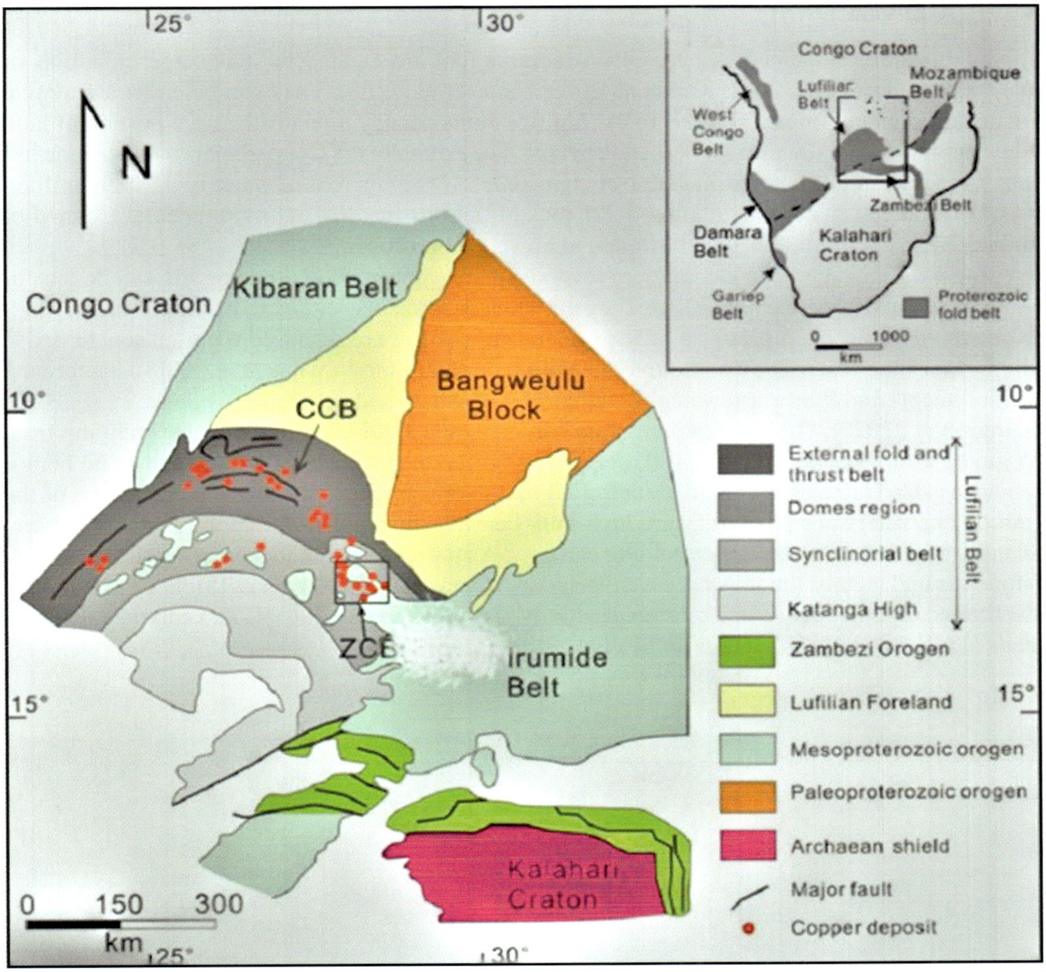

FIGURE 16.3 Geological setting of the Zambian Copperbelt (ZCB) and Northwestern deposits in Zambia.

at some distance and cardinal position from mining activities (Ncube et al., 2012; Ncube & Phiri, 2015). An ecological and health risk study by Kourouma (2018) explicitly dealt with the amount of heavy metals that the Zambian Copperbelt region population is exposed to due to their elevated concentrations in both soils and vegetables.

The increased demand for mineral commodities at global level has resulted in several installations of micro processing plants that add to the mine waste burden (Kambani, 2003; Mwaanga et al., 2019). Much of the waste is disposed in tailings dams, and the mining industry generates huge quantities of tailings each year (Gitari et al., 2018) . The quantity of waste generated depends on the amount of copper content in the host rock. As such, for every ton of copper produced, 350–420 tonnes of waste are generated, sometimes even more (Kaninga et al., 2020). Therefore, this demands restoration of such landscapes into a considerably more natural state with beneficial ecosystems. However, restoring post-mining landscapes is a complex process with varied challenges, including the harsh growing medium for most plants and limited knowledge on the kind of site-specific plants that are suitable for restoration purposes (Gitari et al., 2018; Kaninga et al., 2020). The characteristics of the major mine generated wastelands in Zambia are concisely described in the subsequent sections of this chapter.

16.2 TAILINGS DAMS

Tailings dams are described as hydraulic structures that are created to store large quantities of mine waste (Gitari et al., 2018; Kambani, 2003; Neina et al., 2019). They can be either a retention dam or raised embarkments (Figure 16.4). Tailings dams occupy a relatively large portion of land in the country, especially the Copperbelt Province. For example, the Copperbelt province alone has about 9,125 hectares of land estimated to contain about 77 million tons of waste rock, 1,899 million tons of overburden material and around 791 tons of tailings materials (Environmental Council of Zambia, 2008; Sikaundi, 2013) with limited to no vegetation cover.

A single tailings dam can extend over several square kilometers and dam walls can be tens of meters in height (Schoenberger, 2016). They are often described as either active or passive depending on whether unwanted products of the mineral processing are disposed there or not. In active tailings dams, the material disposed into the dam is in slurry form (Wang et al., 2014), comprising solid sediments and fluid wastewater as well as heavy metals and sulphides (Chileshe et al., 2020; Kaninga et al., 2020). The tailings dam often comprise fine material with grain sizes ranging from 625 μm to 2 mm (Edraki et al., 2014), depending on the parent rock, and they tend to be homogeneous as size fractions are mainly controlled by the efficiency of the concentrator. Some of the common characteristics of tailings are high bulk density, compaction due to the fine soil texture, low infiltration rate and low water holding capacity. As tailings are produced during the extraction of minerals and metals, the soils are juvenile and incohesive (Asensio et al., 2018). They also tend to have limited organic matter, and microbial activity is associated with low pH and enriched in heavy metals (Chileshe et al., 2020; Ginocchio et al., 2017).

FIGURE 16.4 Depiction of one of the degraded mining generated wastelands and relict vegetation (herbaceous component).

FIGURE 16.5 Mine wasteland with waste rock materials on one of mine sites, Zambia.

16.3 WASTE DUMPS

Prior to any production-related mining activity, the overburden materials, which include layers of soil and waste rock, are removed to gain access to underlying ore deposits (Vela-Almeida et al., 2015) and piled within the area (Figure 16.5) that a respective mining house enjoys surface rights. This kind of development mining can strip material down to a depth of 30 m or more and the removed layers are stored either as dumps at the mine sites or elsewhere (Figure 16.2). Ore with too low concentrations considered as being below the cut-off grade is also part of the overburden as it is of no interest to the industry and is stored as waste rock dumps at the mine sites (Broda et al., 2017).

The overburden materials are usually coarse-grained rocks that are heterogeneous and a source of acid mine drainage (AMD), therefore, risking groundwater pollution as the sulfide ores present in the rock react with rainwater (Broda et al., 2017). The waste rocks dumps can cover large areas and are a source of concern for the local environment as they can facilitate AMD (Franks et al., 2011). The overburden is typically nutrient poor and unsuitable for reclamation purposes because of the potential phytotoxicity of the excavated earth materials without amendments (Carrick and Krüger, 2007), even though they be useful in landscape contouring and when covered with layers of suitable soil, possible reclamation attempts can be made (Broda et al., 2017).

The increasing amount of land mined not only in Zambia but also across the region has created an environment that if unsupported by strong mine development and closure policies poses a serious threat to human health and agricultural productivity. Mining generated wastelands present serious pollution hazards to the environment, human health, agricultural productivity and groundwater (Kaninga et al., 2020). The threat to human health usually results from the proximity of mining generated wastelands to human settlements (Mwaanga et al., 2019) and these threats comprise different aspects, including but not limited to groundwater pollution, bioaccumulation into food chains and soil contamination (Ginocchio et al., 2017; Kaninga et al., 2020; Neina et al., 2019; Yohannes et al., 2021).

16.4 ABANDONED AND DECOMMISSIONED MINES IN ZAMBIA

Mine closure implies the process of terminating operations either temporarily or permanently. It marks an important aspect of mine life that goes beyond the closing down of mining and process plant operations (Bainton and Holcombe, 2018). Closure plans must be developed in a way that they comply with the rules that promote sustainable mining practices. There is limited literature regarding mine closure in Zambia. This is partly resulting from the reopening of mines that were at some point closed, such as the Maamba Colleries Limited (Besa et al., 2020). Furthermore, some mining firms have closed down certain mines and opened new ones within the same locality. For instance, Baluba mine in Luanshaya was shut down because of plummeting copper prices and energy challenges while Muliashi mine remained operational despite being in the same locality and owned by the same company (Bainton and Holcombe, 2018; Besa et al., 2020). Another example of previously closed mine where operations have set in is the Kalengwa mine with Moxico Resources Limited currently conducting feasibility studies for possible orebody extensions and efficient mineral processing technologies. Much of the available literature focuses on health-related issues arising from the Kabwe lead-zinc abandoned mine site (Mwaanga et al., 2019; Nyambe and Kawamya, 2005; Peša, 2022; Yohannes et al., 2021). The limited amount of literature could also be attributed to the lack of transparent guidelines on reporting of abandoned mining operations for artisanal, small-scale and large-scale mining. The focus has always been on large-scale mining while artisanal and small-scale mining also leaves behind several unreported excavations in various mining areas of Zambia.

16.5 ENVIRONMENTAL IMPACTS ASSOCIATED WITH MINING

Due to Zambia's long history of mining, there are many historical legacy sites that cause environmental problems. In many historical mining areas, and there are now new operations which make it hard to distinguish between historical and current impacts of mining. The copper industry has dominated the mining scene in Zambia for more than eight decades since the first commercial mine was opened in 1928 (Simutanyi, 2008). A number of environmental impacts are directly linked to operating copper mines. Of late, reports of environmental impacts arising from manganese and coal mining have been observed in the country near facilities for such mines (Nyekele, 2015). The impacts include but are not limited to soil contamination, air pollution and water pollution.

16.5.1 Air and Soil Contamination

Accumulation of metals in soil arising from wind-borne dust particles (from dry tailing dams) and particle fall-out from smelters have been observed in the Copper district of Central Africa. The main soil contaminants that occur in concentrations high enough to constitute a significant hazard to human health, are copper and cobalt (SGAB et al., 2005). Large areas within the Copper District of Zambia are contaminated as a variety of elements occur in elevated concentrations in addition to copper and cobalt as demonstrated by a number of studies (Czech Geological Survey, 2007). A number of studies show the highest concentrations for many elements across the mining towns, namely Kitwe, Mufulira and Chingola, and also to a lesser extent Kalulushi, Chililabombwe and Chambishi (Czech Geological Survey, 2007; Kourouma, 2018).

Both metallurgical processing and wasteland facilities are the main sources of heavy metal pollution in the Copper District of Central Africa. Fallout from the Cu smelter plants and mining generated wastelands has drastically degraded tropical landscapes and soil in the district (Munyemba, 2010). A mosaic of eroded bare soil and short grasslands consisting of metal-tolerant plant species (metallophytes) has replaced the miombo woodland (the dominant open forest of southern Africa). The soils are acidic (mean pH 4.6) and contains elevated Cu, Zn, Co and Pb concentrations (Mpundu, 2010; Shutcha et al., 2015). For example, total Cu soil concentration may be as high as 50,000 mg kg^{-1} in some parts, i.e., 250 times higher than non-contaminated soils from the miombo forest (220 mg kg^{1}) in some parts of the Copper District (Faucon et al., 2011) with some parts ranking among the ten most polluted areas in the world (Narendrula et al., 2012).

Heavy metal contaminated soils represent a secondary source of contamination and are a major environmental issue due to low pH of Ferralsol (Ngongo et al., 2009; Pourret et al., 2015; Mwangi, 2016; Kourouma, 2018), thus increasing heavy metals mobility especially in the rainy season. This has resulted in chronically exposing neighboring human populations to metal-rich dust that represents a serious health hazard. Additionally, it also tends to expose other populations in general

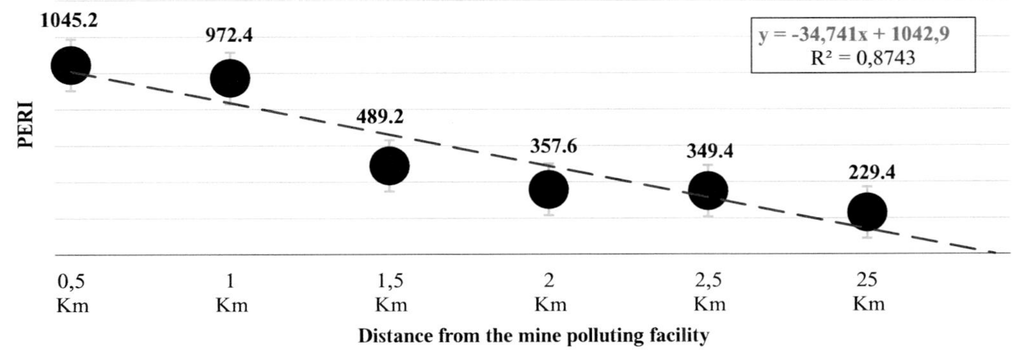

FIGURE 16.6 Variation of the Potential Ecological Risk Index (PERI) with distance from the wasteland facility (Kourouma, 2018).

through either food chain or transport mechanism that results in pollutants being transferred away from the point of source (Kourouma, 2018). Unacceptably high metal concentrations (S, Cr, Cd) have been detected in Kitwe, Zambia, on wood sawdust and tree barks at some distance and cardinal position from mining activities (Ncube et al., 2012; Ncube and Phiri, 2015).

In the case of copper, sub-surface samples show that the surface soil layer contains at least ten times, and quite often more than 50 times, higher concentrations in more or less the whole Copperbelt (Czech Geological Survey, 2007) with potential ecological risk being highly correlated with distance from the wasteland facility (Figure 16.6). Additionally, observations also demonstrate that some soil contaminants can be vertically mobile in the soil profiles with Pb at the rate of 1.36 cm/y (Ettler et al., 2012).

The spatial distribution of metal contaminated vegetables (tomatoes and cabbage) grown at different distances from wasteland facilities shows that the closer the garden is to the facility, the higher the health risk associated with consumption of such vegetables (Figure 16.7). This demonstrate that the farther away the garden is from the wasteland facility, the less the concentration of heavy metals in vegetables.

16.5.2 Water Pollution

The major mining and processing facilities mostly lie within the catchment area of the Kafue River. Kafue's watershed is the most developed in the country and is coming under increasing threat from pollution as well as competition in water utilization (Environmental Council of Zambia, 2008) as it is the main supplier of water to the major towns of the Copperbelt Province. The dissolved copper content has been shown to be considerably higher than the national standards (SGAB et al., 2005). The most significant contributors to metal pollution are the Nchanga, Nkana and Konkola mining operations (SGAB et al., 2005). Even though high concentrations from the mining areas are mostly local, elevated concentrations of dissolved sulfur can be traced all the way down to the confluence with the Zambezi River (Pettersson, 2002). When the water discharge in Kafue is increased during the rainy season, secondary particles are re-suspended and transported downstream in the river system. Mining activities in the Copperbelt also severely affect the waterways through extensive siltation. Concentrations of suspended elements (i.e. small particles) are naturally very low in the Kafue River but because of pollution from the mining industry suspended concentrations of copper and cobalt can be seen down to the junction with the Zambezi River (Pettersson, 2002). The siltation problem is partially a result of erosion from existing tailing dams and waste dumps (contribution of 9,000 tons/year), but the larger impact comes from ongoing mining operations. For example, dewatering of Konkola mine contributes 15,000 tons/year and the Nchanga mining area alone contributes 91,000 tons/year by, for example, discharging tailings directly into the river (SGAB et al.,

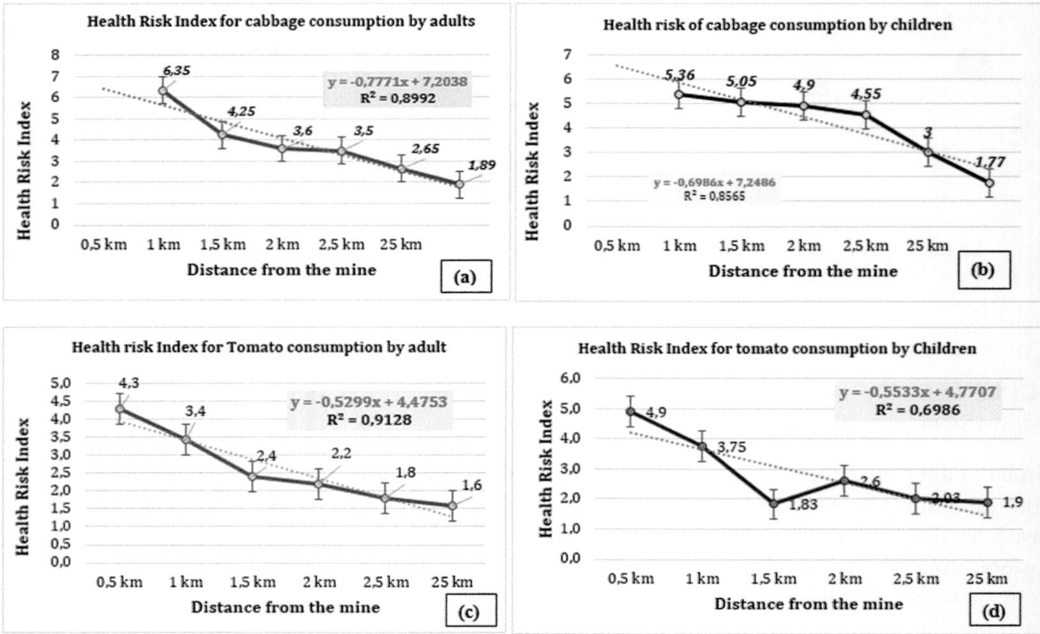

FIGURE 16.7 Health risk index arising from consumption of vegetables grown at different distances from the wasteland facility.

2005). Due to extensive siltation caused by mining activities, there is a continuous buildup in the river channel, and the bed sediment in many places is totally dominated by tailing materials with high metal concentration.

16.6 MANAGEMENT OF ABANDONED MINES AND WASTELANDS

16.6.1 National Legal and Policy Framework

The need to balance environmental requirements and economic activities as the country meets the social needs of its citizens led the government of the Republic of Zambia (GRZ) to adopt the National Conservation Strategy (NCS) in 1985 that resulted in formulation of the principles of sustainable development in the country. Later the NCS transformed into the National Environmental Action Plan (NEAP) in 1992. The NCS and NEAP facilitated development of the Environmental Protection and Pollution Control Act (EPPCA) in 1990 that led to the establishment of the Environmental Council of Zambia (ECZ) in 1992. The EPPCA was replaced in 2011 by the Environmental Management Act No. 12 of 2011. Consequently, this resulted in the name of ECZ changing to Zambia Environmental Management Agency (ZEMA). For mining activities, a parallel environmental management regulatory framework was also established under the Ministry of Mines and Minerals Development with the initial SI being enforced in 1997. Details of each of the regulations and laws are as indicated below:

The Environmental Management Act No. 12 of 2011: It is the supreme environmental management law in Zambia under which all environmental protection and pollution control measures are enforced. Its provisions require that all new projects undergo an environmental impact assessment (EIA) process before implementation. Once the project EIA is approved and implemented, licensing, auditing and compliance inspections follow as administrative tools. These follow-up activities are conducted in accordance with provisions of subsidiary regulations under the same Act. The subsidiary regulations cover activities such as, but not limited to, waste management, pesticides/chemicals management, air pollution prevention, water quality management and hazardous waste management. Key SIs under this Act relevant to mining and environmental management are SI No. 28 and SI No 112.

Environmental Protection and Pollution Control (Environmental Impact Assessment) Regulations, SI No. 28 of 1997. They provide the main framework for conducting EIAs in the country starting from screen up to decision-making. There are two levels of EIA in Zambia, namely an Environmental Project Brief (EPB), which is a basic EIA study for projects that have less significant impacts, and the second stage, an Environmental Impact Statement (EIS), which is a detailed EIA study applicable to projects with much more significant impacts. The EIA regulations also require that alterations and extensions to already existing projects must be subjected to the EIA process subject to activity screening.

Environmental Management (Licensing) Regulations, SI No. 112 of 2013. The SI No. 112 is a consolidation of the following regulations.

- Air and Water Pollution.
- Waste Management.
- Hazardous Waste.
- Pesticide and Toxic Substances.
- Ozone Depleting Substances.

The Mines and Minerals Development Act (MMDA) of 2015: This Act repealed the Mines and Minerals Development Act No. 7 of 2008. The Act requires mining license holders or developers to report on their environmental performance to the Mines Safety Department (MSD), which is the delegated authority for regulating environmental and safety issues relating to all mining and mineral processing operations in Zambia. This regulatory role overlaps with ZEMA; hence, compliance enforcement is conducted jointly where applicable. Prior to issuance of the mining licenses, large-scale mining license applicants are required to prepare Environmental Management Plans (EMPs), a requirement fulfilled through the EIA regulations. The mine license holder is required to prepare EMPs in accordance with Section 35 (d) and Section 80 of the MMDA. These plans are required to demonstrate how the applicants will mitigate adverse impacts that projects may have on the environment. The Act, however, states that EMPs must comply with the provisions of the Environmental Protection and Pollution Control (Environmental Impact Assessment) Regulations, Statutory Instrument No. 28 of 1997. This provision links the mandate of both MSD and ZEMA. The Key SI of this Act relevant to mining and environmental management include SI No. 29 and SI No. 102.

The Mines and Minerals (Environmental) Regulations of 1997 (SI No 29): This SI provides a framework for conducting and reviewing EIA reports for all mining and related projects. It also provides follow-up and compliance activities, such as environmental auditing and compliance inspections. It defines specific issues that must be covered in an EIA report for a mining project and the overall report structure. However, decision making authority for EIA reports is under the Environmental Protection and Pollution Control (Environmental Impact Assessment) Regulations, SI No. 28 of 1997. However, MSD is a consulted before the decision is made.

Mines and Minerals Environmental Protection Fund (EPF) Regulations of 1998 (SI No. 102): The SI provides the mechanism for setting up and operating the EPF as an environmental protection security fund for mining-related projects for post-mining environmental liabilities. The EPF is managed by the Ministry of Mines and Minerals Development (MMMD). It provides for periodic environmental audits and assessment of environmental management costs with reference to the Decommissioning and Closure Phase. Depending on the quality of environmental management by a mining company, a percentage of the security fund (ranging from 5% to 25%) is paid annually to the government.

16.6.2 Mining Closure Planning in Zambia

Globally, there is a well-designed expectation by industry, financiers and practitioners that mine closure planning should be an intrinsic part of the entire mining life cycle (Franks et al., 2011). There is an expectation that mine closure planning should have a pragmatic approach and commence early. This is strongly emphasized from the mitigation and adaptive environmental management in environmental impact assessments, EIA (Morrison-Saunders et al., 2016). Therefore, environmental impact assessments form an integral process in environmental management and have a strong link to mine closure planning. However, policy and principles for mine closure planning are weakly established for Africa compared to other mining continents such as Australia (Bainton and Holcombe, 2018; Morrison-Saunders et al., 2016) . Currently, the major focus for mining in Africa seems to have a bias toward improving governance strategies related to attracting investment and associated opportunities for harnessing mineral resource development (Singer, 2017) with generally less impetus on mine closure.

While mining policies and guidelines are provided at the national level across Africa, there is limited or lack of commitment across the mining industry on mine closure and environmental rehabilitation. In Zambia, the Mines and Minerals Development Act 2008 (MMDA) requires an environmental management plan (EMP), including protection and reclamation of land and water resources to be included in applications for large-scale mining with provisions for environmental impact studies. Furthermore, there is a separate Environmental Management Act 2011 mandating the mines minister to set conditions for the rehabilitation of mining lands. The Mines and Minerals Environmental Regulations under the MMDA need EIAs to specify environmental impact statements (EIS) that address the EMP, plans for mine-site rehabilitation and management, estimated costs and provisions for annual updates. Within the Environmental Management Act, there is a requirement for mining houses to contribute toward environmental restoration under an Environmental Protection Fund (EPF) that is supposed to go toward rehabilitation in an event that the mine closes prematurely. These policies tend to be limited as they do not cover artisanal and small-scale mining operations with regards to internationally acceptable and sustainable mining practices. Additionally, information, such as the application of appropriate strategies (use of appropriate native species), is still scanty as studies dealing in this subject matter are still subjects of investigation.

16.7 REHABILITATION OF MINE WASTELANDS

In the past, rehabilitation of mine wastelands has been undertaken in selected sites in Zambia with the use of species such as *Eucalyptus grandis*, *Mangifera indica*, *Leuceana leucocephala*, *Tithonia diversifolia*, *Cynodon dactylon*, *Vachelia spp*, *Senegalia spp*, *Pericopsis angolensis* and *Jatropha circus* (Figure 16.8) though with very little success (Kambinga and Syampungani, 2012). Two reasons may be advanced for this failure, namely; (i) lack of coordination between the mining houses and researchers, (ii) Lack of involvement of local communities in technology transfer and therefore inability of such innovation to support rural livelihood within close proximity to the mining area.

Other mines have enclosed animals on a tailings with a view that animal dung would eventually improve soil fertility and thereby support plant establishment and development (Figure 16.9). However, this may tend to compound the problem in that grazing of animals on plants with elevated levels of heavy metals may result in bioaccumulation of such heavy metals in animal tissues.

Consequently, most of the current projects undertaken in the Zambian Copperbelt are focusing on (i) the assessment of soil pollution and heavy metals mobility in soils, (ii) the effect of mineral and organic amendment to improve soil quality, (iii) the selection of most suitable plant species

FIGURE 16.8 Some examples of species trials on a Copperbelt tailings dam (Zambia).

FIGURE 16.9 Animals enclosed on a tailings dam in Kitwe, Zambia.

based on their ability to tolerate high concentration of heavy metals in soil and their plants traits and (iv) application of biochar in soil amendments. Field experiments showed encouraging results for the phytostabilization of metal-polluted soils around a smelting plant. However, results from large-scale experiments are lacking because of low collaboration among stakeholders (research institutions, state offices, mining companies) and no involvement of local communities.

16.8 CONCLUSION

The chapter has presented a detailed account on the current status of abandoned mines and mining-generated wastelands with a focus on strategies for restoration of degraded mining landscapes. The study notes that there exists no consolidated information on the occurrence of abandoned mines and mining-generated wastelands in Zambia. Overall restoration efforts in Zambia, as in many mining countries in Africa, have been limited compared to other mining jurisdictions despite the urgent need to prevent further land degradation and land use changes. The major challenge to restoration of degraded mining landscapes in Zambia has been limited knowledge of the degraded landscapes and, therefore, appropriate strategies necessary for phytoremediation and phytostablization.

The use of native flora is important, not only because of the reduced risk of invasive plant species, but also due to easier adaptation to biotic and abiotic factors that may enhance survival rate. Furthermore, native species growing on copper-rich soils tend to develop tolerance mechanisms that regulate metal uptake and, therefore, are, to a certain extent, already adapted to elevated metal concentrations.

However, the pace of restoration research and practice after mining disturbance in Zambia has been slow. Currently, mining houses in Zambia are required to restore the land following mining closure but restoration technologies that can be implemented are lacking and this needs to be addressed from a holistic approach. Biochar appears to be among the most promising amendments that can support plant growth and development on mining- generated wastelands because it increases the nutrients and carbon content of soils. A combination of biochar with native species and other organic amendments is one of the potential steps toward restoration of mine wastelands and in efforts to mitigate challenges of continued leakage of heavy metals from tailings dams.

REFERENCES

Ali, H., Khan, E., & Sajad, M. (2013). Phytoremediation of heavy metals – Concepts and applications. *Chemosphere, 91*(7), 869–881.

Asensio, V., Flórido, F. G., Ruiz, F., Perlatti, F., Otero, X. L., & Ferreira, T. O. (2018). Screening of native tropical trees for phytoremediation in copper-polluted soils. *International Journal of Phytoremediation, 20*(14), 1456–1463. https://doi.org/10.1080/15226514.2018.1501341

Bainton, N., & Holcombe, S. (2018). A critical review of the social aspects of mine closure. *Resources Policy, 59*, 468–478. https://doi.org/10.1016/J.RESOURPOL.2018.08.020

Besa, B., Kabwe, J., Masinja, J., & Banda, W. (2020). Socio-economic impact of mine closure and development of exit strategy for rural mining areas in zambia: A case study of Kalumbila district. *Springer Series in Geomechanics and Geoengineering*, 411–421. https://doi.org/10.1007/978-3-030-33954-8_48

Broda, S., Aubertin, M., Blessent, D., Hirthe, E., & Graf, T. (2017). Improving control of contamination from waste rock piles. *Environmental Geotechnics*, *4*(4), 274–283. https://doi.org/10.1680/envgeo.14.00023

Broughton, D., & Rogers, T. (2010). Discovery of the Kamoa copper deposit, Central Africa Copperbelt, D.R.C. *Society of Economic Geologists Special Publication, 15*(1), 287–298.

Busia, K., &Akong, C. (2017). The African mining vision: perspectives on mineral resource development in Africa. *Journal of Sustainable Development Law and Policy (The)*, *8*(1), 145. https://doi.org/10.4314/jsdlp.v8i1.7

Capistrant, P. L. (2013). Geology of the Enterprise hydrothermal nickel deposit, eastern Kabompo dome, north-western province, Zambia. Unpub. M.Sc. thesis, Colorado School of Mines, 186p.

Carrick, P. J., & Krüger, R. (2007). Restoring degraded landscapes in lowland Namaqualand: Lessons from the mining experience and from regional ecological dynamics. *Journal of Arid Environments*, *70*(4), 767–781. https://doi.org/10.1016/J.JARIDENV.2006.08.006

Chileshe, M. N., Syampungani, S., Festin, E. S., Tigabu, M., Daneshvar, A., & Odén, P. C. (2020). Physicochemical characteristics and heavy metal concentrations of copper mine wastes in Zambia: Implications for pollution risk and restoration. *Journal of Forestry Research*, *31*(4), 1283–1293. https://doi.org/10.1007/s11676-019-00921-0

Cox, D. P., Lindsey, D. A., Singer, D. P., Moring, B. C., & Diggles, M. F. (2007). Sediment-hosted copper deposits of the world: Deposit models and database. *U.S.G.S Open File Report 03-107*, 53pp.

Crundwell, F. K., du Preez, N. B., & Knights, B. D. H. (2020). Production of cobalt from copper-cobalt ores on the African Copperbelt – An overview. *Minerals Engineering*, *156*. https://doi.org/10.1016/j.mineng.2020.106450

Czech Geological Survey, (2007). Environmental-geochemical atlas of the central-northern part of the Copperbelt Province of Zambia. Project of the Development Cooperation Programme of the Czech Republic.

Edraki, M., Baumgartl, T., Manlapig, E., Bradshaw, D., Franks, D. M., & Moran, C. J. (2014). Designing mine tailings for better environmental, social and economic outcomes: A review of alternative approaches. *Journal of Cleaner Production*, *84*(1), 411–420. https://doi.org/10.1016/J.JCLEPRO.2014.04.079

Environmental Council of Zambia (2008). State of the environment in Zambia. Environmental Council of Zambia, Lusaka

Ettler, V., Mihaljevič, M., Křı́bek, B., Majer, V., & Šebek, O. (2011). Tracing the spatial distribution and mobility of metal/metalloid contaminants in Oxisols in the vicinity of the Nkana copper smelter, Copperbelt province, Zambia. *Geoderma*, *164*(1–2), 73–84. https://doi.org/10.1016/J.GEODERMA.2011.05.014

Ettler, V., Cihlová, M., Jarošı́kováa, A., Mihaljeviča, M., Drahota, P., Křı́bek, B., Vaněkc, A., Penı́žek, V., Sracek, O., Klementová, M., Engelf, Z., Kamona, F., & Mapani, B. (2019). Oral bioaccessibility of metal(loid)s in dust materials from mining areas of northern Namibia. *Environment International*, *124*, 205–215.

Faucon, M. P., Parmentier, I., Colinet, G., Mahy, G., Luhembwe, M. N., & Meerts, P. (2011). May rare metallophytes benefit from disturbed soils following mining activity? The case of the Crepidorhopalon tenuis in Katanga (DR Congo). *Restoration Ecology 19*, 333–343.

Festin, E. S. (2020). *Post-mining restoration in Zambia Screening native tree species for phytoremediation potential.*

Festin, E. M., Salk, C., Tigabu, M., Syampungani, S., & Odén, P. C. (2019). Biological traits of tropical trees suitable for restoration of copper-polluted lands. *Ecological Engineering, 138*, 118–125.

Festin, E. S. (2020). Post-mining restoration in Zambia: Screening native tree species for phytoremediation potential. PhD Thesis, Swedish University of Agricultural Sciences.

Franks, D. M., Boger, D. V., Côte, C. M., & Mulligan, D. R. (2011). Sustainable development principles for the disposal of mining and mineral processing wastes. *Resources Policy, 36*(2), 114–122. https://doi.org/10.1016/J.RESOURPOL.2010.12.001

Ginocchio, R., León-Lobos, P., Arellano, E. C., Anic, V., Ovalle, J. F., & Baker, A. J. M. (2017). Soil physicochemical factors as environmental filters for spontaneous plant colonization of abandoned tailing dumps. *Environmental Science and Pollution Research, 24*(15), 13484–13496. https://doi.org/10.1007/s11356-017-8894-8

Gitari, M. W., Akinyemi, S. A., Ramugondo, L., Matidza, M., & Mhlongo, S. E. (2018). Geochemical fractionation of metals and metalloids in tailings and appraisal of environmental pollution in the abandoned Musina Copper Mine, South Africa. *Environmental Geochemistry and Health, 40*(6), 2421–2439. https://doi.org/10.1007/s10653-018-0109-9

Hitzman, M. W., Broughton, D., Selley, D., Woodhead, J., Wood, D., & Bull, S. (2012). The Central AfricanCopperbelt: Diverse stratigraphic, structural, and temporal settings in the world's largest sedimentary copper district. *Society of Economic Geologists, Special Publication, 16*, 487–514.

Humphreys, D. (2010). The great metals boom: A retrospective. *Resources Policy, 35*(1),1–13. https://doi.org/10.1016/j.resourpol.2009.07.002

Kambani, S. M. (2003). Small-scale mining and cleaner production issues in Zambia. *Journal of Cleaner Production, 11*(2), 141–146. https://doi.org/10.1016/S0959-6526(02)00033-1

Kambing'a, M & Syampungani, S. (2012). Performance of Tree Species Growing on Tailings Dam Soils in Zambia: A Basis for Selection of Species for Re-vegetating Tailings Dam. *Journal of Environmental Science and Engineering* B 1, 827–831.

Kaninga, B. K., Chishala, B. H., Maseka, K. K., Sakala, G. M., Lark, M. R., Tye, A., & Watts, M. J. (2020). Review: Mine tailings in an African tropical environment—mechanisms for the bioavailability of heavy metals in soils. *Environmental Geochemistry and Health, 42*(4), 1069–1094. https://doi.org/10.1007/s10653-019-00326-2

Kolala, C., & Dokowe, A. (2021). Economic potential of industrial minerals in Zambia – A review. *Resources Policy, 72*, 101997. https://doi.org/10.1016/J.RESOURPOL.2021.101997

Korouma, J. (2018). Risks Assessment of Soil Heavy Metals Contamination and Implications on Agroecosystems and Human health in Copperbelt Province, Zambia. Msc Dissertation, Copperbelt University, Zambia.

Lydall, M. I., & Auchterlonie, A. (2011). The Democratic Republic of Congo and Zambia: Growing global 'hotspot' for copper-cobalt mineral investment and exploitation. In *6th Southern Africa Base Metals Conference 2011, S. Afr. Inst. Mining and Metal.*, pp. 25–37.

Mendez, M. O., & Maier, R. M. (2008). Phytostabilization of mine tailings in arid and semiarid environment-An emerging remediation technology. *Environmental Health Perspectives, 116,* 278–283.

Morrison-Saunders, A., McHenry, M. P., Sequeira, A. R., Gorey, P., Mtegha, H., & Doepel, D. (2016). Integrating mine closure planning with environmental impact assessment: Challenges and opportunities drawn from African and Australian practice. *Impact Assessment and Project Appraisal, 34*(2), 117–128. https://doi.org/10.1080/14615517.2016.1176407

Mpundu, M. M. (2010). Contaminations des sols en Elements Traces Metalliques a Lubumbashi Katanga/RD Congo. Evaluation des risques de contamination de la chaine alimentarie et choix de solutions de remediation. PhD Thesis, University of Lubumbashi, Democratic Republic of Congo.

Munyemba, F. K. (2010). Quantification et modélisation de la dynamique paysagère dans la région de Lubumbashi: évaluation de l'impact écologique des dépositions issues de la pyrométallurgie. PhD thesis. University of Lubumbashi, DR Congo 255p.

Mwaanga, P., Silondwa, M., Kasali, G., & Banda, P. M. (2019). Preliminary review of mine air pollution in Zambia. *Heliyon, 5*(9), e02485. https://doi.org/10.1016/J.HELIYON.2019.E02485

Mwangi, E. (2016). Classification of Native Metallophytes occurring at Chabeshi Copper Mine Tailings Dam based on their response to heavy metals contaminated soils. Msc Dissertation, Copperbelt University, Zambia.

Mwitwa, J., German, L., Muimba-Kankolongo, A., & Puntodewo, A. (2012). Governance and sustainability challenges in landscapes shaped by mining: Mining-forestry linkages and impacts in the Copper Belt of Zambia and the DR Congo. *Forest Policy and Economics, 25*, 19–30. https://doi.org/10.1016/J.FORPOL.2012.08.001

Ncube, E. (2012). Air pollution on the Copperbelt Province of Zambia: Effects of Sulphur Dioxide on Vegetation and Humans, Kitwe. School of Natural Resources Report, Copperbelt University.

Ncube, E., & Phiri, B. (2015) Concentrations of Heavy Metals in Eucalyptus and Pinus Wood Sawdust and Smoke, Copperbelt Province, Zambia. Maderas. Ciencia y Tecnología, *17*, 585–596.

Neina, D., Ranst, E. van, & Verdoodt, A. (2019). Chemical and mineralogical properties of post-mining sites in two gold mining concessions in Ghana. *Technology and Development, 6*. www.gjstd.org

Ngongo, M. L., Van Ranst, E., Baert, G., Kasongo, E. L., Verdoodt, A., Mujinya, B. B., & Kukalay, J. M. (2009). Guide des Sols en Republique Democratique du Congo, Tome I: Etude et Gestion. Don Bosco, Lubumbashi.

Nyambe, I. A., & Kawamya, V. M. (2005). Approaches to sustainable minerals development in Zambia. *Geological Society Special Publication*, *250*, 73–86. https://doi.org/10.1144/GSL.SP.2005.250.01.08

Nyekele, J. (2015). Assessment of toxic elements in water, sediments, forage and goats in the vicinity of Maamba coal mine, Sinazongwe, Zambia. Msc Dissertation, Copperbelt University, Zambia.

Parra, A., Zornoza, R., Conesa, E., Gómez-López, M., & Faz, A. (2016). Evaluation of the suitability of three Mediterranean shrub species for phytostabilization of pyritic mine soils. *Catena, 136*, 59–65.

Peša, I. (2022). Mining, waste and environmental thought on the Central African copperbelt, 1950–2000. *Environment and History*, *28*(2), 259–284. https://doi.org/10.3197/096734019X15755402985703

Pettersson, U. (2002). A geochemical study of the Kafue River, Zambia, Doctoral thesis, Luleå University of Technology

Pourret, O., Lange, B., Houben, D., Colinet, G., Shutcha, M., & Faucon M. P. (2015). Modeling of Cobalt and copper speciation in metalliferous soils from Katanga (Democratic Republic of Congo). *Journal of Geochemical Exploration 149*, 87–96.

Schoenberger, E. (2016). Environmentally sustainable mining: The case of tailings storage facilities. *Resources Policy*, *49*, 119–128. https://doi.org/10.1016/J.RESOURPOL.2016.04.009

SGAB, SWECO, THOMRO & UNZA. (2005). Preparation of phase 2 of a consolidated environmental management plan – Project summary report. ZCCM Investments Holdings, Copperbelt Environment Project.

Shutcha, M. N., Faucon, M., Kissi, C. K., Collinet, G., Mahy, G., Luhembwe, M. N., Visser, M., & Meerts, P. (2015). Three years of Phytostabilization experiment of bare acidic soil extremely contaminated by copper smelting using plant biodiversity of metal-rich soils in tropical Africa (Katanga, DR Congo). *Ecological Engineering*, *82*, 81–90.

Sikaundi, G. (2013). Copper mining industry in Zambia: environmental challenges. http://unstats.un.org/unsd/environment. Accessed 6 Sept 2023.

Simutanyi, N. (2008). Copper mining in Zambia: The developmental legacy of privatization. *Institute for Security Studies*, *165*, 1–16.

Singer, D. A. (2017). Future copper resources. *Ore Geology Reviews*, *86*, 271–279. https://doi.org/10.1016/J.OREGEOREV.2017.02.022

Syampungani, S., Tigabu, T., Matakala, N., Handavu, F., & Oden, P. C. (2016). Coppicing ability of dry miombo woodland species harvested for traditional charcoal production in Zambia: A win-win strategy for sustainaing rural livelihoods and recovery of woodland ecosystem. *Journal of Forestry Research.* https://doi.org/10.1007/s11676-016-0307-1

Vela-Almeida, D., Brooks, G., & Kosoy, N. (2015). Setting the limits to extraction: A biophysical approach to mining activities. *Ecological Economics*, *119*, 189–196. https://doi.org/10.1016/J.ECOLECON.2015.09.001

Venkateswarlu, K., Nirola, R., Kuppusamy, S., Thavamani, P., Naidu, R., & Megharaj, M. (2016). Abandoned metalliferous mines: Ecological impacts and potential approaches for reclamation. In *Reviews in environmental science and biotechnology* (Vol. 15, Issue 2, pp. 327–354). Springer Netherlands. https://doi.org/10.1007/s11157-016-9398-6

Wang, C., Harbottle, D., Liu, Q., & Xu, Z. (2014). Current state of fine mineral tailings treatment: A critical review on theory and practice. *Minerals Engineering*, *58*, 113–131. https://doi.org/10.1016/J.MINENG.2014.01.018

William, R. J., et al. (2014). Status and ecological effects of the world's largest carnivores. *Science*, *343*, 1241484. https://doi.org/ 10.1126/science.1241484

Wojcik, A. (2021). *World exploration trends 2021 PDAC special edition.*

Yohannes, Y. B., Nakayama, S. M. M., Yabe, J., Toyomaki, H., Kataba, A., Nakata, H., Muzandu, K., Ikenaka, Y., Choongo, K., & Ishizuka, M. (2021). Delta-aminolevulinic acid dehydratase (ALAD) and vitamin D receptor (VDR) genes polymorphisms in children residing in an abandoned leadzinc mine area in Kabwe, Zambia. *Meta Gene*, *27*, 100838. https://doi.org/10.1016/J.MGENE.2020.100838

17 Novel Swampy Forest System as a Strategy to Meet Coal Mine Wastewater Discharge Compliance Requirements in Indonesia

Ihsan Noor, Yudi Firmanul Arifin, Bambang Joko Priatmadi, Akhmad Rizalli Saidy, Charles Lee, Tim Roberts, Irdika Mansur and Lana Saria

17.1 BACKGROUND

Coal mining activities lead to economic development and sources of energy, but these activities can also damage the environment, resulting in widespread pollution of soil, water and air if they are not managed properly (Badhurahman et al., 2020; Worlanyo and Jiangfeng, 2021). Acid mine drainage (AMD) is the biggest and most persistent problem in coal mining throughout the world and in Indonesia. AMD is evident in terms of low pH and high heavy metal content, especially Fe and Mn in water. This water can pose a negative impact when it is discharged directly to public waterways (Gautama et al., 2014; Ojonimi et al., 2021).

The Indonesian government's Law number 3 for 2020 regarding Mineral and Coal Mining requires the holder of a Mining License (IUP) or a Special Mining License (IUPK) to apply good mining practices, which means the proper management and monitoring of environmental issues, including reclamation and mine closure. Based on the Decree issued by the Minister of Energy and Mineral Resources number 1827.K for 2018 regarding guidelines for implementing good mining practices, particularly if a mining activity results in AMD, managing AMD requires conducting a geochemical study. Such a geochemical study includes at least (a) identification of the potential for AMD, (b) prevention of AMD and (c) mitigation of AMD. Another regulation called Government Regulation (PP) number 78, year 2010 concerns Reclamation and Mine Closure and states that the company that has the permit must perform AMD management and revegetation achievement during both the operation and the mine closure stages. There are serious financial implications for a mine operation owner if reclamation and/or post-mining guarantees are not fulfilled.

Pit lakes, which are commonly known as voids, occur as an impact of the excavation of the land surface in open-pit mining activities during coal extraction. After mining activities have ended, AMD can fill up the voids and create problems when the overflow is not managed properly and an acceptable water quality standard cannot be achieved, according to the regulation (Geller et al., 2013; Chernos et al., 2022).

During operations, the mining pit area cannot avoid the problem of AMD formation from rocks on the pit walls after land clearing and overburden removal. The existence of a reaction process of sulfide minerals that eventually enter the void increases acidity by decreasing the pH value of the

DOI: 10.1201/9781315141855-17

water in the void (Acharya and Gehendra, 2020; Blanchette and Lund, 2016). Efforts have been undertaken in Indonesia to treat water that enters a mining pit area by collecting it into a pool on the mine floor (pit sump). The wastewater is then pumped out of the pit for processing. In addition to the mining pit, the overburden (OB) disposal area has the potential to form AMD. Disposal that has been finalized can also potentially form AMD if the OB management process is not carried out properly, namely, if the separation of potential acid-forming (PAF) and non-acid-forming (NAF) materials is not carried out (Dwiki et al., 2015). AMD can lead to a substantial risk that incurs high costs if not identified early and properly managed. To achieve sustainable mine area rehabilitation, managing environmental aspects during active mining operations is very important (Naidu et al., 2019) because better AMD management will, first, lead to successful revegetation rates in post-mining reclamation and, second, avoid public water contamination (Noor et al., 2019).

17.2 AMD MANAGEMENT

The management of AMD can be grouped into two main categories, namely, prevention and remediation. Prevention technology focuses on inhibiting AMD formation by controlling the source, while remediation technology addresses the amelioration of AMD that has occurred before being released into public waters (RoyChowdhury et al., 2015; Skousen et al., 2017). Remediation technology is a curative action and can be grouped into active and passive treatments to improve the runoff water to meet the compliance parameters of the following applicable regulations: first, the central government of the Environment Ministry Decree number 113, year 2003 concerning quality standards for wastewater from mining activities; and, second, the regional regulation of the Governor of South Kalimantan Number 036, year 2008 concerning Wastewater Quality Standards for Mining Activities (Adiloğlu, 2018; Noor et al., 2020a; Jallath et al., 2018). The compliance parameter values are summarized in Table 17.1.

17.3 EXCESSIVE COST OF ACTIVE TREATMENT

Active treatment is generally considered to be expensive compared to passive treatment, especially when mining operations have ceased (Jallath et al., 2018). Active treatment requires several process stages that are more complex and more costly operationally than those of passive treatment (Seervi et al., 2017). However, developing passive treatment processes in Indonesia is a challenge because of the high annual rainfall (as well as maximum daily rainfall), high acidity load and large catchment area. These factors contribute to high flow rates that will eventually undermine the effectiveness and efficiency of a system to process AMD (Ojonimi et al., 2021; Tong et al., 2021).

TABLE 17.1
Threshold Value Compliance Parameters

Compliance Parameters	Unit	Threshold Maximum Value National-1	Threshold Maximum Value Regional-2
pH	-	6–9	6–9
Total Suspended Solids (TSS)	mg L^{-1}	400	200
Iron (Fe)	mg L^{-1}	7.00	7.00
Manganese (Mn)	mg L^{-1}	4.00	4.00
Cadmium (Cd)	mg L^{-1}	-	0.05

Regional-2 = Regulation of Governor of South Kalimantan, Indonesia, Number 038, year 2008
National-1 = Decree of Environmental Ministry of Republic Indonesia, Number 113, year 2003

FIGURE 17.1 In-pit treatment of AMD using the liming process (documentation),

As an example of the active handling of AMD at one coal mining company in South Kalimantan, the company neutralized AMD that had accumulated in a void utilizing the in-pit treatment method (Figure 17.1) and lime mixing machine to treat flowing AMD in a water treatment pond (Figure 17.2). This helped to meet the quality standard parameter values set by the Indonesian government. A field-scale experimental treatment in a void used Ca $(OH)_2$ to neutralize 460,000 m^3 of acidic water, increasing the average pH from 2.8 to 7. The company needed to spend at least USD 0.04 per m^3 of acid water or USD 18,400 on the void (Gautama et al., 2014; Noor et al., 2020a).

FIGURE 17.2 Lime mixing machine run by water flow as part of active treatment (documentation).

17.4 DEVELOPMENT OF PASSIVE TREATMENT

Passive treatment is an AMD processing technique that can be integrated with reclamation activities on post-mining land. One form of passive treatment development is known as "Swampy Forest System", which can be defined as post-mining reclamation. It combines organic matter application and the planting of understory plants from certain types of grasses and woody plants that survive by living on flooded land with low pH values. This concept is based on the development of artificial swampy forests with surface runoff so that the processing of wastewater from coal mining activities and the implementation of reclamation for the purpose of revegetation can be conducted together (Noor et al., 2020a).

The development of passive AMD treatment with a swampy forest system is conducted in several stages. The first stage is the addition of selected organic matter in the form of waste from oil palm plantations, namely, empty oil palm fruit bunches (EFBs), to support efforts to increase the pH of the AMD during the short-term stage (Noor et al., 2020b). The material selection processes involving batch reactors to determine the best material for the next combination in the second stage are illustrated in Figure 17.3.

The second stage incorporates planting local and nonlocal species that were selected as a medium-term strategy to control water acidity levels and achieve heavy metal uptake by plants. Grass species selection experiments determined the types of local grasses, namely, *Eleocharis dulcis* (purun grass) and *Cyperus rotondus* (batibati grass), while the nonlocal grass species are *Typha angustifolia* (typha grass) and *Vetiveria zizaniodes* (vetiver grass) (Noor et al., 2021).

In the third stage, the selected local tree species are planted and include *Melaleuca leucadendra* (galam tree) and *Nauclea subdita* (bangkal tree), while the nonlocal tree species are *Nauclea orientalis* (longkida tree) and *Melaleuca cajuputi* (kayuputih tree). The planting of trees is part of a long-term stage to help streamline AMD processing in a measurable and definite way. Here, the aim is to meet the quality standard values set in the regulations (Noor et al., 2022).

FIGURE 17.3 Batch reactor experiment preparation for the organic matter, grass and tree species selection process (documentation).

17.5 INTEGRATED PROCESS OF RECLAMATION AND AMD PASSIVE TREATMENT

One of the biggest challenges in coal mining activities is determining the best way to restore overburden (OB) in the form of overburden removal during coal extraction. Some coal mining operations are unable to return their overburden to the ex-mining pit area (more commonly known as the in-pit dump area) due to material shortages or inappropriate material balance. This leaves a pit lake in the form of voids that must be managed and requires a large amount of energy, time and money to remediate. To achieve the right material balance, effective methods and planning are needed so that the overburden material for ex-coal mines can be matched and, in the end, reduce the number of voids left behind by saving a certain volume of overburden material (OB), which can be utilized in other areas.

The use of reclamation areas in the form of in-pit dumps for AMD management to manage runoff from voids will assist reclamation activities. Utilization of in-pit dump land for the construction of swampy forest systems as a passive treatment development is an integrated management strategy between the reclamation process and AMD management obligations. This process is applied to the runoff from voids so that AMD management can be carried out effectively, thereby reducing energy, costs and processing time (Noor et al., 2020a).

The swampy forest system is a passive treatment system designed by first connecting runoff water from the voids to a sediment trap (mud trap pond). Then, it flows to several other compartments that have been filled with a combination of organic material media planted with grass and tree species that have been selected for each of these compartments with the aim of improving the quality of the AMD entering the swampy forest system passively and naturally. In this way, the water that comes out of the swampy forest system meets the environmental quality standard values in accordance with regulations. The number of pollutants, low pH values and high contents of heavy metals such as Fe and Mn are subsequently reduced (Noor et al., 2020a).

17.6 EXPERIMENTAL DESIGN OF THE SWAMPY FOREST SYSTEM

The swampy forest system first requires the choice of plants and other materials to passively improve the AMD. The choice of materials is made by assessing each material to establish the best individual role. The first batch reactor experiment is performed to determine which selected organic matter from several types of available organic material around the mining area can increase the pH of acidic water to a neutral pH level. The second batch reactor experiment is performed to determine the type of grass among several types of grass available, including both local and nonlocal grass types. The type of grass selected must be able to aid the deposition and absorption of heavy metals contained in AMD. The third batch reactor experiment is a process to determine the best tree species by considering those species that follow the applicable provisions for the revegetation of ex-mining land. The function of trees will help the process of absorbing heavy metals by choosing the best type of hyperaccumulator and preparing reclamation land that can be planted with suitable wood plants. Details of the process of the swampy forest stage are shown in Figure 17.4 (Noor et al., 2020a).

17.7 LABORATORY SIMULATION OF SWAMPY FOREST SYSTEM

The results of the material selections were used for the subsequent combination experiment, namely, a laboratory simulation experiment intended to develop a passive treatment with a swampy forest system using the concept of a constructed forested wetland. Laboratory simulation experiments were performed by preparing an experimental reactor with a length of 200 cm, a width of 100 cm and a height of 60 cm. The laboratory simulation experiment was a combination of the three materials that were selected in the previous media selection stage. This experiment was designed to

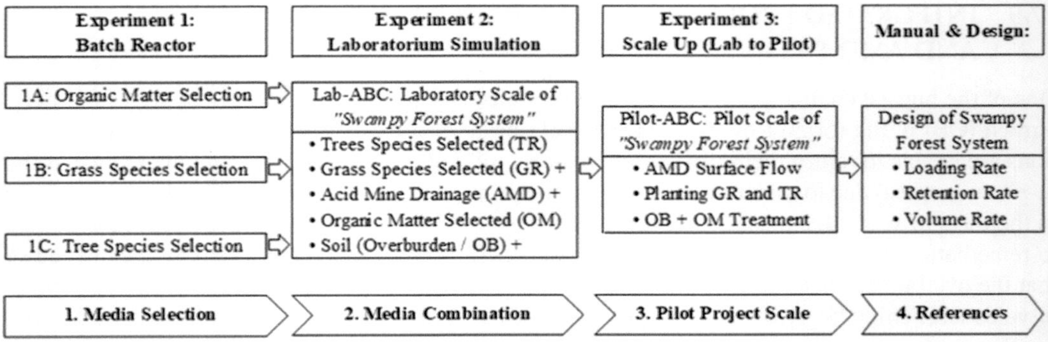

FIGURE 17.4 Swampy forest system research steps.

understand the combination of the three materials, namely, the selected organic matter, grass species and tree species.

The treatment protocol was to place a layer of OB (250 kg) in the bottom layer in each reactor and then plant the selected grass (18 clumps each species) and tree species (three seeds each species) based on the layout detailed in Figure 17.5. Then, treatment was continued by adding cow manure fertilizer (CMF) as an organic fertilizer to support the grass and tree plantings, and finally empty fruit bunches (EFB) were added to the second layer (Noor et al., 2020b). Each reactor was then incubated for four weeks. After the incubation process was completed, the experiment was continued by slowly filling the reactor with surface water (collecting rainwater) until the water level of each reactor had an average height of 15 cm from the top OB layer and allowing it to acclimate for seven days. Then, 25% of the remaining water in each reactor was drained via a tap at the bottom and replaced with AMD for the second acclimation of another seven-day process. This step was repeated a third time by draining 50% of the remaining water and a fourth time by draining 75% of the remaining water. When the total acclimatization period was completed, all (100%) the water in the reactor was replaced again with full AMD to start the treatment period.

After an incubation period of four weeks, the AMD continued to flow into the experimental box and entered the inlet via the surface system, flowing to box-1, then box-2, then box-3, and finally out of the outlet (as a simulation of public water bodies). The measurement of the pH of the water in each reactor started on day 1 (the first day of treatment) and continued until day 5. The pH measurement was carried out in the reactor box with a pH meter. The other parameter measurements included the total suspended solids (TSS) and heavy metal content in the form of Fe and Mn on day 5 of the treatment period. The laboratory simulation experiment is illustrated in Figure 17.5 (Noor et al., 2023).

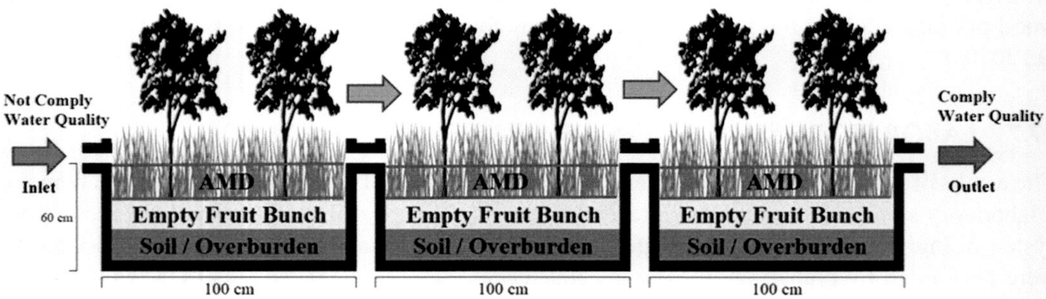

FIGURE 17.5 Cross-section of the laboratory simulation experiment.

TABLE 17.2
Calculations in the Scale-Up Process Are Based on the Basic Data Obtained from the Laboratory Simulation Experiments

Description	Laboratory Simulation Results	Scale Up	Pilot Project Design Calculation in the Field
(Basic data)	(a)	(b)	(c = a x b)
Total volume of water	V_1		$V_2 = (V_1 \times T \times Cf_1)$
Debt of water flow in	D_1	T (times)	$D_2 = (D_1 \times T \times Cf_2)$
Retention time	R_1		$R_2 = (R_1 \times T \times Cf_3)$

Cf2 = CF for water debt and Cf3 = CF for retention time
Note: Cf_1= correction factor (CF) for water volume.

17.8 PILOT PROJECT SCALE IMPLEMENTATION

The third stage was the implementation of the basic data obtained in previous laboratory simulation experiments. The laboratory simulation experiments based on combinations of organic matter, grass species and tree species yielded basic data reference for scaled-up applications in the field. The swampy forest system has demonstrated its ability to change the water quality of water that does not meet the quality standard parameter values to water that is in compliance with applicable regulations (Noor et al., 2020a). The scale-up process from the basic data scale to the pilot project scale is shown in Table 17.2.

This pilot project field experiment was conducted on a one-hectare reclamation area. Four compartments were constructed in a serial system, meaning that the AMD would flow into the inlet of the first compartment and exit to the second compartment and then into the third and fourth compartments. The wastewater then flowed out at the outlet of the fourth compartment to public waterways. Compartment size refers to the design. The length and width ratio is 4:1, resulting in a certain total volume and retention time (Coppini et al., 2019; Cong Manh et al., 2019). This third stage of experimental research validated the basic data obtained from the laboratory simulation experiments, namely, water discharge, retention time and reactor volume of the compartment (Cheng et al., 2018; Hsu et al., 2011).

17.9 CONSTRUCTION OF SWAMPY FOREST SYSTEM

The swampy forest system is a constructed wetland (CW) designed to manage void runoff water, and it consists of an open channel that drains water from the voids, a sediment pond to help improve water quality by deposition of dissolved heavy metals and particles and then a core compartment pond acting as a CW. Pond structures are designed for different functions. At the planning stage, aerial photos are taken at predetermined locations, and the CW shape is designed according to the location conditions. Land arrangements and construction of compartment ponds and embankments are carried out using heavy excavators. After pond construction is completed, the main compartment is filled with organic material, and then certain types of grass and trees that have been chosen are planted. It is ensured that grass and tree species can grow well so that they can cope with the AMD that flows from the voids through open channels into the sediment trap and then into several swampy forest compartment units in sequence. The water in the swampy forest compartment was incubated for 30 days. During the incubation process, the swampy forest is subjected to a reduction process. Under reduction conditions, the low pH will increase and become stable at a value of 6 to 7. The design of each of the swampy forest compartments is described in more detail below, and the layout of the swampy forest system is illustrated in Figures 17.6, 17.7, and 17.8 (Noor et al., 2023).

FIGURE 17.6 Construction of the pilot project of the swampy forest system (documentation)..

FIGURE 17.7 Selected organic materials, grass and trees in the treatment ponds (documentation).

FIGURE 17.8 Water quality monitoring of the pilot project swampy forest system (documentation).

17.10 ONLINE PH MONITORING SYSTEM OF THE PILOT PROJECT SWAMPY FOREST

Mining companies in Indonesia now must obey a new regulation that requires the installation of an online system to monitor water quality from mining activities, so a swampy forest pilot project has been designed for this system. A pH meter was installed at the inlet where water enters the system and at the outlet of the swampy forest system. At the inlet, several water parameters were observed that generally did not meet the quality standards, specifically pH < 4, Fe > 7 mg L^{-1} and Mn > 4 mg L^{-1}. Additionally, other parameters such as TSS and Cd are also monitored although mostly meet the quality standard values. Observations at the outlet revealed that all parameters did meet the quality standards. These results indicate that the swampy forest system changed the noncompliant water quality parameters to meet the values of the quality standard. Figure 17.9 illustrates an example of pH monitoring for 12 weeks with a discharge of 1,000 m^3 per day. Considering that the swampy forest system is a constructed wetland, there are some additional parameters (COD and BOD) to be monitored that must comply with Forestry and Environmental Ministry Regulation Number 3, year 2022 as the new applied regulation for constructed wetland management in Indonesia.

The online monitoring system is one in which the sensor readings are recorded directly onto the server every hour. A comparison of the pH value of the water at the inlet and the outlet shows a stable trend, meaning that when the inlet pH decreases, the outlet pH also decreases. The pH value of the water obtained varied from 0.98 to 3.85 so that an average of approximately 2.42 was obtained.

17.11 CONCLUSION

It is concluded that the swampy forest system can improve the pH range by approximately 2.42 points when the calculation of the average runoff discharge obtained in the field is 1000 m^3 per day. The decline in the pH of water occurs mostly at night, so the recorded pH value also varies, and

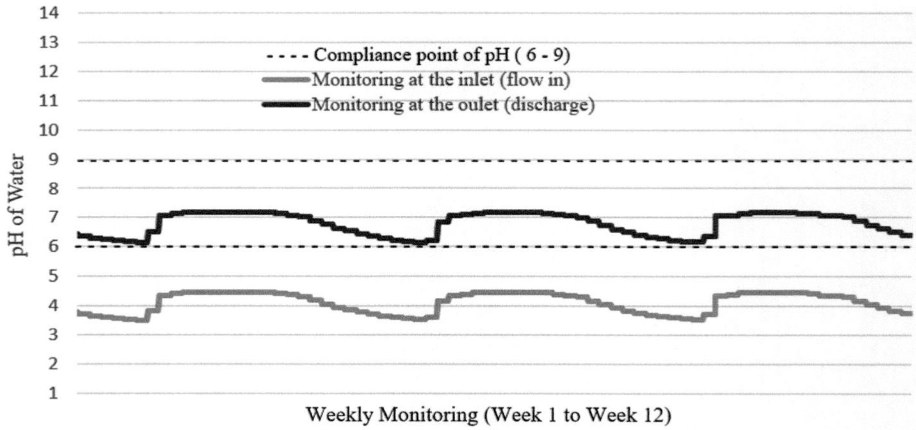

FIGURE 17.9 Data from pH monitoring during week 1 to week 12 at the inlet and outlet of the pilot project of the swampy forest system.

the pH stabilizes again during the day. The swampy forest system, in fact runs well. The range of the increase can be estimated with the same pattern over a long period of time. The development of AMD passive treatment with a swampy forest system for treating runoff from coal mines will greatly help companies manage mining-related liquid waste more effectively and sustainably during the operational period. At the same time, it helps with preparing land for reclamation when mining ceases, especially with respect to Loan-Use Forestry Areas Permit (IPPKH) in Indonesia. These must be returned to the government according to the applicable criteria after the reclamation of the ex-mine is completed.

In the swampy forest system, the role of certain materials, as well as plants and microorganisms, will be very important in reducing the heavy metal content in the excavated soil of coal mines. The swampy forest system is expected to provide important outcomes for mining industry players and government policymakers regarding wetland reclamation efforts in Indonesia. This system will reduce and eliminate heavy metal contamination in AMD and lead to better revegetation of flooded lands. Finally, reducing the use of chemicals and reusing palm oil waste will greatly boost the possibilities for sustainable development.

REFERENCES

Acharya, B. S., & Gehendra, K. (2020). Acid mine drainage from coal mining in the United States – An overview. *Journal of Hydrology, 588*. https://doi.org/10.1016/j.jhydrol.2020.125061

Adiloğlu, S. (2018). Heavy metal removal with phytoremediation. *Advances in Bioremediation and Phytoremediation* (pp. 115–126). Intech Open. https://doi.org/10.5772/intechopen.70330

Badhurahman, A., Gautama, R. S., & Kusuma, G. J. (2020). REE enrichment pattern in acid mine drainage and overburden from coal mine in Indonesia. *Proceedings of the Postponed 14th IMWA Congress-"Mine Water Solutions"*, 9–13 November. International Mine Water Association Congress 5, 1–6. https://www.imwa.info/docs/imwa_2020/IMWA2020_Badhurahman_1.pdf

Blanchette, M. L., & Lund, M. A. (2016). Pit lakes are a global legacy of mining: An integrated approach to achieving sustainable ecosystems and value for communities. *Current Opinion in Environmental Sustainability, 23*, 28–34. http://dx.doi.org/10.1016/j.cosust.2016.11.012

Cheng, G., Li, Q., Su, Z., Sheng, S., & Fu, J. (2018). Preparation, optimization, and application of sustainable ceramsite substrate from coal fly ash/waterworks sludge/oyster shell for phosphorus immobilization in constructed wetlands. *Journal of Cleaner Production, 175*, 572–581. https://doi.org/10.1016/j.jclepro.2017.12.102

Chernos, M., MacDonald, R. J., Straker, J., Green, K., & Craig, J. R. (2022). Simulating the cumulative effects of potential open-pit mining and climate change on streamflow and water quality in a mountainous watershed. *Science of the Total Environment*, *806*(1). https://doi.org/10.1016/j.scitotenv.2021.150394

Cong Manh, N., Van Minh, P., Tri Quang Hung, N., Thai Son, P., & Minh Ky, N. (2019). A study to assess the effectiveness of constructed wetland technology for polluted surface water treatment. *VNU Journal of Science: Earth and Environmental Sciences*, *35*(2). https://doi.org/10.25073/2588-1094/vnuees.4372

Coppini, E., Palli, L., Antal, A., Del Bubba, M., Miceli, E., Fani, R., & Fibbi, D. (2019). Design and start-up of a constructed wetland as tertiary treatment for landfill leachates. *Water Science and Technology*, *79*, 145–155. https://doi.org/10.2166/wst.2019.030

Dwiki, S., Shimada, H., Gautama, R. S., Kusuma, G. J., Sasaoka, T., Koten, F., & Matsumoto, S. (2015). Evaluation of acid mine drainage characterization for predicting post drainage water quality in coal mines. *Inzynieria Mineralna*, *16*(2), 23–28. https://bibliotekanauki.pl/articles/319296.pdf

Gautama, R. S., Novianti, Y. S., & Supringgo, E. (2014). Review on in-pit treatment of acidic pit lake in Jorong Coal Mine, South Kalimantan, Indonesia. In *An interdisciplinary response to mine water challenges* (pp. 645–649). Xuzhou: China University of Mining and Technology Press. https://www.imwa.info/docs/imwa_2014/IMWA2014_Gautama_645.pdf

Geller, W., Schultze, M., Kleinmann, R., & Wolkersdorfer, C. (2013). *Acidic pit lakes: The legacy of coal and metal surface mines*. Environmental Science and Engineering (Subseries: Environmental Science). Springer. https://doi.org/10.1007/978-3-642-29384-9

Hsu, C. B., Hsieh, H. L., Yang, L., Wu, S. H., Chang, J. S., Hsiao, S. C., Su, H. C., Yeh, C. H., Ho, Y. S., & Lin, H. J. (2011). Biodiversity of constructed wetlands for wastewater treatment. *Ecological Engineering*, *37*(10), 1533–1545. https://doi.org/10.1016/j.ecoleng.2011.06.002

Jallath, J. E. S., Romero, F. M., Iturbe Argüelles, R., Cervantes Macedo, A., & Goslinga Arenas, J. (2018). Acid drainage neutralization and trace metals removal by a two-step system with carbonated rocks, Estado de Mexico, Mexico. *Environmental Earth Sciences*, *77*, 86. https://doi.org/10.1007/s12665-018-7248-2

Naidu, G., Ryu, S., Thiruvenkatachari, R., Choi, Y., Jeong, S., & Vigneswaran, S. (2019). A critical review on remediation, reuse, and resource recovery from acid mine drainage. *Environmental Pollution*, *247*, 1110–1124. https://doi.org/10.1016/j.envpol.2019.01.085

Noor, I., Arifin, Y. F., Priatmadi, B. J., & Saidy, A. R. (2020a). Development of the Swampy Forest system for passive treatment of acid mine drainage during post mining land reclamation: A new concept review. *Ecology, Environment & Conservation*, *26*(2), 901–909. http://www.envirobiotechjournals.com/EEC/v26i220/EEC-66.pdf

Noor, I., Arifin, Y. F., Priatmadi, B. J., & Saidy, A. R. (2020b). Oil palm empty fruit bunch as the selected organic matter in developing the Swampy Forest system for passive treatment of acid mine drainage. *Ecology, Environment & Conservation*, *26*(3), 1424–1431. http://www.envirobiotechjournals.com/EEC/vol26i32020/EEC-75.pdf

Noor, I., Arifin, Y. F., Priatmadi, B. J., & Saidy, A. R. (2021). Role of the selected grass species in developing of swampy forest system for passive treatment of acid mine drainage. *Material Science Forum Journal*, *1025*, 273–278. https://doi.org/10.4028/www.scientific.net/msf.1025.273

Noor, I., Arifin, Y. F., Priatmadi, B. J., & Saidy, A. R. (2023). Laboratory simulation of the swampy forest system for the passive treatment of acid mine drainage in coal mine reclamation areas. *Scientific Reports*, *13*(1), 1–10. https://doi.org/10.1038/s41598-023-32990-x

Noor, I., Arifin, Y. F., Priatmadi, B. J., Saidy, A. R., & Mansur, I. (2022). Role of the tree species selected in developing swampy forest system for passive treatment of acid mine drainage. *Technium Sustainability*, *2*(1), 46–53. https://doi.org/10.47577/sustainability.v2i1.5889

Noor, I., Udiansyah, U., Priatmadi, B. J., & Winarni, E. (2019). Evaluation of degraded land management on soil physical properties of coal post mining revegetation land. (In Bahasa: Evaluasi Pengelolaan Lahan Terdegradasi Terhadap Sifat Fisika Tanah Pada Lahan Revegetasi Pasca Tambang Batubara). *EnviroScienteae*, *15*(3), 441. https://doi.org/10.20527/es.v15i3.7438

Ojonimi, T. I., Okeme, I. C., Phiri-Chanda, T., & Ameh, E. G. (2021). Acid mine drainage (AMD) contamination in coal mines and the need for extensive prediction and remediation: A review. *Journal of Degraded and Mining Lands Management*, *9*(1), 3129–3136. https://doi.org/10.15243/jdmlm.2021.091.3129

RoyChowdhury, A., Sarkar, D., & Datta, R. (2015). Remediation of acid mine drainage-impacted water. *Current Pollution Reports*, *1*, 131–141. https://doi.org/10.1007/s40726-015-0011-3

Seervi, V., Yadav, H. L., Srivastav, S. K., & Jamal, A. (2017). Overview of active and passive systems for treating acid mine drainage. *International Advance Research Journal in Science, Engineering and Technology*, *4*(5), 131–137. https://doi.org/10.17148/iarjset.2017.4525

Skousen, J., Zipper, C. E., Rose, A., Ziemkiewicz, P. F., Nairn, R., McDonald, L. M., & Kleinmann, R. L. (2017). Review of passive systems for acid mine drainage treatment. *Mine Water and the Environment*, *36*, 133–153. https://doi.org/10.1007/s10230-016-0417-1

Tong, L., Fan, R., Yang, S., & Li, C. (2021). Development and status of the treatment technology for acid mine drainage. *Mining, Metallurgy and Exploration*, *38*(1), 315–327. https://doi.org/10.1007/s42461-020-00298-3

Worlanyo, A. S., & Jiangfeng, L. (2021). Evaluating the environmental and economic impact of mining for post-mined land restoration and land-use: A review. *Journal of Environmental Management*, *279*, 111623. https://doi.org/10.1016/j.jenvman.2020.111623

18 Application of Risk-Based Management Strategies to Legacy Mines

Dane Lamb, Peter Sanderson, Liang Wang, Balaji Seshadri, Haibin Xu, Mezbaul Bahar, Jianhua Du and Ravi Naidu

18.1 INTRODUCTION

Historical mining operations in Australia operated and ceased within the context of very different social expectations compared to the current day. Modern expectations of mining are that appropriate environmental controls are in place prior to commencing, during and at closure of mining. However, mining of typically small metalliferous ores in the past commenced, operated and ceased with the primary expectation of ore recovery, job creation and economy stimulation. A critical difference in the past was attribution of responsibility for mine closure. Historical mining closure commonly had no clear responsible party. The lack of clear responsibility delegation has resulted in enormous numbers of legacy mine sites in Australia and internationally. Indeed, there are more than 550,000 legacy mine sites in the United States alone (Soucek et al., 2000), >10,000 in Canada, 5,500 in Japan and > 10,000 abandoned metalliferous mines in the United Kingdom (Mayes et al., 2009). Serious human and environmental health impacts have been reported across all inhabited continents, in some cases with likely convincing links between health problems and legacy mine sites (Lamb et al., 2017; Zhuang et al., 2009). Many legacy mine sites around the world are metalliferous and/or pyritic, and, as such, incorporate a range of inorganic contaminants, including high levels of acidity, cyanide, metalloids (arsenic, antimony, chromium) and metals (e.g. cadmium, copper, lead, zinc, iron, aluminium and manganese). In New South Wales, Australia, there are estimated to be between 500 and 20,000 legacy mine sites (Archer and Caldwell, 2004; Grant et al., 2002). The risks from contaminants from metalliferous mines to humans and ecological receptors are largely unknown.

Rehabilitation and "cleanup" of legacy mine sites is typically associated with extreme cost. Governments are invariably left with expensive cleanup costs given that funds or budgets are limited for rehabilitation (Lamb et al., 2017). One of the primary concerns of legacy mine sites is public safety and human health impacts from contamination. Although the presence of contaminants at a given site is of concern, the urgency for rehabilitation and action depends primarily on exposure to human and ecological receptors of contaminants. It is thus important to recognise that although a hazard may exist, the associated risk to human health and the environment from contamination may be low.

Risk-based land management (RBLM) utilises an evidence-based approach on risks associated with contaminants and other stressors from exposure of receptors to contaminants (Naidu et al., 2008a, 2008b). For this reason, it is critical that the science underpinning the elements of a strategy is sound or can be improved. Other important elements include source characterisation, identification of receptors and exposure routes to each receptor. Absolutely critical is the exposure of contaminants to a given receptor, which includes transportation mechanisms, and factors controlling

DOI: 10.1201/9781315141855-18

the bioavailability to receptors. Risk assessment of contaminated sites involves characterisation of potential danger posed by contaminants to ecological and human health.

The growing recognition of the issues surrounding legacy mine sites has resulted in increasing site-specific risk assessments. The majority of risk assessments focus on human health and ecological risk assessments (ERA) rather than a consistent risk-based management strategy put in place for legacy sites. Lim et al. (2008), Zhuang et al. (2009) and Aslibekian et al. (2003) conducted human health risk assessments (HHRAs) using bioaccessibility and plant accumulation data. Similarly, ecological risk assessments were noted for metalliferous abandoned mines (Antunes et al., 2008; Jasso-Pineda et al.; 2007; Min et al., 2013). The incorporation of bioavailability into assessments is indicative of the path forward. Similar to comprehensive rehabilitation, detailed risk assessments (ecological especially) are extremely costly, and a tiered approach will help avoid unnecessary expenditure. Elements of contaminant transport, groundwater contamination, receptor interaction with the site, ecological impacts and human health need to be part of a consistent approach in order to inform management decisions.

A key component of RBLM is to use several stages of assessment within a single site or between sites, allowing prioritisation of funds and assessment of remediation options over time. The methodology makes it possible to identify sites of high risk and selective use of public funds for remediation where possible or indeed necessary (Mayes et al., 2009). In this study, a comparable approach has been applied in New South Wales. Both Mayes et al. (2009) and the current work emphasise prioritisation between sites. In both cases, within a given site further development and prioritisation is demanded. In particular, an approach that underpins remedial options and decisions especially in situations when high capital expenditure is being considered.

This chapter outlines the development of a risk-based framework to manage the environmental risks of metalliferous mine sites throughout New South Wales. The first section outlines the methodology, while the second section uses two sites to demonstrate its application. Finally, the path for making progress is discussed within the context of contaminated site remediation in Australia at the time of writing.

18.2 RISK-BASED MANAGEMENT OF LEGACY MINES

Within New South Wales, a prioritisation scheme previously existed for legacy mine sites. Risk-based management of contaminated sites is an approach that encourages informed use of funds for risk assessment and remediation (Mayes et al., 2009). The Legacy Mines Programme has already conducted an inter-site qualitative ranking based on risk estimated from available data. Significant uncertainty remains within many existing sites due to discrepancies in the data set and level of knowledge about the actual risk posed by contaminants to human and ecological receptors. There was a need to identify high-risk zones using an evidence-based approach. The risk-based approach expanded on existing information by focussing on the exposure of receptors to contaminants with suitable bioavailability and process strategies. For the expansion, a generalised conceptual site model (Figure 18.1) was devised considering the toxicology to human and ecological systems with a focus on aquatic and terrestrial ecotoxicology, human health impacts (soil and water ingestion, dust exposure), surface water impacts, groundwater impacts and erosion.

The proposed methodology involves an intra-site risk assessment tool to help make decisions on remediation and remediation goals. An overview of the proposed methodology is shown in Figure 18.2. The initial proposed methodology included the full process of assessing each legacy mine site (DEFRA, 2011). This includes problem characterisation (including receptor identification), risk assessment and remediation appraisal. The process is iterative over time enabling investment as or when funds become available.

For this chapter, two legacy mine sites – Mole River arsenic mine and Webb's Consol mine - served as examples of implementing the RBLM strategy. Both sites were metalliferous mines with acid mine drainage (AMD) issues which were detected in seepages and receiving streams with

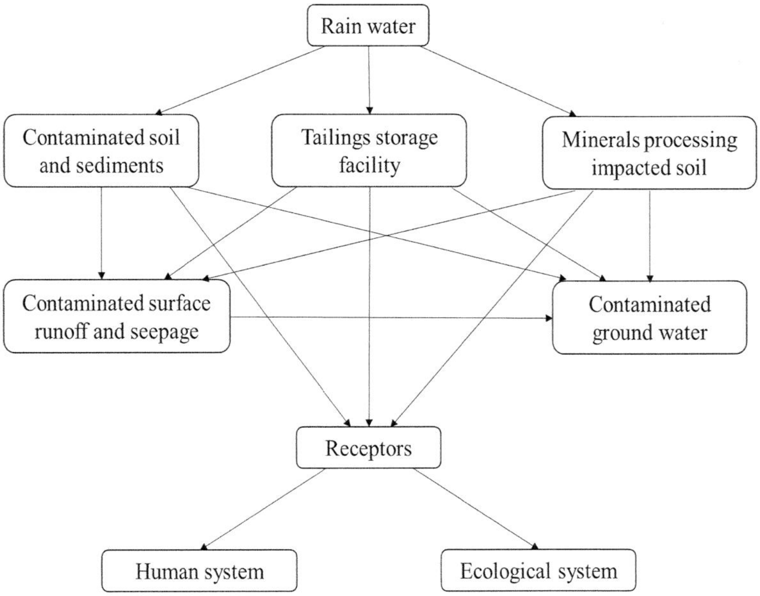

FIGURE 18.1 Generalisation of a mine site conceptual site model.

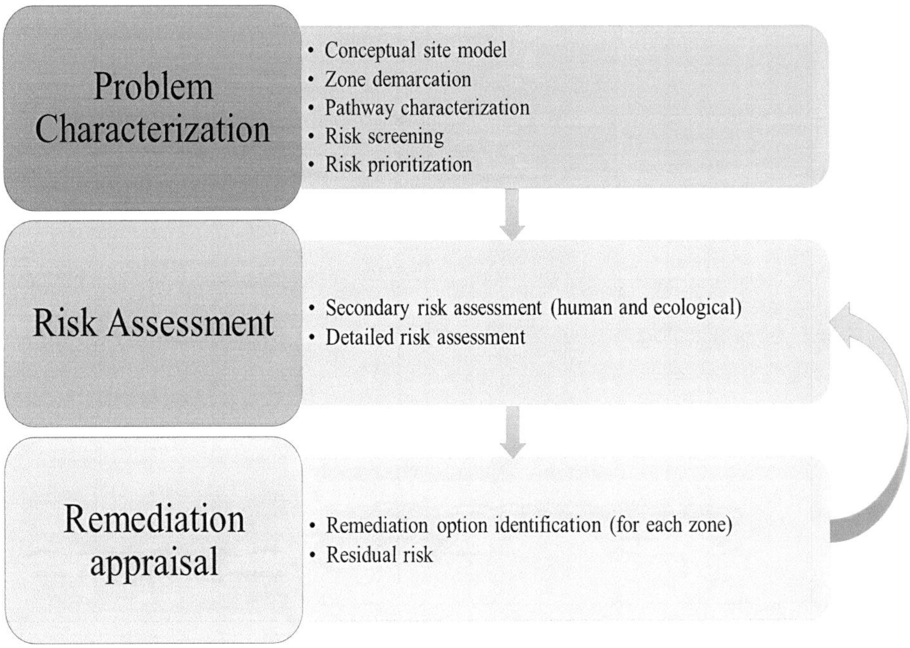

FIGURE 18.2 Outline of the methodology used for this risk-based approach.

extreme total metal(loid) concentrations (Ashley et al., 2004; Ashley and Lottermoser, 1999). Many shafts and processing areas existed at the sites and hence contaminants were dispersed across the sites with the presence of high to extremely high concentrations of metal(loid)s. The main issue included high concentrations of arsenic, lead or zinc with acidity being an added problem.

18.3 THE RBLM FRAMEWORK FOR LEGACY MINES

The work involved ranking mine sites based on identified risk factors using rankCARE™ software developed by Cooperative Research Centre for Contamination Assessment and Remediation of the Environment (crcCARE). It primarily functions to prioritise contaminated sites for making decisions. This easy-to-use application considers a list of weighted "parameters" to define the extent of contamination risk and urgency of action for a given contaminated site/area. Sites can also be split into one or more areas, with parameters applied to each one. A final score is rendered for each area and all data are presented in a graph allowing the areas to be ranked. The data can be added, modified and deleted as the situation changes. The components of the approach are shown in Figure 18.3. Each component is briefly outlined below.

18.3.1 Surface Soil Assessment

Assessment of the surface on the site was undertaken to ascertain areas which most represented risk to human health and the environment. The assessment examined spatial distribution of metals in soils and waste, and also analysed soil properties for determination of properties relevant to erosion. Characterisation of the mineralogy was conducted to generate insights into the mineral phases, which is important to understand metals' mobility in the environment.

18.3.2 Soil Sampling and Analysis

Soil samples were collected by grid-based sampling across the mine sites to a depth of 0.15 m. For the purposes of measurement of total elemental content by X-ray fluorescence (XRF) (Olympus

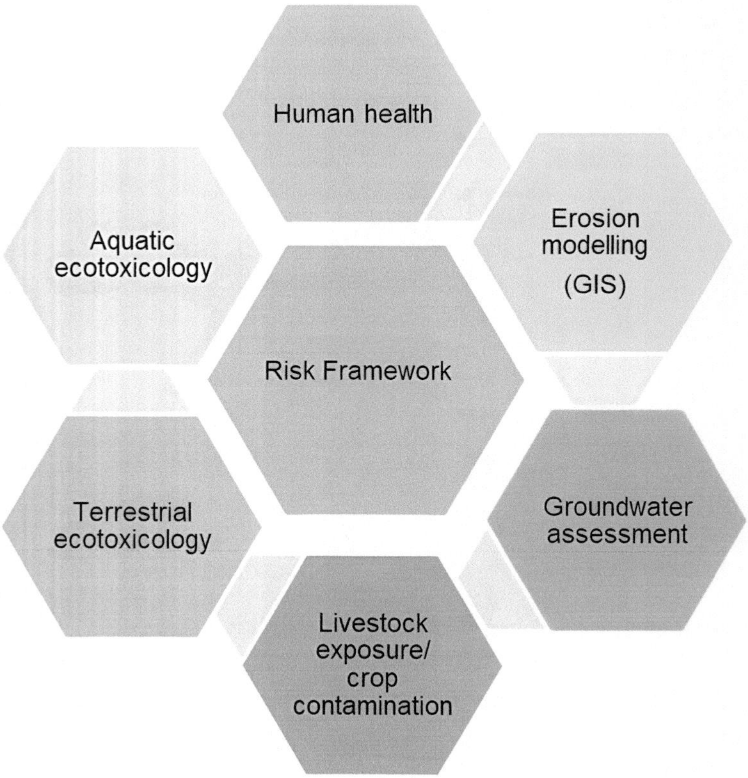

FIGURE 18.3 Components of the RBLM framework.

Delta DP 6000), soil texture analysis and determination of soil properties, samples were taken and put into zip-locked bags. A subsample was pressed into small plastic cups for XRF analysis.

In the laboratory, soil samples were dried at 37°C and sieved to <4 mm with stainless steel sieves. Samples were stored under ambient conditions in plastic containers prior to use. Soil pH and electrical conductivity were measured using standard methods (Rayment and Higginson, 1992). Soil texture was determined on selected samples at the site using the hydrometer method (Gee and Bauder, 1986). Texture measurements are essential for the erosion component. Furthermore, the mineralogy of selected samples was analysed using X-ray diffraction (XRD) (PANalytical Xpert Pro) and scanning electron microscopy (SEM) (Philips XL30 SEM with Oxford ISIS EDS). Here the purpose was to determine mineral phases in samples from soil and waste piles. Selected bulk composite soil samples from each site domain were also collected for terrestrial ecological risk assessments.

18.3.3 Soil Erosion Modelling

The Revised Universal Soil Loss Equation (RUSLE) coupled to Geographic Information System (GIS) was applied to estimate soil loss. The RUSLE is a widely used and accepted model that predicts the average soil erosion rate from a certain area (Fistikoglu and Harmancioglu, 2002; Lu et al., 2006; Tim et al., 1992). The coupled GIS-RUSLE model has the benefits of: (1) being able to map large areas with relative ease; (2) identification of key sources of sediment from within an area; (3) combining RUSLE erosion indices with XRF data, and therefore; (4) identifying key sources of contaminant loss from within a site and region.

The RUSLE equation can be expressed (Tim et al., 1992) as follows:

$$A = R \times K \times L \times S \times C \times P \qquad (1)$$

Where:

A represents the potential long-term average annual soil loss in tonnes per hectare per year (tonnes/ha/yr);

R is the rainfall and runoff factor, equal to the sum of the annual or seasonal energy intensity interaction factors for all storms;

K denotes the soil erodibility factor. It is the average soil loss in tonnes/hectare for soil in a particular structure;

L (length) and S (steepness) factors represent a ratio of soil loss. The steeper and longer the slope, the higher the risk for erosion;

C (cover) and P (practice) factors stand for the land management factor. These two factors can be determined by selecting the on-site vegetation type and tillage/farming method.

The RUSLE parameters were derived for each site:

- R: Using rainfall data from nearby weather stations.
- K: Based on soil texture determinations and soil K value monograph, a kriging interpolation function was applied for mapping the soil K value to the mining site area.
- LS: The slope angle was determined by using the slope function in ARCGIS and the slope length calculated from a Flow Accumulation function.

$$LS\ factor = 0.4^{\frac{FlowAcc \times Resolution}{22.1}} \times 1.4^{\frac{\mathrm{Sin}([Slope]=0.01745)}{0.09}} \times 1.4$$

Where:

FlowAcc – the slope length calculated using the Flow Accumulation Function;

Resolution – the map resolution of the ArcGIS map;

Slope – the slope steepness in degrees calculated using the Slope function.

- C: Land cover management (C) measures the effects of all the interrelated vegetative cover. It is a factor that measures the protection of the soil surface from raindrop and overland flow impacts by vegetative material. The value of C factor can vary from 0 for very well protected soils to 1.5 for finely tilled, ridged surfaces that produce much runoff, leaving it susceptible to rill erosion.
- P: The land management practice (P) measures the effects of management practices such as contouring, terracing and strip cropping used to reduce soil erosion. For areas where there is no support practice the P factor is set to 1.0. For C factor, the land cover information was obtained using the satellite image from Google Maps. Vegetative cover level was manually classified using the image classification function from ArcGIS. The P factor for the mining area was set to 1.0.

Application of the RUSLE model coupled with GIS makes it possible to calculate contaminant yield (kg/year) derived from erosion at the site's areas.

The contaminant yield CY (kg/year) of a given area was calculated as:

$$CY = SY \times Er \times SDR \tag{5}$$

Where SY is estimated from the coupled RUSLE-GIS-based erosion model, as described above. SDR is the sediment delivery ratio and it accounts for differences between eroded sediment and what is delivered to a given stream point:

$$SDR = 10\ r/l$$

Where r is the difference in elevation between a stream point and the average flow path distance to a given stream point; l is the average distance from a land cell to a given stream point. The SDR value may be calculated using many different methods, ranging from relatively simple to quite complicated approaches. We have chosen a simple method based largely on slope and distance (Tim et al., 1992).

Er is the enrichment ratio, which is used to correct for different transport pathways, notably clay and "colloidal" transportation:

$$Er = Cs/Cl \tag{6}$$

Where Cs is the sediment concentration and Cl is the average concentration on a given land unit. The term CY_b is used as a comparison with background soil concentrations using the same input parameters. The term CY_c refers to contaminated yield for contaminated areas.

18.3.4 Surface Water Sampling

Water and sediment samples were collected from surface waters that were affected by erosion and runoff from the mine sites. Samples were collected upstream and downstream from the mine site to ascertain what happened to the surface water quality. Samples were also collected from any surface water running through the mine sites. Water samples were analysed for major cations (Ca, Mg, Na, K), major anions (SO42-, Cl-, NO3-), total cyanide and metal(loid)s (Al, Mn, Fe, As, Ba, Cd, Cr, Cu, Pb, Mn, Ni and Zn).

18.3.5 Groundwater Assessment

The main aim of the groundwater investigation was to assess the potential impact of the legacy mine on local water quality from acid mine drainage (AMD) and contaminants. Groundwater may be extracted and utilised by surrounding land users. Legacy mine sites for which there was much concern for impacted groundwater were identified. A limited number of groundwater wells were installed on the site upgradient and downgradient to establish groundwater flow direction and impact. Following installation of groundwater wells at the identified sites, a groundwater sampling programme was undertaken to determine standing water levels (SWL), field groundwater quality parameters including dissolved oxygen (DO), electrical conductivity (EC), pH, oxygen reduction potential (ORP) and temperature and collection of samples for laboratory analysis of metal content.

18.3.6 Human Health Risk Assessment

Contaminants from legacy mine sites represent a potential risk to residents in the vicinity of the mines or to those accessing these sites. Conceptual site models (CSMs) were devised for each legacy mine site as a basis for identifying sources, pathways and receptors for each site. Generally, a receptor may be exposed to contaminants in the soil, water or dust. Additionally, contaminants may migrate by downslope runoff and erosion, windblown dust, infiltration to groundwater, which can also affect receptors offsite. As such, the assessment of potential contaminant exposure and the associated risk was determined.

A screening level human health risk assessment was conducted to assess exposure to contaminants of concern in soil and water. Mining activities on the site have affected each zone to varying degrees and so exposure and risk have been determined for samples across these zones. Risks were also determined for each surface water and groundwater sampling point. The exposure assessment was conducted according to the approach provided in the NEPM ASC, 2013 for the ingestion pathway using the recommended intake parameters. This was done for the primary exposure pathway ingestion of contaminated soil and water. A basic principle adopted was to determine the risk originating from the highest possible exposure (daily exposure to contaminated media).

The equation for calculating ingestion dose is as follows:

$$\text{Ingested Dose} = C_{media} \text{ x IngR x CF x FI x F x ED/ BW x AT}$$

Where:

Ingested Dose = Average daily potential dose (mg/kg-day)
C_{media} = Concentration of contaminant in medium (e.g., mg/L, mg/kg)
IngR = Ingestion rate (e.g., L/day, mg/day)
CF = Conversion Factor ($1x10^{-6}$ to convert mg to kg)
FI = Fraction ingested from contaminated source (unitless)
F = Exposure frequency (days/year)
ED = Exposure duration (years)
BW = Body weight (kg)
AT = Averaging time (ED*365) and non-threshold exposures (=70 years x 365)

Exposure concentrations for As and Pb in soil were based on *in vitro* measurement of the bioaccessible fraction, which estimates the fraction of a contaminant available for uptake by people. Bioaccessibility was determined by a chemical test validated for arsenic called SBRC (Juhasz et al., 2014). Meanwhile, risk was calculated by comparing the calculated exposed dose to toxicity reference doses (RfD) provided by US EPA IRIS database. The oral reference dose (RfD) is a numerical estimate of daily oral exposure to the human population, including sensitive subgroups such as children, which is not likely to create harmful outcomes during a lifetime. Oral toxicity values are based on the empirical relationship between the occurrence of toxic effects and the amount of chemical ingested.

Toxic risks refer to the non-carcinogenic harms incurred due to the exposure above a certain concentration of a contaminant. These are called threshold exposures. Risk is indicated by a hazard quotient (HQ) > 1. The hazard quotient was calculated based on the total uptake of contaminant divided by the oral RfD. Carcinogenic risks are non-threshold exposures and are expressed in terms of increased incremental cancer risk from exposure over a lifetime. Cancer risk is calculated by daily uptake x cancer slope factor. The upper limit of acceptable cancer risk is 10^{-4} or 1 in 10,000 (US EPA, 1991). By converting the risk to a rank score, this makes it possible to compare between domains on sites and between legacy mines and for prioritisation of risk management of these sites.

18.4 APPLICATION TO MOLE RIVER, NSW

The Mole River arsenic mine site is approximately 450 m above sea level, located about 70 km north of Glenn Innes, New South Wales (Figure 18.4). The area of potential impact has a relatively mild slope, ranging from approximately 475 m to 430 m to the nearby Mole River. The site has previously been identified as having high arsenic (As) concentrations with minor occurrences of antimony (Sb), manganese (Mn) and zinc (Zn) (Ashley and Lottermoser, 1999). Previous studies also recorded downstream contamination in the two adjacent surface water bodies, Sam's Creek and Mole River. Farmers reside both upstream and downstream of the site.

Total concentration determinations with XRF were initially used to identify appropriate demarcations on site. These zones are: main processing area (MP), Inner west (IW), Outer west (OW), North east (NE) and South east (SE). Figure 18.5 shows the distribution of As throughout the site. The XRF data indicates the spatial distribution and magnitude of the contaminant of immediate concern, As. There were occasional incidences of elevated Zn, and to a much lesser degree, Cu, Mn and Sb (Table 18.1). Generally, these minor events were associated with extremely elevated As around the main processing area. Across the site (n=104) the minimum, maximum, median and mean As concentrations were 6.10, 22,700, 144 and 811 mg/kg, respectively. Based on the median As concentration, the areas with the highest to lowest As levels were (value within parenthesis is the median As content (mg/kg)):

Inner west (316)> Main processing (241) > South east (184) > Outer west (127)> North east (122)

Using the Kruskal Wallace test (Zar, 1999), no area was significantly higher in As than another area ($P<0.05$). The spatial map (Figure 18.5), however, shows there were four hot spots identified from XRF measurements on-site. The largest contaminated area was the IW domain. This area revealed very high levels at the most elevated aspect of the site (south end). This area also extends towards the south end of the MP domain, and north down to the western collection dam. There was also a hotspot associated with the eastern collection dam on the east of the site, which extends into the SE zone. Across Sam's Creek to the east, As contamination was observed into the neighbouring farm. Concentrations in the corner of the adjacent eastern property were <500 mg/kg. Arsenic contamination in this area is possibly due to an explosion that occurred in the 1930s. Based on the available As data, the suggestion that "six square miles" (Millar, 1996, cited in Ashley and Lottermoser, 1999) of impact from an incident in the 1930s appears likely to be an exaggeration.

The remaining hotspot occurred in the OW zone (outside of the fenced mining area). There was a high level of As in the exposed (non-vegetated) area just outside the site fence. The elevated As concentrations extend downslope to the edge of the study area. The clear "tailing" of As soil concentrations was also reflected by a change in grass species. This is visible on the site in addition to aerial photographs.

Figure 18.6 shows the interpolated As concentrations from 0–500 mg/kg, 500–3000 and > 3000 mg/kg As across the site. These concentrations represent the generic Health Investigation Levels (HILs) HIL B and D (NEPC, 2013). Arsenic concentrations <500 mg/kg are not likely to present risk to human health based on default trigger values. HIL B represents "residential [land] with

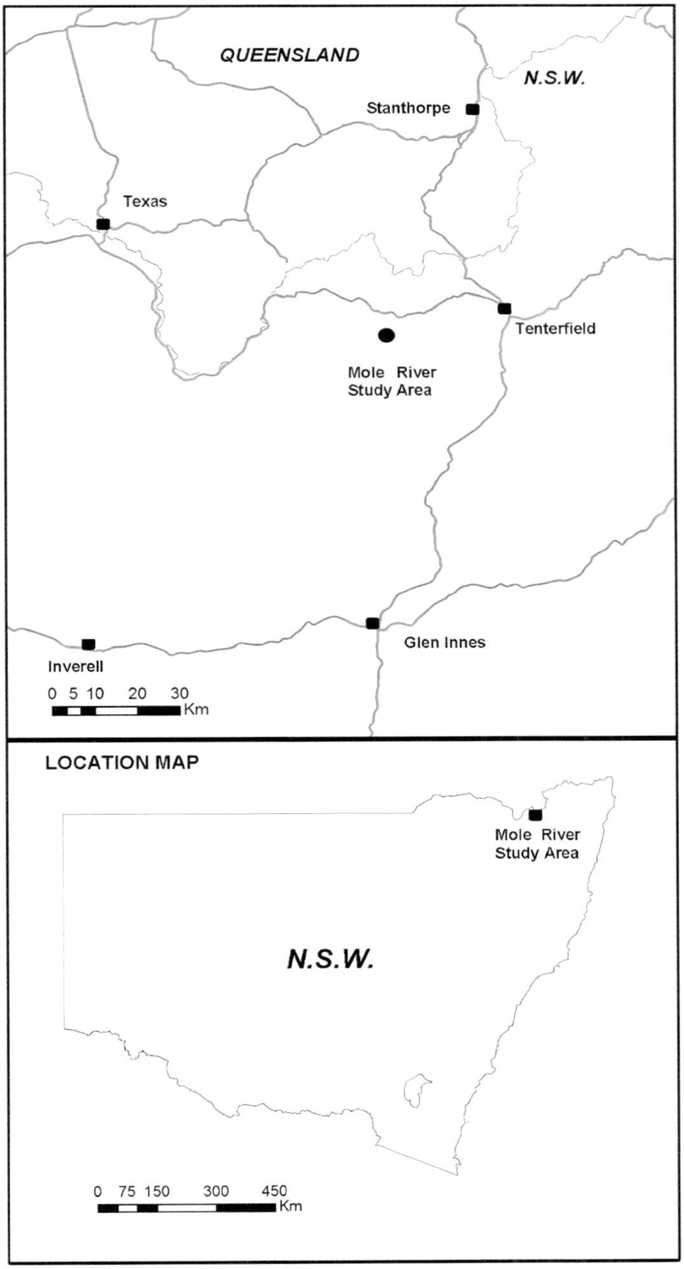

FIGURE 18.4 Location of the Mole River arsenic mine in NSW, Australia.

minimal opportunities' for soil access with fully and permanently paved yard space". HIL D (3000 mg/kg) represents commercial and industrial zones. Cadmium and Pb are of little to no concern based on total determinations. Manganese was significantly higher than background in the south east zone using analysis of variance (P<0.005). There were only 5/104 samples exceeding 30 mg/kg Sb (approximate XRF detection limit) on site. All were located in the MP domain.

FIGURE 18.5 Total As concentrations (mg/kg) determined by a portable XRF device. The concentration ranges are categorised by NEPC EIL values (<160 mg/kg) and other nominal concentration ranges. Data greater than 4000 mg/kg were not included due to limited data points in this range.

18.4.1 Mineralogy of As at Mole River

The arsenic-bearing rock processing area consists of a furnace connected to a chimney by condensing flue, other surrounding facilities including crusher and nitric acid plant. Three representative samples were taken from furnace brickworks, along condensing flue and along the drainage channel from the furnace to the bottom pond. XRD patterns indicated that samples taken from the chimney area consisted of arsenolite (As_2O_3, 2%), together with quartz, gypsum and feldspar minerals (heat-treated albite and orthoclase). Around the drainage channel, unweathered arsenic-bearing mineral arsenopyrite was evident (AsFeS, 3%) as well as 23% of oxidised product gypsum. The remaining minerals in this area consisted of feldspars.

The brick and concrete are covered by white salt crystals granules with a distinguished pink hue as shown in Figure 18.7. XRD spectra indicated that the crystals mainly consist of gypsum ($CaSO_4$, 42%), muscovite (42%), krautite ($MnHAsO_4.H_2O$ 6%), claudetite (As_2O_3, 2%) and realgar (As_4S_4, 8%). Claudetite is an arsenic oxide mineral formed as an oxidation product of arsenic sulphides and is normally colourless or white. Claudetite may be associated with realgar (As_4S_4). The minor realgar occurrences give the salt leached out from furnace brickworks a pink hue. The large mass of gypsum indicates that the oxidisation of arsenic sulphide mineral in the processing area has been ongoing for decades. The formation of multi-colour efflorescent salts is a sign of end-to-final step in the evolution of an acidic mine drainage system as the consumption of the neutralisation potential of the minerals and subsequent outflow of acidic, heavy metal-rich leachates.

TABLE 18.1
Total As, Cu, Mn, Sb, Pb and Zn Concentrations (mg/kg) Broken Down into the Five Zones (Note "bdl" Is Below Detection Limit)

Area	As (mg/kg)	Cu (mg/kg)	Mn (mg/kg)	Sb (mg/kg)	Pb (mg/kg)	Zn (mg/kg)
Main Processing						
Minimum	9.3	10	47	31	13	14
Average	2,560	37.7	368	62.2	39.1	97.1
Quartile 1	34.6	14	308	37	20.7	43.0
Median	241	16	342	68	25.8	48.2
Quartile 3	2,080	22	439	71	49.6	120
Maximum	22,700	181	949	104	107	447
Standard deviation	5,390	50.7	222	29.4	27.0	101
Count	19	13	19	5	19	19
North East						
Minimum	6.1	8	193	<bdl	15.4	20.5
Average	137	19.9	495	<bdl	38.3	91.8
Quartile 1	74.9	11.3	338	<bdl	27.7	49.7
Median	122	16	434	<bdl	37.3	76.5
Quartile 3	227	23	646	<bdl	48	108
Maximum	367	56	995	<bdl	73	239
Standard deviation	103	12.4	214	<bdl	15.2	62.9
Count	22	18	22	<bdl	22	22
South East						
Minimum	49.6	7	360		11.6	46
Average	317	37	1,100	<bdl	34.4	87.2
Quartile 1	68.9	14.2	807	<bdl	17	62
Median	184	31	1,150	<bdl	22.3	76
Quartile 3	314	52	1,480	<bdl	32	91
Maximum	2,050	86	1,700	<bdl	107	178
Standard deviation	534	27.7	463	<bdl	31.0	40.5
Count	13	12	13	<bdl	13	13
Inner West						
Minimum	15.2	127	11	<bdl	13	41.7
Average	1,200	417	27.2	<bdl	28.8	62.0
Q1	71	272	12	<bdl	17.2	47.5
Q2	319	401	17	<bdl	22.8	55
Q3	1,160	542	25	<bdl	25.0	64
Max	11,800	751	138	<bdl	127	156
SD	2,740	190	30.7	<bdl	25.8	26.1
Count	18	18	17	<bdl	18	18
Outer West						
Minimum	28.3	8	195	<bdl	13.7	28
Average	337	35.1	475	<bdl	31.5	77.4
Q1	67.6	10.8	305	<bdl	19.8	53.9
Q2	127	16	414	<bdl	27.6	71
Q3	323	38.3	612	<bdl	34.3	90.5
Max	2,250	194	951	<bdl	97	190
SD	530	43.0	211	<bdl	18.8	35.6
Count	34	26	34	<bdl	34	34

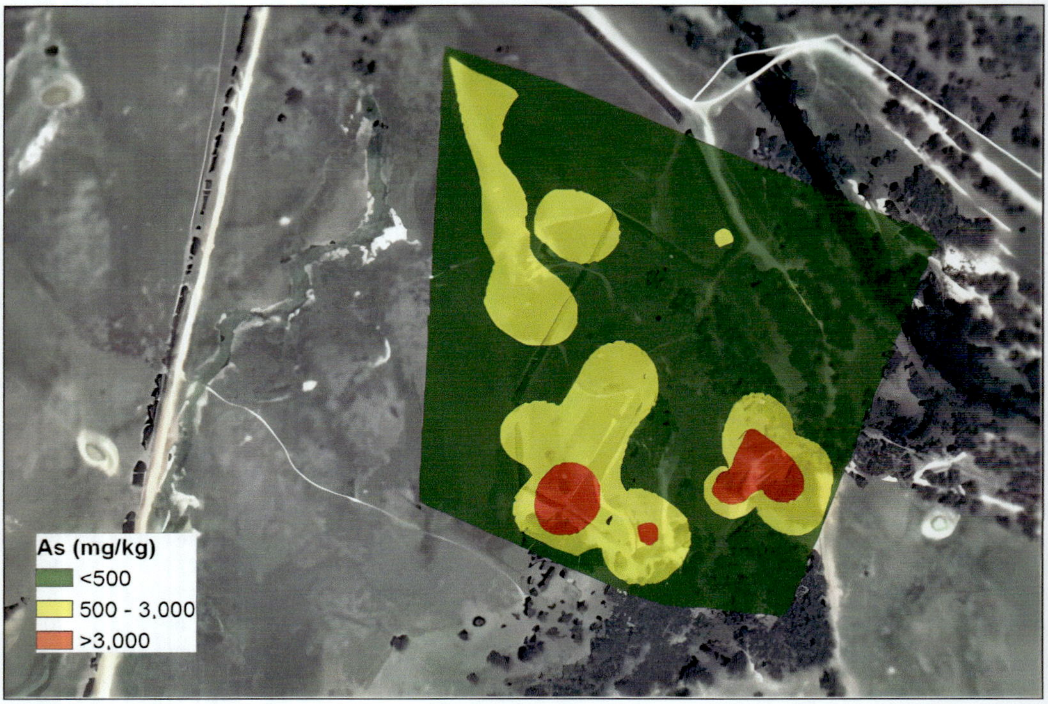

FIGURE 18.6 Total As concentrations (mg/kg) determined by a portable XRF device categorised by HIL B (<500 mg/kg), and HIL D (>3000 mg/kg).

FIGURE 18.7 White and pink crystals are growing on the brickworks of the furnace in processing area.

18.4.2 Erosion

The highest soil erosion was observed in the southern area, corresponding to a soil erosion rate of 10 t ha $^{-1}$ year^{1} as predicted in the Mole River catchment (Figure 18.8). The RUSLE prediction conducted in the Mole River arsenic mine area can be considered as one method that validated the soil erosion results predicted for Mole River catchment, as more reliable data of the soil texture was obtained from the field survey. The good agreement between modelling results in Mole River catchment, Mole River arsenic mine area and previous studies (Lu et al., 2003; Teng et al., 2016) strongly suggests acceptable consistency between predictions and observations.

Arsenic and soil loss are predicted to co-occur with hotspots observed within the XRF maps. The RUSLE predicted erosion losses of As are more specifically observed within exposed soil surfaces. There are several co-relationships between soil loss and predicted soil-As losses. High soil As is likely to reduce plant productivity in these areas. As a result of the lack of vegetation, the erosion rates are predicted to increase, subsequently mobilising more As on the soil surface.

18.4.3 Human Health Risk Assessment

Arsenic contamination of soil, surface water and groundwater may pose a health risk to the residents in the vicinity of the mines or to those who access these sites. Potential receptors include people going to the site or in the vicinity and those using surface water or groundwater downstream of the site. A camp site exists to the north east approximately 150 m from the eastern corner of the site, just off Potters Road. Furthermore, bioaccessibility measurements were conducted on soil samples from each zone in the site area to determine the fraction of contaminants in soil available for human uptake by ingestion. The bioaccessibility of As ranged from <1 – 205 mg/kg, with the highest bioaccessible As around the former workings area in the IW zone and another hotspot in the NW just beyond the fence (Figure 18.9).

Arsenic bioaccessibility followed the total As concentrations across the site, with the exception of areas in the main processing areas and to the east. As indicated in Figure 18.9, As bioaccessibility in these regions is very low. Bioaccessible As is relatively low, so the risk from exposure to As contaminated soil is also relatively low (Table 18.1). The HQ values are all well below 1, although cancer risk is above that considered negligible (10^{-6}) but below that considered unacceptable 10^{-4}. A risk ranking score was calculated for soil exposure, including carcinogenic and non-carcinogenic risk. The risk ranking score assigns a logarithmic score based on HQ and Cancer Scores documented in Table 18.2. The risk scores are all relatively low <1. Within the MP domain, this may be attributable to As mineralogy. The primary arsenic-bearing mineral processed on the site was Arsenopyrite (FeAsS), refined to Arsenic Oxide (As_2O_3) and Arsenic Pentoxide (As_2O_5). The solubility of pure As_2O_3 is relatively low (approximately 20g/L at 25 °C); only a small fraction of the total soil mass consisted of these minerals, which may have been separated out in the <250 μm fraction.

Despite the bioaccessibility data generally following the spatial total concentration trends on-site, the data indicate a relatively low risk from incidental ingestion (Table 18.2). The hazard quotient and cancer risk from incidental soil ingestion are relatively low on-site despite the very high total As concentrations. This emphasises the importance of accounting for bioaccessibility and bioavailability with reference to impacts, rather than generic soil total concentration trigger guidance values.

In contrast, the surface and groundwater associated with the Mole River site indicate significant risks to humans (Table 18.2). The concentration of arsenic in Sam's Creek (the adjacent water course which feeds into Mole River) ranged from 0.019 mg/l upstream to 1.775 mg/l adjacent to the processing area. Arsenic in Sam's Creek downstream from the site was also elevated, ranging from 0.177 to 1.21 mg/l. The Australian drinking water health guideline (NHMRC, 2011) for As is 0.01 mg/l (10 μg/l). In Mole River, the upstream concentration of As was 0.002 mg/l. Downstream As concentrations in Mole River were elevated within 200 m of the site (0.101 mg/l), reducing with distance to levels similar to upstream levels (0.003 mg/l).

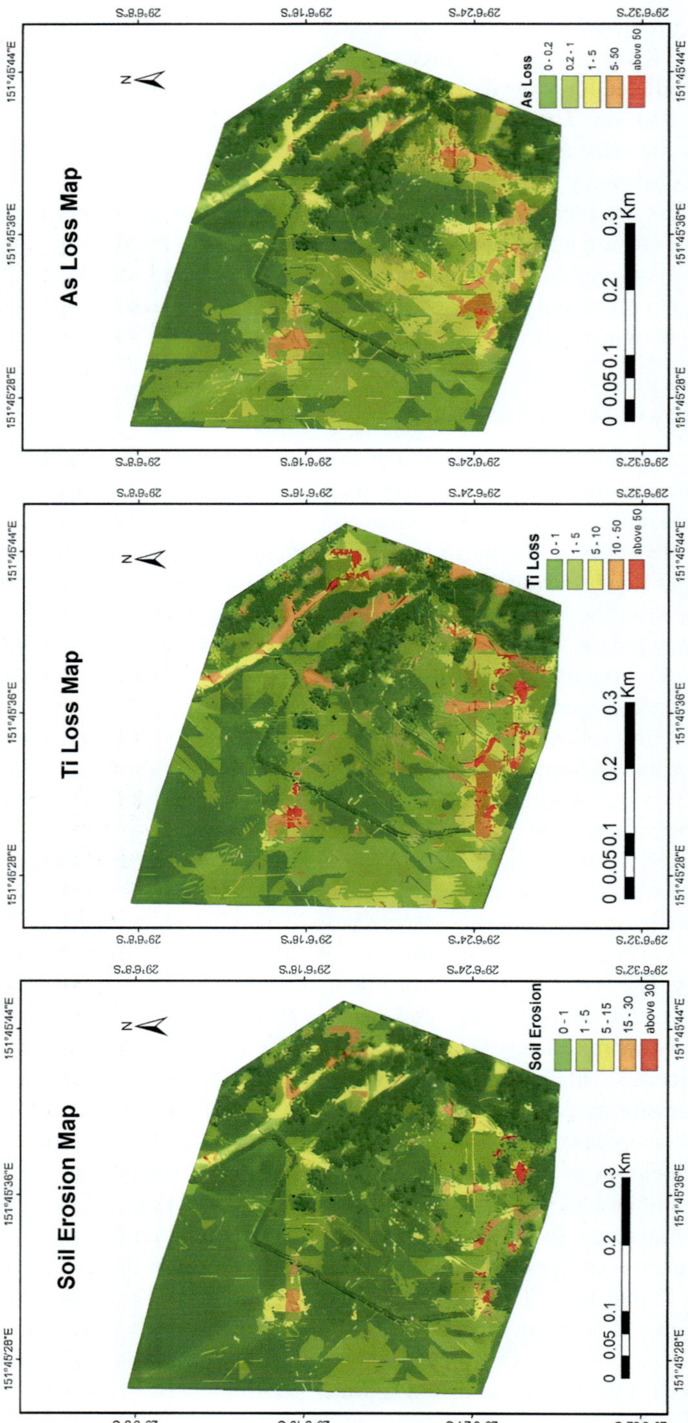

FIGURE 18.8 GIS-RUSLE erosion models results for the Mole River site (Units: soil - tonnes/ha/year. Metals kg/ha/year).

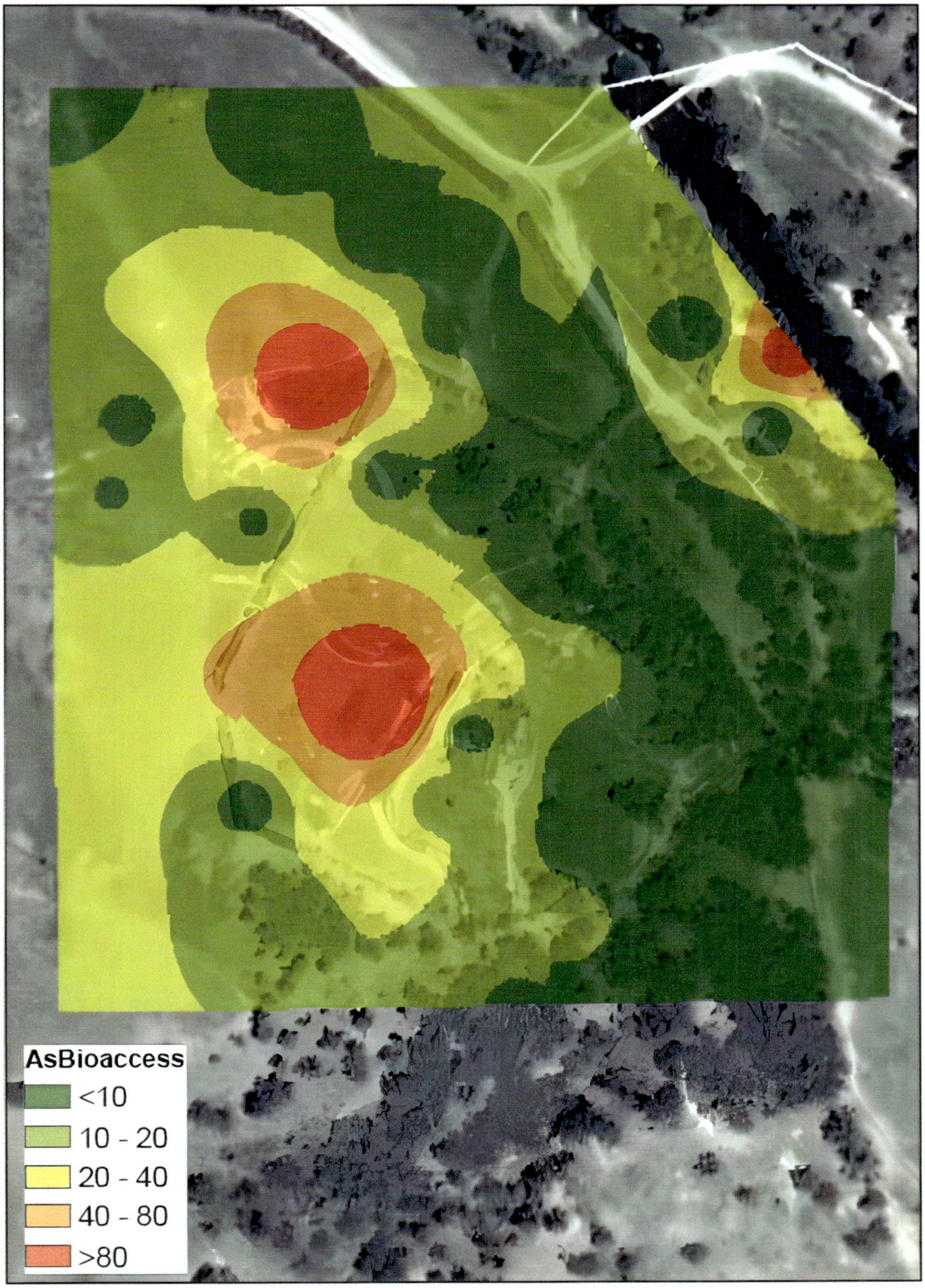

FIGURE 18.9 Bioaccessible As (mg/kg) in soil from mine site areas.

The concentration of As in surface water leads to calculated HQ much greater than > 1 in Sam's Creek downstream from the mine (Table 18.2). The cancer risk is also well above the unacceptable level (10^{-4}). Downstream from the intersection of Sam's Creek with Mole River, the HQ and cancer risks are also quite high for the closest sampling point (192 m). The As concentration declines rapidly with distance and so the risk downstream in Mole River is less (HQ<1). Meanwhile in the

TABLE 18.2
Exposure Assessment for As in Soil, Surface and Ground Water (Surface Water from Sam's Creek (SC) and Mole River (MR))

	As mg/L	As	As Rfd mg/kgbw/day	As HQ	Cancer Risk
Soil – IW domain (average)	n.a.	3.61×10^{-05}	0.003	0.12	5.41×10^{-05}
Soil – MP domain (average)	n.a.	7.51×10^{-06}	0.003	0.03	1.13×10^{-05}
Soil – NE domain (average)	n.a.	3.23×10^{-06}	0.003	0.05	2.45×10^{-05}
Soil – OW domain (average)	n.a.	1.86×10^{-05}	0.003	0.06	2.79×10^{-05}
Soil – SE domain (average)	n.a.	4.39×10^{-06}	0.003	0.01	6.58×10^{-06}
SC Upstream (470 m)	0.019	0.000755	0.0003	2.52	1.13×10^{-03}
SC adj to processing (320 m)	1.7775	0.070622	0.0003	235	1.06×10^{-01}
SC Near road (195 m)	0.177	0.007032	0.0003	23.4	1.05×10^{-02}
SC Near MR intersection (20 m)	1.21	0.048074	0.0003	160	7.21×10^{-02}
MR Upstream (-610 m)	0.002	7.88E-05	0.0003	0.263	1.18×10^{-04}
MR downstream (192 m)	0.101	0.004005	0.0003	13.4	6.01×10^{-03}
MR Downstream - bridge (365 m)	0.003	0.000114	0.0003	0.381	1.72×10^{-04}
Mole River - crossing (8900 m)	0.003	0.000103	0.0003	0.344	1.55×10^{-04}
Groundwater well 01 (MR-GW01)	0.004	0.000159	0.0003	0.530	2.38×10^{-04}
Groundwater well 02 (MR-GW02)	0.953	0.037864	0.0003	126	5.68×10^{-02}
Groundwater well 03 (MR-GW03)	0.023	0.000914	0.0003	3.05	1.37×10^{-03}
Groundwater well 04 (MR-GW04)	0.039	0.00155	0.0003	5.17	2.32×10^{-03}
Groundwater well 05 (MR-GW05)	0.004	0.000159	0.0003	0.530	2.38×10^{-04}

groundwater wells, the As concentration exceeded the drinking water guideline (0.01 mg/l) in three of the wells (GW02, GW03, GW04). Water from these wells also has HQ >1. All of the groundwater wells have a cancer risk > 10^{-4}. The highest risk is associated with GW02, which was a previously installed well near the drainage collection dam. GW03 and GW04 also have elevated As concentrations and risk. The high groundwater well concentrations are a cause for concern around off-site movement of As.

On-site several collection dams have been constructed to collect surface runoff. The As concentrations in these dams were extremely high (1-41 mg/L). The drainage collection dam at the bottom of the site was highly contaminated (41 mg/l As), despite the water dams in the western part of the site also having unsafe levels of As. Clearly, the As content in these dams is also hazardous under chronic exposure scenarios (HQ>1, Cancer risk > 10^{-4}) (Table 18.2). The minimal lethal dose for As lies in the range of 1 mg/kg/day. Consumption of water in Sam's Creek results in an average lifetime dose of 0.07 mg/kgbw/day. The dose from the drainage collection dam exceeds the minimum lethal level (1.627 mg/kgbw/day). Previous investigations (unpublished data) have indicated more than 60 mg/L of As in this dam. This means there is an acute toxicity risk to children and adults should they consume water from this dam. At this level, wildlife may also be endangered by dissolved As levels.

18.5 APPLICATION TO WEBB'S CONSOLS

The Webbs Consols mine produced lead, copper, zinc and silver. It is a good example of the potential environmental impacts that historical mining has left behind in this region. The mine still contains significant historical artefacts with operations starting in the mid-1880s and continuing until the 1950s (Appleton, 2003). Due to the age of the mine, metal recovery from ore was relatively low by modern standards, similarly to its environmental standards. Previous investigations at the site have indicated exceedingly high concentrations of metals and metalloids in soil and surface waters. It should be noted that despite the removal of most contaminated material from the drainage line, XRF analysis in 2016 indicated a high level of As, Pb and Zn contamination across the site (Table 18.3). Total XRF concentrations from Webb's Consols are reported for different domains of the site (Southern area and Castlereagh, Gates Crusher, Main Processing area, Mount Galena and Shaft Four and Unknown Dump). The major contaminants of concern at the site are As, Pb and Zn. Across much of the site, concentrations of As, Pb and Zn are highly elevated relative to control samples. The control sites had average concentrations of 9.6, 65 and 56 mg/kg for As, Pb and Zn, respectively.

The SOD domain generally had smaller total concentrations than other areas on the site. The median As, PbPb and Zn concentrations were 203, 194 and 357 mg/kg, respectively. Paradoxically, the highest concentrations found on site were detected in the SOD area: 366,000, 15,600 and 20,800 mg/kg for As, Pb and Zn, respectively. These measurements were made adjacent to a capping works used to isolate tailings material. The GC domain was moderately contaminated relative to other areas. The median As, Pb and Zn levels were 446, 1050 and 577 mg/kg, respectively, indicating a relatively high Pb and Zn concentration. The source of contamination in this area is the Gates Crusher and fluvial deposition from the site. The highest As concentration was actually found on the Consols Creek bank within the GC domain (46,100 mg/kg). As expected, the "hot spot" was located near the Gates Crusher.

Located around the main Lucky Lucy Shaft and waste pile, and an area west (upslope) of the Lucky Lucy shaft, the Lucky Lucy area is characterised largely by high As (median 596 mg/kg) and moderate to low Pb (median 138 mg/kg) and Zn (median 144 mg/kg). The main processing area has very high concentrations of As, Pb and Zn, although these levels are largely associated with waste rock. The median concentrations were 2,819, 2,948 and 1,032 mg/kg for As, Pb and Zn, respectively.

The GZ domain is the largest and main section of the Webb's Consols site. Arsenic and Pb tend to be of greater significance in this area. Median concentrations were 1,429, 1,310 and 322 mg/kg for As, Pb and Zn, respectively. There are three major hotspots in this area. These include the Mount Galena shaft area, Shaft Four and a small processing plant approximately 30 m north of the Southern Dam. The deposition area of Consols Creek in the Mount Galena domain is generally highly contaminated with As and Pb. The western slope (elevated from Consols Creek) in this area is believed to be uncontaminated based on XRF measurements. On the eastern side of the GZ domain, soils have very high contaminant levels but are well vegetated.

18.5.1 Mineral Composition of Webbs Consols

The mineralogy of the sample taken from the main processing with the semi-quantification result from XRD shows mineral composition of 37% of quartz (SiO_2), 48% of primary rock mineral microcline (Al0.94K0.871Na0.12O8Rb0.008Si3.06), 6% of muscovite (H2Al3K1O12Si3) and arsenic- and lead-bearing minerals: 7% anglesite (O4Pb1S1) and 3% beudantite (H6Al0.46As1.07Fe2.54O14Pb1S0.93). Beudantite forms part of solid solutions containing Pb, As and Fe, and therefore it can resemble scoradite. Similarly, the presence of beudantite could be confirmed and detected by SEM and EDS. The XRD results indicate that around the main shaft area was one of the worst heavy metals–contaminated spots, since the mineralogy consisted of 64% microcline, 15% muscovite, 6% anglesite (O_4PbS), 5% beudantite ($H_6Al_{0.46}As_{1.07}Fe_{2.54}O_{14}PbS_{0.93}$) and 10% scorodite ($Fe(AsO_4)(H_2O)_2$)). The presence of scorodite and beudantite could also be confirmed by EDS and SEM (data not shown). The presence of anglesite was also confirmed by SEM and EDS analysis. The presence of scorodite, anglesite and beudantite is indicative of an oxidised environment following exposure of metal pyritic materials.

XRD of the waste pile located at the Lucky shaft similarly had 12% scorodite, which matched stoichiometrically to the As content in the waste pile. The abundant presence of anglesite and scorodite agrees with previous mineralogical evaluation of the surface environments at Webb's Consols

TABLE 18.3
Statistics for the Total Concentrations of As, Pb and Zn Broken Down into Each Area

Domain	Zn mg/kg	As mg/kg	Pb mg/kg
SOD	Zn	As	Pb
Average	1266	10620	848.9
Maximum	20780	366200	15560
Median	357.0	203.0	194.0
Minimum	0.0	10.3	22.9
N	47	47	47
Q1	105.5	54.5	125.0
Q3	1131	1359	796.0
Standard Deviation	3474	53580	2688
GC	Zn	As	Pb
Average	938.1	4,624	4,765
Maximum	5667	46,130	31,48
Median	577.0	446.0	1052
Minimum	28.8	81.0	32.7
N	25	25	25
Q1	316.0	267.0	483.0
Q3	1,038	1,891	3,721
Standard Deviation	1,166	11,610	5,613
LL	Zn	As	Pb
Average	193	8,670	253
Maximum	1190	218,000	2,330
Median	144	596	138
Minimum	0	19.5	0
N	36	36	36
Q1	101	214	77.0
Q3	235	3,680	312
Standard Deviation	201	5,190	1,368
Main Processing area	Zn	As	Pb
Average	3324	19,090	12,05
Maximum	21,270	231,000	79,41
Median	1032	2819	2948
Minimum	86.0	59.0	48.0
N	33	33	33
Q1	520.0	272.0	487.0
Q3	3009	10,740	13,92
Standard Deviation	4,960	46,050	21,12
GZ	Zn	As	Pb
Average	1,590	8,270	3,960
Maximum	50,040	366,200	157,600
Median	322.0	1,429	1,310
Minimum	27.1	7.7	49.0

(Continued)

TABLE 18.3 (CONTINUED)
Statistics for the Total Concentrations of As, Pb and Zn Broken Down into Each Area

Domain	Zn	As	Pb
	mg/kg	mg/kg	mg/kg
N	58	58	58
Q1	179.8	474.3	666.0
Q3	659.3	5,001	3,850
Standard Deviation	6,648	33,440	13,97
UD	Zn	As	Pb
Average	224.3	36.01	214.4
Maximum	1,345	185	611
Median	85	26	172
Minimum	28	0	41
N	21	21	21
Q1	50	17	92
Q3	225	40	279
Standard Deviation	313.5	40.71	165.6

(Ashley et al., 2004). The lack of Pb incorporated within scorodite in the Lucky Lucy domain is due to the smaller amounts of Pb (i.e. no beudantite), as also shown with XRF measurements.

18.5.2 Erosion Modelling

The rates of As, Pb and Zn movement from erosion reflect areas of high concentrations. In the northern area, the Unknown Dump indicates very little movement of As (up to 6g/year), while Lucky Lucy has a relatively high rate of As movement. Arsenic mobilisation by erosion is very localised around the shaft and waste pile in this domain. The remaining northern areas of the site indicated relatively low mobilisation of As via erosion. Using a portable rainfall simulator on the waste pile, it was shown that rainfall at >100 mm/hour would be required to cause runoff and, hence, erosion. The high porosity of the pile resulted in >90% of incoming rainfall percolating through. Lead and Zn erosion rates in all three northern areas were relatively small in comparison to the southern areas.

The southern area consists of the capped dam, the Castlereagh workings and the Mount Galena domain, which includes Shaft Four. In the GZ domain, the surface erosion maps of this area indicated very high rates of soil loss, with As, Pb and Zn largely isolated to sizable hotspots. Arsenic, Pb and Zn erosion losses from the GZ domain reached up to 9.8 kg/ha/yr, 9.3 kg/ha/yr and 1.4 kg/ha/yr, respectively. Lead and Zn erosion occurred predominantly around the GZ domain and around the Main Shaft Dam, Main Shaft area. Substantial erosion of As was predicted to occur from nearby Shaft Four. Arsenic erosion losses also occur significantly from nearby Shaft four. The SOD domain contributes little to metal losses from the site. This area has lower total metal(loid) concentrations and high vegetation cover. The hot spots in this area are mostly adjacent to the dam.

The estimates of As, Pb and Zn entering the stream (CY_c) were made and converted to risk scores. The two most important areas are LL and the GZ domains. This was partially due to the size of the areas but also the high concentrations and higher soil erosion loss rates. Mount Galena contributes 64%, 58% and 38% of the total estimated rates of As, Pb and Zn loss from the site as a whole. Lucky Lucy is also an important contributor of As (29%), Pb (31%) and Zn (50%). The risk scores reflect this for these two areas.

18.5.3 Human Health – Webb's Consols

Site assessment identified key areas of the mine site that were highly contaminated and potentially impacting the surrounding environment and presenting a risk to human health. The contaminated soils in processing and mine workings areas, remaining waste rock and the lined tailings waste cap represent potential sources for contaminant transport through surface water runoff, infiltration to groundwater and by windblown dust. Erosion and surface water runoff into Consols Creek may be a significant pathway for contaminant transport.

Arsenic and Pb are the primary contaminants of human health risk at Webb's Consols, based on an assessment of contamination at the site. Concentrations of As and Pb are quite high particularly in the main processing area and mine workings areas. The primary Pb and As bearing minerals in the main processing area are anglesite ($PbSO_4$), beudantite $PbFe_3(OH)_6SO_4AsO_4$ and scorodite ($Fe(AsO_4)(H_2O)_2$) comprising, respectively, 6%, 5% and 10% of the waste rock. In the Gates Crusher area, anglesite and beudantite were the Pb and As minerals reported, comprising 7% and 3% of the sample, respectively. Anglesite results mainly from the weathering of Galena. Anglesite exhibits relatively low solubility in water (32 mg/L) and low-medium bioavailability (~10%) (Casteel et al., 2006). As solubility has been observed (as low as < 0.2 mg L^{-1}) beudantite at pH 5 and solubility of scorodite was reported between 0.4 and 3.6 mg L^{-1} (Krause and Ettel, 1989). The bioavailability of scorodite has previously been reported to be around 7% (Makris et al., 2008).

Bioaccessibility measurements were conducted on soil samples from around the site to determine the fraction of contaminants in soil available for human uptake via ingestion. The bioaccessibility of As ranged from 5.5 to 1,470 mg/kg, with the highest bioaccessible As found in the GZ domain (Figure 18.10). The bioaccessibility of Pb is generally higher than for As ranging from 19.5 to 9,680 mg/kg (Figure 18.10). The highest bioaccessible Pb is also in the GZ domain, reflecting the high contamination in that area.

18.5.3.1 Soil

A risk ranking score was calculated for soil exposure, including carcinogenic and non-carcinogenic risk. The sites with the highest bioaccessible As and Pb present the greatest risk to human health (GZ and DSP). These are soils from the Mount Galena domain and downstream from the small processing plant near the top of the site. The HQ was >1 for a number of samples for As and Pb (Figure 18.11). The cancer score exceeds the USEPA 10^{-4} (0.0001) level for unacceptable risk for all samples with the exception of Unknown Dump (UDD), up from the small processing area (USP) and the control, based on calculated lifetime exposure. The samples UDD and USP exceed the USEPA acceptable risk level of 10^{-6}.

18.5.3.2 Surface and Groundwater

The concentration of As in Severn River upstream from the site was 2-5 μg/L. In Consols Creek, the concentrations were elevated and ranged from 81-994 μg/L just before it enters the Severn River. The highest concentration in the surface water was 15 mg/L in the main shaft dam. Concentrations of As decrease in Severn River with distance from the site. At the camp site downstream from the site, the concentration of arsenic was 9 μg/L, which is elevated relative to upstream concentrations. The Australian drinking water health guideline (NHMRC, 2011) for As is 0.01 mg/L (10 μg/L).

The minimal acute lethal dose for As is in the range of 1 mg/kg/day. Consumption of the main shaft dam water results in an average lifetime dose of 0.6 mg/kgbw/day. In Consols Creek, samples exceed HQ>1 for As, Pb, Cd, Mn and Zn (Table 18.4). Severn River samples downstream from the mine are elevated with As and have HQ > 1 (Table 18.4). A risk ranking score based on HQ and cancer scores are given for the surface water samples in Figure 18.12.

The cancer risk exceeds the USEPA 10^{-4} (0.0001) for all water samples based on exposure assessment. The cancer risk scores are 0.00012 – 0.0003 upstream (suggesting elevated background As in the river). In Consols Creek, the cancer risk scores are an order of magnitude higher between

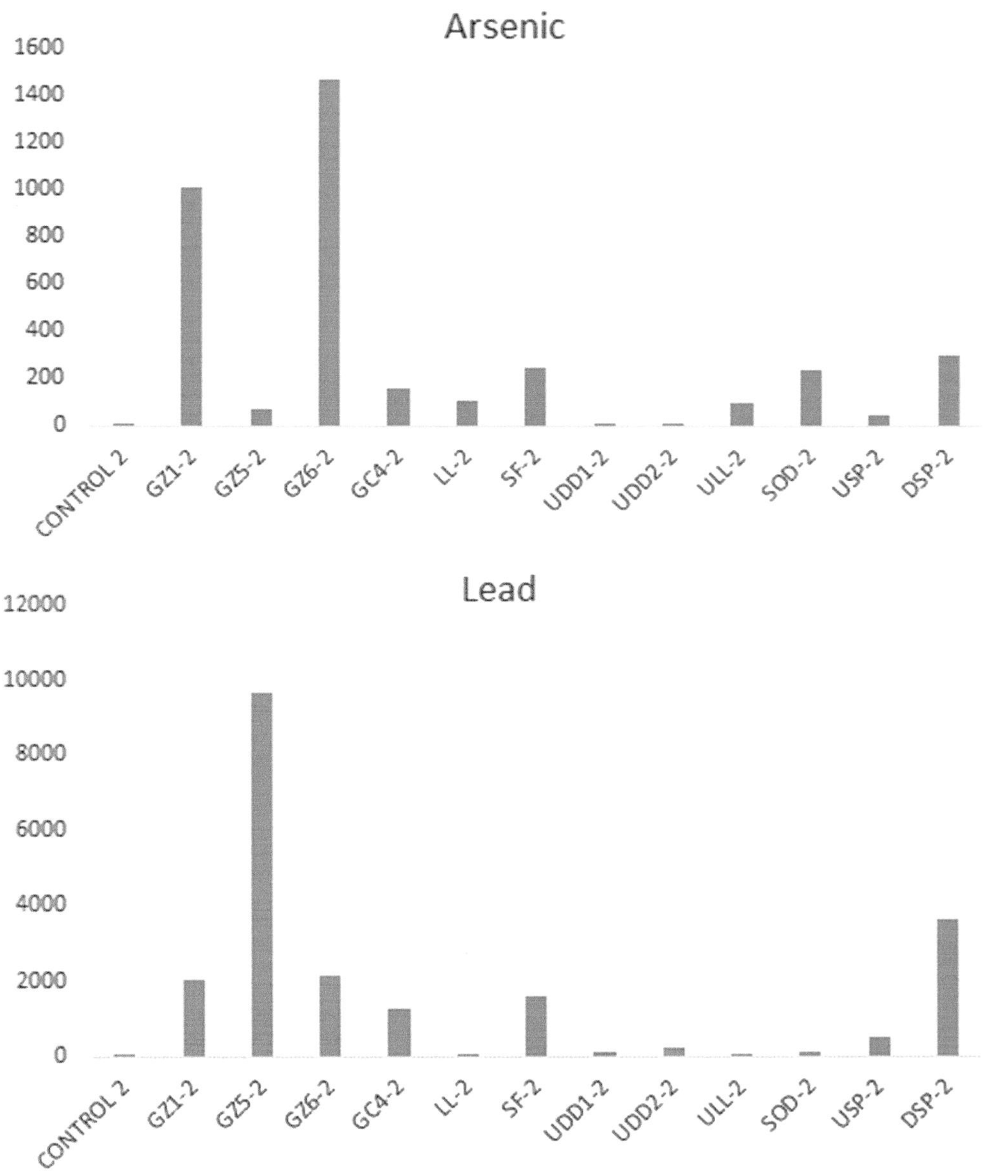

FIGURE 18.10 Bioaccessible As and Pb (mg/kg) in soil from mine site samples.

0.00483 and 0.00793. Downstream from the intersection of Consols Creek with Severn River, the cancer risk scores range from 0.0025 (500m downstream) to 0.00054 (at the camp site). The cancer risk scores have been converted to a ranking score for each sample in Figure 18.12.

As noted, the drinking water guideline for As is 0.01 mg/L. The limit for non-potable groundwater is 0.07 mg/L. Arsenic concentration exceeds the drinking water guideline (0.01 mg/L) in three of the wells (GW02, GW03, GW04) and exceeds the USEPA 10^{-4} (0.0001) guideline for cancer risk. The highest risk was associated with GW02 with 0.279 mg/L As (Table 18.5). Pb and Manganese concentrations in the groundwater are slightly elevated in GW01 and GW02, respectively, though the HQ <1. The highest concentration of Mn in GW02 exceeded the Australian drinking water

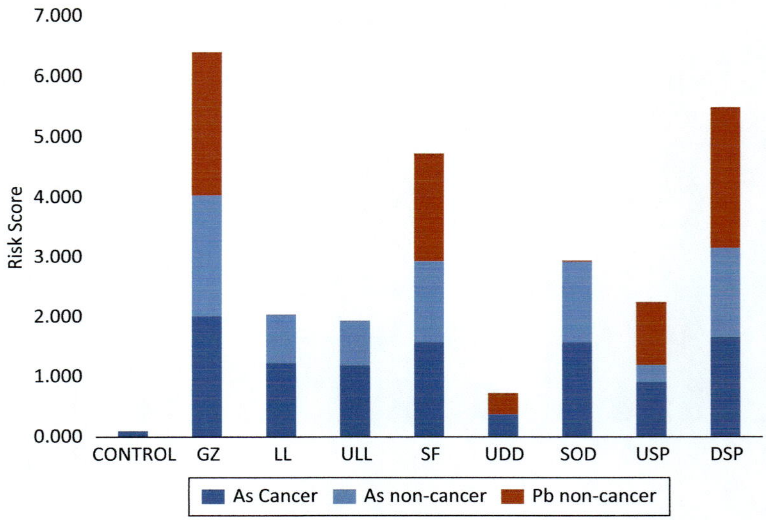

FIGURE 18.11 Cancer and non-cancer risk ranking scores for As and non-cancer ranking for Pb for Webb's Consols soil in different areas of the site. Stacked columns represent the total risk ranking score for soil exposure, individual risk ranking scores represented by the colour according to the figure legend.

TABLE 18.4
Exposure Assessment for Surface Water from Webb's Consols Mine

Sample	As mg/kgbw/day	Pb mg/kgbw/day	As HQ	Pb HQ	As Cancer Risk
Upstream	1.589×10^{-4}	3.97×10^{-5}	0.53	0.04	2.38×10^{-4}
Cap dam	1.987×10^{-3}	3.97×10^{-5}	6.62	0.04	2.98×10^{-3}
Main shaft dam	0.596	2.34×10^{-3}	1986	2.34	8.94×10^{-1}
Consols Creek	4.145×10^{-3}	1.09×10^{-1}	13.82	109	6.22×10^{-3}
Before Severn	3.949×10^{-2}	1.59×10^{-4}	131.64	0.16	5.92×10^{-2}
Unknown dump dam	7.946×10^{-4}	3.58×10^{-4}	2.65	0.36	1.19×10^{-3}
UD outlet into SR	1.589×10^{-4}	3.97×10^{-5}	0.53	0.04	2.38×10^{-4}
SR-CC	3.774×10^{-3}	3.97×10^{-5}	12.58	0.04	5.66×10^{-3}
500 m downstream	1.669×10^{-3}	3.97×10^{-5}	5.56	0.04	2.50×10^{-3}
>2km downstream	3.178×10^{-4}	3.97×10^{-5}	1.06	0.04	4.77×10^{-4}

guideline of 0.5 mg/L. The risk ranking score for groundwater wells at Webb's Consols indicates As is the primary risk driver.

18.5.4 Webb's Consols Ranking

The risk scores calculated from each risk category are summed to yield an overall score. The rank scores for the five areas within Webb's Consols are provided in Figure 18.13. In addition, for each risk category, the areas have been ranked (Table 18.6).

The risk scores clearly indicate that the Mount Galena area represents the highest overall risk based on each category. This is reflected in being ranked as the highest risk for each risk category except for the terrestrial ecological risk category. In this case, it was ranked second.

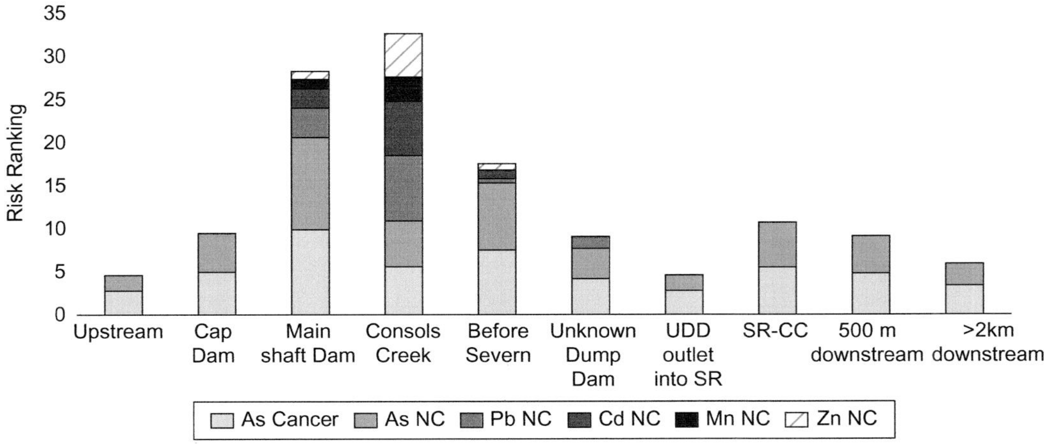

FIGURE 18.12 Risk ranking for Webb's Consols mine based on surface water exposure. Stacked columns represent the total risk ranking score for surface water exposure, individual risk ranking scores represented by the colour according to the figure legend.

TABLE 18.5
Exposure Assessment for Groundwater from Webb's Consols Mine

	GW As mg/L	As Dose $mg/kg_{bw}/day$	HQ	Cancer Risk
GW01	0.001	0.00004	0.13	5.96×10^{-5}
GW02	0.279	0.01108	37.0	1.66×10^{-5}
GW03	0.046	0.00183	6.09	2.74×10^{-5}
GW04	0.026	0.00103	3.44	1.55×10^{-5}

Ranking of groundwater across different zones within one site is difficult as there are few strong rationales for separation. This is especially the case when the groundwater has a high human health risk component. The variation in the groundwater score in this report is due to SPLP extractions. The SPLP extractions provide a surrogate for potential leachability. Due to the high As and Mn content in the Webb's Consols groundwater, all areas have a relatively high-risk score.

The human health risk scores are relatively high across Webb's Consols. The Unknown Dam was the lowest and Mount Galena the highest. Despite the high As concentrations in Lucky Lucy, the human health overall score was low. As discussed above, this is likely limited by the slow rates of dissolution of scorodite.

The ecological risk scores were relatively consistent across all zones at Webb's Consols. Almost all areas had sufficient pore-water concentrations of metals and metalloids to cause a significant reduction (>70%) in plant growth. In contrast to the other risk categories, the highest risk was the Unknown dam. This area had very high levels of dissolved Zn (up to 180 mg/L). This area has sections with very low organic matter, coarse texture and pH.

Erosion risk scores, which are an indication of future ongoing off-site impacts, were highest in Lucky Lucy and Mount Galena zone. The Southern area, Gates Crusher and Unknown Dam all had relatively low erosion risk scores.

Based on the overall rank scores, the risk scores decrease in the following order:

Mount Galena> Southern Area> Lucky Lucy> Gates Crusher> Unknown dam

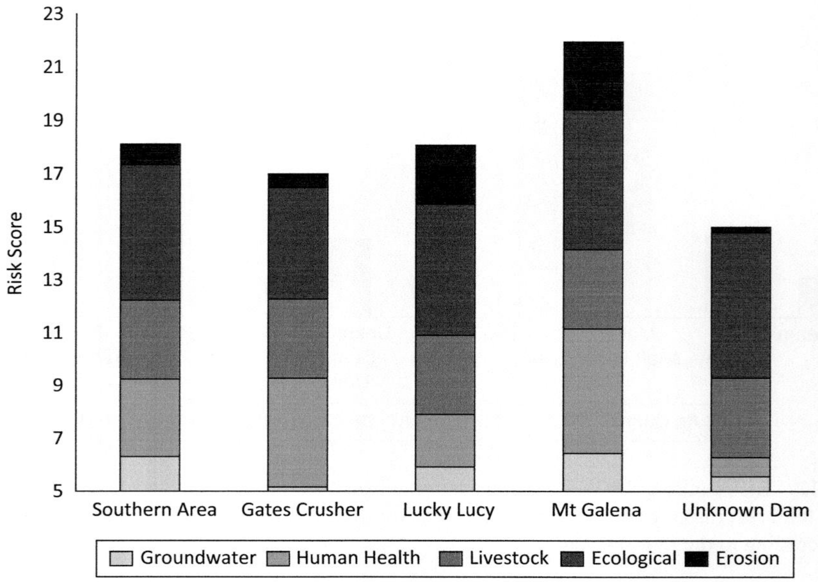

FIGURE 18.13 Risk scores calculated for the five areas of Webbs Consols.

TABLE 18.6
Ranks of Each Area by Risk Category (Ecological Ranks Not Presented)

Rank	Erosion	Human Health	Ecological	Groundwater
1	Mount Galena	Mount Galena	Unknown Dam	Mount Galena
2	Lucky Lucy	Gates Crusher	Mount Galena	Southern Area
3	Southern Area	Southern Area	Southern Area	Lucky Lucy
4	Gates Crusher	Lucky Lucy	Lucky Lucy	Unknown Dam
5	Unknown Dam	Unknown Dam	Gates Crusher	Gates Crusher

The primary area representing risk within the Webb's Consols site based on the ranking systems is Mount Galena, followed by the Southern Area. The Mount Galena area scored high in each of the categories and also represents a major proportion of the site. The ranking process clearly indicates that Webb's Consols is of higher concern, and, therefore, it should be prioritised in terms of remedial action. There are smaller differences between domains of Webb's Consols. As a result of high acidity, low vegetation cover and high bioaccessibility of As, Pb and Zn, Mount Galena area within Webb's Consols is the most highly ranked. The multiple issues within Mount Galena and Webb's Consols imply that several remediation measures are needed. The remediation of the contaminated area in this legacy mine site needs to be carried out either by isolating or by removing the contaminants (heavy metal(loid)s) to safe levels. In general, remediation of mine sites often consists of isolating contaminated materials in pre-existing tailings storage facilities, capping tailings and waste rock piles with clean topsoil, and collecting and treating any contaminated mine water if necessary. Unlike organic chemicals and plastics, metals cannot be degraded chemically or biologically into non-toxic and environmentally neutral constituents. Thus, sites contaminated with toxic metals present a particular challenge for remediation. Soil remediation has been the subject of a significant amount of research work in the past decade; this has resulted in a number of remediation options currently available or being developed. While excavation and ex-situ treatment have been traditionally used as remediation options in the past, the costs involved in such approaches have changed the strategies recently. In-situ options, such as the use of suitable soil amendments and plant-based technology or works including both the above in-situ methods, are being used as promising alternative technologies to traditional options (Dybowska et al., 2006).

Contamination in the Mount Galena zone is diffuse, but there are several hot spots. There are many areas that, based on total concentrations, are heavily contaminated, but support vegetation which has been able to colonise these areas. The hotspots on site are obvious by the lack of vegetation cover as a result of high bioavailability of As, Mn, Pb and Zn.

18.6 CONCLUSIONS

Metalliferous legacy mine sites are generally left with public institutions once mining has ceased. Metalliferous mines leave behind high levels of toxic metals and metalloids, which may or may not pose a risk to local ecosystems or the human food chain. The rehabilitation and "cleanup" of such sites is often associated with extreme cost. Unfortunately, the great majority of these sites did not have established guarantees to ensure adequate mine closure. Instead, public authorities are faced with significant expense with limited funds available for rehabilitation. Methodologies are needed to enable sound management of environmental risk. Remediation of metal(loid)s in itself is sufficiently challenging from a technical point of view. The extra pressure of limited funds adds further challenges.

The approach described in this chapter was to use a risk-based land management approach across several areas to better evaluate the risk of legacy mine sites in New South Wales. The areas included in the work included gastrointestinal bioaccessibility, terrestrial ecotoxicology, GIS-based erosion modelling, preliminary groundwater assessments and aquatic ecotoxicology. Erosion, gastrointestinal bioavailability were included in this chapter for simplicity. Each of these components was calculated for sites and within sites to allow ranking. Sites which rank highly would be considered to be of high risk. Conversely, those with low rank will be considered to be relatively low overall risk. The key component of the approach is to make use of up-to-date methods for evaluating the risk of bioavailable fractions, rather than total concentrations. In metalliferous sites, in particular, this is essential due to the high chance of mineralogical control of aqueous solubility. The approach allows public institutions to remediate impacts sites on the basis of risk, leaving lower-risk sites to be remediated at a later time.

REFERENCES

Antunes, S. C., Castro, B. B., Pereira, R., and Gonçalves, F. (2008). Contribution for tier 1 of the ecological risk assessment of Cunha Baixa uranium mine (Central Portugal): II. Soil ecotoxicological screening. *Science of The Total Environment* **390**, 387–395.

Appleton, J. (2003). A re-assessment of the heritage significance of historical features at Webb's Consols Mine Emmaville, Northern NSW. New South Wales Department of Land and Water Conservation, Unpublished.

Archer, M. J. G., and Caldwell, R. A. (2004). Response of Six Australian plant species to heavy metal contamination at an abandoned mine site. *Water, Air, and Soil Pollution* **157**, 257–267.

Ashley, P. M., and Lottermoser, B. G. (1999). Arsenic contamination at the Mole River mine, northern New South Wales. *Australian Journal of Earth Sciences* **46**, 861–874.

Ashley, P., Lottermoser, B., Collins, A., and Grant, C. (2004). Environmental geochemistry of the legacy Webbs Consols mine, New South Wales, Australia. *Environmental Geology* **46**, 591–604.

Aslibekian, O., and Moles, R. (2003). Environmental risk assessment of metals contaminated soils at silvermines abandoned mine site, co tipperary, Ireland. *Environmental Geochemistry and Health* **25**, 247–266.

Casteel, S. W., Weis, C. P., Henningsen, G. M., and Brattin,W.J. (2006). Estimation of relative bioavailability of lead in soil and soil-like materials using young swine. Environmental health perspectives **114**(8), 1162.

DEFRA. (2011). Guidelines for environmental risk assessment and management. Department of Environment, Food and Rural Affairs.

Dybowska, A., Farago, M., Valsami-Jones, E., and Thornton, I. (2006). Remediation strategies for historical mining and smelting sites. *Science Progress* **89**, 71–138.

Fistikoglu, O., and Harmancioglu, N. B. (2002). Integration of GIS with USLE in assessment of soil erosion. *Water Resources Management* **16**, 447–467.

Gee, G. W., and Bauder, J. W. (1986). Particle-size analysis. In: *Methods of Soil Analysis, Part A* (A. Klute, ed.), pp. 383–411. Madison, WI: American Society of Agronomy.

Grant, C. D., Campbell, C. J., and Charnock, N. R. (2002). Selection of species suitable for legacy mine site rehabilitation in New South Wales, Australia. *Water, Air, and Soil Pollution* **139**, 215–235.

Jasso-Pineda, Y., Espinosa-Reyes, G., González-Mille, D., Razo-Soto, I., Carrizales, L., Torres-Dosal, A., Mejía-Saavedra, J., Monroy, M., Ize, A., Yarto, M., and Díaz-Barriga, F. (2007). An integrated health risk assessment approach to the study of mining sites contaminated with arsenic and lead. *Integrated Environmental Assessment and Management* **3**, 344–350.

Juhasz, A. L., Herde, P., Herde, C., Boland, J., and Smith, E. (2014). Validation of the predictive capabilities of the SBRC-G in vitro assay for estimating arsenic relative bioavailability in contaminated soils. Environmental science & technology **48**(21), 12962–12969.

Krause, E. and Ettel, V. (1989). Solubilities and stabilities of ferric arsenate compounds. Hydrometallurgy **22**(3), 311–337.

Lamb, D. T., Sanderson, P., Wang, L., Kader, M., and Naidu, R. (2017). Phytocapping of mine waste at legacy mine sites in New South Wales. In: *Spoil to Soil: Mine Site Rehabilitation and Revegetation* (M. B. Kirkham, N. Bolan and Y. Ok, eds.), pp. 215–240. Boca Raton: CRC Press.

Lim, H., Lee, J., Chon, H., and Sager, M. (2008). Heavy metal contamination and health risk assessment in the vicinity of the abandoned Songcheon Au–Ag mine in Korea. *Journal of Geochemical Exploration* **96**, 223–230.

Lu, H., Moran, C. J., and Prosser, I. P. (2006). Modelling sediment delivery ratio over the Murray Darling Basin. *Environmental Modelling & Software* **21**, 1297–1308.

Lu, H., Prosser, I. P., Moran, C. J., Gallant, J. C., Priestley, G., and Stevenson, J. G. (2003). Predicting sheetwash and rill erosion over the Australian continent. *Soil Research* **41**, 1037–1062.

Mayes, W. M., Johnston, D., Potter, H. A. B., and Jarvis, A. P. (2009). A national strategy for identification, prioritisation and management of pollution from abandoned non-coal mine sites in England and Wales I: Methodology development and initial results. *Science of The Total Environment* **407**, 5435–5447.

Min, X., Xie, X., Chai, L., Liang, Y., Li, M., and Ke, Y. (2013). Environmental availability and ecological risk assessment of heavy metals in zinc leaching residue. *Transactions of Nonferrous Metals Society of China* **23**, 208–218.

Naidu, R., Pollard, S. J. T., Bolan, N. S., Owens, G., and Pruszinski, A. W. (2008a). Bioavailability: The underlying basis for risk-based land management. In: *Developments in Soil Science* (A. B. M. A. E. Hartemink and N. Ravendra, eds.), Vol. 32, pp. 53–72. Elsevier.

Naidu, R., Semple, K. T., Megharaj, M., Juhasz, A. L., Bolan, N. S., Gupta, S., Clothier, B., Schulin, R., and Chaney, R. (2008b). Bioavailability, definition, assessment and implications for risk assessment. In: *Chemical Bioavailability in Terrestrial Environments* (R. Naidu, ed.), pp. 39–52. Amsterdam: Elsevier.

NEPC. (2013). Schedule B1: Guideline on investigation levels for soil and groundwater. National Environment Protection Council.

NHMRC/NRMMC. (2011). Australian drinking water guidelines paper 6 national water quality management strategy. In National Health and Medical Research Council, National Resource Management Ministerial Council, Commonwealth of Australia, Canberra, NHMRC, N., Ed. Commonwealth of Australia:Canberra.

Rayment, G., and Higginson, F. (1992). *Australian Laboratory Handbook of Soil and Water Chemical Methods*. Melbourne: Inkata Press.

Soucek, D., Cherry, D., Currie, R., Latimer, H., and Trent, G. (2000). Laboratory to field validation in an integrative assessment of an acid mine drainage–impacted watershed. *Environmental Toxicology and Chemistry* **19**, 1036–1043.

Teng, H., Viscarra Rossel, R. A., Shi, Z., Behrens, T., Chappell, A., and Bui, E. (2016). Assimilating satellite imagery and visible–near infrared spectroscopy to model and map soil loss by water erosion in Australia. *Environmental Modelling & Software* **77**, 156–167.

Tim, U. S., Mostaghimi, S., and Shanholtz, V. O. (1992). Identification of critical nonpoint pollution source areas using geographic information systems and water quality modeling. *JAWRA Journal of the American Water Resources Association* **28**, 877–887.

USEPA, U.S.E.P.A. (1991). Integrated Risk Information System (IRIS), Chemical Assessment Summary National Center for Environmental Assessment.

Zar, J. H. (1999). *Biostatistical Analysis*, Fourth Ed. London: Prentice-Hall.

Zhuang, P., McBride, M. B., Xia, H. P., Li, N. Y., and Lia, Z. A. (2009). Health risk from heavy metals via consumption of food crops in the vicinity of Dabaoshan mine, South China. *Science of The Total Environment* **407**, 1551–1561.

19 Mount Morgan Mine Site, Queensland – Long-Term Site Management

Oskar Kadletz, John Frankish and Sam Pegg

19.1 OVERVIEW

Mount Morgan Mine is a large historic gold and copper mine located near Rockhampton in central Queensland. Mining began in 1882 and lasted for 108 years through a combination of underground, open-cut pit mining and tailings retreatment operations. Over the life of the mine, approximately 7.5 million ounces (250t) of gold, 37 tonnes of silver and 387,000 tonnes of copper were extracted.

The legacy of mining the sulphide-rich orebody is a significantly changed landscape with approximately 270 hectares of historic waste rock dumps and tailings, a contaminated water-filled open cut pit and a downstream environment impacted for 20 km. Approximately half of the historic mine site is heritage-listed and includes more than 60 heritage-listed structures. Many are in extremely poor condition.

The Mount Morgan Mine is a complex site, managed by the Queensland Department of Resources, with the focus on site safety and water management. The mining leases have been preserved, and there is an agreement in place between the state and the mining lease holder which limits the mining lease holder's liability for mining legacies and facilitates mineral reprocessing and remediation opportunities. The ongoing management of the mine site costs the Queensland government millions of dollars per year.

Rehabilitation of the mine site is challenging from many perspectives, including the mine's size, complex mineralogy and chemistry, management of the runoff and seepage from the historic tailings, working with the heritage listings over it and the cost of works. However, the site still has large residual mineral resources with the potential to support reprocessing, beneficial landform changes and reducing site water impacts. There is also rising interest in repurposing opportunities, such as renewable energy production. These may be possible with relatively minor site improvements and there is potential to provide ongoing site benefits.

19.2 SITE BACKGROUND AND HISTORY

The Mount Morgan deposit was pegged in 1882. Subsequently, the Mount Morgan Gold Mining Company Limited was formed in 1886 and went into voluntary liquidation in 1927 following a catastrophic underground fire in 1925 that was extinguished by deliberate flooding. In 1929, Mount Morgan Limited was formed. Open-cut mining commenced in 1932 and continued through to 1981. Between 1982 and 1990, tailings re-retreatment was undertaken, and approximately 30 million tonnes of tailings were reprocessed and then deposited in the open-cut pit. Over its 98-year life, the Mount Morgan mine processed more than 145 million tonnes of ore and overburden and produced approximately 100 million tonnes of various waste stockpiles, which were dispersed around the mine site. The location of the stockpiles was generally chosen for operational efficiency, rather than with a site closure plan in mind.

 DOI: 10.1201/9781315141855-19

Under an agreement between the State of Queensland and Mount Morgan Limited, the mining company undertook closure studies and prepared a decommissioning and environmental management plan. It also undertook a series of environmental works, which were completed to government requirements in December 1992. The state government subsequently released Mount Morgan Limited from liability for past mining and ongoing impacts and took over management of the site.

Mount Morgan's mining history is unique, demonstrating the evolution of mining techniques over the century-long span of its operation, and a large proportion of the site has been heritage-listed. Mount Morgan was one of the first mines to undertake open-cut mining in Australia, and it was the deepest man-made excavation in Australia in 1982.

The main driver for the placement of the waste stockpiles was operational convenience; the environmental and community considerations which are applied to today's mine planning were not a requirement at that time (Figure 19.1). Furthermore, the historic waste stockpiles include approximately 40 million tonnes of tailings, which still contain significant quantities of gold.

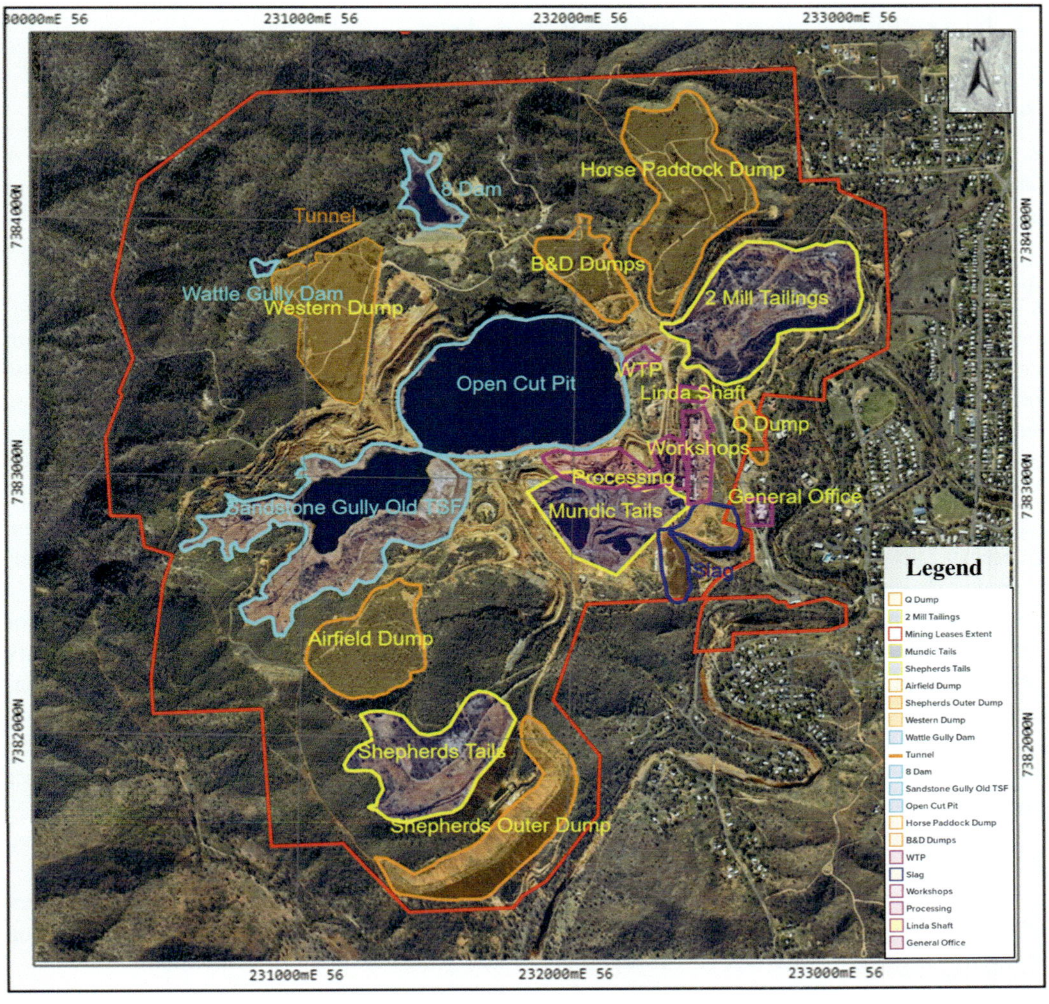

FIGURE 19.1 Map showing the mine layout and major stockpiles.

19.3 SITE IMPACTS AND RISKS

The key management considerations with Mount Morgan Mine are:

- Managing safety and health risks, including those associated with site landforms, site condition and historic infrastructure.
- General site management (including electrical infrastructure, roads, vegetation, fire and pest control).
- Conducting operational management and maintenance of the site and its infrastructure.
- Water management.
  - Reducing the water level in the open cut pit; including minimising rainfall runoff inflows to the pit.
  - Managing acidic seepage from historic stockpiles, and minimising the off-site impact of this seepage.
  - Securing sufficient clean water storage for future demands, including pit water treatment.
- Maintaining Emergency Action Plans for low-likelihood high potential impact scenarios, including a Contingency Plan for mine pit overtopping, and the management and monitoring of subsequent downstream impacts.
- Supporting reprocessing and repurposing opportunities.

19.3.1 Site Safety and Security Management

The government's site priorities focus on mitigating hazards and maintaining safe operations. A comprehensive site safety management system and a suite of operational protocols have been set in place to support site management.

19.3.2 Water Management

The greatest off-site impact is from acid rock drainage (ARD) resulting from rainfall contact with pyrite minerals in the stockpiles and flowing into the adjacent Dee River. A primary hazard at the site is the large volume of acidic water contained in the open-cut pit and other impoundments. Key challenges in site water management include topography and catchment changes that occurred with stockpile placement, as well as managing the inflows to the mine open-cut pit, which contains about 10 gigalitres of acidic water.

Naturally occurring valleys on site were filled with waste rock, tailings and overburden. Preferential flow paths and drainage lines remain within these valleys, resulting in ongoing contaminated groundwater flow and seepage.

A seepage interception and pump back system was installed as part of the works required prior to the state taking over site management in 2003 (Figure 19.2). The system captures low-quality seepage at a number of points on the mine site and returns it to the open-cut pit to minimise contaminated flows offsite. However, this inflow to the pit contributes to a reduction in pit water quality and adds significant extra volume in heavy wet seasons.

As of January 2022., the seepage interception system incorporates nine seepage collection sumps downstream of the major stockpiles and above the Dee River.

Major rainfall events (with return frequencies of less than 1 in 1,000 years) occurred in 2011 and again in 2013, with the open-cut pit overtopping for the first time in 2013 (see Figure 19.3).

To mitigate the overtopping risk and reduce open-cut pit water levels, the state government installed a lime-dosing water treatment plant in 2006. Mechanical evaporation units have also been employed since 2011 to help reduce the open-cut pit water level as quickly as possible.

As of 31 January 2022, the water level in the open cut was 5.65 metres below the spillway.

FIGURE 19.2 Showing the Mount Morgan mine seepage interception system as of January 2022.

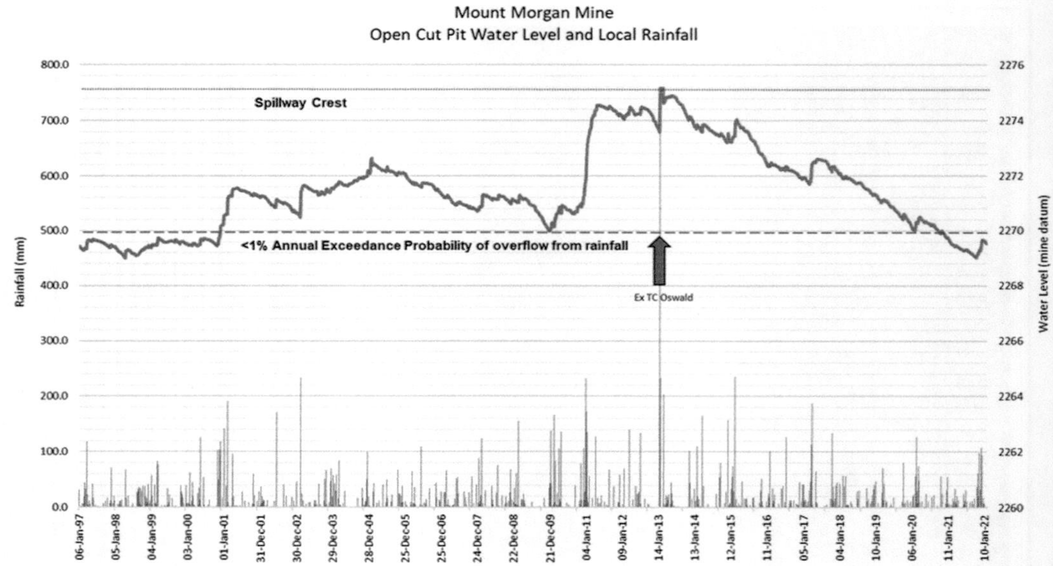

FIGURE 19.3 Rainfall and the mine pit water level over time.

TABLE 19.1
Sulphate Levels from AMD Sources

AMD Source	Approximate Stockpile Volume Million Tonnes	Range of $SO4^{2-}$-Concentration in Runoff (mg/L)	Estimated Total Area (ha)
Waste Rock Dumps	100	15,000 to 55,000	177
Tailings Storages	40	15,000 to 24,000	73
Open Cut Pit Water		16,600	54

Source: Mount Morgan water monitoring records, Department of Resources, Queensland Government
Note – the actual loads for any year increase and decrease with the amount of rainfall for that year.

In 2020 the department undertook an upgrade to a clean water containment dam and associated pumping infrastructure upstream of the mine, to divert clean water runoff away from the open-cut pit. These works have diverted approximately 48% of the total catchment that would otherwise flow into the pit, further reducing the risk of overflow due to rainfall.

The seepage interception system collects between 300 and 2,000 megalitres of seepage per year, depending on the amount of rainfall received over the site. The seepage is pumped back into the open-cut pit. This inflow has a detrimental impact on the mine pit water quality and on the operation of the water treatment plant.

Table 19.1 shows how the open-cut pit water level has been affected by rainfall and the operation of the water treatment plant.

19.3.3 Downstream Environmental Impacts

The Dee River is impacted by ARD for approximately 20 km downstream of the Mount Morgan mine. At this point, the Dee River merges with Fletchers Creek, which has a similar-sized catchment and provides dilution resulting in a marked improvement in river conditions downstream.

In 2004, the government removed three unstable concrete dams adjacent to the mine. The cleanup also removed 400,000m^3 of sulphidic tailings and contaminated water held behind these structures, thereby reducing the ongoing contaminant load to the river. River water quality was further improved with the commissioning of the lime-based water treatment plant. in 2006, enabling treated water to be released from the plant to dilute residual ARD close to the mine site.

19.4 LONG-TERM SITE MANAGEMENT GOALS

The long-term aim is to progressively reduce site risks and impacts and remediate the site in a practical way that is consistent with the state's abandoned mines management policy.

In practical terms the activities can be grouped into four categories:

- Safe – by removing or mitigating hazards on site to prevent potential exposure of the surrounding community.
- Secure – by implementing control measures to limit the level of adverse impacts on the surrounding and downstream environments.
- Durable – by minimising the ongoing maintenance and monitoring requirements for a site (including geotechnical and geochemical stability).
- Productive– by investigating opportunities to commercialise abandoned mines or repurpose the land for a future appropriate use.

In 2003 the state published a preliminary Rehabilitation Plan for the Mount Morgan Minesite[1], which considered several remediation options. The major works identified included removal of the most contaminating stockpiles to the mine pit and clean water diversions to minimise rainfall reporting to the pit. Completion of these works would require hundreds of millions of dollars. The mining leases were preserved when the state took on management of the site in 1993, and an agreement was set in place with the mining lease holders to facilitate reprocessing at the site and cooperation on remediation activities.

It was envisaged that mineral reprocessing would provide the economic capacity to drive site remediation. Under current legislation, mining cannot be undertaken by the government itself, and facilitating industry proposals for reprocessing has raised a variety of challenges.

Experience to date has identified four key considerations for a mining proponent to achieve a successful reprocessing operation at the Mount Morgan mine:

1. Finding a suitable, economic mineral extraction process (applicable to the local mineralisation, that is viable with current metal and reagent prices, and the cost of the mineral extraction plant).
2. Achieving the required statutory resource activity and environmental approvals for an already heavily disturbed site.
3. Achieving private sector investment to build the mineral extraction plant and a new tailings storage facility (requiring the confidence of investors that the mineral extraction process works, and that the private company proposing it can undertake the operation).
4. Limitation of liability for past mining disturbances for the current lease holder (achieved through a Deed of Agreement between the state and the holder of the historic mining leases).

19.5 RESIDUAL MINERAL RESOURCE REPROCESSING

Reprocessing historic tailings to recover gold and possibly other minerals presents opportunities on several fronts. Primarily, it would be the economic impetus for re-commencement of mining on-site, with the subsequent flow-on effects for the regional economy. Second, the reprocessed tailings including pyrite – the source of AMD – would be deposited in a new tailings storage facility built to today's standards, reducing site risks and impacts. Third, the reprocessing would also facilitate progressive site remediation and beneficial landform changes to assist with ongoing site management. Fourth, reprocessing would also enable beneficial use of the mine pit water for part of the reprocessing stream.

Several tailings reprocessing flowsheets have been considered since the mid-2000s to recover gold and potentially other minerals from the tailings. The complexities of the reprocessing chemistry coupled with other site considerations and impacts from the Global Financial Crisis hindered successful implementation.

The current mining lease holder, Heritage Minerals Pty Ltd, is looking to apply proprietary mineral processing and reagent recycling technology to solve these challenges.

REFERENCES

1 – Rehabilitation Plan for the Mount Morgan Minesite Central Queensland – Corinne Unger, Tania Laurencont, Lauren Keliher, Chris McCombe, Greg Bartley; November 2003, Department of Natural Resources and Mines.

20 mineCARE
Managing Contaminated Mine Sites with Advanced Software Tools

Liang Wang, Dane Lamb, Zhaoming Dong, Peter Sanderson, Jianhua Du and Ravi Naidu

20.1 INTRODUCTION

Historical mining operations in Australia, particularly metalliferous legacy mine sites, were conducted under different social expectations compared to modern times. As a result, many metalliferous legacy mine sites were left behind with no clear responsible party for mine closure, resulting in serious human and environmental health risks (Venkateswarlu 2016). These metalliferous legacy mine sites pose significant environmental risks due to the presence of inorganic contaminants, such as acidity, cyanide, metalloids (such as arsenic, antimony and chromium) and metals (such as cadmium, copper, lead, zinc, iron, aluminum and manganese) (Entwistle 2019). These contaminants can persist in the environment for long periods and are known to cause harm to both humans and ecological receptors (Erdogan 2020). In Australia, there are an estimated 500 to 20,000 metalliferous legacy mine sites in New South Wales alone, with many more globally (Unger 2017). These sites can pose serious health risks to nearby communities and wildlife, including respiratory problems, skin irritation and long-term exposure to heavy metals.

To protect human health, ecological receptors and the environment, it is essential to manage metalliferous legacy mine sites with a comprehensive approach that addresses the environmental risks associated with these sites. Nevertheless, rehabilitating and cleaning up legacy mine sites can be extremely expensive, with limited funds or budgets available for rehabilitation, often leaving governments with the burden of the cost. According to the US EPA 2015, the cost for the cleanup of legacy mines across the nation, excluding coal mines, has been estimated to range from $20 billion to $54 billion (SACRAMENTO 2015). In Australia, the state government of Victoria has estimated that rehabilitating the state's abandoned mines could cost up to $1 billion (VAGO 2020). The NSW government allocated a budget of $107.7 million in 2021 for the remediation of the state's abandoned mine sites over the next decade. The primary focus of the funding is on areas where mining ceased between 50 to 100 years ago, such as Captains Flat in the Southern Tablelands. Remediating abandoned mining sites is a complex and costly process that requires significant resources, including time, money and expertise. However, the benefits of remediation are numerous, including improved environmental health, reduced public health risks and the potential for economic benefits such as land reuse and improved land values. Therefore, due to limited resources, remediating all of these sites at once is challenging.

To address this issue, crcCARE has developed mineCARE, an advanced software tool that supports the management and risk assessment of contaminated mine sites. MineCARE uses a graphical statistical analysis app that provides consultants, researchers and government representatives with an intuitive and user-friendly interface to analyze and manage large sets of data. The scoring system in mineCARE evaluates contaminated mine sites based on various factors, such as potential risks to human health and the environment. The scoring system assigns a numerical value to each site,

DOI: 10.1201/9781315141855-20

allowing users to prioritize different sites for remediation based on their level of risk. The factors taken into account in the scoring system include the type and quantity of contaminants present at the site, the proximity of the site to sensitive receptors such as water sources or residential areas and the potential pathways for exposure to contaminants. Users can customize the scoring system to suit their specific needs and priorities, assigning higher weightings to factors such as proximity to residential areas or sensitive ecosystems, depending on the specific goals and objectives of the remediation project.

MineCARE allows for the input of site-specific information, such as the chemical composition of the area, with a focus on heavy metals, as well as any potential impacts on human health or the environment, such as contaminated groundwater that may be flowing toward residential or aquatic areas. On-site surveys and sampling can be conducted to fill in any missing data points or incomplete information. Using this data, the system generates a customized report that compares the user's site profile against average and maximum values, assigning a rating to each risk area. These ratings can then be utilized to identify high-risk areas that require prioritization for remediation efforts.

By providing a transparent and objective means of evaluating contaminated mine sites, the scoring system in mineCARE helps stakeholders make informed decisions about prioritizing sites for remediation. It identifies sites that pose the highest risk to human health and the environment, ensuring that limited resources are allocated in a targeted and effective manner, maximizing the impact of remediation efforts.

20.2 MINECARE SCORING SYSTEM

The mineCARE scoring system is an equation used to calculate the overall score of a contaminated mine site based on its potential risks to human health and the environment. The equation is as follows:

$$\text{Score}_{\text{mineCARE}} = \text{Score}_{\text{Human Impact}} + \text{Score}_{\text{Ecological Impact}} + \text{Score}_{\text{Groundwater Impact}} \quad (20.\ 1)$$

Each component of the equation is calculated based on various factors related to the contamination of the site, such as the concentration and type of contaminants present, the site's proximity to human settlements and the potential for contaminants to migrate to groundwater systems. The score for each component is calculated using a range of methods, including expert knowledge, statistical analysis of field data and numerical models. The scores for each component are then combined to give an overall score for the site, which can be used to prioritize sites for remediation or allocate resources effectively.

20.2.1 Human Impact

The mineCARE scoring system for human impact is shown in Table 20.1. The system uses a set of assessment factors to calculate a score for the potential human impact of a contaminated mine site. The formula takes into account the area of soil affected (F1) and the non-cancer and cancer effects of the soil (F2 and F3, respectively), as well as the groundwater exposure pathway (F4) and the non-cancer and cancer effects of both groundwater (F5 and F6) and surface water (F7 and F8). These factors are weighted differently in the formula based on their relative importance in determining the potential human impact of the contaminated mine site. The scores for each factor are then combined using a formula to calculate the overall score for human impact (Score Human impact).

$$\text{Score}_{\text{Human impact}} = \text{F1} \times (\text{F2} + \text{F3}) + \text{F4} \times (\text{F5} + \text{F6}) + \text{F7} \times (\text{F8} + \text{F9}) \quad (20.\ 2)$$

TABLE 20.1
MineCARE Scoring System – Human Impact

Description	ID	Scores
Area of Soil Contamination	F1	1-4
Non-cancer effects of soil	F2	0-3
Cancer effect of soil	F3	0-3
Groundwater exposure pathway	F4	0 or 1
Non-cancer effects of groundwater	F5	0-3
Cancer-effect of groundwater	F6	0-3
Surface water exposure pathway	F7	0 or 1
Non-cancer effects of surface water	F8	0-3
Cancer-effect of surface water	F9	0-3

20.2.1.1 F1 – Area of Soil Contamination

F1 in the mineCARE scoring system refers to the assessment factor for the area of soil that is contaminated. It is scored on a scale of 1 to 4 based on the extent of the contamination and the presence of ground cover. The descriptions and scores are as follows: 1 for small contaminated area (<50 m^2) with dense ground cover, 2 for large contaminated area (>50 m^2) with dense ground cover, 3 for small contaminated area with no ground cover and 4 for large contaminated area with no ground cover.

20.2.1.2 F2 – Soil Non-Cancer Effects

The mineCARE scoring system prioritizes the assessment of risks to human health by focusing on the most significant exposure pathway: ingestion of contaminated soil and water. While other pathways, such as dust inhalation and skin absorption, can also contribute to exposure, ingestion is considered the most common and direct method of exposure. It occurs during routine activities like eating, drinking and unintentionally consuming contaminated substances. Soil and water, being in direct contact with our surroundings, are more susceptible to contamination from industrial activities, mining operations or pollution. Contaminants can seep into soil and water, increasing the likelihood of ingestion. The digestive system efficiently absorbs substances from ingested soil and water, allowing contaminants to be readily absorbed into the bloodstream. Ingestion often involves repeated exposure over a prolonged period, leading to cumulative exposure to contaminants. Considering these factors, mineCARE prioritizes ingestion as the primary exposure pathway, ensuring that its risk assessment and management strategies address the most prevalent and significant route of exposure.

The amount of soil ingestion can be estimated using Equation 3:

$$e_{s-i} = \frac{c_{s-i} \times d_s \times b_s}{bw} \quad (20.3)$$

Where -

C_{s-i} is the concentration of chemicals in soil in milligrams per kilogram (mg/kg), d_s is the daily soil ingestion in kilograms (kg), b_s is the bioavailability of the chemicals, which is set at 1, and bw is the body weight in kilograms (kg). Three age groups were considered: 0–6 years, 6–18 years and above 18 years. And the estimated daily soil ingestion for the 0–6 year age group is 0.047 grams, while for the 6–18 year and above 18 year age groups it is 0.05 grams.

The non-cancer effect of soil can be determined by using the total hazard quotient (THQ), which takes into account the cumulative exposure of various chemicals through dust and soil (Dong 2023). This approach allows for a more comprehensive evaluation of the potential health risks associated with exposure to multiple contaminants in contaminated soil. The hazard quotient for each chemical is calculated by dividing the total uptake of the chemical by its oral reference dose (RfD), which indicates the daily exposure level of a chemical that is not expected to cause any adverse health effects over a lifetime of exposure. The RfD is based on a comprehensive review of available toxicity data and takes into account factors such as the age and health of the exposed population (Zakaria 2021). By using the THQ approach, it is possible to identify the individual chemicals that contribute the most to the overall risk posed by contaminated soil, and take appropriate actions to minimize exposure. Therefore, the equation for calculating the THQ is given by:

$$THQ = \sum_{i=1}^{n} \frac{e_i}{RfD_i} \tag{20.4}$$

Where -

e_i represents the total uptake of each chemical from soil and dust, and RfD_i represents the oral reference dose for that chemical. It is an estimate of the amount of a chemical that a person can be exposed to on a daily basis over a lifetime without suffering any adverse health effects. The RfD value for each chemical is collected from the Integrated Risk Information System (IRIS) database by US EPA (Thayer 2022). The summation is performed for all the chemicals of concern, denoted by "n". This method enables the assessment of non-cancer effectss of soil by considering the combined effects of all the chemicals in the soil.

The F2 score value is calculated using the formula "1.5 × (log (THQ) + 1)". The calculated score value falls between the range of 0 and 3. For THQ values less than or equal to 0.1, the score value assigned is 0. For THQ values greater than 0.1 but less than or equal to 10, the score value is calculated using the formula "1.5 × (log (THQ) + 1)". For THQ values greater than 10, the score value assigned is 3.

20.2.1.3 F3 - Soil cancer effect

F3 in mineCARE scoring system assesses the potential cancer effects of soil on human health. The Total Cancer Risk (TCR) is calculated using the measured concentration of each chemical and its corresponding cancer risk. The formula for calculating TCR is given as follows:

$$TCR = \sum_{i=1}^{n} C_i \times r_i \tag{20.5}$$

Where the c_i is the measured concentration for each chemical, and r_i is the corresponding cancer risk.

The corresponding cancer risk (r_i) is typically derived from the slope factor or cancer potency factor of a chemical, which is the slope of the dose-response curve for cancer risk at low doses. These values are often provided by regulatory agencies such as the US EPA through their IRIS or other toxicological databases (Thayer 2022). The slope factor or cancer potency factor is used to estimate the risk of cancer associated with exposure to a specific chemical over a lifetime.

The F3 score values are assigned based on TCR calculated for the cancer effect of soil using equation 5. The score value is calculated using the formula "(log ($TCR/10^{-5}$)+1)". The calculated score value falls between the range of 0 and 3. If the TCR value is less than or equal to 10^{-6}, a score of 0 is assigned. For TCR values between 10^{-6} and 10^{-3}, the score is calculated using the given formula. If the TCR value is greater than 10^{-3}, a score of 3 is assigned. This scoring system helps to classify the potential cancer risk associated with the soil, with higher scores indicating a greater risk.

20.2.1.4 F4 to F9 - Groundwater and Surface Water Exposure Pathway

The mineCARE F4 and F7 scores evaluate the risk associated with groundwater and surface water exposure pathways by considering the development of a conceptual site model and the establishment of contaminant-exposure pathway-receptor links. A score of 0 is assigned if the conceptual site model is developed without establishing these links, while a score of 1 is assigned if the links are established. Similar to the soil risk assessment, the groundwater (F5 & 6) and surface water (F8 & 9) risk assessments also take into account both non-cancer and cancer impacts, which are determined using the total hazard quotient (eq. 4) and the corresponding cancer risk (eq. 5). This approach helps to ensure that the potential risks associated with exposure to contaminants through groundwater and surface water pathways are accurately assessed and appropriately managed.

20.2.2 Ecological Impact

The ecological risk assessment is evaluated based on the toxicological studies and erosion models. Toxicological studies are an essential component of the ecological risk assessment, as they provide information about the potential adverse effects of contaminants on ecological receptors. Whereas, erosion investigation is used to estimate the transport of contaminants through soil and sediment and the potential for these contaminants to reach water bodies.

$$\text{Score}_{\text{Ecological Impact}} = \text{F9} + \text{F10} + \text{F11} + \text{F12} \tag{20.6}$$

20.2.2.1 F9 - Distance of Nearby Aquatic Environment Impact

To assess the extent of contamination of nearby water sources, the concentration of hazardous substances in water samples collected downstream, such as rivers, is analyzed and compared with the regulated screening levels (RSL) (Bopp 2019). If the concentration of hazardous substances exceeds the RSL, it indicates that the water source is impacted by the mining activities. In such cases, the distance between the mine site and the water sampling location is used to evaluate the extent of impact. The mineCARE F9 score categorizes the extent of demonstrated impact based on the distance downstream. A score of 0 indicates that there is no impact on the aquatic environment based on toxicological data. A score of 1 is assigned if the impact is within a distance of 100m downstream. Similarly, a score of 2, 3, 4 and 5 is assigned if the impact is between 100–500m, 500–5,000m, 5,000–10,000m and more than 10,000m downstream, respectively. The greater the distance of demonstrated impact, the higher the score.

20.2.2.2 F10 Erosion

F10 in mineCARE is focused on assessing the risk of erosion and sedimentation that can potentially spread contaminants from mining sites into nearby ecosystems. The impact of erosion on a site is determined based on various factors like soil type, topography and rainfall intensity. As demonstrated in Table 20.2, the F10 score can be calculated using two methods: onsite observation or computer-based models like the Revised Universal Soil Loss Equation (RUSLE) and rainfall simulation (Wang 2020).

In mineCARE, an observation-based method is used to assess erosion. This method involves examining the stability of tailings and contaminated materials, as well as the extent of erosion

TABLE 20.2
F10 Score Based on the Erosion Assessments

Observation-Based	RUSLE Modelling	Score
Tailings impoundment and/or contaminated material appears stable (no seepage/erosion)	Very low (<10 tons/ha/year)	1
Evidence of slight erosion of waste / contaminated material	Low (10–20)	2
Evidence of moderate erosion of waste / contaminated material	Moderate (20–30)	3
Tailings / waste is unstable high erosion evident	Severe (>30)	4

observed. The assessment generates a score that ranges from 1 to 4, with each score representing a specific level of stability or instability. A score of 1 indicates that the tailings and contaminated material appear stable, with no signs of erosion or seepage. Conversely, a score of 4 suggests that the tailings or waste is unstable, with high levels of erosion observed. The erosion assessment in mineCARE is critical because it provides a quick and effective way to identify sites that require immediate remediation due to unstable conditions that could lead to contamination of the surrounding environment. Using this method, stakeholders can prioritize their resources and allocate them more efficiently toward addressing the most pressing concerns, ultimately leading to better environmental outcomes.

The calculation of erosion using rainfall simulation or RUSLE model takes into account various input factors to calculate the erosion potential of the site. These input factors include rainfall erosivity factor (R), soil erodibility factor (K), slope length/gradient factor (LS), ground cover and management factor (C) and erosion control practice factor (P) (Wang 2020). Based on the computed soil loss, the F10 score is assigned according to the severity of erosion, as demonstrated in Table 20.2.

20.2.2.3 F11 and F12 – Toxicity Study

In order to evaluate the possible negative effects of contamination, it is imperative to carry out toxicity studies using soil and water samples with the highest levels of contaminants. These studies can be conducted through soil pot and hydroponic experiments, which involve planting and growing crops such as cucumbers in soil or water contaminated with harmful substances. By measuring and recording the growth rate of these plants, an assessment can be made of the extent of contamination and the potential harm it poses (Rahman 2022).

The relative growth rate of the plants is an important indicator of the toxicity of the sample and is measured using the F11 score for soil pot experiments and the F12 score for hydroponic experiments. This score reflects the ability of the plant to grow and develop in the presence of harmful substances. The results of these studies are crucial in determining the potential risks posed by the contamination to both human health and the environment.

The scoring criteria for the relative impact of the contaminant on the relative growth rate (RG) of plants used in the toxicity studies. The relative impact is calculated as (100 – RG) of the plants. The score assigned to each range of the relative impact is as follows: if the relative impact is less than 10%, the score is 0; if the relative impact is between 10% and 50%, the score is calculated as 0.05 * (100-RG); if the relative impact is greater than 50%, the score is 3.

20.2.3 Groundwater Impact

The groundwater impact assessment for mineCARE is based on six factors (G1-G6) and their associated scores.

$$\text{Score}_{\text{Groundwater Impact}} = \sum_{1}^{6} G_i \tag{20.7}$$

The significance of this assessment is to evaluate the potential impact of mining activities on the groundwater quality and to identify the level of risk posed to human health and the environment.

G1, This parameter considers the levels of contaminants in the groundwater. The score assigned depends on whether the concentration of contaminants is below or above the regulated screening level. If the concentration of contaminants is below the regulated screening level, a score of 0 is assigned. In contrast, if the concentration of contaminants is higher, a score of 1 is given.

G2, assesses the synthetic precipitation leaching procedure (SPLP) values (Araya 2020). The SPLP test determines the potential for contaminants to leach from a material under specific environmental conditions. The score assigned depends on whether the SPLP values are below or above the regulated screening level. If the SPLP values are below the regulated screening level, a score of 0 is assigned. Conversely, if the SPLP values are higher, a score of 1 is given.

G3, takes into account the depth to groundwater table in an unconfined aquifer below ground surface. The assessment in G3 assigns a score ranging from 1 to 3 based on the depth of the groundwater table. A score of 1 is assigned if the groundwater table is less than 5 meters below the ground surface (bgs), a score of 2 is assigned for depths between 5 and 10 meters bgs and a score of 3 is assigned for depths greater than 10 meters bgs. The score is indicative of the level of contamination in the groundwater, as a deeper groundwater table suggests that contaminants have infiltrated deeper into the underground. Therefore, a higher score suggests a higher risk to human health and the environment.

G4: This parameter evaluates off-site impact and liability issues based on the direction and velocity of groundwater flow toward sensitive receptors. The score assigned for this category ranges from 1 to 4 and is determined by the direction and velocity of groundwater flow toward sensitive receptors. A score of 1 is given when the groundwater flow direction is toward a sensitive receptor and the velocity is less than 5 meters per year. A score of 2 is assigned when the groundwater flow direction is toward a sensitive receptor and the velocity is greater than 5 meters per year but less than 20 meters per year. A score of 3 is given when the groundwater flow direction is toward a sensitive receptor and the velocity is greater than 20 meters per year but less than 30 meters per year. Finally, a score of 4 is assigned when the groundwater flow direction is toward a sensitive receptor and the velocity is greater than 30 meters per year.

G5, The mineCARE assessment also considers the proximity of the mining site to the nearest drinking groundwater borehole or residence. It should be noted that there may be illegal groundwater wells in the area, so it is assumed that residents have access to a groundwater well unless evidence shows otherwise. A score ranging from 1 to 4 is assigned depending on the distance between the nearest drinking groundwater borehole or residence and the mining site. If the distance is less than or equal to 5,000 meters, a score of 1 is assigned. If the distance is less than or equal to 2,.000 meters, a score of 2 is given. If the distance is less than or equal to 1,000 meters, a score of 3 is assigned. Finally, if the distance is less than or equal to 500 meters, a score of 4 is given. The closer the mining site is to the nearest drinking groundwater borehole or residence, the higher the potential risk of contamination.

G6, assesses the applicability of institutional control. Institutional control refers to the use of administrative or legal measures to manage and control environmental risks at a site. The applicability of institutional control is evaluated in G6 of the mineCARE method, with a score assigned based on the level of control that can be applied. If stringent institutional control measures can be applied at the site, such as implementing strict monitoring and reporting requirements or conducting regular site inspections, a score of 1 is assigned. Examples of stringent institutional control measures could include installing barriers to prevent migration of contaminants, implementing treatment systems to remove contaminants from groundwater or monitoring groundwater quality at regular intervals. If institutional control measures can be applied but are difficult to follow up or enforce, a score of 2 is assigned. For example, this may be the case when the site is in a remote location, making regular monitoring and inspection difficult, or when access to the site is restricted due to safety concerns. If no institutional control measures can be applied, a score of 3 is assigned.

20.3 STATISTICAL ANALYSIS

20.3.1 Normal Distribution and Box Chart

MineCARE offers a comprehensive set of tools for analyzing the risk associated with mining activities. One such tool is the normal distribution and box chart, which provides a graphical representation of the overall score distribution within the mineCARE system. The normal distribution plot, as illustrated in Figure 20.1, shows the mean score (μ) and standard deviation (σ) of the data. This allows users to assess the overall trend of the data and identify any outliers that may represent extreme cases.

In addition to the normal distribution plot, mineCARE also provides a box chart (Figure 20.2) that offers a more detailed view of the data. The box chart displays the 25th, 50th and 75th percentiles of the data, along with the highest and lowest values within a specified range (usually 1.5 times the interquartile range). This provides users with a quick snapshot of the range of scores within the data set and allows them to compare the score of each individual case with the broader statistical data. By comparing the score of each case with the statistical data provided by the normal distribution and box chart, users can gain a deeper understanding of the severity of each case by comparing the characteristics of new cases with those of existing cases in the database.

20.3.2 Ranking

In addition to the normal distribution and box chart, mineCARE also includes a ranking scatter plot that enables the user to rank the existing cases based on their mineCARE score (Figure 20.3). The mineCARE score is an indicator of the potential risks to humans and the environment. The higher

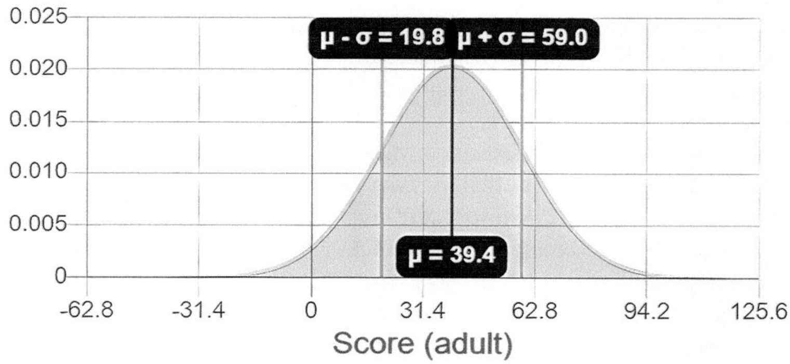

FIGURE 20.1 MineCARE statistic tool: normal distribution.

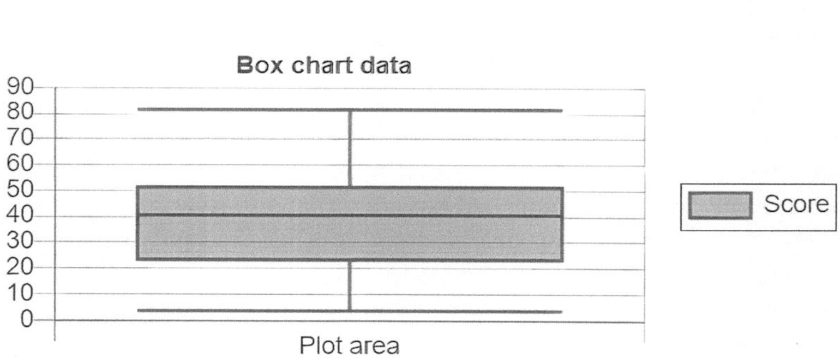

FIGURE 20.2 MineCARE statistic tool: box chart.

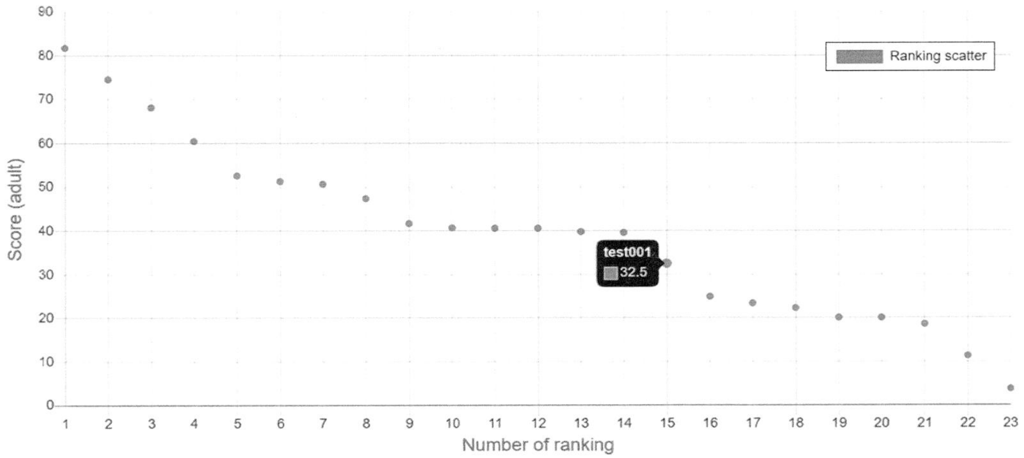

FIGURE 20.3 MineCARE ranking scatter plot.

the score, the higher the potential risks. By ranking the cases, the authorities can easily identify those with higher scores, which require immediate attention and priority actions to mitigate the risks.

The ranking scatter plot provides a visual representation of the mineCARE score for each case, with the x-axis representing the ranking order and the y-axis representing the mineCARE score. This allows the user to quickly identify cases that are of higher concern based on their position in the plot. For instance, cases that fall in the upper right quadrant of the plot have both high ranking and high mineCARE scores, indicating that they pose the highest potential risk and should be given priority attention.

20.3.3 Radar Chart and Traffic Light Indication

MineCARE provides detailed sub-scores for each case, including human impact (adult and child), ecological impact and groundwater impact, in addition to the overall score. These sub-scores are presented using a radar chart, which also shows the mean and maximum sub-scores in the mineCARE database. This allows users to compare the severity of each section with that of other cases in the database. Additionally, mineCARE uses traffic light colors to indicate the severity of each case, with green indicating that all sub-scores are below average, amber indicating that one or more sub-scores are above average and red indicating that all sub-scores are above average. Figure 20.4 demonstrates how the radar chart and traffic light colors are used in mineCARE. By providing this detailed information, mineCARE allows users to have a comprehensive understanding of the risk associated with each case, including the specific areas of affect that require attention. The radar chart and traffic light colors provide a visual representation of the data, making it easier for users to identify high-priority cases that require immediate action.

20.4 DISCUSSION

20.4.1 Sampling

In order to utilize the mineCARE system, it is important for users to provide detailed information regarding the chemical composition of the soil, groundwater (if applicable) and surface water (if applicable) at the contaminated mine site. To ensure that the highest risk areas are accurately identified, users should input the highest recorded concentrations of each chemical.

Hotspots within a contaminated mine site can be located in various ways. One approach is to review historical reports of the site, which may contain information on processing areas, tailings or other

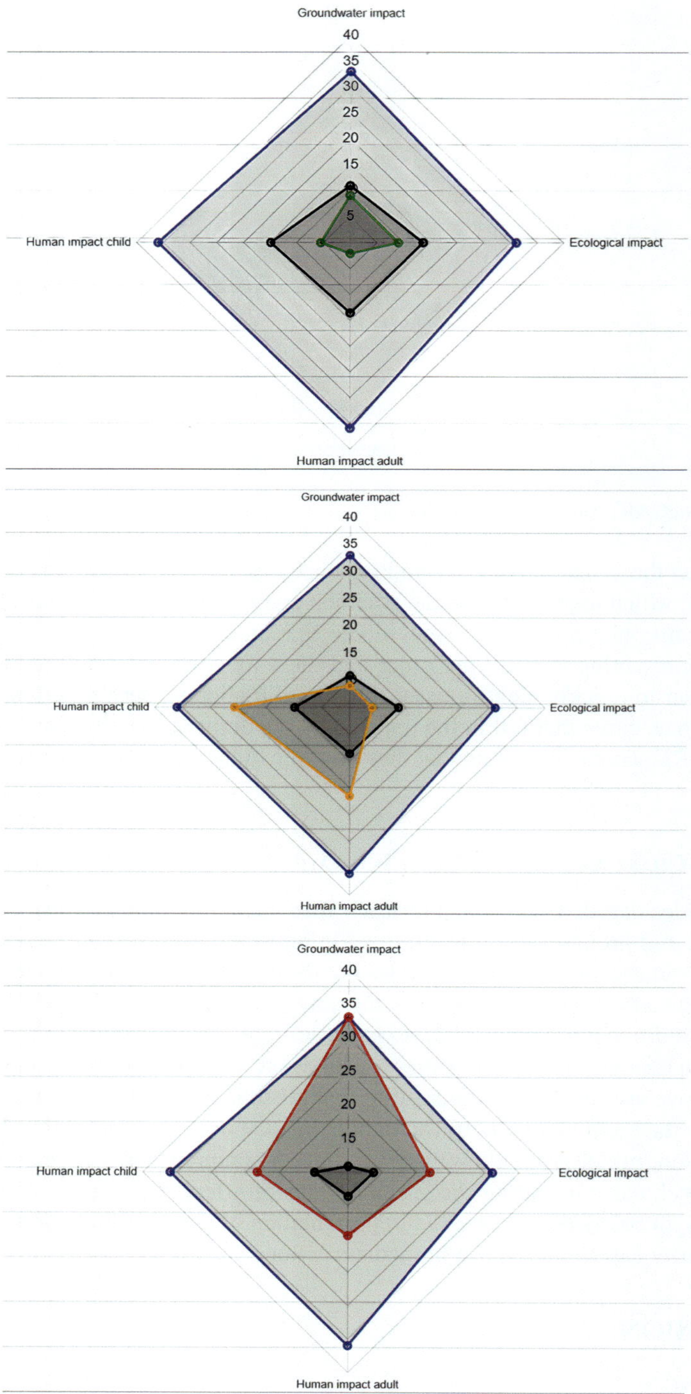

FIGURE 20.4 MineCARE radar chart with traffic light indication for each case. The blue line represents the maximum scores in the system, while the black line shows the average scores. The green, amber and red lines indicate the scores for each individual case.

areas where contamination is likely to occur. Another method is to conduct a walkover survey of the site using a handheld X-ray fluorescence (XRF) analyzer (Ravansari 2020). The resulting data can then be used to quickly generate a spatial distribution map of contaminant levels across the site, providing valuable information for assessing the severity of contamination and locating the hotspots. To ensure comprehensive coverage of the impacted area, it is recommended that users conduct field XRF measurements in a grid with approximately 50-meter intervals. As an example, Figure 20.5 illustrates

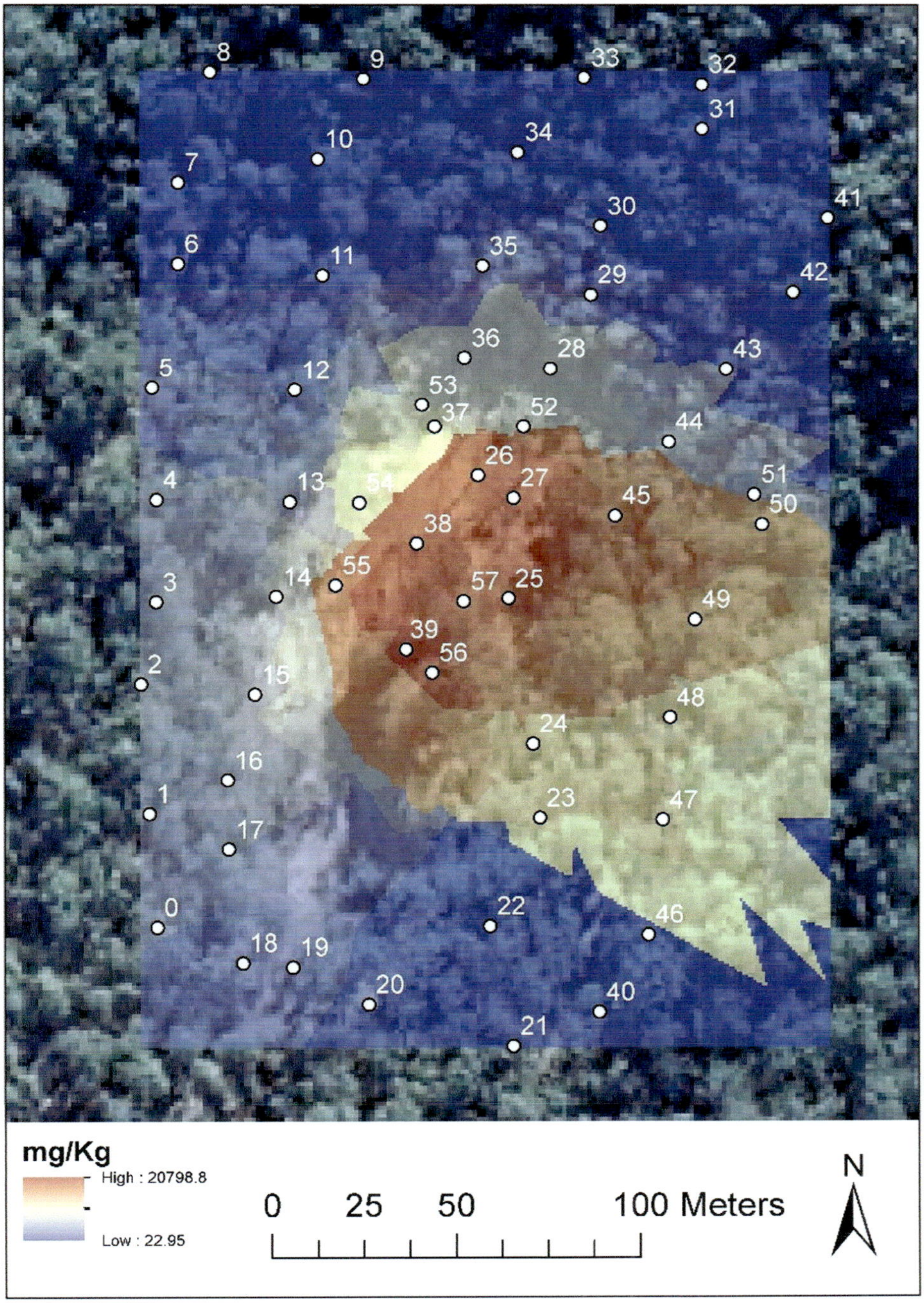

FIGURE 20.5 A contour map for copper distribution using a handheld XRF.

the distribution of copper levels at a legacy copper mining site that was generated using a handheld XRF analyzer.

After identifying hotspots, the hydrological conditions must be considered to determine groundwater access for further examination. To assess surface water contamination, potential affected waterways must be identified by conducting a thorough site assessment and mapping the flow of water from the contaminated site. Samples of both the creek and the river water should be collected at key points along the waterways, including the point where the creek or river meets the contaminated site, and downstream points to determine the extent of contamination and any potential impacts on aquatic life and human health. Additional sampling may also be necessary at other locations and depths along the waterways to fully assess the scope of contamination.

20.4.2 Erosion Assessment

In mineCARE, erosion can be calculated using two methods: onsite observation or computer-based models like the Revised Universal Soil Loss Equation (RUSLE). Figure 20.6 illustrates a scenario of a legacy mine site situated on a steep hillside. The mining area appears to be exposed with no vegetation covers, which can result in erosion of contaminated soil due to water runoff. The contaminated soil is then likely to be carried away, potentially causing contamination of nearby water sources.

The Revised Universal Soil Loss Equation (RUSLE) is a mathematical model that predicts the amount of soil loss caused by water erosion or wind erosion. It is a widely used tool for soil erosion assessment and management in agriculture, forestry and land-use planning.

The RUSLE model estimates soil erosion by taking into account five factors: rainfall and runoff erosivity, soil erodibility, slope length and steepness, crop management and conservation practices. The model uses these factors to calculate the annual soil loss per unit area of land, which is expressed as tons of soil per acre or metric tons per hectare.

In Australia, the mean annual rainfall (MAR) information can be collected from a nearby public weather station. The erosivity factor (R) can be calculated using the equation (20.8):

$$R = 17.67 \log(\text{MAR}) - 27.24 \tag{20. 8}$$

Australian SPADE (Spatially-enabled Probabilistic Approaches for Downscaled Environments) provides general soil profile including soil erosion. eSPADE is an online tool from the NSW Department of Planning that provides detailed soil information for NSW. It offers soil profiles and maps based on the Soil and Land Information System (SALIS) system, which groups soils by properties. Users can search by location or map and access data like type, texture, pH and soil erodibility.

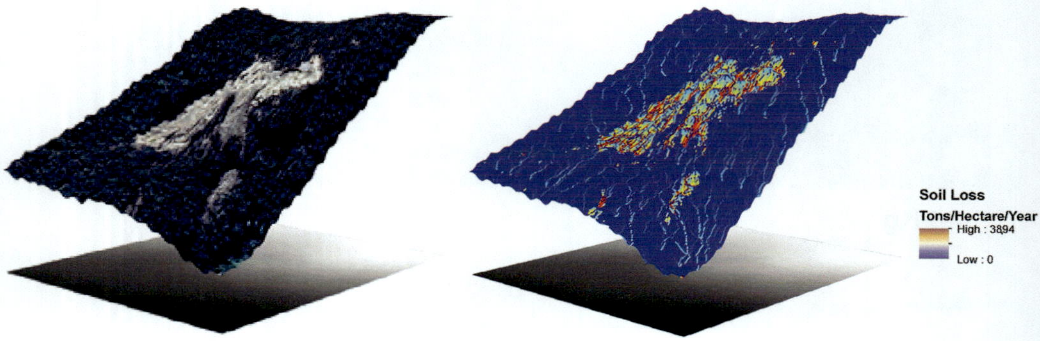

FIGURE 20.6 Erosion assessment based on satellite image and RUSLE model.

Slope length and steepness can be calculated by the spatial analysis function in ArcGIS with the Digital Elevation Model (DEM) data from ELVIS (Elevation Information System), Australian National Elevation Data Framework. To determine the information about land vegetation or ground cover, satellite images such as Google Earth can be utilized. Once the satellite images are obtained, the likelihood classification function in ArcGIS can be employed to calculate the extent of vegetation or ground cover present in the study area. The likelihood classification function uses statistical algorithms to analyze pixel values in the satellite images, which enables it to classify areas based on their likelihood of containing vegetation or ground cover.

20.4.3 Toxicity Study

A toxicity study can be conducted using soil and groundwater samples obtained from a contaminated site. In this type of study, a pot experiment can be set up using cucumber seeds as the test plant. The soil sample is added to the pots along with 30 cucumber seeds, and the growth and development of the plants are observed over a specified period of time. A hydroponic study can be conducted using contaminated groundwater as the source of the nutrient-rich water. Another 30 seeds can be planted in a hydroponic system, such as a nutrient film technique or deep water culture, using the contaminated groundwater as the growing medium. The plants can be monitored for growth, and the concentrations of contaminants in the plants and water can be measured over time.

In this experiment, a controlled environment can be created using a greenhouse, and the growth rate of the plants, as well as any observable abnormalities or deformities, can be recorded and analyzed. To ensure accurate results, several replicates of each experiment can be conducted, and a control group can be established using uncontaminated soil or water. This control group will help to eliminate any external factors that may affect plant growth and ensure that the observed effects are due to the contaminants in the soil or water.

20.5 CONCLUSION

Metalliferous legacy mine sites pose environmental and health risks due to inorganic contaminants that can persist for extended periods. However, rehabilitating these sites is expensive and challenging, with limited resources available for remediation. To address this issue, crcCARE developed mineCARE, an advanced software tool that supports the management and risk assessment of contaminated mine sites.

The mineCARE scoring system combines three impact scores, human, ecological and groundwater, to provide an overall score that can be used to prioritize remediation efforts. The system calculates the scores using advanced technologies, statistical analysis, and numerical models, providing a comprehensive and accurate assessment of the risks posed by contaminated mine sites. In addition, mineCARE offers statistical analysis tools, such as normal distribution and box chart, ranking scatter plot, radar chartand traffic light indication, to identify cases requiring immediate attention, prioritize actions and understand the severity of each case. The software provides a comprehensive approach for managing contaminated mine sites, including identifying hotspots, assessing erosion, conducting toxicity studies and accurate sampling of soil, groundwater and surface water.

To utilize mineCARE effectively, historical reports and walkover surveys using handheld XRF analyzers can help locate hotspots, and hydrological conditions must be considered when assessing groundwater and surface water contamination. Erosion assessment can be calculated through onsite observation or computer-based models like RUSLE, and toxicity studies can be conducted using soil and groundwater samples with cucumber seeds as a test plant in a pot experiment or hydroponic system.

ACKNOWLEDGMENTS

The authors express their gratitude to the Cooperative Research Center for Contamination Assessment and Remediation of the Environment (CRC CARE) and the University of Newcastle for their support in making this research possible. The funding for this research was provided by CRC CARE Pty Ltd. The research was carried out at the Global Center for Environmental Remediation (GCER) at the University of Newcastle.

REFERENCES

Araya, G., Toro, N., Castillo, J., Guzmán, D., Guzmán, A., Hernández, P., Jeldres, R. I. and Sepúlveda, R. (2020). "Leaching of oxide copper ores by addition of weak acid from copper smelters." *Metals* **10**(5): 627.

Bopp, S. K., Kienzler, A., Richarz, A.-N., van der Linden, S. C., Paini, A., Parissis, N. and Worth, A. P. (2019). "Regulatory assessment and risk management of chemical mixtures: Challenges and ways forward." *Critical Reviews in Toxicology* **49**(2): 174–189.

Dong, Z., Cui, K., Liang, J., Guan, S., Fang, L., Ding, R., Wang, J., Li, T., Zhao, S. and Wang, Z. (2023). "The widespread presence of triazole fungicides in greenhouse soils in Shandong Province, China: A systematic study on human health and ecological risk assessments." *Environmental Pollution* **328**: 121637.

Entwistle, J. A., Hursthouse, A. S., Marinho Reis, P. A. and Stewart, A. G. (2019). "Metalliferous mine dust: Human health impacts and the potential determinants of disease in mining communities." *Current Pollution Reports* **5**: 67–83.

Erdogan, I. G., Fosso-Kankeu, E., Ntwampe, S. K. O., Waanders, F. B. and Hoth, N. (2020). Management of metalliferous solid waste and its potential to contaminate groundwater. *Recovery of Byproducts from Acid Mine Drainage Treatment*. E. Fosso-Kankeu, Wolkersdorfer, C., Burgess, J. Wiley, 1–21.

Rahman, S. U., Nawaz, M. F., Gul, S., Yasin, G., Hussain, B., Li, Y. and Cheng, H. (2022). "State-of-the-art OMICS strategies against toxic effects of heavy metals in plants: A review." *Ecotoxicology and Environmental Safety* **242**: 113952.

Ravansari, R., Wilson, S. C. and Tighe, M. (2020). "Portable X-ray fluorescence for environmental assessment of soils: Not just a point and shoot method." *Environment International* **134**: 105250.

Sacramento, C. (2015). "EPA: National cost of leaking-mine cleanup could surpass $50 billion." 2023, from https://www.cbsnews.com/sacramento/news/epa-national-cost-of-leaking-mine-cleanup-could-surpass-50-billion/.

Thayer, K. A., Angrish, M., Arzuaga, X., Carlson, L. M., Davis, A., Dishaw, L., Druwe, I., Gibbons, C., Glenn, B., Jones, R., Phillip Kaiser, J., Keshava, C., Keshava, N., Kraft, A., Lizarraga, L., Persad, A., Radke, E. G., Rice, G., Schulz, B. and Shaffer, R. M. (2022). "Systematic evidence map (SEM) template: Report format and methods used for the US EPA Integrated Risk Information System (IRIS) program, Provisional Peer Reviewed Toxicity Value (PPRTV) program, and other 'fit for purpose' literature-based human health analyses." *Environment International* **169**: 107468.

Unger, C. (2017). Legacy issues and abandoned mines. *Mining in the Asia-Pacific*. T. O'Callaghan, Graetz, G. Springer.

VAGO. (2020). *Rehabilitating Mines*. Victorian Government Printer, 142.

Venkateswarlu, K., Nirola, R., Kuppusamy, S., Thavamani, P., Naidu, R. and Megharaj, M. (2016). "Abandoned metalliferous mines: Ecological impacts and potential approaches for reclamation." *Reviews in Environmental Science and Bio/Technology* **15**(2): 327–354.

Wang, L., Cheng, Y., Lamb, D. and Naidu, R. (2020). "The application of rapid handheld FTIR petroleum hydrocarbon-contaminant measurement with transport models for site assessment: A case study." *Geoderma* **361:** 114017.

Zakaria, Z., Zulkafflee, N. S., Mohd Redzuan, N. A., Selamat, J., Ismail, M. R., Praveena, S. M., Tóth, G. and Abdull Razis, A. F. (2021). "Understanding potential heavy metal contamination, absorption, translocation and accumulation in rice and human health risks." *Plants* **10**(6): 1070.

21 Derelict Mine Ownership – Past, Present and Future

Tim Roberts, E. Pottie, S. Lucas and P. Stevens

21.1 INTRODUCTION

Mining of the earth's natural resources has been a necessary function of societies across the globe since the dawn of civilisation; however, once extracted, mined sites often leave enduring socio-cultural and ecological legacies. The ecological impacts they leave behind are just one part of the broader problem of globally degraded landscapes, and for every action within complex social-ecological systems, there are consequences (Commoner, 1972). The issue is significant globally, with derelict mines accumulating over many centuries. In Australia, with a relatively short mining history, there are more than 50,000 derelict mines, beginning with the first gold rushes (Figure 21.1) and continuing today in the current intensified resource extraction.

Derelict mines are any lands that have been used for, or affected by, mining operations and which have been abandoned such that an "owner", title or leaseholder no longer exists (Ministerial Council on Mineral and Petroleum Resources and Minerals Council of Australia, 2010). Ongoing ecological impacts arising from derelict mines are present across the globe in the form of Acid Mine Drainage (AMD), bioaccumulation, permanent scars upon the landscape, loss of habitat and biodiversity and the lethal buildup of concentrated gases, in many cases spelling biological death for surrounding ecosystems, and the possibility of far reaching offsite impacts.

The inextricably linked nature of social and ecological systems mean that changes in one domain will necessarily impact the other (Walker and Salt, 2006), such that changes to the ecological landscape can manifest in a sense of "loss of place" and feelings of "Solastalgia" (Albrecht, 2005). Ownership of both the mines and the subsequent impacts are therefore linked: a derelict mine is one without an owner to take responsibility for ongoing remediation of social and ecological impacts. So who is responsible? While governments generate wealth from mining through royalties and taxing, the communities who lived there previous to, during and post-mining are left with the often-negative impacts.

This chapter addresses the issue of ownership and the discrepancies of definition inherent within derelict mines. Case studies used will highlight the way in which ecological harm can manifest itself in social distress. Neath in the Hunter Valley, NSW, and Redbank mines highlight the variances in impacts arising from the mining process and the difficulty in ascribing responsibility. Possibilities for adaptive reuse will be explored, as well as value adding.

21.2 OWNERSHIP

It is important to address the nomenclature surrounding the issue, as variances in derelict mine terminology can be part of the problem. Indeed, an international workshop in 2001 identified the lack of a clear definition and absence of any standards of rehabilitation as a central cause of inaction and progress on derelict mines (Pepper, Roche and Mudd, 2014). According to the NSW Mining Act, a derelict mine is "any land that ...a) was used for, or has been affected by, mining operations or prospecting operations, and b) has been abandoned...whether or not it is possible to identify or

DOI: 10.1201/9781315141855-21

FIGURE 21.1 Stocqueler, Edwin, 1855, Australian gold diggings, ca.1855, retrieved March 9, 2017, from http://nla.gov.au/nla.obj-134299327

locate the landholder or the holder of an authorisation under which operations ...were carried out" (Mining Act, 1992).

Alternatively, the Ministerial Council on Mineral and Petroleum Resources (MCMPR), in the Strategic Framework for Managing Abandoned Mines, defines derelict mines as "mines where mining leases or titles no longer exist, and responsibility for rehabilitation cannot be allocated to any individual, company or organisation responsible for the original mining activities" (MCMPR, 2010: 6, emphasis own). Whether or not an individual or company can be identified as responsible, the state government owns everything below the topsoil of existing properties (Cottle, 2013), and as the authority responsible for assigning titles, leases and rights, should then be responsible for the impacts that arise from the assigning of those rights. The injustices arising from derelict mine impacts are not just inflicted by careless mining corporations; rather, they are inflicted by the complacency of authorities who are in a position of power and fail to avert or prevent the harm (Venkatapuram, Bell and Marmot, 2010). While freehold landowners do not have mineral rights or the capacity to prevent access to minerals by third parties approved by the state, they are still deemed responsible for derelict mines on freehold land (Unger, Lechner, Kenway, Glenn and Walton, 2015). So private buyers should remember "Caveat emptor" (buyer beware!).

Indeed, even the term "ownership" is a contested term. Despite derelict mines no longer having owners, the adverse ecological and resulting social impacts are "owned" by the communities and individuals who inhabit the mine region. Ownership of mine sites is threefold and includes:

- the government, which owns the resources below the topsoil (Cottle, 2013), and is responsible for assigning rights, approvals, setting conditions and legislation, as well as collecting royalties and taxation.
- the holders of these assigned land, water and mineral entitlements (typically private land leased to mining companies).
- the communities associated with the land before, during and after mining.

However, ownership understood as proprietorship does not adequately address its nuances. Social systems are inextricably linked to and within ecological systems, and the sense of ownership extends beyond the legal understanding of exclusive proprietorship and the right to capitalise upon the land as one pleases within the legislation (Solomon, Katz and Lovel, 2008). It extends to the cultural and spiritual connection Indigenous people have had with the land for thousands of years prior to European Invasion, and which continues today with many groups being displaced from their lands and Country (Solomon et al., 2008). For the Wonnarua people who inhabit Country around the Upper Hunter region, their connection to the land is illustrated through the traditional stories passed down through generations that tell of the ancestral beings who created the sentient landscape, of which humans are a part (Connor, Higgenbotham, Freeman and Albrecht, 2008). Still in operation today, these traditional concepts of "place" mean that land and Country are a responsibility that requires care, values which sit uneasily with the economic narrative of place (Solomon et al., 2008).

Non-Indigenous peoples also hold a sense of place and value it beyond solely economic purposes, and historic legacies, cultural, environmental, aesthetic and intrinsic values are ascribed to the land. Environmental philosopher Glenn Albrecht coined a new term, "Solastalgia" to describe the feeling he encountered when coal mining took over the paddocks and hills of the Hunter Valley (Albrecht, 2005). When the landscape is changed by external forces such as mining or natural disaster, there is a loss felt and a longing for this sense of place. Expressing the distress of the residents, Solastalgia is "the homesickness you have when you are still at home" (Albrecht, 2005). It is clear, then, that ideations of ownership extend beyond simply the capitalist understanding of exclusive proprietorship. Rather, the enduring ecological legacies are owned by the communities associated with the landscape and manifested in a variety of ways within the social domain.

While ownership in the eyes of the law denotes proprietorship, this chapter also considers ownership as taking responsibility for outcomes. As government and industry profit from resource extraction, it is just that they should take responsibility for the legacies created. However, despite an estimated 50,000 abandoned mines across Australia, there is no national inventory of these mines, their risks and impacts, much less a national strategy for approaching their rehabilitation (Pepper et al., 2014; Unger et al., 2015). There is a common perception that derelict mine legacies are a feature of the past, from a time when regulation failed to hold operators accountable; however, the reality is that current mine operations are creating a legacy for future generations due to insufficient demands by government on the industry (Pepper et al., 2014).

The expanded scale and intensity of current mine operations in the Hunter Valley NSW will result in numerous voids and pit lakes across the landscape (Hancock, Wright and De Silva, 2004). Despite this scar on the landscape, there is no legislated requirement for void backfilling in Australia in regards to rehabilitation of the site. It is therefore the role and responsibility of government to strengthen its demands upon industry to ensure that future legacies are not being created through current mining operations.

21.3 CASE STUDIES IN AUSTRALIA

21.3.1 Case Study: Neath, Hunter Valley, NSW

The Neath Colliery began in 1906 on the Greta Coal Seam Measures that are located to the north-east of Cessnock in the Hunter Valley (NSW) and produced coal until 1960. From 1957 until 1988, the site was used to wash coal, whose history is causing the AMD issues of the present day. The high sulphur content of the Greta coal seam, 5.2%, recorded up to as high as 10.5%, increases AMD in prevalence and intensity (Davies and Tonks, 1993). The waste pile at the Neath colliery is 1.3 kilometres long, up to five metres thick and 300 metres wide. Despite rehabilitation, over the years, the capping on the waste dump has since eroded and the site is now leaching toxic acid mine drainage in the form of sulphuric acid, iron and aluminium into local waterways, at an estimated Olympic swimming pool size every few weeks. The result is a toxic legacy on the surrounding ecosystem such that the creek has turned a burnt orange colour, with a highly acidic pH reading of 2.5, killing local aquatic, plant and animal life.

While the cleanup of abandoned mine sites could be the Crown's biggest contamination liability, there is a lack of connection between community expectation of government responsibility, and the government's action towards these sites. Inadequate funds under the state government's Derelict Mines Programme and priority setting means that the AMD sites surrounding Cessnock have been rated at low-to-medium risk, resulting in monitoring of the sites rather than cleanup. This sits uneasily with local people, who as part of the Kurri Kurri Landcare group, decided to take matters into their own hands, as residents are left to live with and "own" the impacts. After spending eight years attempting to neutralise the acid by applying hydrated lime, the group was ordered to stop by the Department of Lands in December 2008, arguing the group was further exacerbating the problem. Disagreement between community expectations of government responsibility highlights the contested and politicised nature of derelict mine rehabilitation. While the Crown has ownership of the site itself, residents are left to live with the adverse ecological legacy, and they are in a sense the true owners of its impacts. Communities have connections to the landscape and its ecological values, and ecological impacts necessarily have knock-on effects for social systems, as is the case at Neath.

21.3.2 Redbank Copper Mine, Northern Territory

In Australia's Northern Territory, there is an estimated $1 billion in derelict mining liabilities as most mines predate the requirement for rehabilitation security bonds or the 1% levy (NT Government, 2016a). Redbank Copper Mine, situated in the North East of the Northern Territory, 30 km from the Queensland (QLD) border, despite its remote location, has left a prominent sociocultural and ecological legacy (Barton, 2011). Discovered in 1916, an estimated 800 tonnes of copper oxide were singlehandedly mined, hand sorted and transported by horse to the coast by William Masterson until 1961 (Northern Territory Government, 2016b). The site has since had a number of owners; however, operations were intermittent due to declining economic viability, with no significant production since 1996 (Northern Territory Government, 2016b). The site was bought by Redbank Mines Pty Ltd in 2006; however, operations were deemed uneconomic and the site was placed in care and maintenance. In 2015, the Environmental Protection Authority brought charges against the operators, with 26 counts of breaches of their Waste Discharge Licence. However, the charges were dropped in 2016 and the mineral leases were relinquished to the NT Mines Department, which is now taking over responsibility for rehabilitation of the site due to the previous owners being unable to finance the process and years of inaction.

The mine has been leaching copper sulphide into local waterways for the past two decades; contaminating the ecosystem such that Hanrahan's Creek, 500 metres from the mine, has turned an unnatural, vibrant blue-green. Copper concentrations in the creek are measured at 377 milligrammes per litre, and the toxicity has been detected 40 kilometres from the mine at the QLD border, where Hanrahan's Creek turns into Settlement Creek, feeding internationally important wetlands (EcOz Environmental Services, 2009: 169). Part of Redbank's enduring legacy is therefore serious off-site impacts. Hanrahan's Creek itself is biologically inert, deeply troubling to the traditional owners of the site, the Garawa people, who used to fish and swim there before it was poisoned (Roche, Judd and Bista, 2016). A recorded Aboriginal sacred site, Moonlight Falls just 2 kilometres away, is bleached white from the copper. This legacy is owned by the Garawa people, but the spiritual-cultural impacts are not captured in risk assessments, which fail to adequately acknowledge the importance of Country (Roche et al., 2016).

Despite being a relatively small mine site, its toxic legacy is enormous, with an estimated rehabilitation price tag of anywhere between $10 and $100 million. The true cost is unknown as there has been no in-depth research undertaken. As one of the worst legacy mines in the Northern Territory, it is highly politicised and has received priority under the Legacy Mine Unit due to ministerial commitments regarding the sites (Fawcett, 2016). The NT Environmental Protection Agency (EPA) has since defined management failure as the root cause of this legacy, citing "poor due diligence and disclosure, lack of continuity and long-term planning, flawed environmental impact assessment and poor regulation" (Roche et al., 2016). Despite very little production occurring throughout the life of Redbank, there is a disproportionate cost borne by the traditional owners of Country, in the form of spiritual and cultural loss.

The impacts are not merely confined to the site itself, landholders downstream are concerned about their cattle drinking the contaminated water, and the contamination continues to the QLD border, where it threatens ecological damage to wetlands. If a site like Redbank cannot be fixed, future mining operations are a conscious decision to sacrifice ecosystems like Hanrahan's creek, or greater effort needs to be made to protect them (Toohey, 2013). However, with the government admitting that poor regulation during the 1990s is to blame, as exemplified in the meagre security bond for the site set at $150,000, there is a need to strengthen regulation to prevent this occurring again.

While there are many negative legacies left by derelict mines, for those sites not heavily polluted by AMD there may be positive values, such as voids suitable for domestic waste disposal, secondary mining opportunities, tourism and alternative land uses (Unger et al., 2015). The Eden Project in Cornwall (UK) is a well-known example of alternative land use (Jones et al., 2001). A clay quarry until 1995, the 2.2 ha site now comprises three biomes housing plants from all over the world, acting as an educational resource capable of fulfilling its mission to promote sustainability and the understanding of the vital relationship between plants, people and resources.

A fascinating reuse is occurring at the closed Kidston Gold Mine in North Queensland. The first of its kind in the world, the Kidston Project will convert the two disused pits into a hydroelectric storage plant, essentially a "giant battery" (Canaccord Genuity, 2016). In what will be the third largest hydroelectric storage project in Australia, the 250-megawatt plant will have 1,500-megawatt hour's storage capacity, with an inclusion of a 50-megawatt solar farm with the potential to power the storage plant's water pumps (Canaccord Genuity, 2016). Surplus energy from the 50 MW solar farm, capable of powering 27,500 homes per year, will be used to pump water from the lower reservoir to the higher reservoir, which will then be released to generate power for commercial sale during periods of high demand and low solar radiation (Canaccord Genuity, 2016). Australian Renewable Energy Agency has allocated $4 million to the project, and the ASX listed company Genex Power has been awarded a 20-year Revenue Support Deed consisting of a one-way price contract, guaranteeing the sale of electricity above a set floor price for a 20-year period (Genex Power, 2017). With construction starting this year, the project is a visionary opportunity to transform a site of finite resource extraction into renewable energy production, and it sets an example of the value that can be added to mine voids.

Another form of adaptive reuse can be found in the Woodlawn bioreactor near Goulburn, NSW, which has transformed a previous copper and zinc open-cut mine. Since 2004 to date, over 3.8 million tonnes of waste have been used to generate green energy, accepting 20% of Sydney's putrescible waste (Veolia, 2017). As part of the rehabilitation of the mine site, the construction of a mechanical biological treatment plant designed to separate organics from mixed household waste will allow for the creation of compost (Veolia, 2017). The site also includes aquaculture and horticulture, using the waste heat from the energy generation process for fish farming, a working farm applying nutrient and grazing rotation and a 50 MW wind farm (Veolia, 2017). These three examples highlight the opportunities that some unused mine sites can offer. Indeed, there is potential for these sites to reverse some of their damaging history as generators of greenhouse gases through the provision of renewable energy. These examples have added value to the site and set a precedent for what can be done.

21.4 CONCLUSIONS

Derelict mines have left enduring ecological and sociocultural legacies across Australia, from the first gold rushes to the current expanded and intensified mining operations. The interlinked nature of social-ecological systems means that changes in one system necessarily impact the other, and they can be seen in the sadness arising from "loss of place" in the communities left to live with the adverse ecological impacts of mining legacies. Ownership of these legacies is a contested and opaque issue, as in different cases either the Crown or the landholder may own the site, but, in most cases, the communities own the impacts.

While industry profited from the resource extraction, and governments collected royalties and taxes throughout the life of the mine, communities disproportionately bear the legacy burden. The sheer scale of the problem means that rehabilitation is usually dependent on the site being categorised as high risk by state government, as funding for the nation's mining legacy falls far short of that required. As a result, it is important that today's mining operations are not creating legacies for future generations, which requires greater responsibility and ownership of outcomes on the part of government and industry. There are, however, positive opportunities for adaptive reuse, as can be seen in the Kidston Project, the Eden Project and the Woodlawn bioreactor, and all opportunities for adaptive reuse should be explored when dealing with derelict mines.

Ultimately, in dealing with derelict and operational mines, the response taken is indicative of societal values. Prioritising short-term economic gain over the capacity of the earth to continue to sustain us fails to consider the long-term perspective that "Earth can live without us. We cannot live without Earth" (Korten, 2014). Humans are just one component of the complexity of the living earth, and our capitalist narrative of exponential growth is ultimately destructive to the Earth that sustains us (Korten, 2014). Adequately responding to the mining legacy may, in fact, require a global paradigm shift in many of our attitudes to consumption (of all things).

REFERENCES

Albrecht, G. (2005) Solastalgia, a new concept in human health and identity. *Philosophy Activism Nature*, 3, 41–44.

Barton, P. (2011) Acid mine drainage from Redbank mine. Retrieved from http://www.mininglegacies.org/mines/northern-territory/redbank/

Canaccord Genuity. (2016) Genex power (ASX: GNX). Retrieved from http://www.genexpower.com.au/uploads/6/6/1/2/6612684/canaccord_genuity_-_genex_power_-_nov_2016.pdf

Commoner, B. (1972) *The Closing Circle*. London: Jonathan Cape Ltd, 14–48.

Connor, L., Higginbotham, N., Freeman, S., & Albrecht, G. (2008) Watercourses and discourses: Coalmining in the Upper Hunter Valley, New South Wales. *Oceania*, 78(1), 78–90.

Cottle, D. (2013) Land, life and labour in the sacrifice zone: The socio-economic dynamics of open cut coal mining in the Upper Hunter Valley, New South Wales. *Rural Society*, 22(3), 208–216.

Davies, E., & Tonks, E. E. (1993) The development of the Greta coal measures of New South Wales. In *Proceedings of The AusIMM Centenary Conference*, Adelaide, 30 March–4 April 1993, 353–365.

EcOz Environmental Services. (2009) Redbank copper – EIS 2009. Retrieved from http://www.redbankcopper.com.au/images/PDF/environmental/part_4.pdf

Fawcett, M. (2016) Redbank copper mine – Case study. Presented at Proceedings of Dealing with Derelict Mines 2016 – CRC Care, to be published.

Genex Power. (2017) ASX announcement. Retrieved from http://www.genexpower.com.au/uploads/6/6/1/2/6612684/1.pdf

Hancock, G., Wright, A., & De Silva, H. (2004) Long-term final void salinity prediction for a post-mining landscape in the Hunter Valley, NSW, Australia. *Hydrological Processes*, 19, 387–401.

Jones, A. C., Hamilton, D., Purvis, M., & Jones, M. (2001). Eden project, Cornwall: design, development and construction. *Structural Engineer*, 79(20), 30–36.
Korten, D. (2014) A new story for a new economy to find our human place in a living universe, 1–29. Retrieved from http://www.yesmagazine.org/happiness/a-new-story-for-a-new-economy
MCMPR. (2010) Strategic framework for managing abandoned mines in the minerals industry. Ministerial Council on Mineral and Petroleum Resources and Minerals Council of Australia, Canberra. Retrieved from https://industry.gov.au/resource/Mining/Documents/StrategicFrameworkforManagingAbandonedMines.pdf
Mining Act No 29, Sect. 242A (NSW). Retrieved from http://www.legislation.nsw.gov.au/#/view/act/1992/29
Northern Territory Government, Department of Primary Industry and Resources. (2016a) About legacy mines. Retrieved from https://dpir.nt.gov.au/mining-and-energy/mine-rehabilitation-projects/about-legacy-mines
Northern Territory Government, Department of Mines and Energy. (2016b) Redbank mine. Retrieved from https://dpir.nt.gov.au/mining-and-energy/mine-rehabilitation-projects/about-legacy-mines
Pepper, M., Roche, C. P., & Mudd, G. M. (2014) Australia's mining legacy challenge, Life of Mine 2014 Conference, Paper Number 87. Retrieved from www.mpi.org.au/wp-content/uploads/2013/12/Final-LofM-Mining-Legacies.docx
Roche, C., Judd, S., & Bista, S. (2016) Ground truths: Taking responsibility for Australia's mining legacies, Mineral Policy Institute. Retrieved from http://www.mpi.org.au/wp-content/uploads/2016/06/Ground-Truths-2016-web.pdf
Solomon, F., Katz, E., & Lovel, R. (2008) Social dimensions of mining: Research, policy and practice challenges for the mining industry in Australia. *Resources Policy*, 33, 142–149.
Toohey, P. (2013) How the Redbank Copper mine poisoned Hanrahan's Creek. Retrieved from http://www.news.com.au/national/how-the-redbank-copper-mine-poisoned-hanrahans-creek/news-story/38e9f541ba0437beba5630ff5772df3a
Unger, C. J., Lechner, A. M., Kenway, J., Glenn, V., & Walton, A. (2015) A jurisdictional maturity model for risk management, accountability and continual improvement of abandoned mine remediation programs. *Resources Policy*, 43, 1–10.
Venkatapuram, S., Bell, R., & Marmot, M. (2010) The right to sutures: Social epidemiology, human rights, and social justice. *Health and Human Rights*, 12(2), 3–17.
Veolia. (2017) Woodlawn bioreactor. Retrieved from http://www.veolia.com/anz/our-services/services/municipal-residential/recovering-resources-waste/woodlawn-bioreactor
Walker, B., & Salt, D. (2006) *Resilience Thinking: Sustaining Ecosystems and People in a Changing World.* Washington, DC: Island Press.

Index

C

D

J

K

L

M

N

Q

R